Study Guide and Partial Solutions Manual for
Mendenhall/Beaver/Beaver's

Introduction to

Probability and Statistics

Eleventh Edition

Barbara M. Beaver *and* **Robert J. Beaver**
University of California, Riverside

THOMSON

BROOKS/COLE

Australia • Canada • Mexico • Singapore • Spain • United Kingdom • United States

Printed in Canada
1 2 3 4 5 6 7   05 04 03 02 01

Printer: Transcontinental Printing

ISBN: 0-534-39520-1

For more information about our products, contact us at:
Thomson Learning Academic Resource Center
1-800-423-0563

For permission to use material from this text,
contact us by:
Phone: 1-800-730-2214
Fax: 1-800-731-2215
Web: http://www.thomsonrights.com

Asia
Thomson Learning
5 Shenton Way #01-01
UIC Building
Singapore 068808

Australia
Nelson Thomson Learning
102 Dodds Street
South Street
South Melbourne, Victoria 3205
Australia

Canada
Nelson Thomson Learning
1120 Birchmount Road
Toronto, Ontario M1K 5G4
Canada

Europe/Middle East/South Africa
Thomson Learning
High Holborn House
50/51 Bedford Row
London WC1R 4LR
United Kingdom

Latin America
Thomson Learning
Seneca, 53
Colonia Polanco
11560 Mexico D.F.
Mexico

Spain
Paraninfo Thomson Learning
Calle/Magallanes, 25
28015 Madrid, Spain

CONTENTS

PREFACE

The study of statistics differs from the study of many other college subjects. A student must not only absorb a set of basic concepts and applications, but also precede this with the acquisition of a new language.

We think with words. Hence, understanding the meanings of words used in the study of a subject is an essential prerequisite to the mastery of concepts. In many fields, this poses no difficulty. Often, terms in the physical, social, and biological sciences have been encountered in the curricula of the public schools, in the news media, in periodicals, and in everyday conversation. In contrast, few students encounter the language of probability and statistical inference before embarking on an introductory college-level study of the subject. Many consider the memorization of definitions, theorems, and the systematic sequence of steps necessary for the solution of problems to be unnecessary. Others are oblivious to the need. The consequences for both types of students are disorganization and disappointing achievement in the course.

This study guide and student solutions manual attempts to lead you through the language and concepts necessary for a mastery of the material in *Introduction to Probability and Statistics*, 11th edition by William Mendenhall, Robert J. Beaver and Barbara M. Beaver (Duxbury, 2003).

A study guide with answers is intended to be an individual student study aid. The subject matter is presented in an organized manner that incorporates continuity with repetition. Most chapters bear the same titles and order as the textbook chapters. Within each chapter, the material both summarizes and explains again the essential material from the corresponding textbook chapter. This allows you to gain more than one perspective on each topic and, we hope, enhances your understanding of the material.

At the appropriate points in each chapter, you will encounter a set of Self-Correcting Exercises in which problems relating to new material are presented. Terse, stepwise solutions to these problems are found directly following the exercise set. You can refer to these at any intermediate point in the solution of each problem, or use them as a stepwise check on any final answer. The Self-Correcting Exercises not only provide the answers to specific problems but also reinforce the stepwise logic required to arrive at a correct solution to each problem.

Additional sets of exercises can be found at the end of each chapter. These exercises are provided for students who feel that further individual practice is needed in solving the kinds of problems found with each chapter. At this point, having been given stepwise

solutions to the Self-Correcting Exercises, you are now presented
with only final answers to problems. When your answer disagrees
with that given in the study guide, you should be able to find your
error by recalculating and comparing your solution with the
solutions to similar Self-Correcting Exercises. If the answer given
disagrees with your only in decimal accuracy, it can be assumed that
this difference is due only to rounding error at various stages in the
calculations.

In the back section of this study guide you will find detailed
solutions for approximately twenty-five percent of the exercises that
appear in the text – these are the test exercises whose number is
coded in color. These solutions are intended to give you yet another
study aid as you proceed through the course of study.

When the study guide is used as a supplement, the textbook
chapter should be read first. Then you should study the
corresponding chapter within the study guide. Key words, phrases,
and numerical computations have been left blank so that you can
insert a response. The answers are given in the page margins, on
the same line in which the blank occurs. These should be covered
until you have supplied a response for each blank. Bear in mind
that in some instances more than one answer is appropriate for a
given blank. It is left to you to determine whether your answer is
synonymous with the answer given with the margin.

If you happen to find an error in the study guide portion of this
text or in the solutions to the text exercises, please bring it to our
attention.

Barbara M. Beaver
Robert J. Beaver

Introduction:
An Invitation to Statistics

Almost daily we hear announcements or read reports concerning the results of research in such fields as genetics, in which certain genes are associated with specific forms of cancer or disease, or political science, whereby the results of polls regarding government or economic issues are reported, or perhaps human nutrition and its impact on the aging process.

Have you ever wondered how researchers are able to reach and corroborate results that lead to such wide-ranging conclusions? How are they able to generalize results concerning perhaps a few hundred individuals to millions of American citizens? What kinds of information are collected? How is the information analyzed? Who writes the computer programs used in the analysis?

Statistics is a branch of applied mathematics that affects many aspects of our lives. Like any discipline, it has its own jargon and procedures. One of the big hurdles for the student new to statistics is to become conversant with the terms and ideas used by statisticians. We invite you to learn the language and procedures of statistics step by step, thereby enabling you to understand and use these powerful data analytic tools.

THE POPULATION AND THE SAMPLE

One of the basic concepts in the language of statistics is sampling, whereby a specified number of individuals or units are selected in some preassigned way from a much larger body of individuals or units. The set of measurements made for the selected individuals is referred to as the
_____ ; the set of measurements for all individuals or units of
interest to the investigator is called the _____.

sample
population

A **population** is the set that represents all the measurements of interest to the investigator.

A **sample** is a subset selected from the population of interest.

The terms *sample* and *population* have two meanings. If 100 people are selected for inclusion in a telephone poll, the word *sample* is used to refer to both the people selected and their responses to a given question. Statisticians refer to the measurements as the sample, and they describe the people or objects about which the measurements are taken as *experimental* _____ or *elements of the*

_____ .

units
sample

Consider the following situations in which sampling is used in the study of a population:

1. The preferences of all eligible voters in one state are of interest to the sponsors of a state proposition. Rather than a poll of the entire list of eligible voters, questionnaires are sent to a selected group.
2. The FDA has given permission for clinical trials to assess the efficacy of a new drug used in treating Alzheimer's disease.

The following characteristics are common to both of these situations:

data
measurements

could not

1. A sample is taken from a much larger body of _____ .
2. On each element in the sample, one or more _____ are made.
3. The measurements obtained (could, could not) be predicted in advance.
4. Under certain guidelines concerning the sampling procedure, it is expected that the conclusions drawn from the study will apply to the larger body of data.

One difference between these two situations is that the population in the first example exists in fact and its elements are all the eligible voters in the state. The underlying population in the second example exists as a concept; that is, it consists of measurements on all individuals with Alzheimer's disease who could (or would) be treated with the new drug. In this case, the elements of the population (do, do not) exist in fact, and we are sampling a theoretical population.

do not

DESCRIPTIVE AND INFERENTIAL STATISTICS

Whether we sample from a population or conduct a complete census of the population, our first concern is to find a way to organize the data. In either case, we can use graphical or numerical descriptive measures. This branch of statistics is called _____ *statistics*.

descriptive

We can use pie charts, bar graphs, line graphs, and other visual approaches to describe the set of measurements. In addition, or as an alternative, we can describe the set of data by reporting the largest and smallest measurements, or the most frequent value in the set, or the arithmetic average of the observations, or some other outstanding characteristic of the data.

Some of the reasons we might prefer to sample rather than census a population are costs, time, and errors associated with handling large data sets. Furthermore, obtaining a measurement may involve destroying the experimental unit—for example, determining the breaking strength of a length of steel cable or the force required to crush a shipping container. For these and possibly other reasons, we may have only a sample from the population of interest. After analyzing the sample information, we would like to be able to answer questions concerning the population; that is, we want to use the sample information to infer something about the characteristics of the population. This branch of statistics is called
_____ *statistics*.   inferential

Inferential statistics consists of procedures used to make inferences about population characteristics from information contained in a sample drawn from that population.

The **objective of inferential statistics** is to make _____ in the   inferences
form of predictions, decisions, or conclusions about the characteristics of
a population from information contained in the sample. In addition to
the inference itself, we would like a measure of its reliability. How confident are we that the inference we make is correct?

ACHIEVING THE OBJECTIVE OF STATISTICS: THE NECESSARY STEPS

How do we go about making inferences about populations from sample data? We can organize this procedure into five logical steps.

1. **Specify the questions to be answered and identify the population of interest.** The questions to be answered must be properly stated, and the _____ to be sampled must be clearly specified. Sampling   population
only North Dakota dairy farms with the objective of determining the
vitamin D concentration in milk produced by American dairies would
be incomplete because only a subset of the population of interest has
been sampled. Any resulting inferences would not be expected to apply
to the United States in general.
2. **Decide how to select the sample.** The sample contains a quantity of information upon which an inference about the population can be based.

information
design

sample

accurate

reliability

Our aim is to determine the most economical procedure for getting a specified quantity of _____. This is called the *sampling plan*, or the _____ *of the experiment.*

3. **Select the sample and analyze the sample information.** In this step we must extract the information contained in the _____. Methods appropriate for analyzing the data collected using the sampling plan you selected in step 2 will be introduced in subsequent chapters of the text.

4. **Use the information from step 3 to make an inference about the population.** There may be more than one way to make the inference that you desire, but in general the sampling procedure limits which of these techniques are appropriate in any one situation. From among the appropriate techniques, we choose the one that produces the most _____ inference.

5. **Determine the goodness of the inference.** How accurate is a sales projection of 10,000 units? Is the margin of error plus or minus 10 units? 100 units? 1000 units? Is it reliable enough to use in setting production goals? Before we can answer this last question, we must assess the _____ of the inference. Every statistical inference must be accompanied by a measure of reliability that tells you how much confidence you can place in the inference and thereby assess its practical value.

As you proceed through the text and this study guide, you will learn more words and definitions and concepts that are useful in understanding and applying statistical procedures. It will become clear that statistical procedures usually consist of commonsense steps that you may have discovered for yourself if given enough time and patience. You will also find that statistical results must agree with what common sense says is reasonable. If this is not the case, the results are suspect, and the investigator and the statistician need to review the five steps that went into producing the inference in question.

Chapter 1
Describing Data with Graphs

1.1 VARIABLES AND DATA

Whether a set of measurements is a sample or an entire population, you need to be able to describe the data set in a clear and understandable form. This leads us to some new definitions.

A **variable** is a characteristic that changes or varies over time or area, and/or for different individuals or objects under consideration.

Height and weight are variables that change from individual to individual. Years of formal schooling, eye color, marital status, and state of residence are also variables, characteristics that _____ from individual to individual.

vary

An **experimental unit** is the individual or object on which a measurement is taken.

A single **measurement** or **data value** results when a variable is measured on an individual or an experimental unit.

When a variable is measured over time, or for different experimental units, a set of measurements is generated. If a measurement is generated for every experimental unit in the entire collection, the resulting data set constitutes a _____. Any smaller subset of measurements is a sample.

population

A **population** is the set of all measurements of interest to the investigator.

A **sample** is a subset of measurements selected from the population of interest.

Example 1.1

Six vehicles are selected from the vehicles that are issued campus parking permits, and the following measurements are recorded:

Vehicle	Type	Make	Car-pool?	One-way commute distance (miles)	Age of vehicle (years)
1	Car	Nissan	No	23.6	6
2	Car	Toyota	No	17.2	3
3	Truck	Toyota	No	10.1	4
4	Van	Dodge	Yes	31.7	2
5	Motor-cycle	Harley-Davidson	No	25.5	1
6	Car	Chevrolet	No	5.4	9

Discuss the experimental units and the variables used to generate this set of measurements.

Solution

The experimental units are the vehicles that are issued campus parking permits. There are _____ variables measured on each of the six vehicles selected: type of vehicle, make of vehicle, whether or not the vehicle is used in a carpool, the one-way commute distance, and the age of the vehicle. If we consider the commuting distance for all vehicles with campus parking permits as the population of interest, then these six distances represent a _____ from this population.

   When we measure whether the vehicle is used in a carpool, the measurement belongs to one of two categories: yes and no. Unlike commute distance, this variable is not numerically valued, and it results in assigning the observation to one of two categories. Along the same lines, vehicle type and vehicle make are variables that produce categorical data with four and five observed categories, respectively. The variable age of vehicle, like commute distance, is numerically valued and not categorical. Since we have measured five variables on each experimental unit, a measurement actually

five

sample

consists of five observations. The measurement on the first vehicle is
(Car, _____, No, 23.6, 6).

Nissan

1.2 TYPES OF VARIABLES

Variables that give rise to nonnumerical data in which the observations
are categorized according to similarities or differences in kind are called
_____ variables. Hair or eye color, sex, geographic regions, and
religious affiliation are examples of qualitative variables.

qualitative

A **qualitative variable** gives rise to observations that are
categorized according to likenesses or differences in kind.

Qualitative variables give rise to categorical data.
  When the variable used to measure a characteristic produces a
numerical observation, the variable is said to be a _____ variable.
The number of vehicles that pass a given point on an inter-state highway,
the number of broken taco shells in a package of 12 shells, and a person's
height, weight, and annual salary are examples of quantitative variables.

quantitative

A **quantitative variable** gives rise to numerical observations
that represent an amount or quantity.

  Quantitative variables can be further categorized according to the
range of numerical values that a measurement can assume. For example,
the number of customers in a checkout line, the number of boating accidents along a 50-mile stretch of the Mississippi River during the summer
of 1998, and the number of cases of chickenpox in your county during
the past year are variables that take on values associated with the counting numbers: 0, 1, 2,.... Because these variables assume a countable or
discrete number of values, they are called _____ variables.

discrete

Variables that assume a countable or discrete number
of values are called **discrete variables**.

  In contrast, measurements on variables such as height, weight, time,
and volume can assume values that correspond to the infinite number of
points on a line interval. Variables of this type are called _____
variables. A variable is continuous if between any two values of the
variable it is (sometimes, always) possible to find a third.

continuous

always

Example 1.2
Identify the following variables as qualitative or quantitative. If the variable is quantitative, determine whether it is discrete or continuous.

1. The number of specialty cakes sold daily over the counter at the the corner delicatessen.
2. The time required to complete a questionnaire.
3. Your choice of color for a new refrigerator.
4. The number of brothers and sisters you have.
5. The yield of wheat in kilograms from a 1-hectare area in a wheat field.

Solution
Variable 3 is qualitative because a measurement results in a category described as a color. The remaining four variables are _____ because they describe an amount or a quantity. Specifically, variables 1 and _____ are discrete because the values that they may assume are the counting numbers: 0, 1, 2, ... Variables 2 and _____ are continuous; if two people require 2 and 2.5 minutes to complete the questionnaire, it is always possible to find a third person who takes between 2 and 2.5 minutes to complete the questionnaire. Using the same reasoning, we would classify variable 5 as a continuous variable.

   In Example 1.1, five different variables were measured for each of the vehicles. In Example 1.2 five different variables were measured, but this time in different situations and for different experimental units.

quantitative

4

5

When an observation results in a single measurement on one variables, the data is called **univariate data.**

When an observation results in measurements on two different variables, the data is said to be **bivariate data.** When an observation results in measurements on three or more variables, the data is said to be **multivariate data.**

It is important to know the kind of data that you collect because the descriptive methods you use will depend on the type of data you have.

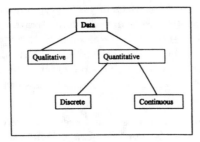

Figure 1.1 Types of Data

Self-Correcting Exercises 1A

1. A market analyst wishes to determine which factors exert the most influence on a purchaser's decision process in selecting his own or her own car. Describe the population of interest to the analyst, and indicate what kind of information the analyst might consider collecting.

2. A medical researcher is interested in determining whether a new drug is more effective than those currently available for treating degenerative arthritis. Identify the population of interest to the researcher. Should severity and time since onset of this disease enter into any proposed sampling plan? Why or why not?

3. An agricultural economist is interested in determining the revenue loss to cotton growers in the Imperial Valley of California due to crop infestation by the pink bollworm. Identify the population of interest to the economist. Would his findings apply equally well to cotton growers in Texas?

4. Identify the following variables as quantitative or qualitative. If the variable is quantitative, indicate whether it is continuous or discrete.
   a. The cost of a specified model of Toyota sedan for sale in Southern California in 1999.
   b. The response from an applicant for a credit card when asked if he or she has an outstanding bank loan.
   c. The 1999 domestic sales for an electronic firm.
   d. The number of potential voters in a sample of 100 potential voters who would vote to dismantle affirmative action procedures.

Solutions to SCE 1A

1. The population consists of the measurements taken on all possible purchasers of a new car. Possible variables of interest to the market analyst might be cost, favored colors and styles, available accessories, mileage ratings, and so on.

2. The population of interest consists of yes/no type measurements (drug is/is not more effective than the current treatment), taken on all people who currently have (or at some time might contract) degenerative arthritis. Since the drug will have different effects depending on the severity and time onset of the disease, these factors should be considered when choosing a sample from the population of interest.

3. The population consists of the revenue losses measured for all growers in the Imperial Valley of California. Unless the growers in Texas have losses which behave exactly as those suffered by the California growers, it is unwise to use the findings for the California growers to make inferences about losses in Texas.

4. a. Cost is a quantitative discrete variable, since the numerical cost can take values which can be listed as {$.00,,.01,,.02, ...}.
   b. The response {yes, no} is a qualitative variable.
   c. Sales is a monetary variable, which is again quantitative discrete.
   d. Number of voters is a quantitative discrete variable, taking the integer values given by {0, 1, 2, 3, ...}.

1.3 GRAPHS FOR QUANTITATIVE AND QUALITATIVE VARIABLES

Statistical tables are used to summarize data in an organized fashion. The form of the table will vary with the kind of variables being summarized. Statistical tables for _____ variables usually consist of a list of categories and the number of observations in each category or the amount measured for each category. In selecting the categories to be used, we must be careful to ensure (1) that an observation can belong to one and only one category and (2) that every observation has a category to which it can be assigned. The following tables can be used to summarize qualitative variables:

qualitative

CLASSIFICATION OF UNIVERSITY STUDENTS

Undergraduate	Graduate	Other
Freshman	Masters level	
Sophomore	Doctoral level	
Junior	Professional	
Senior		

CLASSIFICATION OF TEXTBOOKS

Natural science
Social science
Literature
Music
Other

A **pie chart** is a circular graph that is used to show how a quantity is distributed into several _____. Each piece of the pie represents the proportion of the whole that is in that category. A **bar chart** can also be used to show the amounts or the frequencies in the categories, with the _____ of the bar proportional to the amount or frequency in each category.

parts

height

Example 1.3
The five top-grossing North American concert tours from 1985 to 2000 are listed in the following table. Total gross is in millions.

Artist	Total Gross		Proportion
The Rolling Stones (1994)	$121.	2	.25
Pink Floyd (1994)	103.	5	.21
The Rolling Stones (1989)	98.	0	.20
The Rolling Stones (1997)	89.	3	.18
Tina Turner (2000)	80.	2	.16
Total	$492.	2	1.00

Source: *The World Almanac and Book of Facts, 2002*, p. 277

Construct a pie graph and a bar graph for this data set.

Solution
To construct the pie graph in Figure 1.2, assign to each category a sector of a circle, with the angle of the sector equal to the proportion in the category times 360°.

Tours	Angle
The Rolling Stones (1994)	.25 × 360° = 90.0°
Pink Floyd (1994)	.21× 360° = 75.6°
The Rolling Stones (1989)	.20× 360° = 72.0°
The Rolling Stones (1997)	.18× 360° = 64.8°
Tina Turner (2000)	.16× 360° = 57.6°
Total	360°

Pie Chart of Tours

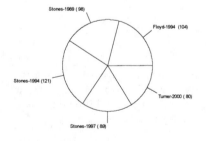

Figure 1.2

Figure 1.3

In constructing the bar chart in Figure 1.3, we plotted the actual dollar amount in each category rather than the percentage associated with each category.

Example 1.4
The table that follows lists the year 2000 deposits in U. S. banks that were insured by the Federal Deposit Insurance Corporation (FDIC).

Class	Deposits (billions)
National banks	$2250
State banks	1032
Not Federal Reserve members	894
Savings and Loans	738
Total	$4914

Source: *The World Almanac and Book of Facts*, 2002, p. 108

Use a pie graph and a bar graph to display these data.

Solution
In constructing the pie graph in Figure 1.4, we need to find the sector angles that correspond to each of the bank categories. For example, the sector angle for national banks is

$$\frac{2250}{4914} \times 360^0 = 164.8^0$$

Pie Chart of Banks

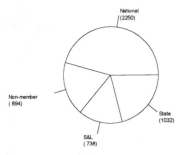

Figure 1.4

In this example, we are given the amounts of the deposits in each of the bank categories rather than a frequency. Therefore, in the bar chart in Figure 1.5, the vertical axis represents the deposits, and the height of each bar is determined by the amount of deposits for each category of bank.

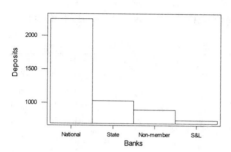

Figure 1.5

Notice that both charts indicate that _____ banks have the | national
largest deposits among the banks listed, but the bar chart allows us to
determine the _____ of deposits for each category of bank. Since | amount
there is no inherent ordering to the categories, they can be displayed in
any order in both the pie chart and the bar chart. Therefore, the shape
of the bar graph does not enter into its interpretation.

A **line chart** is used to display the change in the value of a variable
over _____. The time units are displayed on the horizontal axis | time
and the value of the variable on the vertical axis. The line chart in
Figure 1.6 tracks the sales of Wal-Mart Stores, Inc. This chart is very
effective in showing the slow increase in sales during the early 1980s
and the rapid _____ in sales in the early 1990s. The associated | rise
charts that show the rise in the number of stores and the number of
employees are equally effective in showing the overall increase in the
size of the company as well as its sales.

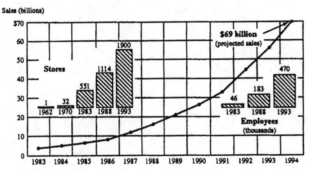

Walmart's third quarter sales for 2000 numbered $134,773 billions. The
rapid rise of the 90's has continued into the 21st century.

Figure 1.6

Example 1.5
The Consumer Price Index (CPI) for the years 1995-2001 follows.

Year	CPI	
1955	80.	2
1960	88.	7
1965	94.	5
1970	116.	3
1975	161.	2
1980	248.	8
1980	322.	2
1985	322.	2
1990	391.	4
1995	456.	5
2001	529.	1

Source: *The World Almanac and Book of Facts*, 2002, p. 103.

Construct both a line chart and a bar chart for the CPI.

Solution

Notice in Figures 1.7 and 1.8 that the Consumer Price Index exhibits a slow growth during the late 50's through the 60's with increases less than 8 points every five years. However, a drastic change took place in the 70's with five-year increases of 21.8 in 1970, 44.9 in 1975, and 87.6 in 1980. The CPI increased an average of about 70 points every five years thereafter. Both the line and bar charts vividly depict the information contained in the data. In this case, the points on the line chart and the bars in the bar chart are ordered in time, and the shapes of the line and the bar chart are relevant to their interpretation.

are

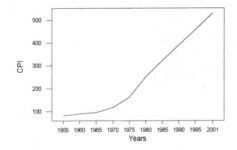

Figure 1.7

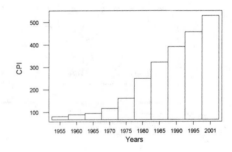

Figure 1.8

Self-Correcting Exercises 1B

1. The five leading causes of death of Americans are given in the table
    that follows.

Cause of Death	Number (in 100,000)
1. Heart disease	7. 3
2. Cancer	5. 5
3. Stroke	1. 6
4. Lung disease	1. 2
5. Accidents	1. 0

Source: *The World Almanac and Book of Facts*, 2002, p. 881

a. Use a bar chart to display these data.
b. Use a pie chart to display these data.
c. Which chart best conveys the information in the table?

2. The following data, supplied by the Bureau of Labor Statistics, gives
    the total civilian labor force from 1995 to 2000.

Years	Employed ($\times 10^6$)	Unemployed ($\times 10^6$)
1995	124. 9	7 .4
1996	126. 7	7 .2
1997	129. 6	6 .7
1998	131. 4	6 .2
1999	133. 5	5 .9
2000	134. 3	5 .7

Source: *The World Almanac and Book of Facts*, 2002, p. 139

a. Construct a bar chart to display the number employed during the
   1995-2000 period.
b. Construct a bar chart to display the number unemployed during
   this period.
c. Use a line chart to display the data in parts a. and b.
d. Which graphical procedure better presents the information
   contained in these data?

3. The 2002 purchases by a car rental agency are given in the table that
    follows.

Car Model	Number purchased
Chevrolet Cavalier	45
Ford Focus	30
Ford Taurus	60
Pontiac Grand Am	15
Toyota Camry	30

a. Display these data using a pie chart.
b. Present these data using another type of graph.

4. The investment portfolio of pension funds for employees of a manufacturing company for 1992 and 2002 are listed in the next table.

Type of Asset	1992		2002	
Common stock	$6000,	000	690,	000
Preferred stock	120,	000	115,	000
Industrial bonds	120,	000	345,	000
Government bonds	300,	000	690,	000
Real estate mortgages	60,	000	460,	000

Construct two pie charts to depict the company's portfolio composition, one for 1992 and one for 2002.

Solutions to SCE 1B

1. The bar chart and pie chart are shown below. Although either chart conveys the tabled information quite well, the bar chart allows you to read the actual numbers of deaths, while the pie chart does not.

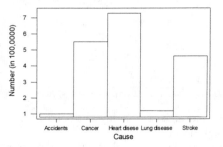

Pie Chart of Cause

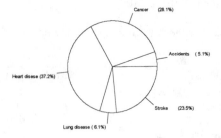

2. a

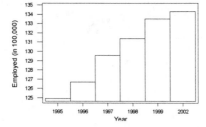

b.

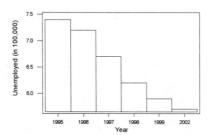

c.

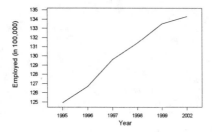

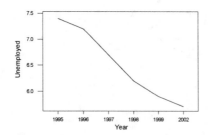

d. Although the vertical scales are not the same for the two line charts, taken together, they do show first, that there has been a significant increase in the number of employed, and second, a related decrease in the number of unemployed. Both charts present the data equally well, and the choice of chart would depend upon your preferences.

3. a.

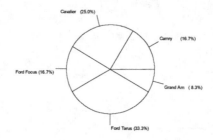

Pie Chart of Model

b.

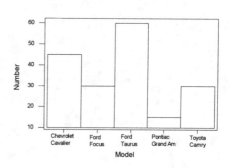

4.

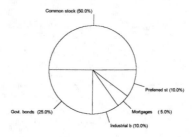

Pie Chart of Assets for 1992

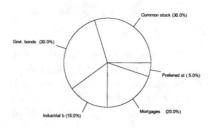

Pie Chart of Assets for 2002

1.4   OTHER GRAPHICAL TECHNIQUES FOR QUANTITATIVE VARIABLES

Dotplots

You may wish to display a set of quantitative data without losing the actual values of the measurements. This can be done by using a dotplot in which the value of an observation is plotted as a point on a horizontal axis. Dotplots are useful in providing a picture of how the observations are distributed along the horizontal axis. Figures 1.9a and 1.9b display dotplots for (a) a sample of n = 10 observations and (b) a sample of n = 100 observations.

Dotplot for n = 10 Observations

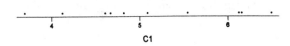

Figure 1.9a

Dotplot of n=100 Observations

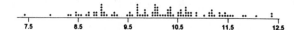

Figure 1.9b

Notice that dotplots of large sets of data help to show where the data pile up and give an indication of how the data are spread along the axis.

Stem and Leaf Plots

A stem and leaf plot is another simple and informative method for displaying data graphically. It belongs to an area of statistics called exploratory data analysis. The stem and leaf plot presents a graph-like picture of the data, while allowing the experimenter to retain the actual observed values of each data value.

In creating a steam and leaf display, we must choose part of the original measurement as the *stem* and the remaining part as the *leaf*. Consider for example a set of four measurements:

624, 538, 465, 552

stem; leaf

46

5

We could use the digit in the hundreds place as the stem, and the remainder (the digits at or to the right of the tens place) as the leaf. In this case, for the observation 624, the digit 6 would be the (stem, leaf) and the digits 24 would be the (stem, leaf). We could also choose to use the digits at or to the left of the tens place as the stem and the remainder as the leaf. Then for the observation 465, the digit(s) _____ would be the stem and the digit(s) _____ would be the leaf. The choice of the stem and leaf coding depends on the nature of the observations at hand.

The stems and leaves are now used as follows:

1. List all the stem digits vertically, from lowest to highest.
2. Draw a vertical line to the right of the stem digits.
3. For each data point, place the leaf digit of that point in the row corresponding to the correct stem.
4. The stem and leaf display may be made more visually appealing by reordering the leaf digits from lowest to highest within each stem row.

Example 1.10

The following data are the 2001 state gasoline tax in cents per gallon for the 50 United States and the District of Columbia.

AL	18.0	IL	19.0	MT	27.0	RI	29.0
AK	8.0	IN	15.0	NE	22.8	SC	16.0
AZ	18.0	IA	20.0	NV	8.0	SD	22.0
AR	19.5	KS	20.0	NH	19.5	TN	20.0
CA	18.0	KY	16.4	NJ	10.5	TX	20.0
CO	22.0	LA	20.0	NM	18.5	UT	24.5
CT	32.0	ME	19.0	NY	29.3	VT	20.0
DE	23.0	MD	23.5	NC	21.2	VA	17.5
DC	20.0	MA	21.0	ND	21.0	WA	23.0
FL	13.1	MI	19.0	OH	22.0	WV	25.3
GA	7.5	MN	20.0	OK	17.0	WI	25.4
HI	16.0	MS	18.4	OR	24.0	WY	14.0
ID	25.0	MO	17.0	PA	25.9		

Source: *The World Almanac and Book of Facts, 2002,* p. 229

Construct a stem and leaf display for the data.

Solution

From an initial survey of the data, the largest observation is 32.0; the smallest observation is 7.5. The data in this example are recorded in tenths of a cent. Suppose that we were to choose the leading digit – that is, the digit in the tens place – as the stem. There would be only _____ stems for the data – namely, 0, 1, 2, and 3. This would not provide a very good visual description of the data.

four

Therefore, we choose to use the digits at and to the left of the ones place as the stem. This is also called the "integer part" of the observation. The leaf will be the remaining portion of the observation – that is, those digits that fall to the (left, right) of the decimal point.

right

The stems from 7 to 32 are listed vertically below and the leaves are entered in the correct row.

Example 1.10

Stem-and-Leaf Display: Gasoline Tax

```
Stem-and-leaf of Gasoline   N  = 51
Leaf Unit = 0.10

    1      7 5
    3      8 00
    3      9
    4     10 5
    4     11
    4     12
    5     13 1
    6     14 0
    7     15 0
   10     16 004
   13     17 005
   18     18 00045
   23     19 00055
   (8)     20 00000000
   20     21 002
   17     22 0008
   13     23 005
   10     24 05
    8     25 0349
    4     26
    4     27 0
    3     28
    3     29 03
    1     30
    1     31
    1     32 0
```

From the stem and leaf display, we may make the following observations:

1. The lowest state gasoline tax is _____ cents per gallon. The highest gasoline tax is _____ cents per gallon.

   7.5
   32.0

2. There are (no, one) extreme values (values that are much higher or much lower than all the others).

   no

3. The data (are, are not) approximately mound-shaped.

   are

4. Most of the gasoline taxes fall between _____ cents per gallon and _____ cents per gallon.

   12
   26

An alternate stem and leaf display can be found if you allow Minitab to select the stems. The following stem and leaf display results when this is the case.

Stem-and-Leaf Display: Gasoline Tax

```
Stem-and-leaf of Gasoline   N  = 51
Leaf Unit = 1.0

   1      0 7
   3      0 88
   4      1 0
   5      1 3
   7      1 45
  13      1 666777
  23      1 8888899999
 (11)     2 00000000111
  17      2 2222333
  10      2 445555
   4      2 7
   3      2 99
   1      3
   1      3 2
```

You can see that in this case stems are used more than once. The rule is to reuse a stem five times: once for twos and threes, again for fours and fives, again for sixes and sevens, again for eights and nines, and finally for zeros and ones. The decimal fraction is dropped and only the tens and ones places in the observations are used. For example, the smallest entry with a stem of 0 and a leaf of 7 represents Georgia's tax of _____ cents. The last entry represents Connecticut's tax of _____ cents.

7.5

32.0

Of the two displays, the second loses some of the information due to the dropping of the last digit in the numbers, but it does a better job of displaying a defining shape of this data set. You might describe the shape as mound-shaped, a shape that is commonly encountered in data analysis.

Both of the stem and leaf displays have a column to the left of the stems that gives the cumulative number of observations starting from both the beginning and end of the set. When the stem class that contains the middle observation in the set if encountered, the number of observation in that class is shown in parentheses. In the second display, there are 7 ordered observations up to and including the number 15, while there are 8 ordered observations that have values of 24 or greater. The middle ordered observation would be in the 26-th position and this number is one of the 12 observations in the stem class 1 for eights and nines.

Example 1.11
The following data represent the planned rated capacities in mega-watts (millions of watts) for the world's 23 largest hydroelectric plants.

20,000	6,000	4,000
18,200	5,328	3,600
13,320	5,225	3,409
10,830	5,020	3,300
10,300	4,678	3,200
7,260	4,500	3,000
6,400	4,500	2,715
6,000	4,150	

Source: *World Almanac and Book of Facts, 2002*, p. 612.

Construct a stem and leaf display for the data.

Solution

We will take the leading digits (the digits at and to the left of the thousands place) to be the stems. To simplify the presentation, take the digit in the hundreds place as the leaf. For example, the observation 13, 320 will have stem 13 and leaf 3. The observation 5,225 will have stem _____ and leaf _____. The stem and leaf display is given below.

5; 2

Stem-and-Leaf Display: Megawatts

```
Stem-and-leaf of Megawatt   N  = 23
Leaf Unit = 1000


   6     0 233333
  (8)    0 44444555
   9     0 6667
   5     0
   5     1 00
   3     1 3
   2     1
   2     1
   2     1 8
   1     2 0
```

2000; 8000

The stem and leaf display indicates a high concentration of capacities between 2000 and 5000 megawatts, with almost all of the data falling between _____ and _____ megawatts. There are four unusually large values.

Sometimes the available stem choices result in a display that contains too many stems (and very few leaves within a stem) or too *few* stems (and many leaves within a stem). In this situation, we may divide the too few stems by stretching them into two or more lines depending on the leaf values with which they will be associated. However, this option tends to complicate the stem and leaf procedure and hence minimizes the major advantage of simplicity in the stem and leaf display.

Interpreting Graphs with a Critical Eye

Graphical displays are very efficient ways to provide you with a quick visual summary of a data set. However, unless you examine the scales used for the graph, you may be led to conclusions that are incorrect. A simple way to distort the information in a graph is to stretch or shrink one or both of the axes.

The number of United States farms from 1950 to 2000 is given in the following table.

Number (Millions)	5.4	4.0	3.0	2.5	2.3	2.2
Year	1950	1960	1970	1980	1990	2000

Source: *The World Almanac and Book of Facts, 2002*, p. 130.

These data are graphed in two different ways in Figure 1.10

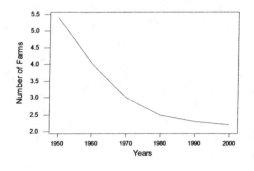

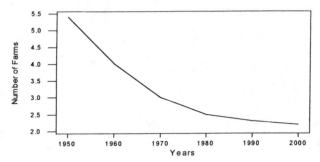

Figure 1.10

In examining graphs for one or more sets of data, you should look at the following aspects of the graph.

Notice that by stretching the horizontal axis and shrinking the vertical axis, what appears as a sharp decline in the number of farms over the period from 1950 to 2000 can be made to look like a slow, gradual decline over that same period.

- Check the vertical and horizontal axes so that you are sure of the scales of measurements.
- Assess the location of the distribution. Where would you place the center of the distribution? If two distributions are involved, do they have the same center?
- Determine the shape of the distribution. Is there one peak in the distribution? If so are the measurements equally distributed to the left and right of this peak? If not, do the measurements tail out to the right or left of the peak?
- Are there any unusually large or small observations that might be possible outliers?

Distributions can be described according to their shape.

A distribution is **symmetric** if the left and right halves of the distribution when divided at the peak value form mirror images.

A distribution is **skewed to the right** if a greater proportion of the measurements lie to the right of the peak values. Distributions that are **skewed to the right** often contain a few unusually large measurements.

A distribution is **skewed to the left** if a greater proportion of the measurements lie to the left of the peak value. Distributions that are skewed left often contain a few unusually small measurements.

Refer to the two stem and leaf plots that follow.

Stem-and-Leaf Display: Y

```
Stem-and-leaf of Y          N  = 200
Leaf Unit = 1.0

     3    3 455
     5    3 67
    11    3 888899
    16    4 00111
    35    4 2222222223333333333
    56    4 444444445555555555555
    84    4 66666666777777777777777777777
  (34)    4 8888888888888999999999999999999999
    82    5 000000000000001111111111
    60    5 2222222222222222233333333
    35    5 4444444445555555555
    16    5 6667777
     9    5 88889
     4    6 0111
```

Figure 1.11a Stem and leaf display of an approximately symmetric
distribution.

Stem-and-Leaf Display: Y

```
Stem-and-leaf of Y          N  = 200
Leaf Unit = 1.0

     1    0 1
    17    0 2222223333333333
    46    0 444444444455555555555555555555
    86    0 66666666666666666666677777777777777777777
  (36)    0 888888888888888888899999999999999999
    78    1 0000000000000000000001111111111111111
    43    1 22222222222222223333
    23    1 444444445555
    11    1 6777
     7    1 9
     6    2 0001
     2    2 23
```

Figure 1.11b Stem and leaf display of a distribution that is skewed
right

Notice that in the distribution in Figure 1.11a, the date are approximately symmetrically distributed about the value 49. The distribution in Figure 1.11b, however, is not symmetric, but has a larger
proportion of the observations to the right of the peak value of 7, and
would therefore be categorized as being skewed to the

right   _____.

1.5 RELATIVE FREQUENCY HISTOGRAMS

The methods in earlier sections provide useful data summaries for categorical data such as political affiliation or for discrete variables such as the number of children in a family, or the number of vehicles owned. Bar and pie charts allow you to display the frequencies or amounts that correspond to each category or to the frequency of the observed values of a discrete variable. You have seen that line charts are useful for plotting time series, data taken at equally spaced intervals of time. You have also used a dotplot to show how the observations on a continuous variable were distributed along a horizontal axis. We would like to find a way to group observations on a continuous variable that would allow us to produce a bar chart similar to one for categorical or discrete data.

An alternative type of chart for a grouped continuous variable is called a **relative frequency histogram**, in which an arbitrary number of adjacent nonoverlapping intervals are defined, and the fraction or proportion of the observations in each interval is plotted like a bar chart. The number of intervals should be between 5 and _____, with large   20
data sets requiring more intervals. These intervals define classes that are similar to the categories used for qualitative data. Hence, the intervals must be defined so that every observation has a class to which it can be assigned, and no observation can belong to more than one class. The next example will show you how to construct a relative frequency histogram.

Example 1.8
The following data are the numbers of correct responses on a recognition test consisting of 30 items, recorded for 25 students:

25	29	23	27	25
23	22	25	22	28
28	24	17	24	30
19	17	23	21	24
15	20	26	19	23

1. First find the highest score, which is _____, and the low-   30
    est score, which is _____. These two scores indicate that   15
    the measurements have a range of 15.
2. To determine how the scores are distributed between 15 and 30,
    we divide this interval into subintervals of equal length. The interval from 15 to 30 could be divided into from 5 to 20 subintervals, depending on the number of measurements available. Wishing to obtain about 7 subintervals, we determine a suitable width by dividing $30 - 15 = 15$ by 7. The integer _____ seems to be   2
    a satisfactory subinterval width for these data.
3. Since the smallest value is 15, we can form subintervals beginning with 15 up to *but not including* 17, from 17 up to *but not including* 19, and so on. Hence we would use the boundary points 15, 17, 19,

21, 23, 25, 27, 29, and 31. In this way, each measurement belongs to one and only one subinterval or class.

4. You can now proceed to tabulate the number of measurements in each of the defined subintervals, and record the class frequencies.

Class	Class Boundaries	Tally	Class Frequency	Relative Frequency
1	15 to < 17	|	1	.04
2	17 to < 19	||	2	.08
3	19 to < 21	|||	3	.12
4	21 to < 23	|||	3	.12
5	23 to < 25	₩₩ ||	7	.21
6	25 to < 27	||||	4	.16
7	27 to < 39	|||	3	.12
8	39 to < 31	||	2	.08

5. The number of measurements that fall in a class is called the class frequency. Of the total number of measurements, the fraction that falls in a class is called the _____ frequency.

relative

As a check on your tabulation, remember that for k classes:

n

a. The sum of the frequencies equals _____.

1

b. The sum of the relative frequencies equals _____.

6. With the data tabulated, you can now construct a relative frequency histogram to describe these data by plotting relative frequency against the classes.

A frequency or relative frequency histogram is easily found using Minitab's graph menu. Notice that the two histograms that follow are identical except for the vertical scale.

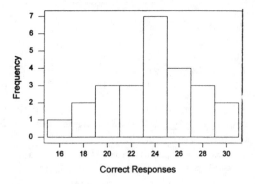

Figure 1.12a

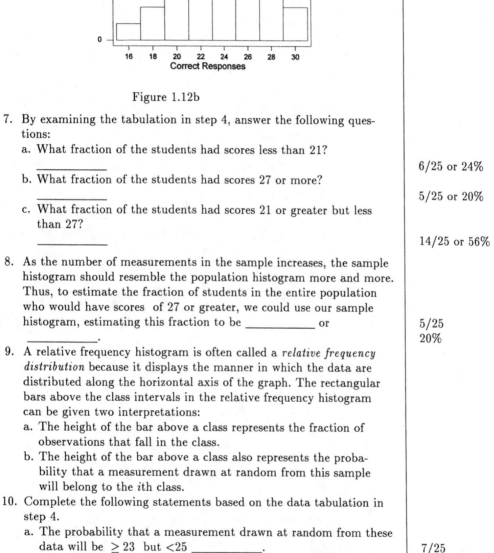

Figure 1.12b

7. By examining the tabulation in step 4, answer the following questions:

a. What fraction of the students had scores less than 21?

_____   6/25 or 24%

b. What fraction of the students had scores 27 or more?

_____   5/25 or 20%

c. What fraction of the students had scores 21 or greater but less than 27?

_____   14/25 or 56%

8. As the number of measurements in the sample increases, the sample histogram should resemble the population histogram more and more. Thus, to estimate the fraction of students in the entire population who would have scores of 27 or greater, we could use our sample histogram, estimating this fraction to be _____ or   5/25

_____.   20%

9. A relative frequency histogram is often called a *relative frequency distribution* because it displays the manner in which the data are distributed along the horizontal axis of the graph. The rectangular bars above the class intervals in the relative frequency histogram can be given two interpretations:

a. The height of the bar above a class represents the fraction of observations that fall in the class.

b. The height of the bar above a class also represents the probability that a measurement drawn at random from this sample will belong to the *i*th class.

10. Complete the following statements based on the data tabulation in step 4.

a. The probability that a measurement drawn at random from these data will be ≥ 23 but <25 _____.   7/25

b. The probability that a measurement drawn at random from these data will be greater than or equal to 19 is _____.   22/25

c. The probability that a measurement drawn at random from these these data will be less than 25 is _____.   16/25

Example 1.9

The following data are the response times in seconds for $n = 25$ first-grade children to arrange three objects by size:

5.2	3.8	5.7	3.9	3.7
4.2	4.1	4.3	4.7	4.3
3.1	2.5	3.0	4.4	4.8
3.6	3.9	4.8	5.3	4.2
4.7	3.3	4.2	3.8	5.4

Construct a relative frequency histogram for these data.

Solution

Verify the entries in the table.

Class	Class Boundaries	Tally	Class Frequency	Relative Frequency
1	2.3 to < 3.0	I	1	.04
2	3.0 to < 3.5	III	3	.12
3	3.5 to < 4.0	IIII I	6	.24
4	4.0 to < 4.5	IIII II	7	.28
5	4.5 to < 5.0	IIII	4	.16
6	5.0 to < 5.5	III	3	.12
7	5.5 to < 6.0	I	1	.05

Complete the following statements based on the preceding tabulation and Figure 1.13.

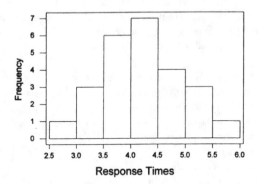

Figure 1.13

1. The probability that a measurement drawn at random from this sample is greater than or equal to 4.5 is _____.

   8/25

2. The probability that a measurement drawn at random from this sample is less than 3.5 is _____.

   4/25

3. An estimate of the probability that a measurement drawn at random from the sampled population would be in the interval from 3.5 or greater but less than 4.5 is _____.

   13/25

In conclusion, the steps necessary to construct a frequency distribution are the following.

1. Determine the number of classes depending on the uniformity and the number of observations. It is usually best to have from _____ to _____ classes. You can use the following table as a guide for selecting an appropriate number of classes. You may use more classes or fewer classes than given in the table if it makes the graph more descriptive or easier to construct.

5; 20

Sample Size	25	50	100	200	500
Number of Classes	6	7	8	9	10

2. Determine the class width by dividing the range by the number of _____ and adjusting the resulting quotient to obtain a convenient figure. With the exception of the first and last classes, all classes should be of _____ width.

classes

equal

3. Locate the class boundaries. Class boundaries should be chosen using the *left-inclusion* and *right-inclusion* rule. With these boundaries, it will (impossible, likely) for a measurement to fall on a class boundary.

impossible

Self-Correcting Exercises 1C

1. In problem 2 in SCE-1B the civilian work force for the period 1995-2000 was given as follows.

Years	Employed (× 10^6)		Unemployed (× 10^6)	
1995	124.	9	7	.4
1996	126.	7	7	.2
1997	129.	6	6	.7
1998	131.	4	6	.2
1999	133.	5	5	.9
2000	134.	3	5	.7

Source: *The World Almanac and Book of Facts*, 2002, p. 139

a. Suppose you wished to make the rise in the number of employed appear as large and dramatic as possible. Construct a bar chart that would accomplish this goal.

b. Use this same approach to make the drop in unemployment over this same period look as large as possible.

2. The average weekly unemployment benefit amount (to the nearest dollar) for each of the 50 states (state programs only) follows.

AL	$159	IN	$222	NE	$188	SC	$190
AK	190	IA	238	NV	222	SD	181
AZ	163	KS	247	NH	217	TN	189
AR	210	KY	225	NJ	290	TX	227
CA	160	LA	182	NM	180	UT	213
CO	256	ME	202	NY	247	VT	216
CT	258	MD	212	NC	231	VA	204
DE	215	MA	293	ND	210	WA	281
FL	220	MI	244	OH	236	WV	198
GA	212	MN	290	OK	214	WI	233
HI	284	MS	157	OR	233	WY	207
ID	209	MO	186	PA	264		
IL	252	MT	188	RI	253		

Source: *The World Almanac and Book of Facts, 2002,* p. 139

a. Use these data to construct a dotplot. Comment on the shape of this distribution.

b. Present these data using a stem and leaf plot. Would your comment on the shape of this distribution differ from that in part a?

c. Use these data to construct a relative frequency histogram. Compare this display with the stem and leaf plot. Are these any apparent differences?

3. Examine the dotplots for the two data sets that follow.

	Set 1	
5	4	4
5	6	4
6	5	3
6	5	7
6	7	4
5	5	6
5	6	2
5	7	5
7		

	Set 2	
4	5	3
5	11	3
4	4	12
2	3	4
9	7	2
6	5	7
4	10	4
8	3	8
4		

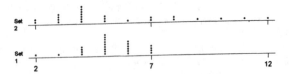

   a.   How would you characterize the shape of the distribution of the first data set?

   b.   How would you characterize the shape of the distribution of the second data set?

   c.   Does either of the data sets appear to contain outliers?

4.   Construct a stem and leaf plot for each of the data sets in Exercise 3. Would you change your description of the two data sets? Are any of the points described as outliers?

5.   Construct a relative frequency histogram for each of the data sets in Exercise 3. How do the histograms compare with the stem and leaf plot in Exercise 4?

Solutions to SCE 1C

1. a. See the bar charts in SCE 1B. To make the rise in the number of employed look large, you should stretch the scale on the vertical axis.

   b. To make the drop in unemployment look large, again stretch the scale on the vertical axis.

2.   Since very few of the values actually repeat, the dotplot does not allow the "stacking" of data points to display the distribution's shape. The stem and leaf plot shows the distribution to be mound-shaped with heavy tails.

Unemployment Insurance Benefits (Dollars)

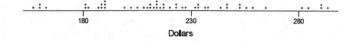

Dollars

Stem-and-Leaf Display: Dollars

```
Stem-and-leaf of Dollars    N  = 50
Leaf Unit = 1.0

    2    15  79
    4    16  03
    4    17
   11    18  0126889
   14    19  008
   18    20  2479
   (9)   21  002344567
   23    22  02257
   18    23  13368
   13    24  477
   10    25  2368
    6    26  5
    5    27
    5    28  14
    3    29  003
```

c.   Since the range of the data is $R = 297 - 157 = 140$ for $n = 50$ observations, we choose to use ten intervals of length 15, beginning at 155. the table below shows the relative frequencies for each class.

Interval	f	f/n
155 to < 170	4	.08
170 to < 185	3	.06
185 to < 200	7	.14
200 to < 215	10	.20
215 to < 230	8	.16
230 to < 245	6	.12
235 to < 260	6	.12
260 to < 275	1	.02
275 to < 290	2	.04
290 to < 305	3	.06

The relative frequency histogram shows a somewhat skewed shape.

Unemployment Insurance Benefits (in Dollars)

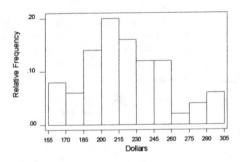

3.   a-b. The second set is skewed to the right with a long right tail indicating a few unusually large observations. The first set is skewed slightly to the left, but is not as extreme as the second set.

c.   The most extreme observations in the right tail of Set 2 might be considered outliers.

4.   The stem and leaf plots generated my Minitab are shown below. The outliers are trimmed by Minitab and are displayed at the top (LO) or bottom (HI) of the plots.

Stem-and-Leaf Display: Set 1 and Set 2

```
Stem-and-leaf of Set 1      N  = 25
Leaf Unit = 0.10
        LO  20, 30,

     6    4 0000
    (9)   5 000000000
    10    6 000000
     4    7 0000

Stem-and-leaf of Set 2      N  = 25
Leaf Unit = 0.10

     2    2 00
     6    3 0000
    (7)   4 0000000
    12    5 000
     9    6 0
     8    7 00
     6    8 00
     4    9 0
     3   10 0
     2   11 0

        HI  120
```

5. The relative frequency histograms are shown below. The shapes remain nearly the same.

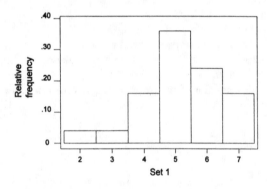

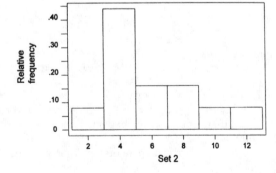

KEY CONCEPTS

I.   *How Data are Generated*
1. Experimental units, variables, measurements
2. Samples and populations
3. Univariate, bivariate, and multivariate data

II. *Types of Variables*
1. Qualitative or categorical
2. Quantitative
   a. Discrete
   b. Continuous

III. *Graphs for Univariate Data Distributions*
1. Qualitative or categorical data
   a. Pie charts
   b. Bar charts
2. Quantitative data
   a. Pie and bar charts
   b. Line charts
   c. Scatterplots or dotplots
   d. Stem and leaf plots
   e. Relative frequency histograms
3. Describing data distributions
   a. Shapes—symmetric, skewed left, skewed right, unimodal, bimodal
   b. Proportion of measurements in certain intervals
   c. Outliers

Exercises

1. The United States exports and imports (in billions of dollars) for the period 1994-2000 are given in the following table.

Year	1994	1995	1996	1997	1998	1999	2000
Exports	$513	$585	$626	$689	$682	$696	$782
Imports	$683	$743	$795	$871	$912	$1,025	$1,281

Source: *The World Almanac and Book of Facts, 2002,* p. 220

a. Display the imports for this period using a bar chart.
b. Display the exports for this period using a bar chart.
c. Use a line chart to track exports and imports from 1994 to 2000.
d. Are there any overall conclusions that you can draw from these charts?

2. The most popular colors for compact and sports cars in a recent year are given in the table that follows.

Color	Percentage
Silver	22.3
Black	14.4
White	11.4
Light Brown	9.9
Medium/Dark Green	9.8
Medium Red	8.3
Bright Red	7.6
Blue	7.1
Teal	2.6
Other	6.8

Source: *The World Almanac and Book of Facts, 2002*, p. 227

Use an appropriate graphical display to present these data.

3. The fuel efficiency of new passenger cars made in the United States over the last 35 years is given in the next table. Use a line chart to track the performance efficiency during this period of time. Is there a trend? Any spectacular changes in this time series?

Year	mpg
1960	15.5
1965	15.4
1970	14.1
1975	15.1
1980	22.6
1985	26.3
1990	26.9
1995	28.3
1999	28.1

Source: *The World Almanac and Book of Facts, 2002*, p.226.

4. The total retail advertising expenditures during a recent year (in trillions of dollars) are given in the following table.

Medium	Expenditures
Newspapers	$4,006
Magazines	804
Television	3,064
Radio	34
Cable	505

a. What graphical techniques are available to present these data?
b. Use the technique that would best dos[;au the salient aspects of these data.

5. The following data are the ages of employees in years of a recently organized small manufacturing firm.

51	44	19	55	42	36	50	52	46
23	27	21	26	33	28	41	32	30
49	32	44	41	34	32	35	34	
30	21	37	18	29	23	23	27	
25	35	31	59	39	38	43	28	

a. Construct a stem and leaf plot for these data. How would you describe the shape of this distribution?

b. Construct a relative frequency histogram for these data using a class width of 5 beginning with the lower limit of the first interval equal to 15. Does the histogram provide any information that is not already evident in the stem and leaf plot?

6. The ages (in months) at which 50 children were first enrolled in a preschool are listed below.

38	40	30	35	39
47	35	34	43	41
32	34	41	30	46
55	39	33	32	32
42	50	37	39	33
40	48	36	31	36
36	41	43	48	40
35	40	30	46	37
45	42	41	36	50
45	38	46	36	31

a. Present a stem and leaf plot for this set of data.

b. Construct a frequency distribution for these data beginning with the lower boundary of the first class equal to 30 and a class width of 5 months.

c. Construct a relative frequency histogram for these data based upon the frequency distribution constructed in part b.

d. Compare the two graphical displays in parts a. and c. Are there any significant differences that would cause you to choose one as the better method to display these data?

e. What proportion of the children were aged 35 months or more, but less than 45 months of age when first enrolled in preschool?

f. If one child were selected at random from this group of children, what is the chance that the child was aged 40 or more but less than 50 months when first enrolled in preschool?

7. The annual rates of profit (in percentages) on stockholders' equity after taxes for 32 industries are given as follows.

10.6	10.8	14.8	10.8
12.5	6.0	10.7	11.0
14.6	6.0	12.8	10.1
7.9	5.9	10.0	10.6
10.8	16.2	18.4	10.7
10.6	13.3	8.7	15.4
6.5	10.1	8.7	7.5
11.9	9.0	12.0	9.1

a. Present a stem and leaf plot of these data. How would you describe the shape of the resulting distribution? Are there one or more outliers?

b. Present a relative frequency distribution of these data using a class width of 2 percentage points, with the lower boundary of the first class at 5.5.

c. Use the relative frequency distribution in part b. to construct a relative frequency histogram for these data.

d. What proportion of the observations were equal to or greater than 7.5, but less than 15.5?

e. If one observation is drawn from this set at random, what is the chance that it will be greater than or equal to 13.5?

Chapter 2
Describing Data with Numerical Measures

2.1 DESCRIBING A SET OF DATA WITH NUMERICAL MEASURES

The chief advantage of a graphical method is its visual representation of data. Many times, however, we are restricted to reporting our data verbally; a graphical method of description cannot be used. The greatest disadvantage to a graphical method of describing data is its unsuitability for making inferences because it is difficult to give a measure of goodness for a graphical inference. Therefore, we turn to *numerical descriptive measures.* We seek a set of numbers that characterizes the frequency distribution of the measurements and at the same time will be useful in making inferences.

We will distinguish between numerical descriptive measures for a population and those associated with a set of sample measurements. A numerical descriptive measure calculated from all the measurements in a population is called a (statistic, parameter). Those numerical descriptive measures calculated from sample measurements are called _____.

parameter
statistics

Numerical descriptive measures are classified into two important types:

1. Measures of **central tendency** locate in some way the "center" of the data or frequency distribution.
2. Measures of **variability** measure the "spread" or dispersion of the data or frequency distribution.

Using measures of both types, the experimenter is able to create a concise numerical summary of the data.

2.2 MEASURES OF CENTER

We first consider two of the more important measures of center that attempt to locate the center of the frequency distribution.

The **mean** of a set of n measurements $x_1, x_2, \ldots, x_n$ is defined to be the sum of the measurements divided by n. The symbol $\bar{x}$ is used to designate the sample mean, whereas the Greek letter μ is used to designate the population mean.

The sample mean has very desirable properties as an inference maker. In fact, we will use $\bar{x}$ to estimate the population mean, μ. To indicate the sum of the measurements, we will use the Greek letter Σ (sigma). Then Σx_i will indicate the sum of all the measurements that have been denoted by the symbol x. Using this summation notation, we can define the sample mean by formula as

$$\bar{x} = \frac{\sum\limits_{i=1}^{n} x_i}{n}$$

The numbers above and below the Greek letter Σ denote the values of i for which the summation is performed; that is,

$$\sum_{i=1}^{n} x_i = x_1 + x_2 + \cdots + x_n$$

We will write Σx_i to mean "the sum of all the x measurements. Using this notation we write the formula for the sample mean as follows:

Sample mean: $\bar{x} = \dfrac{\Sigma x_i}{n}$

Population mean: μ.

Example 2.1
Use a dotplot to display the $n = 5$ measurements 2, 5, 7, 10, 11, 13. Find the sample mean of these observations and compare its value to what you would consider to be the center of the observations in the dotplot.

Solution

The center of the dotplot seems to be between 7 and 8. To calculate the mean, we find

48

$$\sum x_i = \underline{\hspace{3cm}}$$

48

$$\bar{x} = \frac{\sum x_i}{n} = \frac{\underline{\hspace{2cm}}}{6}$$

8

$$\bar{x} = \underline{\hspace{3cm}}$$

In addition to being an easily calculated measure of central tendency, the mean is also easily understood by all users. The calculation of the mean utilizes all the measurements and can always be found exactly.

One disadvantage of using the mean to measure central tendency is well known to any student who has had to pull up one low test

is

score: the mean (is, is not) greatly affected by extreme values. For example, you might be unwilling to accept, say, an average property value of $200,000 for a given area as a good measure of the middle property value if you knew that (1) the property value of a residence owned by a millionaire was included in the calculation and (2) excluding this residence, the property values ranged from $60,000 to $95,000. A more realistic measure of central tendency in this situation might be the property value such that 50% of the property values are less than this value and 50% are greater.

The **median** of a set of n measurements $x_1, x_2, \ldots, x_n$ is the value of x that falls in the middle when the measurements are arranged in order of magnitude. When n is odd, the median is the measurement with rank $(n+1)/2$. When n is **even**, the median is the simple average of the measurements with ranks $n/2$ and $(n/2)+1$—that is, the average of the two middle measurements.

Example 2.2

Find the median of the following set of measurements:

5, 3, 2, 7, 4

Solution

1. Arranging the measurements in order of magnitude, we have

2, 3, 4, 5, 7
    ↑

the middle observation, marked with an arrow, is in the center of the ordered measurements. Hence the median is

4

$$m = \underline{\hspace{3cm}}.$$

2. The median will be the _____ ordered value because
   $(n + 1)/2 = 6/2 = 3$. Hence, the median is _____.

Example 2.3
Find the median of the following set of measurements:

   10, 8, 13, 14, 9, 8

Solution
1. Arranging the measurements in order of magnitude, we have

   8, 8, 9, 10, 13, 14
   ↑

   The center of this set of measurements lies between 9 and 10, and
   would be taken to be 9.5.

2. Since $n = 6$ is even, the median will be the average of the
   _____ and _____ ordered values. Hence,

$$\text{median} = \frac{\underline{} + \underline{}}{2}$$

$$= \underline{}$$

   as shown in step 1.

third, fourth

9; 10

9.5

Example 2.4
Find the mean and median of the following data:

   5, 7, 8, 10, 10, 11, 13, 14
         ↑
         m

Solution

1. $\displaystyle\sum_{i=1}^{8} x_i = \underline{} \qquad \bar{x} = \frac{\Sigma x_i}{n} = \frac{\underline{}}{8} = \underline{}$

78; 78; 9.75

2. To find the median, we note that the measurements are already
   arranged in order of magnitude and that $n = 8$ is even. Therefore,
   the median will be the average of the fourth and fifth ordered values:

$$\text{median} = \frac{10 + 10}{2} = \underline{}$$

10

   In Example 2.4, the mean and median were reasonably close numerical values as measures of central tendency. However, if the measurement $x_9 = 30$ were added to the eight measurements given, the
recalculated mean would be $\bar{x} = $ _____, but the median would
remain at 10, reflecting the fact that the median is a positional average
unaffected by extreme values.

12

   Measures of center do not always produce the same value. When
the distribution of the data is symmetric, then the mean and the median
are equal as shown in Figure 2.2.

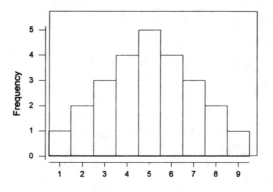

Figure 2.2

inflated

is not affected

larger

When there are one or more large values in the data set and the distribution exhibits skewing, the mean, which depends upon the values of each observation is _____ by these large values. However, the median, which measures the center using the middle value of the observations when arranged from small to large, (is, is not affected) by large values in the right tail of the distribution, as shown in Figure 2.3. In situations like this, the mean will always be (larger, smaller) than the median. If a distribution is strongly skewed, the median is the appropriate measure of the center of a distribution.

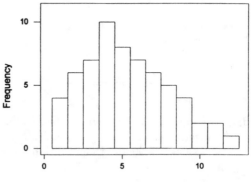

Figure 2.3

Another measure of the center of a set of observations is the value that occurs most frequently in the set, or if the data have been grouped into a frequency table or histogram, the category with the largest frequency could be used to locate the center of the set. This measure of center is called the mode.

The **mode** is the category with the highest frequency or the
most frequently occurring value of x. When the data have
been grouped using a frequency table or histogram, the class
with the highest frequency is called the **modal class**.

For the data in Figure 2.2 the modal class is 5, while for the data in
Figure 2.3, the modal class is 4. It is possible for a set of data to
exhibit two or more modes. When this happens, the indication is that
there is a mix of two or more populations. For example, in measuring
the length of fish taken from a lake in one season, you may see two
modes when constructing a histogram for these data. The two modes are
no doubt due to the mixture of the populations of male and female fish.
In this case, we would report that the data are *bimodal*. Distributions
with two or more modes should be examined more carefully to determine whether one or more characteristics are available that would be
useful in determining which individuals belong to the populations
represented in the sample.

Self-Correcting Exercises 2A

1. Find the mean, median and mode of the following observations.

    2, 5, 4, 6, 4, 3, 6, 7, 4, 3, 5

2. a. Find the mean, median and mode of the following observations
    and compare their values.

    5, 7, 3, 5, 6, 8, 5, 6,4, 6, 25

   b. Eliminate the last observation $x = 25$ and then find the mean,
      median and mode. How do these values compare with those
      found using the full data set?
   c. How do possible outliers (such as 25) affect the values of these
      three measures of center?

3. Suppose that as a prospective buyer you wish to compare the prices
   of homes in several parts of town.
   a. If the homes in one area of town consisted of tract homes with
      about the same square feet of living space, what measure would
      you use in finding the typical cost of a home in this area?
   b. If the homes in another area of town consisted of homes which
      were built by individual contractors with varying square feet
      of living area including several large custom homes, what
      measure might you wish to use to find the typical cost of a home
      in this area?

4. The following data are the ages (in months) at which $n = 50$ children were first enrolled in a preschool.

38	40	30	35	39
47	35	34	43	41
32	34	41	30	46
55	39	33	32	32
42	50	37	39	33
40	48	36	31	36
36	41	43	48	40
35	40	30	46	37
45	42	41	36	50
45	38	46	36	31

a. Find the mean and median for these data.

b. Construct a frequency histogram for these data using 30 as the lower limit of the first class, and a class width of 5 months. (See Chapter 1, Exercise 6) What is the modal class? What is its endpoint?

c. Compare the values of the mean, median and the midpoint of the modal class.

d. From the histogram, comment on the shape of the distribution. Do the values of the mean and the median support your answer concerning the shape of the distribution?

5. The average weekly unemployment benefit amount for the 50 US states are given to the nearest dollar.

$159	$222	$188	$190	$284
190	238	222	181	209
163	247	217	189	252
210	225	290	227	157
160	182	180	213	186
256	202	247	216	188
258	212	231	204	233
215	293	210	281	264
220	244	236	198	253
212	290	214	233	207

Source: *The World Almanac and Book of Facts, 2002*, p.159

a. Find the mean and median for these data.

b. Construct a histogram and find the midpoint of the modal class.

c. Compare these three measures of center. Using only the measures of center, would you expect this distribution to be fairly symmetric or skewed? Is your conclusion supported b y the shape of the histogram in part b?

Solutions to SCE 2A

1. a. Arrange the set of data in order of ascending magnitude.

    2, 3, 3, 4, 4, 4, 5, 5, 6, 6, 7

    median $= 4$; $\bar{x} = \Sigma x_i/n = 49/11 = 4.45$; mode $= 4$.

2. a. Arrange the set of data in order of ascending magnitude.

    3, 4, 5, 5, 5, 6, 6, 6, 7, 8, 25

    median $= 6$; $\bar{x} = \Sigma x_i/n = 80/11 = 7.27$; modes $= 4$ and 6.
    b. If the value $x = 25$ is removed, median $= (5+6)/2 = 5.5$; $\bar{x} = \Sigma x_i/n = 55/10 = 5.5$; modes $= 5$ and 6. The mean is smaller.
    c. The mean is affected by the outlier, while the median and mode are not.

3. a. If all homes will cost about the same amounts, and there are no unusually high or low costing homes in the tract, use the sample mean.
    b. Since the custom homes may be much more expensive than the other homes, the average or mean cost may not be a good measure of center. You should use the median cost.

4. a. The data are arranged in order of ascending magnitude below.

30	30	30	31	31	32	32	32	33	33
34	34	35	35	35	36	36	36	36	36
37	37	38	38	39	39	39	40	40	40
40	41	41	41	41	42	42	43	43	45
45	46	46	46	47	48	48	50	50	55

    The median for $n = 50$ observations is the average of the 25th and 26th ordered observations or

    $$\text{median} = \frac{39 + 39}{2} = 39$$

    and $\bar{x} = \dfrac{\Sigma x_i}{n} = \dfrac{1954}{50} = 39.08$.

    b. The frequency histogram is shown below. The modal class is 35 to < 40 with midpoint $(35 + 40)/2 = 37.5$.
    c. Notice that the mean is shifted slightly to the right of both the median and the midpoint of the modal class.
    d. The distribution in part b is slightly skewed to the right, which explains why the mean is shifted to the right of the median and the midpoint of the modal class.

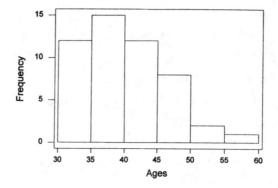

5. a.   The data are arranged in order of ascending magnitude.

157	189	212	225	252
159	190	212	227	253
160	190	213	231	256
163	198	214	233	258
180	202	215	233	264
181	204	216	236	281
182	207	217	238	284
186	209	220	244	290
188	210	222	247	290
188	210	222	247	293

The median for n = 50 observations is the average of the 25th and 26th ordered observations or

$$\text{median} = \frac{215 + 216}{2} = 215.5$$

and

$$\bar{x} = \frac{\Sigma x_i}{n} = \frac{10998}{50} = 219.96$$

b.   Using the relative frequency histogram from Exercise 2, SCE 1C, the modal class is 200 to < 215 which has midpoint 207.5.

c.   Since the mean is shifted to the right of the median and the modal class, the distribution would exhibit skewing to the right. See the histogram in Exercise 2, SCE 1C.

2.3 MEASURES OF VARIABILITY

Having found measures of central tendency, we next consider measures of the variability or dispersion of the data. A measure of variability is necessary because a measure of central tendency alone does not adequately describe the data. Consider the two histograms in Figure 2.4.

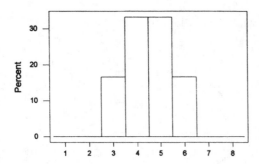

Figure 2.4a

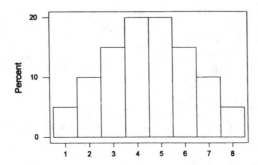

Figure 2.4b

Both sets of data have a mean equal to _____. However, the second set of measurements displays much more variability about the mean than does the first set.

4.5

In addition to a measure of central tendency, a measure of variability is indispensable as a descriptive measure for a set of data. A manufacturer of machine parts would want very (little, much) variability in her product in order to control oversized or undersized parts, whereas an educational testing service would be satisfied only if the test scores showed a (large, small) amount of variability in order to discriminate among people taking the examination.

little

large

We have already used the simplest measure of variability, the range.

The **range** of a set of measurements is the difference between the largest and smallest measurements.

Example 2.5

Find the range for each of the following sets of data:

Set I:	23	73	34	74
	28	29	26	17
	88	8	52	49
	37	96	32	45
	81	62	23	62

8; 88

Range = 96 − _____ = _____

Set II:	8.8	6.7	7.1	2.9
	9.0	0.2	1.2	8.6
	6.3	6.4	2.1	8.8

0.2; 8.8

Range = 9.0 − _____ = _____

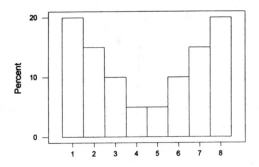

Figure 2.5a

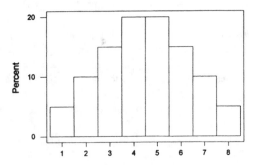

Figure 2.5b

Although the range is a simply calculated measure of variation, it alone is not adequate. Both distributions in Figure 2.5 have the same range, but they display different variability.

We base the next important measure of variability on the dispersion of the data about their mean. Consider two sets of measurements. The first set consists of N measurements and represents the entire population of interest to the experimenter. The second set is a sample of n measurements taken from this population. Define μ to be the population mean. For a finite population consisting of N measurements,

$$\mu = \frac{\Sigma x_i}{N}$$

Although the indexes above and below the summation sign are omitted, the sum is found by summing over all N observations in the population.

$$\sum_{i=1}^{N} x_i = x_1 + x_2 + \cdots + x_N$$

Similarly, we define

$$\bar{x} = \frac{\Sigma x_i}{n} = \frac{x_1 + x_2 + \cdots + x_n}{n}$$

to be the sample mean. Finally, we define the quantity $(x_i - \mu)$, or $(x_i - \bar{x})$ (depending on whether the set of measurements is a population or a sample), as the ith deviation from the mean. Large deviations indicate (more, less) variability of the data than do small deviations. We could utilize these deviations in different ways.

more

1. If we attempt to use the average of the N (or n) deviations, we find that the sum of the deviations is _____. To avoid a zero sum, we could use the average of the absolute values of the deviations. This measure, called the **mean deviation**, is difficult to calculate, and we cannot easily give a measure of its goodness as an an inference maker.

zero

2. A more efficient use of data is achieved by averaging the sum of squares of the deviations. For a finite population of N measurements, this measure is called the **population variance.**

The **population variance** is defined as the average squared deviation from the mean and is given by

$$\sigma^2 = \frac{\Sigma(x_i - \mu)^2}{N}$$

Large values of σ^2 indicate (large, small) variability, whereas (large, small) values indicate small variability.

large
small

Since the units of σ^2 are not the original units of measurement, we can return to these units by defining the standard deviation.

The **standard deviation** σ is the positive square root of the variance; that is,

$$\sigma = \sqrt{\sigma^2} = \sqrt{\frac{\sum (x_i - \mu)^2}{N}}$$

For a sample of size n, it would seem reasonable to define the sample variance in a similar way, using $\bar{x}$ in place of μ and n in place of N. However, we choose to modify this definition slightly.

Since our objective is to make inferences about the population based on sample data, it is appropriate to ask whether the sample mean and variance are good estimators of their population counterparts, μ and σ^2. The fact is that $\bar{x}$ is a good estimator of μ, but the quantity $\sum (x_i - \bar{x})^2 / n$ appears to (underestimate, overestimate) the population variance σ^2 when the sample is small.

The problem of underestimating σ^2 can be solved by dividing the sum of squares of deviations by $n - 1$ rather than n. We then define

$$s^2 = \frac{\sum (x_i - \bar{x})^2}{n - 1}$$

as the sample variance. The sample standard deviation is then

$$s = \sqrt{s^2} = \sqrt{\frac{\sum (x_i - \bar{x})^2}{n - 1}}$$

Example 2.6
Calculate the sample mean, variance, and standard deviation for the following data:

4, 2, 3, 5, 6

Solution
Arrange the measurements in the following way, first finding the mean, $\bar{x} =$ _____ .

x_i	$x_i - \bar{x}$	$(x_i - \bar{x})^2$
4	0	0
2	−2	4
3	−1	1
5	1	1
6	2	4
$\sum x_i = 20$ $\sum (x_i - \bar{x}) = $ _____		$\sum (x_i - \bar{x})^2 = $ _____

underestimate

4

0; 0

After finding the mean, complete the second column and note that its sum is zero. The variance is

$$s^2 = \frac{\sum (x_i - \bar{x})^2}{n-1} = \frac{}{4} = \underline{}$$

10; 2.5

The standard deviation is

$$s = \sqrt{2.5} = \underline{}$$

1.581

The calculation of $s^2 = \sum (x_i - \bar{x})^2/(n-1)$ requires the calculation of the quantity $\sum (x_i - \bar{x})^2$. To facilitate this calculation, we introduce the identity

$$\sum (x_i - \bar{x})^2 = \sum x_i^2 - \frac{\left(\sum x_i\right)^2}{n}$$

the proof of which is omitted. This computation requires the following:

1. We need the ordinary arithmetic sum of the measurements, $\sum x_i$.

2. We also need the sum of the squares of the measurements, $\sum x_i^2$.

   To calculate this, we *first square* each measurement and *then sum* these squares.

3. To calculate $\left(\sum x_i\right)^2$, we *first sum* the measurements and *then square* this sum.

Example 2.7
Calculate s^2 for Example 2.6.

Solution
Display the data in the following way, finding $\sum x_i$ and $\sum x_i^2$.

x_i	x_i^2
4	16
2	4
3	9
5	25
6	36
$\sum x_i = \underline{}$	$\sum x_i^2 = \underline{}$

20; 90

1. We first calculate

$$\sum (x_i - \bar{x})^2 = \sum x_i^2 - \frac{\left(\sum x_i\right)^2}{n}$$

$$= 90 - \frac{(20)^2}{5}$$

80

$$= 90 - \underline{\hspace{2cm}}$$

10

$$= \underline{\hspace{2cm}}$$

2. Then

10; 2.5

$$s^2 = \frac{\sum (x_i - \bar{x})^2}{n-1} = \frac{\underline{\hspace{1.5cm}}}{5-1} = \underline{\hspace{2cm}}$$

Example 2.8
Calculate the mean, variance, and standard deviation of the following data: 5, 6, 7, 5, 2, 3.

Solution
Display the data in a table.

x_i	x_i^2
5	25
6	36
7	49
5	25
2	4
3	9
$\sum x_i = \underline{\hspace{1.5cm}}$	$\sum x_i^2 = \underline{\hspace{1.5cm}}$

28; 148

4.67

$$\bar{x} = \frac{\sum x_i}{n} = \frac{28}{6} = \underline{\hspace{2cm}}$$

$$\sum (x_i - \bar{x})^2 = \sum x_i^2 - \frac{\left(\sum x_i\right)^2}{n}$$

$$= 148 - \frac{(28)^2}{6}$$

$$= 148 - 130.67$$

17.33

$$= \underline{\hspace{2cm}}$$

17.33; 3.467

$$s^2 = \frac{\sum (x_i - \bar{x})^2}{n-1} = \frac{\underline{\hspace{1.5cm}}}{6-1} = \underline{\hspace{2cm}}$$

The standard deviation is the square root of the variance. Therefore

$$s = \sqrt{s^2} = \sqrt{3.467} = 1.862$$

When working with the variance or standard deviation, remember the following points.

- The larger the value of s^2 or s, the greater the variability of the set of measurements.
- If s^2 or s is equal to zero, all of the measurements have the same value.
- The standard deviation s is computed in order to have a measure of variability in the same units as the observations.

Self-Correcting Exercises 2B

1. Given the $n = 5$ observations 2, 4, 3, 4, 5, calculate
   a. the sample mean $\bar{x}$.
   b. the sample variance and standard deviation using the definition formula.
   c. the sample variance and standard deviation using the short-cut formula. How do these values compare with those found in part b?
   d. If you have a statistical calculator, use it to find $\bar{x}$ and s. Com- these values with those found in parts b and c.

2. Given the $n = 9$ observations 7, 9, 10, 6, 8, 7, 8, 9, 8, calculate
   a. the range.
   b. the mean, variance, and the standard deviation.
   c. find the ratio of the range divided by the standard deviation. The range is approximately how many standard deviations?

3. Refer to the data in SCE 2A, Exercise 4.
   a. Find the variance and standard deviation for these data.
   b. Find the range of these data.
   c. Compare the range and the standard deviation by finding the ratio of the range to the standard deviation. Approximately how many standard deviations is the range?

Solutions to SCE 2B

1. $\sum x_i = 18; \sum x_i^2 = 70; n = 5$

   a. $\bar{x} = \dfrac{\sum x_i}{n} = \dfrac{18}{5} = 3.6$

   b. $s^2 = \dfrac{\sum (x_i - \bar{x})^2}{n-1} = \dfrac{(2-3.6)^2 + \ldots + (5-3.6)^2}{4}$

   $$= \dfrac{(-1.6)^2 + (.4)^2 + (-.6)^2 + (.4)^2 + 1.4^2}{4}$$

$$= \frac{5.2}{4} = 1.3 \text{ and } s = \sqrt{1.3} = 1.14$$

c. $s^2 = \dfrac{\sum x_i^2 - \dfrac{\left(\sum x_i\right)^2}{n}}{n-1} = \dfrac{70 - \dfrac{18^2}{5}}{4} = \dfrac{5.2}{4} = 1.3$ and

   $s = \sqrt{1.3} = 1.14$. Answers are the same.

2.   a. R $= 10 - 6 = 4$.

   b. $\bar{x} = \dfrac{\sum x_i}{n} = \dfrac{72}{9} = 8$;

   $s^2 = \dfrac{\sum x_i^2 - \dfrac{\left(\sum x_i\right)^2}{n}}{n-1} = \dfrac{588 - \dfrac{72^2}{9}}{8} = \dfrac{12}{8} = 1.5$;

   $s = \sqrt{1.5} = 1.225$.

   c. R/$s = 4/1.225 = 3.27$. The range is between 3 and 4 standard deviations.

3.   a. $s^2 = \dfrac{\sum x_i^2 - \dfrac{\left(\sum x_i\right)^2}{n}}{n-1} = \dfrac{78{,}118 - \dfrac{1954^2}{50}}{49} = \dfrac{1755.68}{49} = 35.8302$;

   $s = \sqrt{35.8302} = 5.99$.

   b. Using the sorted data in the solution for Exercise 4, SCE 2A, R $= 55 - 30 = 25$.

   c. R/$s = 25/5.99 = 4.2$. The range is slightly more than 4 standard deviations.

2.4   ON THE PRACTICAL SIGNIFICANCE OF THE STANDARD DEVIATION

Having defined the mean and standard deviation, we now introduce two theorems that will use both these quantities in more fully describing a set of data.

Tchebysheff's Theorem: Given a number k greater than or equal to 1 and a set of n measurements $x_1, x_2, \ldots, x_n$, <u>at least</u> $(1 - 1/k^2)$ of the measurements will lie within k standard deviations of their mean.

The importance of this theorem is that it applies to any set of measurements. It applies to a population using the population mean μ and the population standard deviation σ, and it applies to a sample from a given population using $\bar{x}$ and s, the sample mean and sample standard deviation. Since this theorem applies to _____ set of

any

measurements, it is of necessity a conservative theorem. It is therefore very important to stipulate that _____ $(1 - 1/k^2)$ of the measurements will lie within k standard deviations of their mean.

at least

Complete the following chart for the values of k given:

k	Interval $\bar{x} \pm ks$	Interval Contains at Least the Fraction $(1 - 1/k^2)$
1	$\bar{x} \pm s$	_____
2	$\bar{x} \pm 2s$	_____
3	$\bar{x} \pm 3s$	_____
10	$\bar{x} \pm 10s$	_____

0
3/4
8/9
99/100

Summarizing the results in the last table we can restate the theorem in the following way.

- At least none of the measurements lie in the interval from $\mu - \sigma$ to $\mu + \sigma$.
- At least 3/4 of the measurements lie in the interval from $\mu - 2\sigma$ to $\mu + 2\sigma$.
- At least 8/9 of the measurements lie in the interval from $\mu - 3\sigma$ to $\mu + 3\sigma$.

Although the first statement is not very helpful, the remaining two intervals provide valuable information about the set of measurements. You should also be aware that we do not need to use values of k that are integers. For example, taking $k = 3.5$, the proportion of the measurements falling into the interval $\mu = 3.5\sigma$ to $\mu + 3.5\sigma$ is at least

$$1 - \frac{1}{(3.5)^2} = 1 - .0816 = .9184$$

or approximately 92%.

Example 2.9
The mean and variance of a set of $n = 20$ measurements are 35 and 25, respectively. Use Tchebysheff's Theorem to describe the distribution of these measurements.

Solution
Collecting pertinent information we have

$$\bar{x} = 35, \quad s^2 = 25, \quad s = \sqrt{25} = 5$$

1. At least 3/4 of the measurements lie in the interval $35 \pm 2(5)$, or from _____ to _____.

25; 45

2. At least 8/9 of the measurements lie in the interval $35 \pm 3(5)$, or from _____ to _____.

20; 50

3. At least 15/16 of the measurements line in the interval $35 \pm 4(5)$, or from _____ to _____.

15; 55

Example 2.10
If the mean and variance of a set of $n = 50$ measurements are 42 and 36, respectively, describe these measurements using Tchebysheff's Theorem.

Solution
Pertinent information: $\bar{x} = 42$, $s^2 = 36$, and $s = 6$.

1. At least 3/4 of the measurements lie in the interval $42 \pm 2(6)$, or from _____ to _____.

30; 54

2. At least 8/9 of the measurements lie in the interval $42 \pm 3(6)$, or from _____ to _____.

24; 60

3. At least 15/16 of the measurements lie in the interval $42 \pm 4(6)$, or from _____ to _____.

18; 66

Another rule for describing the variability of a set of data cannot be used for all data, but it can be used when the data exhibit a piling up in the center of the distribution. Such data sets are often referred to as "mound-shaped." The closer the distribution of the data is to the distribution in Figure 2.6, the more accurate will be your results using this rule.

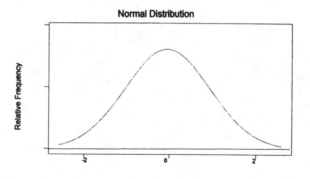

Figure 2.6

Empirical Rule: Given a distribution of measurements that is approximately bell-shaped, the interval
1. $\mu \pm \sigma$ contains approximately 68% of the measurements.
2. $\mu \pm 2\sigma$ contains approximately 95% of the measurements.
3. $\mu \pm 3\sigma$ contains almost all (approximately 99.7%) of the measurements.

This rule holds reasonably well for any set of measurements that has a distribution that is mound-shaped. Bell-shaped or mound-shaped is taken to mean that the distribution has the properties associated with the normal distribution, whose graph is given Figure 2.6.

Example 2.11

A random sample of 100 oranges was taken from a grove and individual weights were measured. The mean and variance of these measurements were 7.8 ounces and 0.36 (ounces)^2, respectively. Assuming the measurements produced a mound-shaped distribution, describe these measurements using the Empirical Rule.

Solution

First find the intervals needed.

k	$\bar{x} \pm ks$	$\bar{x} - ks$	to	$\bar{x} + ks$	
1	$\bar{x} \pm s$	_____		_____	7.2; 8.4
2	$\bar{x} \pm 2s$	_____		_____	6.6; 9.0
3	$\bar{x} \pm 3s$	_____		_____	6.0; 9.6

Then approximately

1. _____% of the measurements lie in the interval from _____ to _____ .

   68
   7.2; 8.4

2. _____% of the measurements lie in the interval from _____ to _____ .

   95
   6.6; 9.0

3. _____% of the measurements lie in the interval from _____ to _____ .

   100 (or 99.7)
   6.0; 9.6

   When n is small, the distribution of measurements (would, would not) be mound-shaped and the Empirical Rule (would, would not) be appropriate in describing these data. Since Tchebysheff's Theorem applies to any set of measurements, it can be used regardless of the size of n.

   would not
   would not

Example 2.12

The following data are the response times in seconds for $n = 25$ first-grade children to arrange three objects by size given in Example 1.9.

5.2	3.8	5.7	3.9	3.7
4.2	4.1	4.3	4.7	4.3
3.1	2.5	3.0	4.4	4.8
3.6	3.9	4.8	5.3	4.2
4.7	3.3	4.2	3.8	5.4

a. Find the mean and standard deviation of these data.
b. Use Tchebysheff's Theorem and the Empirical Rule to describe the distribution of these observations.

Solution

The sum and sum of squares needed to calculate the mean and standard deviation are

$$\sum x_i = 104.9 \text{ and } \sum x_i^2 = 454.81$$

Then

$$\bar{x} = \frac{\sum x_i}{n} = \frac{104.9}{25} = 4.1960$$

and

$$s^2 = \frac{\sum x_i^2 - \frac{\left(\sum x_i\right)^2/n}{n-1}}{n-1} = \frac{454.81 - \frac{(104.9)^2/25}{25}}{24} = \frac{14.6496}{24}$$

$$= .6104 \text{ with } s = \sqrt{.6104} = .781.$$

k	Interval $\bar{x} \pm s$	Frequency	Relative Frequency
1	3.42 – 4.98	17	.68
2	2.63 – 5.76	24	.96
3	1.85 – 6.54	25	1.00

Does Tchebysheff's Theorem apply? Yes–notice that the relative frequencies in the three intervals in the table all exceed the proportions given for Tchebysheff's Theorem. In addition, the observed relative frequencies for these three intervals are very close to the 68%, 95% and almost 100% figures given by the Empirical Rule.

2.5 A CHECK ON THE CALCULATION OF s

For mound-shaped or approximately normal data, we can use the range to check the calculation of s, the standard deviation. According to Tchebysheff's Theorem and the Empirical Rule, at least 3/4 and more likely 95% of a set of measurements will be in the interval $\bar{x} \pm 2s$. Hence, the sample range R should approximately equal $4s$, so that

$$s \approx \frac{R}{4}$$

This approximation requires only that the computed value be of the same order as the approximation.

Example 2.13
Check the calculated value of s for the first set of data given in Example 2.5.

Solution
For these data, the range is $96 - 8 =$ _____, and

$$s \approx \frac{88}{4} = \underline{\hspace{2cm}}$$

88

22

Comparing 22 with the calculated value, 25.46, we (would, would not) have reason to doubt the accuracy of the calculated value.

> would not

In referring to the second set of data in Example 2.5, which consists of 12 measurements, we find that the range is $9.0 - .2 = 8.8$. Hence, an approximation to s using $R \approx 4s$ yields

$$s \approx \frac{8.8}{4} = \underline{\hspace{2cm}}$$

> 2.2

When compared with the calculated value 3.21, this approximation is not as close as the approximation for the first set of data.

Since extreme measurements are more likely to be observed in (large, small) samples, we can adjust the approximation to s by dividing the range by a divisor that depends on the sample size n. A rule of thumb to use in approximating s by using the range is given in the following table:

> large

n	Divide Range by
5	2.5
10	3
25	4
100	5

Example 2.14
Use the range approximation to check the calculated value of s, 3.21, for the second set of data in Example 2.5.

Solution
We know that $R = 8.8$; hence, for $n = 12$ measurements, we use the approximation

$$s \approx \frac{R}{3} = \frac{8.8}{3} = \underline{\hspace{2cm}}$$

> 2.93

which more closely approximates the calculated value of s, 3.21, than did the earlier approximation, 2.2.

Example 2.15
Use the range approximation to check the calculation of s for the data given in Example 2.6.

Solution
For the five measurements, 4, 2, 3, 5, 6, the range is $6 - 2 = \underline{\hspace{2cm}}$. Therefore, an approximation to s is

> 4

$$s \approx \frac{R}{2.5} = \frac{4}{2.5} = \underline{\hspace{2cm}}$$

> 1.6

which closely agrees with the calculated value of 1.581.

Self-Correcting Exercises 2C

1. A set of observations consists of the values:
   10, 5, 1, 10, 7, 3, 5, 2, 3, 8.
   a. Use the range approximation to estimate the value of s, the sample standard deviation. (*Hint:* use the appropriate divisor as given in the table at the end of Section 2.5.)
   b. Find the standard deviation, s. How does it compare to the estimate you found in part a?
   c. Construct a stem and leaf plot for these $n = 10$ observations. Is the data mound-shaped?
   d. Could you use Tchebysheff's Theorem to describe these data? The Empirical Rule? Why or why not?

2. Suppose you are told that the mean and standard deviation of a sample of $n = 500$ observations were $\bar{x} = 50$ and $s = 10$. You know nothing else about the shape of the distribution for these data.
   a. What can be said about the proportion of observations between 40 and 60?
   b. What can be said about the proportion of observations between 30 and 70?
   c. What can be said about the proportion of observations smaller than 30?

3. Suppose now you are told that the data in Exercise 2 are mound-shaped.
   a. What can be said about the proportion of observations between 40 and 60?
   b. What can be said about the proportion of observations between 30 and 70?
   c. What can be said about the proportion of observations smaller than 30?

4. Refer to Exercise 4, SCE-2A.
   a. Find the mean and standard deviation of these data.
   b. Construct a histogram and describe the shape of these data.
   c. Find the actual proportion of observations within the intervals $\bar{x} \pm s$, $\bar{x} \pm 2s$, and $\bar{x} \pm 3s$. How do these observed proportions compare with those given by Tchebysheff's Theorem and the Empirical Rule?

Solutions to SCE 2C

1. a. $s \approx R/3 = (10 - 1)/3 = 3$

   b. $s^2 = \dfrac{\sum x_i^2 - \dfrac{\left(\sum x_i\right)^2}{n}}{n - 1} = \dfrac{386 - \dfrac{54^2}{10}}{9} = \dfrac{94.4}{9} = 10.4889;$

$s = \sqrt{10.4889} = 3.239$, which is close to the estimate in part a.

c. The data is not mound-shaped, as shown by the stem and leaf plot. You could use Tchebysheff's Theorem, but <u>not</u> the Empirical Rule to describe the data.

1	1 0
2	2 0
4	3 00
4	4
(2)	5 00
4	6
4	7 0
3	8 0
2	9
2	10 00

2. a. Since nothing is known about the shape of the distribution, you must use Tchebysheff's Theorem to describe the data. The interval 40 to 60 represents $\bar{x} \pm s \Rightarrow 50 \pm 10$. Since $k = 1$, you can say only that *at least none* of the measurements are in this interval.

b. The interval 30 to 70 represents $\bar{x} \pm 2s \Rightarrow 50 \pm 20$. Since $k = 2$, you can say that *at least 3/4* of the measurements are in this interval.

c. If at least 3/4 of the measurements are between 30 and 70, at most 1/4 of the measurements are outside this interval. Since you know nothing about the shape of the distribution, all of these measurements might be less than 30.

3. a. Using the Empirical Rule, approximately 68% of the measurements will be between 40 and 60.

b. Approximately 95% of the measurements will be in the interval $\bar{x} \pm 2s \Rightarrow 50 \pm 20$ or 30 to 70.

c. From b., there are 5% of the measurements outside the interval from 30 to 70. Since a mound-shaped distribution is symmetric about the mean, $1/2(5\%) = 2.5\%$ will be less than 30.

4. a. From Exercise 3, SCE 2B, $\bar{x} = 35.83$ and $s = 5.99$.

b. From Exercise 4, SCE 2A, the histogram is slightly skewed to the right.

c. $\bar{x} \pm s \Rightarrow 35.83 \pm 5.99$ or 29.84 to 41.82 contains $35/50 = .70$ or 70% of the measurements. $\bar{x} \pm 2s \Rightarrow 35.83 \pm 11.98$ or 23.85 to 47.81 contains $45/50 = .90$ or 90% of the measurements. $\bar{x} \pm 3s \Rightarrow 35.83 \pm 17.97$ or 17.86 to 53.80 contains $49/50 = .98$ or 98% of the measurements. Since the distribution is not quite mound-shaped, the proportions are not exactly as described by the Empirical Rule.

2.6 MEASURES OF RELATIVE STANDING

Occasionally, we wish to know the position of a measurement relative to others in the set. The mean and standard deviation of the set of measurements can be used to calculate a sample z-score used to measure the relative standing of a measurement in a data set. The z-score is defined as follows:

The **sample z-score** corresponding to a measurement x is

$$z\text{-score} = \frac{x - \overline{x}}{s}$$

---

mean

A z-score measures the distance between a measurement and the sample _____ measured in units of standard deviation.

    The magnitude of a z-score takes on meaning when used in conjunction with Tchebysheff's Theorem and the Empirical Rule. At least 3/4 and more likely 95% of the measurements in a set lie within

two

_____ standard deviations of the mean; therefore, z-scores between -2 and $+2$ are highly likely. At least 8/9 and more likely all

three

of the measurements lie within _____ standard deviations of the mean. Therefore, z-scores larger than 2 but less than 3 in absolute value are not very likely, and z-scores larger than 3 in absolute value are very unlikely.

    A set of measurements may contain values that for some reason lie far from the middle of the distribution in either direction. These values lie in the tails of the distribution and are called **outliers**. Sometimes outliers are generated when a mistake is made in recording the data; for example, a number may be misread or mistyped. Perhaps the environmental conditions under which an experiment is performed changed drastically for a short period of time and gave

outliers

rise to one or more _____. **Hence, if a z-score exceeds 2 in absolute value, we would suspect that the measurement is an outlier.** In this case, the experimenter should check to see whether a faulty measurement has been obtained.

Example 2.15
Consider the following sample of $n = 10$ observations.

    10, 27, 24, 13, 22, 17, 19, 23, 20, 49

The value $x = 49$ looks much larger than the other observations in the set and may be an outlier. Calculate the z-score corresponding to $x = 49$.

Solution
You can verify that for these observations, $\overline{x} = 22.4$ and $s = 10.65$. The z-score corresponding to $x = 49$ is

$$\text{z-score} = \frac{x - \bar{x}}{s} = \frac{49 - 22.4}{10.65} = 2.50 \;.$$

Since the z-score does not exceed 3 in absolute value, we cannot say with any certainty that the value $x = 49$ is an outlier. However, the z-score $= 2.50$ is larger than 2, and not highly likely: perhaps this observation should be checked for a possible recording error, or an error in the measurement itself.

The median is defined as the measurement that falls in the middle when all measurements are arranged in increasing magnitude. The median is also defined to be the value of x, such that at most 50% of the measurements are _____ than x and _____ are greater.   less; half
Therefore, in addition to providing a measure of the center of a set of measurements, the median allows the user to assess the position of one particular measurement in relation to the others in the set. For example, if the median family income in a particular state was $26,500, then half the families in the state had incomes less than $26,500 and half had incomes _____ than $26,500.   greater

A **percentile** is another measure of relative standing that is most often used for large data sets.

Let $x_1, x_2, \ldots, x_n$ be a set of measurements arranged in order of increasing magnitude. The **pth percentile** is that value of x such that at most $p\%$ of the measurements are less than x and at most $(100 - p)\%$ are greater than x.

Example 2.16
An elementary school child has scored in the 91st percentile on the SAT tests. What does this mean?

Solution
Scoring in the 91st percentile implies that at most 91% of the children who took the examination scored lower than this child, and at most 9% scored higher.

Example 2.17
A set of test scores are approximately mound-shaped with mean 73 and standard deviation 10. If a student scored 93 on the test, into what percentile would his score fall?

Solution
The value $x = 93$ lies two standard deviations above the mean. Hence, using the Empirical Rule and the fact that a mound-shaped distribution is symmetric about its mean, we can draw the following conclusions:

1. The percentage of scores falling below 73 is _____%.   50
2. The percentage of scores falling between 53 and 93 is approximately

95; 47.5

47.5; 97.5

2.5

98

median; lower
upper

95%. Hence, the percentage falling between 73 and 93 is
_____/2 = _____%, and the percentage of scores
below 93 is then

$$50 + \text{_____} = \text{_____}\%$$

3. Then, 97.5% of the scores lie below $x = 93$ and _____%
   lie above $x = 93$.

Rounding to the nearest percent, the score $x = 93$ lies in the
_____th percentile.

Percentiles provide an excellent way to measure relative standing
and are used frequently in the presentation of test scores, sociological
data, and medical data. However, the user must be careful when
using percentiles with small sets of data. In this situation, the exact
percentile may fall between two data points, and any value that falls
between these two points can be taken as the percentile value. Most
often, however, percentiles are encountered in the description of large
sets of data.

By the definition of a percentile, the 50th percentile is also called
the _____. The 25th percentile is called the (lower, upper)
quartile, and the 75th percentile is called the (lower, upper) quartile.

The upper and lower quartiles, when located along with the
median on the horizontal axis of a frequency distribution representing the measurements of interest as in Figure 2.7, divide that frequency distribution into four parts, with each containing an equal
number of measurements. The upper and lower quartiles are formally
defined and calculated as follows:

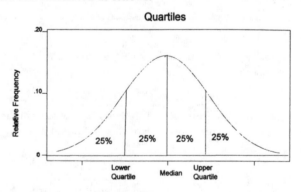

Figure 2.7

Let $x_1, x_2, \ldots, x_n$ be a set of n measurements arranged in
order of increasing magnitude. The **lower quartile**, Q_1, is
a value of x such that at most 1/4 of the measurements are
less than x and at most 3/4 are greater than x. The **upper
quartile**, Q_3, is a value of x such that at most 3/4 of the
measurements are less than x and at most 1/4 are greater.

When a small set of data is considered, the above definition some-times admits many numbers that would satisfy the criteria necessary for Q_1 and Q_3. For this reason, we will avoid this inconsistency by calculating sample quartiles in the following way:

Let $x_1, x_2, \ldots, x_n$ be a set of n measurements arranged in order of increasing magnitude. The **lower quartile**, Q_1, is the value of x in the $.25(n+1)$ position, and the **upper quartile**, Q_3, is the value of x in the $.75(n+1)$ position.

When $.25(n+1)$ and $.75(n+1)$ are not integers, the quartiles are found by interpolation, using the values in the two adjacent positions. (This definition of sample quartiles is consistent with that used in the MINITAB package.)

Example 2.18
Find the upper and lower quartiles for the following set of measurements:

   3, 8, 7, 1, 1, 12, 13, 9, 3, 2, 10

Solution
The measurements are first arranged in order of increasing magnitude:

   1, 1, 2, 3, 3, 7, 8, 9, 10, 12, 13

Since $n = 11$, the lower quartile, Q_1, will be in position

   $.25(n+1) = .25(11+1) = 3$

and the upper quartile, Q_3, will be in position

   $.75(n+1) = .75(12) = 9$

Since the quartile positions (are, are not) both integer-valued, the quar-tiles can be read directly as   are

   $Q_1 = 2$ and $Q_3 = 10$

Example 2.19
Find the upper and lower quartiles for the following prices:

   $2.15, 3.50, 6.80, 4.29, 1.67, 2.20, 1.59, 2.98

Solution
The measurements are first arranged in order of increasing magnitude:

   $1.59, 1.67, 2.15, 2.20, 2.98, 3.50, 4.29, 6.80

Since $n = 8$, the lower quartile, Q_1, is found in position

   $.25(n+1) = .25(9) = 2.25$

and the upper quartile, Q_3, is found in position

$$.75(n+1) = .75(9) = 6.75$$

Therefore,

$$Q_1 = 1.67 + .25(2.15 - 1.67) = 1.67 + .12 = 1.79$$

and

$$Q_3 = 3.50 + .75(4.29 - 3.50) = 3.50 + .59 = 4.09$$

We should note that there are many possible values that will satisfy the definition of a quartile. We choose to use the above convention in order to avoid complicated arithmetic calculation. You should not be surprised to find that other references may choose to calculate quartiles in a different way.

A measure of spread or dispersion that is resistant to outliers is the range of the measurements between the first and third quartile. This measure is called the **interquartile range.**

The **interquartile range** for a set of measurements is the difference the upper and lower quartiles.

$$\textbf{IQR} = Q_3 - Q_1$$

The interquartile range is used in exploratory data analysis as a surrogate measure for the standard deviation. In Example 2.18 the $IQR = Q_3 - Q_1 = 10 - 2 = 8$, while in Example 2.19 the $IQR = 4.09 - 1.79 = 2.30$, indicating that the data in Example 2.18 is (more, less) variable than the data in Example 2.19.

more

Many of these descriptive measures are easily found using the descriptive statistics menu in the Minitab package,including the arithmetic mean, the median, the standard deviation, the maximum and minimum values and other measures that we have not discussed. One of these measures is the **trimmed mean**, which is the arithmetic mean of the middle 90% of the observations after the largest 5% and the smallest 5% are excluded (or trimmed). A printout of the Minitab descriptive statistics output for the data in Example 2.12 is given in Figure 2.8.

Descriptive Statistics

Variable	N	Mean	Median	TrMean	StDev	SE Mean
Times	25	4.196	4.200	4.204	0.781	0.156

Variable	Minimum	Maximum	Q1	Q3
Times	2.500	5.700	3.750	4.750

Figure 2.8

Notice in Figure 2.8 that the median = 4.200 and trimmed mean = 4.204 have very similar values, which is what we would expect, since both ignore the extremely large and extremely small observations. In addition, the arithmetic mean $\bar{x} = 4.195$ is also close to these two values and the first and third quartiles are approximately equally distant from the center of the distribution. Taken together, these measures indicate that the distribution is fairly (<u>symmetric,</u> <u>nonsymmetric</u>) as seen in the histogram in Figure 2.9 and the stem and leaf plot in Figure 2.10

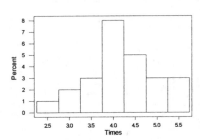

Figure 2.9

Stem-and-Leaf Display: Times

```
Stem-and-leaf of Times
N   = 25
Leaf Unit = 0.10

    1     2  5
    4     3  013
   10     3  678899
  (7)     4  1222334
    8     4  7788
    4     5  234
    1     5  7
```

Figure 2.10

2.7 The Five-Number Summary and Boxplot

The median and the upper and lower quartiles divide a set of data into four sets, each containing the same number of observations. If we add the largest number (maximum) and the smallest (minimum) to this group, we have a set of numbers that provide a quick summary of the data distribution.

The five-number summary consists of the smallest number, the lower quartile, the median, the upper quartile and the largest, presented in order from smallest to largest

$$Min \qquad Q_1 \qquad median \qquad Q_3 \qquad Max$$

Bu definition, one-fourth of the measurements in a data set lie between each of the four adjacent pairs of numbers.

The five-number summary can be used to make a simple graph called a boxplot to visually describe the data. From a boxplot you can detect skewness in the distribution as well as any outliers.

To construct a boxplot:

- Calculate the median, the upper and lower quartiles and the IQR.
- Draw a horizontal line representing the scale of measurement.
- Above the horizontal line form a box whose left and right ends are Q_1 and Q_2.
- Draw a line through the box at the median.

The box is shown in Figure 2.11.

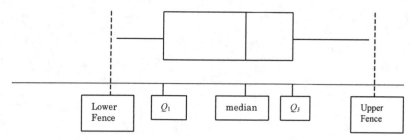

Figure 2.11

Detecting outliers—observations that are beyond:

- Lower fence: $Q_1 - 1.5$ *(IQR)*

- Upper fence: $Q_3 + 1.5(IQR)$

The upper and lower fences are shown in Figure 2.11, but they are not usually drawn on the boxplot. Any measurement outside the lower and upper fences is deemed an **outlier**. Measurements inside the fences are not considered unusual.

To finish the boxplot:

- Mark any outliers with an asterisk (*) on the graph.

- Extend horizontal lines called "whiskers" from the ends of the box to the smallest and largest observations that are not outliers.

Example 2.20

Construct a boxplot for the following data (Example 2.15).

10, 27, 24, 13, 22, 17, 19, 23, 20, 49

Solution

To find the median and the quartiles, you must first order the $n = 10$ values from smallest to largest.

10, 13, 17, 19, 20, 22, 23, 24, 27, 49

The median will be in position $.5(n + 1) = .5(10 + 1) = 5.5$ and will therefore consist of the average of the 5th and 6th observations.

$$m = \frac{20 + 22}{2} = \underline{\hspace{2cm}}.$$

21

The lower and upper quartiles will be in positions

$.25(n + 1) = .25(11) = 2.75$ and $.75(n + 1) = .75(11) = 8.25$, respectively.

Therefore

$$Q_1 = 13 + .75(17 - 13) = 13 + 3 = \underline{\hspace{2cm}}$$

16

and

$$Q_3 = 24 + .25(27 - 24) = 24 + .75 = \underline{\hspace{2cm}}.$$

24.75

The interquartile range is then

$$IQR = Q_3 - Q_1 = 24.75 - 16 = 8.75.$$

Using the interquartile range we can now find the inner and outer fences.

$$Q_1 - 1.5(IQR) = 16 - 1.5(8.75) = 16 - 13.125 = 2.875 \text{ or } 2.88$$

and

$$Q_3 + 1.5(IQR) = 24.75 + 1.5(8.75) = 24.75 + 13.125 = 37.975 \text{ or } 37.98.$$

The smallest and largest values that are inside the upper and lower fences are 10 and 27. They are the smallest and largest values that are not deemed to be outliers. Next, extend horizontal lines from the ends of the box to the values 10 and 27.

The last thing to do is identify any outliers and indicate their positions with asterisks. The value $x = 49$ lies beyond the upper fence and is therefore considered to be an $\underline{\hspace{2cm}}$.

outlier

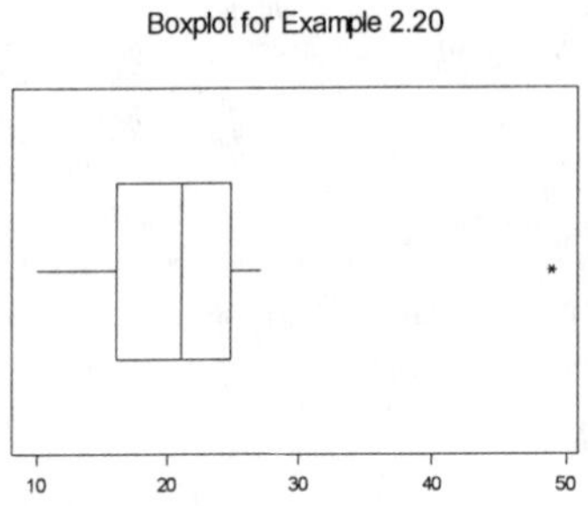

Boxplot for Example 2.20

Figure 2.12

Self-Correcting Exercises 2D

1. Find the median, the lower and upper quartiles and the inter-quartile range for the following data: 4, 0, 5, 3, 6, 2, 5, 9, 5, 3.

2. Refer to the data in Exercise 1.
   a. Calculate $\bar{x}$ and s.
   b. Calculate the z-score for the smallest and largest observations in the set. Is either of these observations unusually small or large?

3. Construct a boxplot for the data in Exercise 1. Are there any suspected outliers? Any extreme outliers?

4. Refer to the data in SCE-2A, Exercise 5.
   a. Find the mean, standard deviation, median and lower and upper quartiles.
   b. Construct a boxplot for these data. Comment on the distribution of these data. Is the distribution relatively symmetric? Are there any outliers?
   c. Find the proportion of observations falling into the intervals $\bar{x} \pm s$, $\bar{x} \pm 2s$, and $\bar{x} \pm 3s$. Compare these proportions with those given by Tchebysheff's Theorem and the Empirical Rule. Do the results here support your answer in part b?

Solutions to SCE 2D

1. Arrange the data in order of ascending magnitude:

   0, 2, 3, 3, 4, 5, 5, 5, 6, 9

   The positions of the median, lower and upper quartiles are

   $$.5(n+1) = .5(11) = 5.5$$
   $$.25(n+1) = .25(11) = 2.75$$
   $$.75(n+1) = .75(11) = 8.25$$

   Then $m = (4+5)/2 = 4.5$;

$$Q_1 = 2 + .75(3 - 2) = 2.75;$$
$$Q_3 = 5 + .25(6 - 5) = 5.25$$
and $IQR = Q_3 - Q_1 = 2.5.$

2. a. $\bar{x} = \dfrac{\sum x_i}{n} = \dfrac{42}{10} = 4.2;$

$$s^2 = \frac{\sum x_i^2 - \dfrac{\left(\sum x_i\right)^2}{n}}{n - 1} = \frac{230 - \dfrac{42^2}{10}}{9} = 5.9556;$$

$$s = \sqrt{5.9556} = 2.44.$$

   b. For $x = 0$, z-score $= \dfrac{x - \bar{x}}{s} = \dfrac{0 - 4.2}{2.44} = -1.72.$

   For $x = 9$, z-score $= \dfrac{x - \bar{x}}{s} = \dfrac{9 - 4.2}{2.44} = 1.97.$

   Neither value is unusually large or small.

3. *Lower and Upper Fences*

$Q_1 - 1.5(IQR) = 2.75 - 1.5(2.5) = -1$
$Q_3 + 1.5(IQR) = 5.25 + 1.5(2.5) = 9$
The largest and smallest values that are not outliers are 0 and 6.

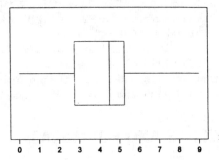

4. a. From Exercise 5a, SCE 2A, $\bar{x} = 181.8$ and $m = 174$. Using the
       sorted data in the solution to that exercise, the positions of the
       upper and lower quartiles are:

$$.25(n + 1) = .25(51) = 12.75$$
$$.75(n + 1) = .75(51) = 38.25$$

   Then

$$Q_1 = 160 + .75(162 - 160) = 161.5;$$
$$Q_3 = 199 + .25(202 - 199) = 199.75$$
and $IQR = Q_3 - Q_1 = 38.25.$ Also,

$$s = \sqrt{\frac{\sum x_i^2 - \dfrac{\left(\sum x_i\right)^2}{n}}{n - 1}} = \sqrt{\frac{1{,}699{,}440 - \dfrac{9090^2}{50}}{49}} = 30.93.$$

b. *Inner Fences*:

$Q_1 - 1.5IQR = 161.5 - 1.5(38.25) = 104.125$

$Q_3 + 1.5IQR = 199.75 + 57.375 = 257.125$

Outer Fences

$Q_1 - 3IQR = 161.5 - 3(38.25) = 46.75$

$Q_3 + 3IQR = 199.75 + 114.75 = 314.5$

Adjacent values are 121 and 253; $x = 270$ is a suspect outlier. The distribution is skewed to the right.

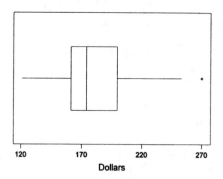

c. $\bar{x} \pm s \Rightarrow 181.8 \pm 30.93$ or 150.87 to 212.73 contains 35/50 $= .70$ or 70% of the measurements. $\bar{x} \pm 2s \Rightarrow 181.8 \pm 61.86$ or 119.94 to 243.66 contains 47/50 $= .94$ or 94% of the measurements. $\bar{x} \pm 3s \Rightarrow 181.8 \pm 92.79$ or 89.01 to 274.59 contains 50/50 $= 1.00$ or 100% of the measurements. Although the distribution is not quite mound-shaped, the proportions match quite well with those described by the Empirical Rule.

KEY CONCEPTS AND FORMULAS

I. *Measures of the Center of a Data Distribution*
   1. Arithmetic mean (mean) or average
      a. Population: μ

      b: Sample of n measurements: $\bar{x} = \dfrac{\sum x_i}{n}$

   2. Median; position of the median $= .5(n + 1)$
   3. Mode
   4. The median may be preferred to the mean if the data are highly skewed.

II. *Measures of Variability*
   1. Range: $R = \text{largest} - \text{smallest}$
   2. Variance

      a. Population of N measurements: $\sigma^2 = \dfrac{\sum (x_i - \mu)^2}{N}$

      b. Sample of n measurements: $s^2 = \dfrac{\sum (x_i - \bar{x})^2}{n - 1}$

b. (Cont'd): $s^2 = \dfrac{\sum x_i^2 - \dfrac{\left(\sum x_i\right)^2}{n}}{n-1}$

3. Standard deviation

   a. Population: $\sigma = \sqrt{\sigma^2}$

   b. Sample: $s = \sqrt{s^2}$

4. A rough approximation for s can be calculated as: $s \approx R/4$. The divisor can be adjusted depending on the sample size.

III. *Tchebysheff's Theorem and the Empirical Rule*

1. Use Tchebysheff's Theorem for any data set, regardless of its shape or size.

   a. At least $1 - \dfrac{1}{k^2}$ of the measurements lie within k standard deviations of the mean.

   b. This is only a lower bound; there may be more measurements in the interval.

2. The Empirical Rule can be used only for relatively mound-shaped data sets. Approximately 68%, 95%, and "almost all" of the measurements are within one, two, and three standard deviations of the mean, respectively.

IV. *Measures of Relative Standing*

1. Sample z-score: $z = \dfrac{x - \bar{x}}{s}$

2. pth percentile: $p\%$ of the measurements are smaller and $(100 - p)\%$ are larger.

3. Lower quartile, Q_1; position of $Q_1 = .25(n+1)$

4. Upper quartile, Q_3; position of $Q_3 = .75(n+1)$

5. Interquartile range; $IQR = Q_3 - Q_1$

V. *The Five-Number Summary and Boxplots*

1. The five-number summary:

   Min Q_1 Median Q_3 Max.

   One-fourth of the observations lie between any two adjacent pairs of numbers.

2. Box plots are used for detecting outliers and shapes of distributions.

3. Q_1 and Q_3 form the ends of the box. The median line is in the interior of the box.

4. Upper and lower fences are used to find outliers.

   a. Lower fence: $Q_1 - 1.5(IQR)$

   b. Upper fence: $Q_3 + 1.5(IQR)$

5. Outliers are located on the boxplot with an asterisk (*).

6. Whiskers are connected to the box from the largest and smallest values that are not outliers.

7. Skewed distributions usually have a long whisker *in the direction* of the skewness, and the median line is drawn *away* from the direction of the skewness.

Exercises

1. The following data are the miles per gallon (mpg) for each of 20 medium-sized cars selected from a production line during the month of March.

23.1	21.3	23.6	23.7
20.2	24.4	25.3	27.0
24.7	22.7	26.2	23.2
25.9	24.7	24.4	24.2
24.9	22.2	22.9	24.6

a. What are the maximum and minimum miles per gallon? What is the range of the data?

b. Construct a relative frequency distribution for these data by using class intervals of length 1 mpg, with the lower limit of the first class equal to 19.5. Remember to *use the left inclusion, right exclusion rule* in assigning observations to classes.

c. What proportion of the observations are greater than or equal to 22.5, but less than 26.5?

d. What proportion of the observations are greater than or equal to 26.5?

e. Find the mean and the standard deviation of these data.

f. Arrange the data from small to large. Find the z-scores for the largest and smallest observation in the set. Would you consider them to be outliers? Why or why not?

g. What is the median? Find the lower and upper quartiles. What is the 90th percentile for these data?

2. Construct a boxplot for the data in Exercise 1. Are there any suspected or extreme outliers? Does this conclusion agree with your results in Exercise 1, part f?

3. An IRS employee randomly sampled 25 tax returns and for each recorded the number of exemptions claimed by the taxpayer. The data are shown below.

2	0	2	1	3
3	1	1	3	2
1	0	7	0	2
0	0	0	4	1
4	3	2	2	5

a. What is the range of these data?

b. Calculate the median, the lower and upper quartiles.

c. Would a stem-and-leaf display be appropriate for these data? Explain.

d. Calculate $\bar{x}$, s^2, and s for these data.

e. Construct a relative frequency histogram for these data using classes of width 1.0. (You might begin with −0.5.)

f. Assuming the above sample is representative of all the taxpayers in the United States,

    i. what is the probability a taxpayer will claim no more than 1
       exemption?
    ii. what is the probability a taxpayer will claim 3 or more
       exemptions?
  g. Can the Empirical Rule be applied to these data? Explain.

4. A strain of "long-stemmed roses" has an approximate normal distribution with a mean stem length of 15 inches and standard deviation
2.5 inches.
  a. If one accepts as "long-stemmed roses" only those roses with a
    stem length greater than 12.5 inches, what percentage of such
    roses would be unacceptable?
  b. What percentage of these roses would have a stem length between
    12.5 and 20 inches?
Hint: Using the symmetry of the normal distribution, 1/2 of 68% of
the measurements lie one standard deviation to the left *or* to the
right of the mean, and 1/2 of 95% of the measurements lie two
standard deviations to the left *or* to the right of the mean.

5. The heights of 40 cornstalks ranged from 2.5 feet to 6.3 feet. In
presenting these data in the form of a histogram, suppose you had
decided to use .5 foot as the width of your class interval.
  a. How many intervals would you use?
  b. Give the class boundaries for the first and the last classes.

6. A machine designed to dispense cups of instant coffee will dispense
on the average μ ounces, with standard deviation $\sigma = .7$ ounce.
Assume that the amount of coffee dispensed per cup is approximately mound-shaped. If 8-ounce cups are to be used, at what value
should μ be set so that approximately 97.5% of the cups filled will
not overflow?

7. A pharmaceutical company wishes to know whether an experimental
drug being tested in its laboratories has any effect on systolic
blood pressure. Fifteen subjects, randomly selected, were given
the drug, and the systolic blood pressures recorded (in millimeters)
were

172	148	123
140	108	152
123	129	133
130	137	128
115	161	142

  a. Approximate s using the method described in Section 2.5.
  b. Calculate $\bar{x}$ and s for the data.
  c. Find values for the points a and b such that at least 75% of the
    measurements fall between a and b.
  d. Would Tchebysheff's Theorem be valid if the approximated s
    were used in place of the calculated s?
  e. Would the Empirical Rule apply to these data?

8. Refer to Exercise 7. Construct a stem-and-leaf display for the data.

9. Refer to Exercise 7. Are there any outliers?
   a. Use the z-score technique.
   b. Use the box plot technique.

10. Toss two coins 30 times, recording for each toss the number of heads observed.
    a. Construct a histogram to display the data generated by the experiment.
    b. Find $\bar{x}$ and s for your data.
    c. Do the data conform to Tchebysheff's Theorem? Empirical Rule?

11. The following data represent the social ambivalence scores for 15 people as measured by a psychological test. (The higher the score, the stronger the ambivalence.)

9	8	15	17	10
14	11	4	12	13
10	13	19	11	9

    a. Using the range, approximate the standard deviation s.
    b. Calculate $\bar{x}$, s^2, and s for these data.
    c. What fraction of the data actually lies in the interval $\bar{x} \pm 2s$?

12. A lumbering company interested in the lumbering rights for a certain tract of slash pine trees is told that the mean diameter of these trees is 14 inches with a standard deviation of 2.8 inches. Assume the distribution of diameters is approximately normal.
    a. What fraction of the trees will have diameters between 8.4 inches and 22.4 inches?
    b. What fraction of the trees will have diameters greater than 16.8 inches?

13. If the mean duration of television commercials on a given network is 1 minute and 15 seconds with a standard deviation of 25 seconds, what fraction of these commercials would run longer than 2 minutes and 5 seconds? Assume that duration times are approximately normally distributed.

14. The following are the prices charged (in cents) for a 16 ounce can of stewed tomatoes at 25 supermarkets in a large metropolitan area.

80	81	89	84	78
92	79	82	89	85
78	91	80	76	80
79	90	82	85	79
81	82	79	85	84

a. Find the range of these data.
b. Construct a relative frequency histogram. Are the data approximately mound-shaped?
c. Calculate the mean and standard deviation.
d. What fraction of the data actually lies in the interval $\bar{x} \pm s$? In the interval $\bar{x} \pm 2s$? Do these results agree with the Empirical Rule?

Chapter 3
Describing Bivariate Data

3.1 BIVARIATE DATA

You may measure two or more variables in your investigation. Multivariate data results when three or more variables are measured. When two variables are measured on each experimental unit, the resulting data are called **bivariate data**. For example, you may be interested in a person's height as well as weight, or you may be interested in the yield of fruit per tree and the age of the tree. It is not difficult to think of the many situations in which you might wish to measure two, or possibly three or more, variables on each experimental unit.

When you have bivariate data, you may wish to describe each variable using graphical or numerical methods. However, you may also wish to describe the relationship between these two variables. How will you do this? The answer will depend upon the kinds of variables that you measure. For example, if you record two qualitative variables such as a person's gender and hair color, you might summarize the data using a two-way table. A typical table might look like this.

Hair Color	Black	Brown	Red	Blonde	Other	Total
Male	15	21	2	10	2	51
Female	18	28	8	23	2	79

In contrast, you may wish to measure the type of residence and the number of family members for 100 families. Your data might be tabulated this way.

Family Number	Type of Residence			
	Apartment	Duplex	Single Residence	Totals
1	8	10	2	20
2	15	4	14	33
3	9	5	24	38
4 or more	6	1	28	35
Totals	38	20	68	126

You may wish to determine if there is a relationship between two variables such as weight and cholesterol level when these two variables are measured for males between 40 and 50 years of age. Another person may wish to assess the relationship between a college student's high school SAT scores and the student's college GPA. In both cases, the two variables measured would be quantitative and you would need some way to determine if one variable increases or decreases with changes in the other variable. Each of these questions can be answered in turn. However, the way in which we display the data and investigate relationships between two variables will depend upon whether one or both of the variables are qualitative or quantitative.

3.2 GRAPHS FOR QUALITATIVE VARIABLES

When at least one of the two variables is qualitative, you can use simple or comparative **pie charts, line charts or bar charts** to display and describe the data. One common situation is to have one quantitative variable measured for two or more groups (a qualitative variable).

Example 3.1
Use the following data on the source of support for several charitable US organizations to construct both a stacked bar chart and two comparative pie charts.

| | Amounts (in millions of dollars) | | |
Organization	Private	Other	Total
1. Salvation Army	$1,397	$1,321	$2,718
2. YMCA	693	2,913	3,606
3. American Red Cross	678	1,727	2,405
4. American Cancer Society	620	52	672
5. American Heart Assn.	358	78	436
Total	$3,746	$6,091	$9,837

Source: *The World Almanac and Book of Facts, 2002*, p.117.

Solution
The stacked bar chart for the data in Figure 3.1 is constructed by plotting the total for each of the organizations and then shading the portion of the bar that corresponds to the private donations.

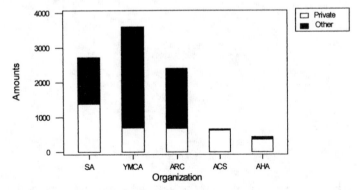

Figure 3.1

The pie charts for private and others sources of income in Figures 3.2a and 3.2b are constructed using the angles calculated as

$$\frac{\text{Number in the category}}{\text{Total}} \times 360°$$

The angles are given in the next table.

| Organization | Amounts (in millions of dollars) | | | |
	Private		Other	
1. Salvation Army	37.3%	134°	21.7%	78°
2. YMCA	18.5%	67°	47.8%	172°
3. American Red Cross	18.1%	65°	28.4%	102°
4. American Cancer Society	16.6%	60°	0.9%	3°
5. American Heart Assn.	9.6%	34°	1.3%	5°
Total	100%	360°	100%	360°

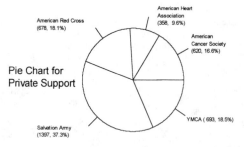

Figure 3.2a. Private Support

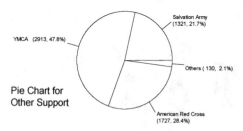

Figure 3.2b. Other Support

The stacked bar chart reveals that the first three organizations receive a substantial portion of their income from other sources, while the remaining two receive almost all of their income from private sources. The two pie charts convey that apart from the Salvation Army, the remaining four organizations have about the same private support, while in terms of other support, the American Cancer Society and the American Heart Association receive almost none.

Comparative Graphs

Two or more sets of data may be compared using appropriate graphs, provided that the data sets have a basis for comparison. For example, identical categories may be measured at several points in time or for different populations. These kinds of comparisons might be made plotting two or more line charts on the same axes, or using a bar chart where the same _____ for different points in time are placed side by side;   categories
another option might be to stack the bars for each category on top of each other. Pie charts that use the same categories might be placed side by side or, for example, one pie chart might show sources of income for a given period, while another might show expenditures for that same period.

Example 3.2
The data given in the table that follows are the receipts and outlays (in billions of current dollars) of the US government from 1993 to 2000. Use a comparative line chart and a side-by-side bar chart to display these data.

	1993	1994	1995	1996	1997	1998	1999	2000
Receipts	1.15	1.26	1.35	1.45	1.58	1.72	1.83	2.03
Outlays	1.41	1.46	1.52	1.56	1.60	1.65	1.70	1.79

Source: *The World Almanac and Book of Facts, 2002*, p.113

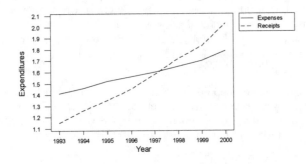

Figure 3.3

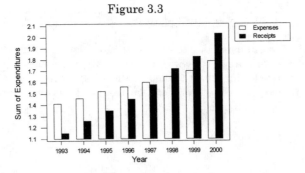

Figure 3.4

Figures 3.3 and 3.4 both show the constant rise in both expenses and receipts from 1993 to 2000. The line chart shows that expenses were well above receipts until 1997 when the US budget appears to be almost balanced. From 1997 through 2000, the government took in more money than it spent.

Self-Correcting Exercises 3A

1. The amounts in the supplementary insurance trust fund from 1970 to 1995 are given in the table that follows.

| Year | Income (in millions) | | |
	Premiums	Government	Other
1970	$936	$928	$12
1975	1,887	2,330	105
1980	2,928	6,932	415
1985	5,524	17,898	1,155
1990	11,494	33,210	1,434
1995	19,244	36,988	1,937

Source: *The World Almanac and Book of Facts, 1997*, p. 712

a. Construct a bar chart displaying the premiums paid for the periods 1970 through 1995.

b. Construct a comparative bar chart for the income from premiums and the government for the periods 1970 through 1995. How would you describe the changes in the source of funds from premiums and the government?

c. Construct side-by-side pie charts showing the three sources of income for the years 1975 and 1995. What conclusions can you draw from these pie charts?

2. The percentage of men 65 years and older in the labor force steadily declined over the last century from 63.1% in 1900 to 18.6% in 2000. The percentage of women 65 or older in the work force has barely changed at all with 8.3% in 1900 and 10.0% in 2000.

% ≥ 65	54.0	41.8	41.4	30.5	24.8	19.3	17.6	18.6
Years	1930	1940	1950	1960	1970	1980	1990	2000

a. Construct a bar chart to represent these data. What salient characteristics of the chart would you wish to mention?

b. Construct a line chart to display these data. Are there any characteristics that are more apparent in this chart than in the bar chart of part a?

3. The number of endangered species listed by the US Department of
    Interior as of August 2001 is given below.

Group	US only	Foreign only
Mammals	63	251
Birds	78	175
Reptiles	14	64
Amphibians	10	8
Fishes	70	11
Crustaceans	18	0
Insects	33	4

Source: *The World Almanac and Book of Facts, 2002*, p. 167.

a. Use a side-by-side bar chart to compare the number of
    endangered species listed by the US versus the number listed
    by foreign countries.
b. Is there a better way to present these same data? Use an
    alternative method for displaying the data that you feel is
    better than the presentation in part a.
4. Transplant surgery is more successful than ever, thanks to
    improved surgical techniques, a better understanding of the body's
    immune system, and the development of drugs that combat organ
    rejection. The following table gives the survival rates of patients
    by the organ transplanted.

Organ	Patient Survival Rate	Number Performed
Heart	78%	2,359
Heart–Lung	53%	69
Kidney	96%	8,592
Liver	75%	3,925
Lung	63%	863
Pancreas	87%	1,026

a. Construct a bar chart showing the patient survival rates for
   each type of transplant. Does this convey the information in the
   table?
b. Can you present this information using comparative pie charts?
   What would be the basis for comparison?
c. Construct a stacked bar chart showing the number of survivors
   as part of the bar showing the number of transplants.
d. Which presentation conveys the information in the chart most
   clearly? Support your answer.

Solutions to SCE 3A

1. a.

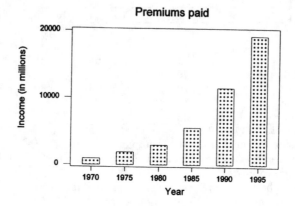

b.

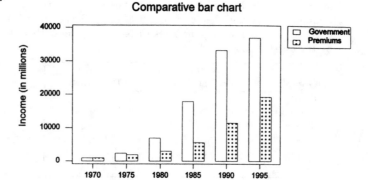

Although both sources of income are increasing, the premiums paid are increasing at a faster rate.

c.

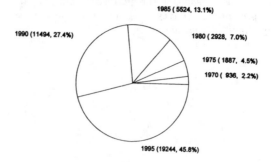

Government

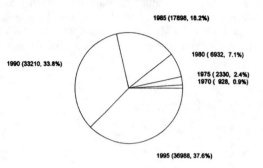

1985 (17898, 18.2%)

1980 (6932, 7.1%)

1975 (2330, 2.4%)
1970 (928, 0.9%)

1990 (33210, 33.8%)

1995 (36988, 37.6%)

Other

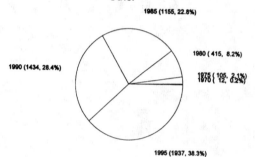

1985 (1155, 22.8%)

1980 (415, 8.2%)

1975 (105, 2.1%)
1970 (12, 0.2%)

1990 (1434, 28.4%)

1995 (1937, 38.3%)

2.   a.

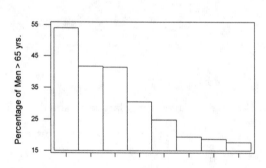

b.

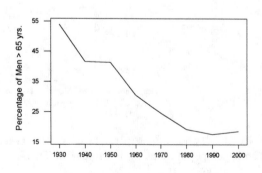

Both charts convey the same basic information. The percentage of men over 65 in the workforce has shown a steady decrease apart from the almost identical percentages for 1940 and 1950, the years that included WWII and the Korean war.

3. a.

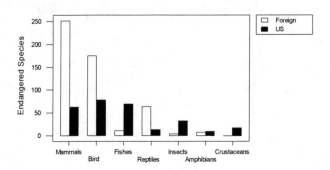

b.

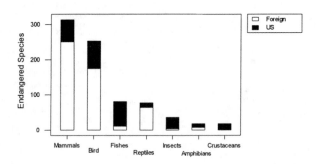

4. a. Bars are reordered from largest to smallest survival rates.

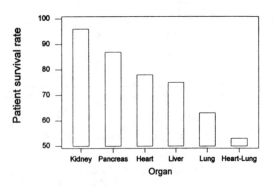

b. No. The two columns in the table are not two different categories or populations to be compared. There is no basis for comparison.

c-d. To create a stacked bar chart, you must find the number of
patients in each category who survived the transplant, and the
number who died. The number who survived is found by
multiplying the second and third columns together, and the
number who died is found by subtraction. The rounded values
are shown below.

Organ	Survived	Died
Heart	1840	519
Heart-Lung	37	32
Kidney	8248	344
Liver	2944	981
Lung	544	319
Pancreas	893	133

The stacked bar chart below, reordered by number of transplants, gives the best description of the data.

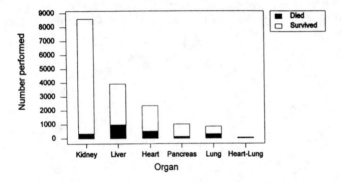

3.3 SCATTERPLOTS FOR QUANTITATIVE DATA

When both variables are quantitative, one is plotted on the horizontal
axis and the other on the vertical axis. One variable is usually identified
as x and the other as y. The pairs of observations (x_i, y_i),
$i = 1, 2, ..., n$ when plotted give rise to a two-dimensional graph called
a **scatterplot**, a generalization of the dotplot in Chapter 1.

An important use of a scatterplot is to determine what if any relationship or pattern might be exhibited in the scatterplot.

- What type of pattern do you see? Is there a constant upward or
  downward trend that follows a straight line pattern? Do the
  plotted points follow a curved pattern? Is there no apparent
  pattern at all?
- How strong is the pattern? If the points were tightly clustered
  within the pattern the pattern would be deemed strong, while if
  the points are widely dispersed about the pattern, it would be
  considered weak.

- Are there any unusual observations? Although they might be more difficult to identify for bivariate data, the set may contain outliers, points that lie far from the rest of the data points. Do the points cluster into one or more groups? If so, is there a reason or underlying variable that is responsible for the grouping?

Example 3.3

The number of passengers x (in millions) and the revenue y (in billions of dollars) for the top ten US airlines in 2000 are given in the next table.

x	86.2	83.9	105.6	56.8	45.1	59.8	72.6	26.4	19.9	13.5
y	20.2	19.4	15.9	11.4	9.9	9.3	5.6	3.5	2.3	2.2

Source: *The World Almanac and Book of Facts, 2002.* pp. 117, 217.

Construct a scatterplot for these data.

Solution

Label the horizontal axis x and the vertical axis y. Plot the points using the coordinates (x, y) for each of the ten pairs. The scatterplot is shown in Figure 3.5.

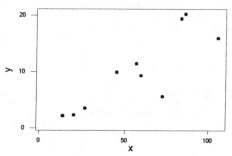

Figure 3.5

upward

The pattern here is an (<u>upward, downward</u>) trend, indicating that the number of passengers and the airline revenues increase together. It appears that there is an underlying straight line relationship between x and y.

Example 3.4

Groups of ten-day-old chicks were randomly assigned to seven groups in which the diet for each group was supplemented with differing amounts of biotin. The data that follow give the biotin intake and the average weight gain per day for each of the groups. Construct a scatterplot and discuss any possible patterns that appear to be present.

Biotin intake, x	.14	2.01	6.06	6.34	7.15	9.65	12.50
Weight gain, y	8.0	17.1	22.3	24.4	26.5	23.4	23.3

Solution
Let the amount of biotin be represented by x and weight gain by y. The plot is given in Figure 3.6, which follows. There appears to be a curved pattern to the points rather than the straight line pattern that you saw in the last example. The pattern also seems quite (weak, strong). With the exception of points 4 and 5, the remaining points seem to lie on the same curved line.

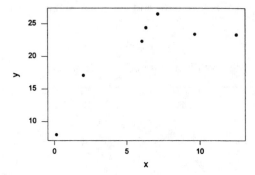

Figure 3.6

3.4 NUMERICAL MEASURES FOR QUANTITATIVE BIVARIATE DATA

One pattern that is commonly observed is one in which x and y exhibit an increasing or decreasing trend whereby the points seem to be scattered above and below, but within a fixed distance from a straight line. When this is the case, we say that x and y exhibit a *linear relationship*. There is a numerical measure that will detect a linear relationship between x and y and also measure the strength of the relationship.

Example 3.5
Of two personnel evaluation techniques available, the first requires a 2-hour test-interview while the second can be completed in less than an hour. The scores for each of the $n = 15$ individuals who took both tests are given in the next table.

Applicant	(Test 1) x	(Test 2) y
1	75	38
2	89	56
3	60	35
4	71	45
5	92	59
6	105	70
7	55	31
8	87	52

strong

Applicant	(Test 1) x	(Test 2) y
9	73	48
10	77	41
11	84	51
12	91	58
13	75	45
14	82	49
15	76	47

Produce a scatterplot for these data and comment on any patterns present.

Solution
The plotted points are given in Figure 3.7 that follows.

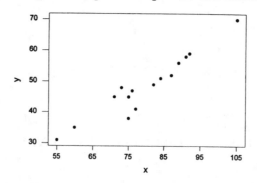

Figure 3.7

It appears that x and y exhibit a linear relationship, whereby y increases (decreases, increases) as x increases.

You could describe each variable, x and y, individually by finding $\bar{x}$, $\bar{y}$, s_x and s_y. However, none of these descriptive measures provides any information about the linear relationship between x and y. A simple measure that does provide this information is called the **correlation coefficient**, denoted by r and defined as

$$r = \frac{s_{xy}}{s_x s_y}$$

The quantities s_x and s_y are the standard deviations of x and y. The quantity s_{xy} is called the **covariance** between x and y and defined as

$$s_{xy} = \frac{\sum (x_i - \bar{x})(y_i - \bar{y})}{n-1}.$$

The calculational form for the covariance is similar to the computing formula for s and is given by

$$s_{xy} = \frac{\sum x_i y_i - \frac{(\sum x_i)(\sum y_i)}{n}}{n-1}.$$

How does the covariance measure a linear relationship between x and y? Consider the cross products given by $(x_i - \bar{x})(y_i - \bar{y})$. When the value of x is larger than its mean, the deviation $(x_i - \bar{x})$ will be positive, and negative if x is smaller than its mean. Similarly, when y is larger than its mean, $(y_i - \bar{y})$ will be positive, and negative if y is smaller than its mean. Notice in Figure 3.8a that when the data points lie mainly in the first and third quadrant, the sum of crossproducts will be positive, and s_{xy} will be positive. In Figure 3.8b most of the points lie in quadrants 2 and 4 in which case the sum of crossproducts will be negative and s_{xy} will be negative. In Figure 3.8c the points are randomly scattered across all four quadrants and the sum of the crossproducts will be approximately zero and s_{xy} will also be approximately zero.

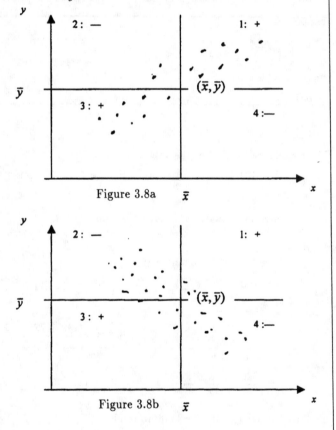

Figure 3.8a

Figure 3.8b

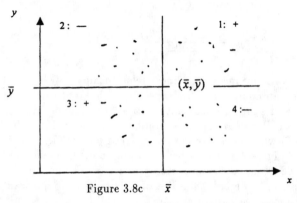

Figure 3.8c $\bar{x}$

Scientific calculators and statistical packages all provide the value of the correlation coefficient when the data are properly entered and the appropriate commands given. Figure 3.9 displays the results of using Minitab to calculate these quantities.

Covariances

	x	y
x	161.838	
y	122.190	101.381

Correlations (Pearson)

Correlation of x and y = 0.954, P-Value = 0.000

Figure 3.9

The covariance table gives the values of

$$s_x^2 = 161.838, \qquad s_y^2 = 101.381 \qquad \text{and} \qquad s_{xy} = 122.190$$

Example 3.5
Find the covariance between x and y for Example 3.4.

Solution
You will need to find the quantities:

$$\sum x_i = 1192 \qquad \sum y_i = 725 \qquad \text{and} \qquad \sum x_i y_i = 59{,}324$$

Then

$$s_{xy} = \frac{\sum x_i y_i - \dfrac{\left(\sum x_i\right)\left(\sum y_i\right)}{n}}{n-1} = \frac{59{,}324 - \dfrac{(1192)(725)}{15}}{14}$$

$$= \frac{59{,}324 - 57{,}613.333}{14} = \frac{1710.667}{14} = 122.190$$

a value that agrees with the Minitab result.

Example 3.6
Use the variances and covariance given in Figure 3.9 to calculate the correlation coefficient.

Solution
The correlation coefficient is found using the formula

$$r = \frac{s_{xy}}{s_x s_y}$$

Therefore

$$r = \frac{(122.190)}{\sqrt{(161.838)(101.381)}}$$

$$r = \frac{122.190}{(128.091)} = \underline{\hspace{2cm}}$$

.954

Our calculated value agrees with that produced by Minitab.

How can you assess how weak or how strong is the linear relationship between x and y? The value of the correlation coefficient lies between -1 and $+1$.

$$-1 \leq r \leq +1$$

When r is close to one, there is a strong positive linear relationship between x and y. When r is close to -1, there is a strong negative linear relationship between x and y. When r is close to 0, there is no linear relationship between x and y.

For the data in Example 3.5, the correlation coefficient was equal to $r = .954$, a value that is close to 1. This would indicate that there is a (weak, strong) linear relationship between x and y.

strong

Another way to describe the linear relationship between two quantitative variables is to determine a best-fitting line that describes the linear relationship. A straight line is described algebraically as

$$y = a + bx$$

and results in the graph given in Figure 3.10.

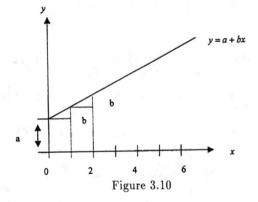

Figure 3.10

From the graph you can see that a is where the line cuts or intersects the y-axis: a is called the y-intercept. From Figure 3.10 you can also see that when x increases one unit, y increases by an amount b. The quantity b is called the **slope** of the line and determines whether the line is increasing ($b > 0$), decreasing ($b < 0$) or horizontal ($b = 0$).

The best fitting line, which is called the **regression** or **least squares line**, is found by minimizing the squared differences between the data points and the line itself as shown in Figure 3.11.

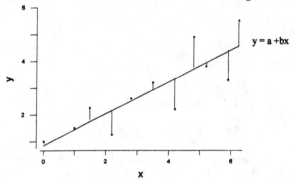

Figure 3.11

The calculational forms for a and b are:

$$b = r\left(\frac{s_y}{s_x}\right) \quad \text{and} \quad a = \bar{y} - b\bar{x}.$$

Since b and r have the same sign, when r is positive, so is b; when r is negative, so is b; and when r is close to 0, so is b.

Example 3.7
Find the best-fitting line for the following data. The value of y is the sales price (in \$1000) of a residence and x is the square footage of heated floor space (in 1000 sq. ft.). Plot the points and the line on the same graph.

x	1.5	2.1	1.7	1.5	1.	2.4
y	89	109	101	91	102	113

Solution
In order to find the least squares line, you will need to calculate the values for the descriptive statistics that follow.

$$\bar{x} = 1.850, \quad \bar{y} = 100.833, \quad s_x = .3564 \quad s_y = 9.5167$$
$$\text{and} \quad r = .9644.$$

You can now find the slope and intercept,

25.75

$$b = r\left(\frac{s_y}{s_x}\right) = .9644\left(\frac{9.5167}{.3564}\right) = \underline{\hspace{2cm}} \quad \text{and}$$

$$a = \bar{y} - b\bar{x} = \underline{\hspace{2cm}} - 25.75(1.85) = \underline{\hspace{2cm}}$$

100.833; 53.20

The plotted line with the points superimposed is shown in Figure 3.12.

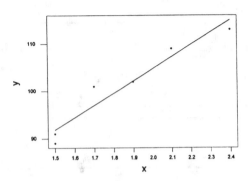

Figure 3.12

When should you describe the linear relationship between two variables using the correlation coefficient r, and when should you use the
regression line $y = a + bx$? Although the point is a technical one, correlation is used when the values of x and y are both random, as when you
randomly select an individual and measure the person's height and
weight (both height and weight would be random variables). On the
other hand, if an experiment involved fixing the values of x and then
measuring y, as in setting the temperature of a reaction and measuring
y, the quantity of reactant produced, then x would be fixed and y
would be random. In this case, the regression line would be the appropriate way to measure the linear relationship between x and y.

Self-Correcting Exercises 3B

1. A set of bivariate data consisting of measurements on two variables
   x and y, follows.

   (1, 1), (2, 3), (2, 4), (3, 5), (3, 6), (4, 8), (4, 9)

   a. Draw a scatterplot to display these data. Does there appear to be
      an underlying relationship between x and y? How would you
      describe it?
   b. Calculate the correlation coefficient. Would you consider the
      correlation to be weak or strong? Why?
   c. Calculate the best-fitting line for these data. How would you
      assess its fit?

2. Consider the set of data shown in the following table.

x	1	2	3	4	5	6
y	5.5	4.2	3.6	2.1	1.3	.4

a. Draw a scatterplot to describe these data. Is there a relationship between x and y? How would you describe the
relationship?

b. Use the following Minitab printouts to calculate the correlation coefficient for these data.

Descriptive Statistics

Variable	N	Mean	Median	TrMean	StDev	SE Mean
x	6	3.500	3.500	3.500	1.871	0.764
y	6	2.850	2.850	2.850	1.917	0.783

Variable	Minimum	Maximum	Q1	Q3
x	1.000	6.000	1.750	5.250
y	0.400 5.500	1.075 4.525		

Covariances

	x	y
x	3.50000	
y	-3.57000	3.67500

c. Use the Minitab printouts in part b to find the best-fitting
line to describe these data.

3. The following is a comparison of the average sentence and
average time served (in months) for various types of offenses.

Type of offense	Sentence (x)	Time served (y)
All Violent	88	39
Homicide	180	84
Rape	116	61
Robbery	92	40
Sexual Assault	81	39
Assault	61	28
Other	67	29

Source: *The World Almanac and Book of Facts, 2002.* p. 765.

a. Produce a scatterplot for these data. Are there any patterns
in the data? Use the Minitab printout to calculate the
correlation coefficient.

Descriptive Statistics: x, y

Variable	N	Mean	Median	TrMean	StDev	SE Mean
x	7	97.4	85.0	97.4	40.56	15.3
y	7	45.71	39.00	45.71	20.06	7.58

Variable	inimum	Maximum	Q1	Q3
x	61.0	180.0	67.0	116.0
y	28.00	84.00	29.00	61.00

Covariances: x, y

	x	y
x	1644.952	
y	803.976	402.571

b. Use the information obtained in part a and the Minitab printouts to find the best-fitting line for these data. How would you interpret the slope in terms of the length of a criminal's sentence and the length of time served?

Solutions to SCE 3B

1. a. The scatterplot shows a strong positive linear relationship between x and y.

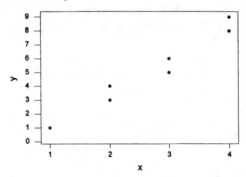

b. $s_{xy} = \dfrac{\sum x_i y_i - \dfrac{(\sum x_i)(\sum y_i)}{n}}{n-1} = \dfrac{116 - \dfrac{19(36)}{7}}{6} = 3.047619;$

$s_x = 1.112697; \ s_y = 2.79455;$ and $r = \dfrac{s_{xy}}{s_x s_y} = .9801$

Since the correlation is very close to 1, it is considered very strong.

c. Calculate $b = r\,\dfrac{s_y}{s_x} = .9801\left(\dfrac{2.79455}{1.112697}\right) = 2.4615$ and

$a = \bar{y} - b\bar{x} = 5.14286 - 2.4615(2.714) = -1.538.$ The equation of the best-fitting line is $y = -1.538 + 2.46x.$ It should fit very well through the points.

2. a. The scatterplot shows a strong negative linear relationship between x and y.

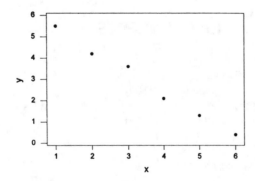

b. Use the **Covariance** portion of the printout to calculate

$$r = \frac{S_{xy}}{s_x s_y} = \frac{-3.57}{\sqrt{3.5(3.675)}} = -.9954.$$

c. Calculate $b = r\dfrac{S_x}{s_y} = -.9954\left(\dfrac{1.917}{1.871}\right) = -1.0199$ and

$a = \bar{y} - b\bar{x} = 2.85 - (-1.0199)(3.5) = 6.42$. Rounding to two decimal places, the equation of the best-fitting line is $y = 6.42 - 11.02x$.

3. a. The scatterplot shows a strong positive linear relationship.

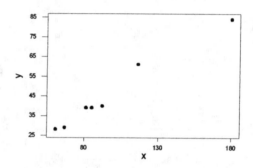

The correlation coefficient is found to be

$$r = \frac{S_{xy}}{s_x s_y} = \frac{803.976}{40.56(20.06)} = .9881 .$$

c. Calculate $b = r\dfrac{S_y}{s_x} = .9881\left(\dfrac{20.06}{40.56}\right) = .4887$ and

$a = \bar{y} - b\bar{x} = 45.71 - .4887(97.43) = 1.904$. The equation of the best-fitting line is $y = 1.90 + .49x$. Notice that for each increase in one year's sentence, the actual time served increases by roughly half (.4887) a year.

KEY CONCEPTS

I. *Bivariate Data*
    1. Both qualitative and quantitative variables
    2. Describing each variable separately
    3. Describing the relationship between the two variables
II. *Describing Two Qualitative Variables*
    1. Side-by-side pie charts
    2. Comparative line charts
    3. Comparative bar charts

    a. Side-by-side

    b. Stacked

  4. Relative frequencies to describe the relationship between the two variables

III. *Describing Two Quantitative Variables*

  1. Scatterplots

    a. Linear or nonlinear pattern

    b. Strength of relationship

    c. Unusual observations: clusters and outliers

  2. Covariance and correlation coefficient

  3. The best-fitting regression line

    a. Calculating the slope and y-intercept

    b. Graphing the line

    c. Using the line for prediction

Exercises

1. The following data represent the total sales and the domestic sales for an electronics firm over the last five years.

Sales (in $100,000)					
Year	1	2	3	4	5
Domestic	550	600	650	690	700
Total	600	650	800	850	960

  a. Construct a bar chart to depict total sales as a function of time.

  b. Construct a stacked bar chart to compare domestic sales as a fraction of the total sales as a function of time.

2. The data that follows represents the average amount of tuition and required fees for public and private four-year colleges and universities.

Academic year	Schools	
	Public	Private
1995 – 1996	$2848	$12,243
1996 – 1997	2987	12,881
1997 – 1998	3110	13,344
1998 – 1999	3229	13,973
1999 – 2000	3349	14,588
2000 – 2001	3506	15,531

Source: *The World Almanac and Book of Facts, 2002*, p. 236

  a. Use a bar chart to display the costs of four-year college tuition and fees during the period 1995-2001.

  b. Use a line chart to display this same information. Which presentation better displays the information in the data?

3. The number of public elementary and secondary schools with and without computers in 2001 are given in the table that follows.

Grade Levels	Number of Schools (thousands)	
	Computers	No Computers
Elementary	48.3	3.0
Middle/Junior High	13.6	1.3
Senior High	16.3	1.6
Special Education	1.7	0.7
Total	79.9	6.6

Source: *The World Almanac and Book of Facts, 2002.* p. 232.

a. Use a bar chart to depict the number of grade level classifications that have computers available.

b. Use a bar chart to depict the number of grade level classifications that do not have computers.

c. Use either a side-by-side or stacked bar chart to relate the numbers of schools with and without computers for the several grade level classifications given. Which of these two options provide the better display of the data?

M-Age	M-Systolic	M-Diastolic
16	116	74
19	118	78
17	108	84
16	132	80
20	134	80
19	126	74
20	114	76
19	110	88
20	130	68
16	120	70
20	106	70
17	120	86
19	128	70
17	126	66
18	118	80
17	140	70
15	120	70
20	120	70
20	140	98
15	110	60
16	100	80
20	130	64
17	110	70
15	170	70
18	130	88

a. Produce a histogram of the systolic blood pressure readings. Use the histogram to estimate the center and the range of the distribution. Comment on the shape of the distribution.
b. Repeat the instructions in part a. for diastolic blood pressures. Comment on the shape of the distribution. Howe does this distribution compare with the systolic blood pressures?
c. Construct a scatterplot of the systolic and diastolic blood pressure readings. Can you discern a pattern in the readings? If so, what is the pattern?
d. Use your calculator or a statistical package to find the correlation coefficient for systolic and diastolic blood pressure reading for these males. Would you consider this correlation to be weak or strong? Why?
e. Find the best-fitting line that describes the relationship between systolic (y) and diastolic (x) blood pressures for males in this age group.
f. Plot the points and the line on the same graph. Would you evaluate this relationship as weak or strong? Why?

5. The age, systolic and diastolic blood pressure for $n = 25$ females between the ages of 15 and 20 are given in the following table.

Age	Systolic	Diastolic
16	100	56
19	140	76
19	110	84
19	116	76
19	100	70
16	120	73
18	108	46
19	88	60
20	116	94
18	98	70
16	124	86
18	90	70
17	110	70
20	100	70
17	110	72
16	104	60
19	100	55
19	100	64
20	110	76
15	110	70
20	120	70
19	120	70
17	110	80
16	100	60
19	88	68

a. Produce a histogram of the systolic blood pressure readings. Use the histogram to estimate the center and the range of the distribution. Comment on the shape of the distribution.

b. Repeat the instructions in part a. for diastolic blood pressures. How do these two distributions compare?

c. Construct a scatterplot of the systolic and diastolic blood pressure readings. Can you discern a pattern in the readings? If so, what is the pattern?

d. Use your calculator or a statistical package to find the correlation coefficient between systolic and diastolic blood pressure readings for males. Is this a relatively strong relationship? Why?

e. Find the best-fitting line that describes the relationship between systolic (y) and diastolic (x) blood pressures for females in this age range.

f. Comment on the comparison of both systolic and diastolic blood pressures for males and females in the age-range of 15-20. Would you expect this same relationship to hold for older groups of males and females?

6. The heights and weights of 15 randomly selected and apparently healthy adults are given in the following table.

Height (inches)	Weight (pounds)
60	105
61	127
62	140
64	137
66	158
66	165
68	169
69	168
69.5	182
70	145
70	185
71	168
71	189
72	200
73	224

a. Produce a scatterplot for these data and describe any apparent relationships that you discern from the plot.

b. Calculate the correlation coefficient relating height and weight for these data. How strong would you describe a linear relationship between height and weight?

c. Produce the best-fitting line describing the relationship between height (x) and weight (y). What can you conclude about the difference in weight for a one-inch increase of height? Do you think that height determines weight? Is it possible to use what you have learned in the course at this point to be able to give a definitive answer to this question?

7. The birth rates per-thousand for the 11 Atlantic states and the District of Columbia in two recent years are given in the following table.

Year 1	Year 2
15.4	14.6
14.8	13.7
13.0	12.5
14.7	14.3
14.3	14.2
17.0	15.9
14.6	14.1
11.8	11.6
14.4	14.2
13.9	13.6
15.4	15.8
13.7	13.3

a. Produce a scatterplot for these data. Are there any apparent trends?

b. Find the correlation coefficient of the birth rates for these two years. How would you interpret the strength of the linear relationship between the birth rates for these two years?

c. If you were to find the best-fitting regression line describing the relationship of birth rates for these two years, would you be able to describe the relationship between birth rates for the remaining states? Would you be able to predict the birth rate for any state for year two based on its birth rate for year 1? Why or why not?

Chapter 4
Probability and Probability Distributions

4.1 THE ROLE OF PROBABILITY IN STATISTICS

We have stated that our aim is to make inferences about a population based on sample information. However, in addition to making the inference, we need to assess how good the inference will be.

Suppose that you are interested in estimating the unknown mean of a population of observations to within two units of its actual value. If an estimate is produced based on the sample observations, what is the chance that the estimate is no farther than two units away from the true but unknown value of the mean?

If an investigator has formulated two possible hypotheses about a population and only one of these hypotheses can be true, when the sample data are collected she must decide which hypothesis to accept and which to reject. What is the chance that she will make the correct decision?

In both these situations, we have used the term *chance* in assessing the goodness of an inference. But chance is just the everyday term for the concept statisticians refer to as _____. Therefore, some elementary results from the theory of probability are necessary in order to understand how the accuracy of an inference can be assessed.

probability

In the broadest sense, the probability of the occurrence of an event A is a measure of one's belief that the event A will occur in a single repetition of an experiment. One interpretation of this definition that finds widespread acceptance is based on empirically assessing the probability of the event A by repeating an experiment n times and observing n_A/n, the relative frequency of the occurrence of event A. When n, the number of repetitions, becomes very large, the fraction n_A/n will approach a number we will call $P(A)$, the probability of the occurrence of the event A.

4.2 EVENTS AND THE SAMPLE SPACE

When the probability of an event must be assessed, it is important to be able to visualize under what conditions that event will be realized. An *experiment* is the process by which an observation or measurement is obtained.

An **event** is an outcome of an experiment.

When an experiment is run repeatedly, a population of observations results. A _____ is any set of observations taken from this population. In this context, an observation may be a measurement, such as a person's height, or it may be a description or a category, such as "dead" or "alive," "male" or "female." A simple event is one that cannot be decomposed.

sample

A **simple event** is defined as one of the possible outcomes on a single repetition of the experiment. *One and only one* simple event can occur on a single repetition of an experiment.

Simple events are denoted by the letter E with a subscript. *Events* consist of a collection of two or more _____ events. Events are denoted by capital letters such as A, B, C, and so on.

simple

An **event** is a collection of one or more simple events.

There are experiments wherein when one event occurs, a second cannot. If, for example, when a coin is tossed and the event "a head is observed on the upper face" occurs, then the event "a tail is observed on the upper face" cannot occur.

Two events are **mutually exclusive** if, when one occurs, the other cannot, and vice versa.

Example 4.1
An experiment involves ranking three applicants, X, Y, and Z, in order of their ability to perform in a given task. List the possible simple events associated with this experiment.

Solution
Using the notation (X, Y, Z) to denote the outcome that X is ranked first, Y is ranked second, and Z is ranked third, we get six possible outcomes or simple events associated with this experiment:

X,Y
Z,Y,X

E_1; E_4

E_1: (X,Y,Z), E_2: (Y,X,Z), E_3: $(Z,$_____$)$,
E_4: (X,Z,Y), E_5: (Y,Z,X), E_6: $($_____$)$

If A is the event that applicant X is ranked first, then A will occur if simple event _____ or _____ occurs.

Example 4.2
The financial records of two companies are examined to determine whether each company showed a profit (P) or not (N) during the last quarter.

a. List the simple events associated with this experiment.
b. List the simple events that make up the event B, "exactly one company showed a profit."

Solution
a. The simple events consist of the *ordered* pairs

E_1: (P,P), E_2: (P,N),

E_3: (N,P), E_4: $($_____$)$

b. Event B consists of the simple events _____ and

_____.

N, N

E_2
E_3

The set of all simple events associated with an experiment is called the *sample space*. A Venn diagram is a pictorial representation of the sample space in which events are depicted as portions of the sample space. The totality of simple events is the sample space and is denoted by S.

The set of all simple events is called the **sample space** and denoted by S.

Example 4.3
An oil wildcatter has just enough resources to drill three wells. Each well will either produce oil or be dry. List the simple events associated with this experiment and construct a Venn diagram to represent the sample space S.

Solution
We can find the simple events using a *tree diagram* in which each successive "branching" of the tree corresponds to a step necessary to generate the simple events (Figure 4.1). Since there are three oil wells, there are three successive branchings. Fill in the missing entries in Figure 3.1. A typical outcome is (Oil, Dry, Dry), which we abbreviate as (O, D, D).

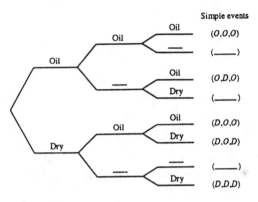

Figure 4.1

The simple events are given by the following *ordered triplets*:

E_1: (O,O,O), E_5: (D,O,O),

E_2: (O,O,D), E_6: (D,O,D),

E_3: (O,D,O), E_7: (_____),

E_4: (O,D,D), E_8: (_____)

1. In this experiment, the event "observe exactly two dry wells" is a (simple, compound) event because it is composed of the simple events _____, _____, and _____. On the other hand, the event "observe no dry wells" is a (simple, compound) event because it is composed of exactly one simple event—namely, _____.

2. Complete the following Venn diagram corresponding to this experiment by assigning the eight simple events to the eight points enclosed by the closed curve.

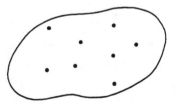

3. Let A be the event "observe no dry wells" and let B be the event "observe at least two wells producing oil." Represent A and B in a Venn diagram in the space below.

Dry; (O,O,D)

Dry

Oil; (D,D,O)
Dry

D,D,O

D,D,D

compound
E_4; E_6; E_7
simple

E_1

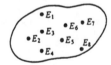

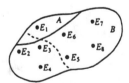

An alternative way to display simple events is to use a **probability table**, as shown in Figure 4.2 in which the population consists of male and female undergraduate students.

Sex	Class			
	Freshman (1)	Sophomore (2)	Junior (3)	Senior (4)
Male	(M, 1)	(M, 2)	(M, 3)	(M, 4)
Female	(F, 1)	(F, 2)	(F, 3)	(F, 4)

Figure 4.2

The entries would consist of the actual cell probabilities or fraction of students in each of the cells in the body of the table.

4.3 CALCULATING PROBABILITIES USING SIMPLE EVENTS

When an experiment and its sample space have been defined, the next step is to assign probabilities to the simple events. If an experiment is repeated a large number of times, n, and the Event A is observed with frequency n_A, then $P(A)$, the probability that the event will occur on a single repetition of the experiment is defined as

$$P(A) = \lim_{n \to \infty} \left(\frac{Frequency}{n} \right) = \lim_{n \to \infty} \left(\frac{n_A}{n} \right)$$

and approximated by the fraction $\frac{n_A}{n}$.

Using the relative frequency definition of probability, we see that for any event A,

0; 1

$$\underline{\hspace{2cm}} \leq P(A) \leq \underline{\hspace{2cm}}$$

If the event A never occurs, then $P(A) = 0$; if the event A always occurs, then $P(A) = \underline{\hspace{2cm}}$. Therefore, the closer its value is to 1, the more likely A is to occur.

For example, in the toss of a coin, if a head appears on the upper face (event H), then a tail (event T) cannot. Therefore, the

are

events H and T are mutually exclusive. Simple events (are, are not) mutually exclusive; therefore, the probabilities associated with simple events satisfy the following conditions:

1. _____ $\leq P(E_i) \leq$ _____   0; 1

2. $\sum\limits_{\text{all } i} P(E_i) =$ _____   1

When it is possible to write down the simple events associated with an experiment and to assess their probabilities, we can find the probability of an event A by summing the probabilities of the simple events contained in _____. A

The **probability of an event** A is found by summing the probabilities of the simple events in A:

$$P(A) = \sum_{\text{all } E_i \text{ in } A} P(E_i)$$

If there are N simple events, all of which are *equally likely*, and n_A of the simple events are contained in the event A, then $P(A) = n_A/N$. However, in general, simple events will not have equal probabilities.

Example 4.4
Suppose that two coins are tossed and the upper faces are recorded. Suppose further that the coins are not fair and the probability that a head results on either coin is greater than one half. The following probabilities are assigned to the simple events:

Simple Event	Outcome	Probability
E_1	HH	.42
E_2	HT	.18
E_3	TH	.28
E_4	TT	.12

a. Verify that this assignment of probabilities satisfies the conditions

$$0 \leq P(E_i) \leq 1 \quad \text{and} \quad \sum_{i=1}^{4} P(E_i) = 1$$

b. Find the probability of the event A, "the toss results in exactly one head and one tail."

c. Find the probability of the event B, "the toss results in at least one head."

Solution

a. We need only to verify the second condition because observation shows that the assigned probabilities satisfy the first condition.

Hence

$$\sum_{i=1}^{4} P(E_i) = P(E_1) + P(E_2) + P(E_3) + P(E_4)$$

.28; .12

$$= .42 + .18 + \underline{\hspace{2cm}} + \underline{\hspace{2cm}}$$

1

$$= \underline{\hspace{2cm}}$$

is

The second condition (is, is not) satisfied.

b. The event A, "exactly one head and one tail," consists of the simple events E_2 and E_3. Therefore,

$$P(A) = P(E_2) + P(E_3)$$

.18; .28

$$= \underline{\hspace{2cm}} + \underline{\hspace{2cm}}$$

.46

$$= \underline{\hspace{2cm}}$$

c. The event B, "at least one head," consists of the simple events E_1, E_2, and E_3. Therefore,

$$P(B) = P(E_1) + P(E_2) + P(E_3)$$

.42

$$= \underline{\hspace{2cm}} + .18 + .28$$

.88

$$= \underline{\hspace{2cm}}$$

Example 4.5

In a shipment of four radios, R_1, R_2, R_3, and R_4, one radio is defective (say, R_3). If a dealer selects two radios at random to display in his store, what is the probability that exactly one of the radios is defective?

Solution

If we disregard the order of selection of the two radios to be displayed, the possible outcomes are

$$E_1: (R_1, R_2), \quad E_4: (R_2, R_3),$$

R_4

$$E_2: (R_1, R_3), \quad E_5: (R_2, \underline{\hspace{1.5cm}}),$$

R_3, R_4

$$E_3: (R_1, R_4), \quad E_6: (\underline{\hspace{1.5cm}})$$

If the radios are selected at random, all combinations should have the same chance of being drawn. Therefore, we assign $P(E_i) =$

1/6

$\underline{\hspace{2cm}}$ to each of the six simple events.

The event D, "exactly one of the two radios selected is defective," consists of the simple events E_2, E_4, and $\underline{\hspace{2cm}}$.

E_6

Therefore,

$$P(D) = P(E_2) + P(E_4) + P(E_6)$$

$$= \tfrac{1}{6} + \tfrac{1}{6} + \tfrac{1}{6}$$

1/2

$$= \underline{\hspace{2cm}}$$

Calculating the Probability of an Event: Simple Event Approach

The simple event approach to finding $P(A)$ has five steps:

Step 1. Define the experiment.

Step 2. List all the simple events. Test to make certain that none can be decomposed.

Step 3. Specify which simple events lie in A.

Step 4. Assign appropriate probabilities to the simple events. Make sure that

$$\sum_{\text{all } i} P(E_i) = 1$$

Step 5. Find $P(A)$ by summing the probabilities for all simple events in A.

Example 4.6

A taste-testing experiment is conducted in a local grocery store. Two brands of soft drinks are tasted by a passing shopper, who is then asked to state a preference for brand C or brand P. Suppose that four shoppers are asked to participate in the experiment and that all four choose brand P. Under the assumption that there is no difference between the two brands, what is the probability of the event A, "all four shoppers choose brand P"?

Solution

Step 1. The experiment consists of observing the preferences of each of the four shoppers for either brand P or brand C.

Step 2. The sample space contains the following _____ simple | 16
events, which can be found using a tree diagram:

E_1: (_____), E_7: $(CCPP)$, E_{12}: $(CCCP)$, | $PPPP$

E_2: $(CPPP)$, E_8: $(CPCP)$, E_{13}: $(CCPC)$,

E_3: $(PCPP)$, E_9: $(CPPC)$, E_{14}: $(CPCC)$,

E_4: $(PCCP)$, E_{10}: $(PCCP)$, E_{15}: $(PCCC)$,

E_5: (_____), E_{11}: (_____), E_{16}: (_____), | $PPPC$; $PCPC$;

E_6: $(PPCC)$ | $CCCC$

Step 3. $A = \{($_____$)\}$. | $PPPP$

Step 4. The requirement that there be no difference between the two brands implies that a probability of _____ should be as- | $1/16$
signed to each simple event.

Step 5. There is(are) _____ simple event(s) in A. Hence, $P(A)$ | one

$=$ _____. | $1/16$

Modified Simple Event Approach

It may be that the list of simple events is quite long. But if equal probabilities are assigned to the simple events, all that is actually required is that you know precisely the number N of events in S and the number n_A of simple events in the event A. Then $P(A) = n_A/N$. If, however, a list is not made of the simple events, you must take care that no simple event in A is overlooked.

Example 4.7

A dealer who buys items in lots of ten selects two of the ten items at random and inspects them thoroughly. He accepts all ten if there are no defectives among the two inspected. Suppose that a lot contains two defective items.

a. What is the probability that the dealer will nonetheless accept all ten items?
b. What is the probability that he will find both of the defective items?

Solution

a. Let A be the event that the dealer accepts the lot.

Step 1. The experiment consists of selecting two items at random from ten items.

Step 2. The sample space consists of *unordered* pairs of the form $(G_1 G_2)$ or $(G_7 D_1)$. The number of simple events is $N = 45$; the simple events are listed below:

$(G_1 G_2)$	$(G_1 G_6)$	$(G_1 D_2)$	$(G_2 G_6)$	$(G_2 D_2)$
$(G_1 G_3)$	$(G_1 G_7)$	$(G_2 G_3)$	$(G_2 G_7)$	$(G_3 G_4)$
$(G_1 G_4)$	$(G_1 G_8)$	$(G_2 G_4)$	$(G_2 G_8)$	$(G_3 G_5)$
$(G_1 G_5)$	$(G_1 D_1)$	$(G_2 G_5)$	$(G_2 D_1)$	$(G_3 G_6)$
$(G_3 G_7)$	$(G_4 G_6)$	$(G_5 G_6)$	$(G_6 G_7)$	$(G_7 D_1)$
$(G_3 G_8)$	$(G_4 G_7)$	$(G_5 G_7)$	$(G_6 G_8)$	$(G_7 D_2)$
$(G_3 D_1)$	$(G_4 G_8)$	$(G_5 G_8)$	$(G_6 D_1)$	$(G_8 D_1)$
$(G_3 D_2)$	$(G_4 D_1)$	$(G_5 D_1)$	$(G_6 D_2)$	$(G_8 D_2)$
$(G_4 G_5)$	$(G_4 D_2)$	$(G_5 D_2)$	$(G_7 G_8)$	$(D_1 D_2)$

Step 3. The event A consists of the simple events that contain two good items. In this case, $n_A = 28$.

Step 4. Since the selection is made at random, each simple event should be assigned the same probability, equal to

_____ .

Step 5. There are 28 simple events in A. Hence, $P(A) =$

_____ .

1/45

28/45

b. Let B be the event that the dealer finds both defective items in the random selection.

Step 3. The event B consists of the single simple event
(_____). Hence, $n_B = $ _____.

Step 5. Therefore, $P(B) = n_B/N = 1/45$.

$D_1 D_2$; 1

Notice that it was very tedious to list the $N = 45$ simple events in this situation. Moreover, it is easy to overlook simple events when their number becomes large. In many instances, the counting rules presented in the next section can be used to find N and n_A without listing the individual simple events. The interested reader is referred to that section.

Self-Correcting Exercises 4A

1. The owner of a camera shop receives a shipment of five cameras from a camera manufacturer. Unknown to the owner, two of these cameras are defective. Suppose that the owner selects two of the five cameras at random and tests them for operability.
    a. Define the experiment.
    b. List the simple events associated with this experiment.
    c. Define the following events in terms of the simple events in part b.
        A: both cameras are defective.
        B: neither camera is defective.
        C: at least one camera is defective.
        D: the first camera tested is defective.
    d. Find $P(A)$, $P(B)$, $P(C)$, and $P(D)$ by using the simple event approach.

2. A lot containing six items is comprised of four good items and two defective items. Two items are selected at random from the lot for testing purposes.
    a. List the simple events for this experiment.
    b. List the simple events in each of the three following events:
        A: at least one item is defective.
        B: exactly one item is defective.
        C: no more than one item is defective.
    c. Find $P(A)$, $P(B)$, and $P(C)$ using the simple event approach.

3. A public relations office proposed the following experiment to assess public attitudes toward large corporations. A person is shown four photographs $(1, 2, 3,$ and $4)$ of crimes that have been committed and asked to select what he or she considers to be the two worst crimes. Although all four crimes shown are robberies involving about the same amount of money, 1 and 2 are robberies committed against private citizens, while 3 and 4 are robberies committed against corporations. If the person shows no bias in his or her selection, find the probabilities associated with the following events:

         A: the selection includes pictures 1 and 2.
         B: the selection includes picture 3.
         C: both *A* and *B* occur.
         D: either *A* or *B* or both occur.

4. In quality control of taste and texture, it is common to have a taster compare a new batch of a food product with one having the desired properties. Three new batches are independently tested against the standard and classified as having the desired properties (H) or not having the desired properties (N).
   a. List the simple events for this experiment.
   b. If in fact all three new batches are no different from the standard, all simple events in part a. should be equally likely. If this is the case, find the probabilities associated with the following events:
        A: exactly one batch is declared as not having the desired properties.
        B: batch number one is declared to have the desired properties.
        C: all three batches are declared to have the desired properties.
        D: at least two batches are declared to have the desired properties.

Solutions to SCE 4A

1. a. One camera is drawn at random from five, after which a second camera is chosen. Each is tested and found to be defective or nondefective.
   b. Since it is important whether the first or second camera is defective (see event *D*), there are 20 simple events to be listed. The first element in the pair denotes the first camera chosen.

$$E_1: G_1G_2 \qquad E_6: G_2D_1 \qquad E_{11}: G_2G_1 \qquad E_{16}: D_1G_2$$
$$E_2: G_1G_3 \qquad E_7: G_2D_2 \qquad E_{12}: G_3G_1 \qquad E_{17}: D_2G_2$$
$$E_3: G_1D_1 \qquad E_8: G_3D_1 \qquad E_{13}: D_1G_1 \qquad E_{18}: D_1G_3$$
$$E_4: G_1D_2 \qquad E_9: G_3D_2 \qquad E_{14}: D_2G_1 \qquad E_{19}: D_2G_3$$
$$E_5: G_2G_3 \qquad E_{10}: D_1D_2 \qquad E_{15}: G_3G_2 \qquad E_{20}: D_2D_1$$

   c. *A*: $\{E_{10}, E_{20}\}$

     B: $\{E_1, E_2, E_5, E_{11}, E_{12}, E_{15}\}$

     C: $\{E_3, E_4, E_6, E_7, E_8, E_9, E_{10}, E_{13}, E_{14}, E_{16}, E_{17}, E_{18}, E_{19},$
       $E_{20}\}$

     D: $\{E_{10}, E_{13}, E_{14}, E_{16}, E_{17}, E_{18}, E_{19}, E_{20}\}$

d. $P(A) = 2/20 = 1/10$ $P(B) = 6/20 = 3/10$
$P(C) = 14/20 = 7/10$ $P(D) = 8/20 = 2/5$.

2. Denote the four good items as G_1, G_2, G_3, G_4 and the two defectives as D_1 and D_2.

   a. $E_1: G_1G_2$ $E_5: G_1D_2$ $E_9: G_2D_2$ $E_{13}: G_4D_1$

   $E_2: G_1G_3$ $E_6: G_2D_3$ $E_{10}: G_3G_4$ $E_{14}: G_4D_2$

   $E_3: G_1G_4$ $E_7: G_2G_4$ $E_{11}: G_3D_1$ $E_{15}: D_1D_2$

   $E_4: G_1D_1$ $E_8: G_2D_1$ $E_{12}: G_3D_2$

   b. "At least one defective" implies one or two defectives, while "no more than one defective" implies zero or one defective.

   A: $\{E_4, E_5, E_8, E_9, E_{11}, E_{12}, E_{13}, E_{14}, E_{15}\}$

   B: $\{E_4, E_5, E_8, E_9, E_{11}, E_{12}, E_{13}, E_{14}\}$

   C: $\{E_1, E_2, E_3, E_4, E_5, E_6, E_7, E_8, E_9, E_{10}, E_{11}, E_{12}, E_{13}, E_{14}\}$

   c. Each simple event is assigned equal probability; that is, $P(E_i) = 1/15$.
   $P(A) = 9/15 = 3/5$, $P(B) = 8/15$, $P(C) = 14/15$.

3. The experiment consists of choosing two photographs from a total of four, and the six simple events are
$E_1: 1, 2$ $E_2: 1, 3$ $E_3: 1, 4$ $E_4: 2, 3$ $E_5: 2, 4$ $E_6: 3, 4$

   Then: A: $\{E_1\}$; B: $\{E_2, E_4, E_6\}$; C: no simple events;

   D: $\{E_1, E_2, E_4, E_6\}$;

   $P(A) = 1/6$; $P(B) = 3/6 = 1/2$; $P(C) = 0$; $P(D) = 4/6 = 2/3$.

4. a. $E_1: HHH$ $E_3: NHN$ $E_5: HHN$ $E_7: NHH$

   $E_2: HNN$ $E_4: NNH$ $E_6: HNH$ $E_8: NNN$

   b. $P(A) = 3/8$; $P(B) = 4/8 = 1/2$; $P(C) = 1/8$;
   $P(D) = 4/8 = 1/2$.

4.3 USEFUL COUNTING RULES (Optional)

There are three basic counting rules that are useful in counting the number of simple events N that arise in many experiments. When all the N simple events are equally likely, the probability of an event A can be found without listing the simple events if N, the number of events in S, and n_A, the number of events in A, can be counted because in this case $P(A) = n_A/N$. This is often important because N and n_A can become quite large.

The **mn** Rule. Suppose a procedure can be completed in two stages. If the first stage can be done in m ways and the second stage in n ways after the first stage has been completed, then the number of ways of completing the procedure is mn (m times n).

Example 4.8
An experiment involves ranking three applicants in order of merit. How many ways can the three applicants be ranked?

Solution
The process of ranking three applicants can be done in two stages:

Stage 1: Select the best applicant from the three.
Stage 2: Having selected the best, select the next best from the remaining two applicants.

The ranking of the remaining applicant will automatically be third. The number of ways of accomplishing stage 1 is _____. When stage 1 is completed, there are _____ ways of accomplishing stage 2. Hence, there are $(3)(2) =$ _____ ways of ranking three applicants.

3
2
6

Example 4.9
A lot of items consists of four good items (G_1, G_2, G_3, and G_4) and two defective items (D_1 and D_2).

a. How many different samples of size two can be formed by selecting two items from these six?
b. How many different samples will consist of exactly one good and one defective item?
c. What is the probability that exactly one good and one defective item will be drawn?

Solution
a. Selecting two items from six items corresponds to the two-step procedure of (1) picking the first item and (2) picking the second item after picking the first. Hence, $m =$ _____, $n =$ _____, and the number of ordered pairs is $N = mn =$ $(6)(5) =$ _____.

6
5
30

b. Selecting one good and one defective item can be done in either of two ways:

1. The *defective* item can be drawn *first* in $m =$ _____ ways and the *good* item drawn *second* in $n =$ _____ ways. Hence, there are $mn =$ _____ ways of selecting a

2
4
8

defective item on the first draw and a good item on the second draw.

2. However, the *good* item can be drawn *first* in $m =$ __________ ways and the *defective* item drawn second in $n =$ __________ ways, so that there are $mn =$ __________ ways in which a good item is drawn first and a defective item is drawn second. 4 2 8
3. Combining the results above, there are exactly $8 + 8 =$ __________ samples that will contain exactly one defective and one good item. 16

c. Let A be the event that exactly one good and one defective item are drawn. From part a, $N =$ __________, and from part b, $n_A =$ __________. Hence, 30 16

$$P(A) = \frac{n_A}{N} = \frac{16}{30} = __________$$ 8/15

Generalized *mn* Rule. If an operation can be performed in k stages with n_1 ways of performing the first stage, n_2 ways of performing the second stage, ... and n_k ways of performing the k-th stage, then the number of ways that the operation can be performed is $n_1 n_2 \ldots n_k$.

For example, if you can choose one of five different colors of paint, one of two interior decors and one of three models of car, then you have $5(2)(3) = 30$ different possible selections to choose from.

An ordered arrangement of r distinct objects is called a **permutation**. The number of permutations that consist of r objects selected from n objects is given by the formula

$$P_r^n = \frac{n!}{(n-r)!} = n(n-1)(n-2)\cdots(n-r+1)$$

with $0! = 1$. Notice that P_r^n consists of r factors, commencing with n.

Example 4.10

In how many ways can three different office positions be filled if there are seven applicants who are qualified for all three positions?

Solution

Notice that assigning the same three people to different office positions would produce different ways of filling the three positions. Hence, we need to find the number of permutations (*ordered arrangements*) of three people selected from seven. Therefore,

5; 210

$$P_3^7 = \frac{7!}{4!} = (7)(6)(\underline{\hspace{2cm}}) = \underline{\hspace{2cm}}$$

Example 4.11
A corporation will select two sites from ten available sites under consideration for building two manufacturing plants. If one plant will produce flashbulbs and the other cameras, in how many ways can the selection be made?

Solution
We are interested in the number of permutations of two sites selected from ten sites because if two sites, say 6 and 8, were chosen, and the flashbulb plant was built at site 6 while the camera plant was built at site 8, this would result in a different selection than would occur if the camera plant was built at site 6 and the flashbulb plant at site 8. Therefore, the number of selections is

9; 90

$$P_2^{10} = (10)(\underline{\hspace{2cm}}) = \underline{\hspace{2cm}}$$

Suppose that in Example 4.11 the two sites that were selected were to have identical plants producing the same merchandise. In that case, the selection (S_1, S_2) would be the same as the selection (S_2, S_1). These two selections represent the number of permutations of two items equal to $2! = 2(1) = 2$. In this case you would consider the number of combinations of two sites as the actual number of possible selections. This would be found by dividing the number of permutations $P_2^{10} = \frac{10!}{8!}$ by $2!$. In this case the number of combinations would be equal to $\frac{10!}{2! \, 8!} = 45$. This leads to the next definition.

A selection of r objects from n distinct objects without regard to their ordering is called a **combination**. The number of combinations that can be formed when selecting r objects from n objects is given as

$$C_r^n = \frac{n!}{r! \, (n-r)!} \qquad \text{with } 0! = 1$$

Example 4.12
How many different 5-card hands can be dealt from an ordinary deck of 52 cards?

Solution
Since it is the value of the 5 cards and not the order in which they were dealt that will differentiate one 5-card hand from another, the number of distinct 5-card hands is

$$C_5^{52} = \frac{52!}{5!\,47!} = \frac{(52)(51)(50)(49)(48)}{(5)(4)(3)(2)(1)} = \underline{\hspace{2cm}}$$

2,598,960

Notice that C_5^{52} is the same as C_{47}^{52}. In general,

$$C_r^n = \underline{\hspace{2cm}}$$

C_{n-r}^n

Example 4.13

An experimenter must select three animals from ten available animals to be used as a control group. In how many ways can the control group be selected?

Solution

Since the order of selection is unimportant, the number of unordered selections is

$$C_3^{10} = \frac{(10)(9)(8)}{(3)(2)(1)} = \underline{\hspace{2cm}}$$

120

Example 4.14

Refer to Example 4.7. A dealer tests two items randomly chosen from a lot of ten items and accepts the lot if the two items are not defective. If a lot contains two defective items, use the counting rules to find the probability that the dealer accepts the lot and the probability that both defectives are found. Define:

      A: No defectives are found
      B: Two defectives are found

Solution

In order to calculate $P(A)$ and $P(B)$ it is necessary to find N, n_A, and n_B.

1. Since a simple event is an unordered pair of the form $(G_1 G_2)$ or $(G_1 D_1)$, the total number of simple events is

$$N = C_2^{10} = \frac{10!}{2!\,8!} = \frac{(10)(9)}{(2)(1)} = 45$$

2. The number of ways to draw no defectives is the same as the number of ways of drawing _____ good items (from a total of _____ good items). Hence,

two
eight

$$n_A = C_2^8 = \frac{8!}{2!\,6!} = \frac{(8)(7)}{(2)(1)} = \underline{\hspace{2cm}}$$

28

3. The number of ways to draw two defective items (from a total of two defective items) is

$$n_B = C_2^2 = \frac{2!}{0!\,2!} = \underline{\hspace{2cm}}$$

1

4. Using the results of steps 1, 2, and 3, we get

$$P(A) = \frac{n_A}{N} = \frac{28}{45}$$

$$P(B) = \frac{n_B}{N} = \frac{1}{45}$$

Notice that this method of solution is much less tedious than the solution used in Example 4.7.

Self-Correcting Exercises 4B

1. Refer to Exercise 1, Self Correcting Exercises 4A.
   a. Using counting rules, count the number of simple events in the experiment.
   b. Count the number of simple events in the events defined in part c of Exercise 1, Self-Correcting Exercises 4A.
   c. Find $P(A)$, $P(B)$, $P(C)$, and $P(D)$. Compare your answers with part d of Exercise 1, Self-Correcting Exercises 4A.

2. Five people are being considered for three awards, and no person can receive more than one award.
   a. In how many ways can these awards be given?
   b. If three of these people are city officials, in how many ways could the awards be given to these officials?
   c. If all candidates are equally qualified for the three awards, what is the probability that the awards will be presented to the three city officials?

3. A sociologist is interested in drawing a random sample of six individuals from a group consisting of ten males and ten females.
   a. How many different samples of size six are possible?
   b. How many samples would consist of all males? All females?
   c. How many samples would consist of three males and three females?
   d. If all samples are equally likely, find the probability that the sample would consist of all persons of the same sex.
   e. Find the probability that the sample contains three males and three females.
   f. Would this type of random sampling insure with a high probability that the sample proportion of males and females reflects the 50% proportion in the population?

Solutions to SCE 4B

1. a. Since the pairs are ordered (see part b, Exercise 1, Self-Correcting Exercises 4A), the number of simple events is
   $$P_2^5 = (5)(4) = 20.$$

b. Event A: $P_2^2 = (2)(1) = 2$; event B: $P_2^3 = (3)(2) = 6$; event C:

$2P_1^2P_1^3 + P_2^2 = 2(2)(3) + 2 = 14$ since the defective item may be

chosen either first or second; event D: $P_1^2P_1^3 + P_2^2 = 6 + 2 = 8$ since once the first camera is found to be defective, the second can be either defective or good.

c. See Exercise 1d, Self-Correcting Exercises 4A.

2. Since the awards are different, order is important.

a. $P_3^5 = 5!/2! = 60.$

b. $P_3^3 = 3! = 3(2) = 6.$

c. $P(\text{city officials receive three awards}) = 6/60 = 1/10$

3. Since a random sample of size 6 does not involve order, we have the following:

a. $C_6^{20} = 38{,}760.$

b. Drawing 6 males from a total of 10 males can be done in C_6^{10} $= 210$ ways. Similarly, 6 females can be drawn in $C_6^{10} = 210$ ways.

c. Three females or 3 males can be drawn in $C_3^{10} = 120$ ways.

Using the mn rule, 3 men *and* 3 women can be drawn in $C_3^{10}C_3^{10}$ $= 120^2 = 14{,}400$ ways.

d. $P(\text{all persons of same sex}) = P(\text{all men}) + P(\text{all women})$ $= (210 + 210)/38{,}760 = .0108.$

e. $P(3 \text{ males and } 3 \text{ females}) = 14{,}400/38{,}760 = .3715.$

f. Yes, 37% is a fairly high probability.

4.5 EVENT COMPOSITION AND EVENT RELATIONS

When one is attempting to find the probability of an event A, it is often useful and convenient to express A in terms of other events whose probabilities are known or perhaps easily calculated. Composition of events occurs in one of the two following ways or a combination of these two:

Intersections. The **intersection** of two events A and B is the event consisting of those simple events that are in both A and B. The intersection of A and B is denoted by $A \cap B$.

Unions. The **union** of two events A and B is the event consisting of those simple events that are in either A or B or both A and B. The union of A and B is denoted by $A \cup B$.

Example 4.15

In each of the Venn diagrams that follow, express symbolically the event represented by the shaded area. In each case the sample space S comprises all the simple events within the rectangle.

1.

$A \cap B$

Symbol _____

2.

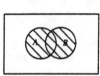

$A \cup B$

Symbol _____

3.

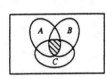

$A \cap B \cap C$

Symbol _____

4.

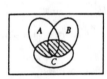

$(A \cap C) \cup (B \cap C)$

Symbol _____

Example 4.16

In each of the Venn diagrams below, shade in the event symbolized.

1. Symbol: $A \cup B$

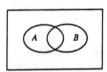

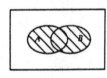

2. Symbol: $B \cap C$

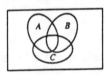

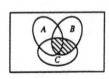

3. Symbol: $(A \cap E_1) \cup (A \cap E_2)$

4. Note that $E_1 \cup E_2 = S$ and $(A \cap E_1) \cup (A \cap E_2) = $ _____ .

A

Example 4.17

Consider an experiment that can result in one of ten simple events with probabilities as given in the table:

E_i	E_1	E_2	E_3	E_4	E_5	E_6	E_7	E_8	E_9	E_{10}
$P(E_i)$	.01	.05	.04	.20	.40	.03	.02	.15	.05	.05

Define the following events:

$A = \{E_1, E_2, E_3\}$

$B = \{E_1, E_3, E_4, E_5\}$

$C = \{E_4, E_5, E_6, E_7\}$

List the simple events and calculate the probability of occurrence for the following events:

1. $A \cap B$	2. $A \cap C$	3. $B \cup C$	4. $A \cup C$
5. A	6. B	7. C	8. $A \cup B \cup C$

Solution

1. The event $A \cap B$ consists of the simple events in both A and B. Hence, $A \cap B = \{E_1, E_3\}$. Since the probability of an event is calculated by summing the probabilities associated with each simple event contained in the event,

$$P(A \cap B) = P(E_1) + P(\underline{\hspace{1.5cm}})$$

E_3

$$= .01 + \underline{\hspace{1.5cm}} = \underline{\hspace{1.5cm}}$$

.04; .05

2. The events A and C contain _____ common simple events. Hence, $P(A \cap C) = $ _____ .

no

0

3. $B \cup C = \{E_1, E_3, E_4, E_5, E_6, E_7\}$ and

$$P(B \cup C) = P(E_1) + P(E_3) + P(\underline{\hspace{1cm}}) + P(E_5) + P(\underline{\hspace{1cm}})$$

E_4; E_6

$$+ P(E_7)$$

$$= .01 + .04 + \underline{\hspace{1.5cm}} + .40 + \underline{\hspace{1.5cm}} + .02$$

.20; 03

$$= \underline{\hspace{2cm}}$$

.70

4. $A \cup C = \{E_1, E_2, E_3, E_4, E_5, E_6, E_7\}$ and

$$P(A \cup C) = \sum_{i=1}^{7} P(E_i) = .75$$

5. $P(A) = .01 + \underline{\hspace{2cm}} + .04 = \underline{\hspace{1.5cm}}$.

.05; .10

6. $P(B) = .01 + .04 + \underline{\hspace{2cm}} + .40 = \underline{\hspace{1.5cm}}$.

.20; .65

7. $P(C) = \underline{\hspace{2cm}} + .40 + .03 + .02 = \underline{\hspace{1.5cm}}$.

.20; .65

8. $A \cup B \cup C = \{E_1, E_2, \underline{\hspace{1cm}}, \underline{\hspace{1cm}}, E_5, E_6, E_7\}$. Hence,

E_3; E_4

$$P(A \cup B \cup C) = \sum_{i=1}^{7} P(E_i) = \underline{\hspace{2cm}}$$

.75

Notice that $A \cup B \cup C = A \cup C$ in this example.

Mutually exclusive events are also called **disjoint** events because they contain no simple events in common.

When events A and B are **mutually exclusive** or **disjoint**, then

- $P(A \cap B) = 0$

- $P(A \cup B) = P(A) + P(B)$

For example, in tossing two fair coins with events A: no heads, B: exactly one head, and C: exactly two heads, let $P(A) = 1/4$, $P(B) = 2/4$, and $P(C) = 1/4$. Since A, B, and C are disjoint, the event $A \cup B$, that there is either zero or one head in the toss, has probability $P(A \cup B) = P(A) + P(B) = 1/4 + 2/4 =$ _____,

3/4

0

while the event $A \cap B$ has probability $P(A \cap B) =$ _____.

The event relationship that follows often simplifies probability calculations:

The event consisting of all those simple events in the sample space S that are not in the event A is defined as the **complement** of A and is denoted by A^C.

1

It is always true that $P(A) + P(A^C) =$ _____. Therefore, $P(A) = 1 - P(A^C)$. If $P(A^C)$ can be found more easily than $P(A)$, this relationship greatly simplifies finding $P(A)$.

Example 4.18
If three fair coins are tossed, what is the probability of observing at least one head in the toss?

Solution

1. Let A be the event that there is at least one head in the toss of

no

three coins. A^C is the event that there are _____ heads in the toss of three coins.

2. There are $n = 8$ possible outcomes for this experiment:

$$(TTT) \quad (HTT) \quad (THT) \quad (TTH)$$
$$(HHH) \quad (THH) \quad (HTH) \quad (HHT)$$

so that each simple event is assigned a probability of

1/8

_____.

1/8; 7/8

3. A^C consists of the single simple event (TTT); $P(A^C) =$ _____ and $P(A) = 1 - P(A^C) =$ _____.

When A and B are mutually exclusive, their intersection $A \cap B$ contains no simple events. Notice that the events A and A^C (are, are not) mutually exclusive.

are

4.6 CONDITIONAL PROBABILITY AND INDEPENDENCE

Two events may be related so that the probability of the occurrence of one event depends on whether a second event has occurred. Let A be the event "a defective item is produced" on a production line, and let B be the event "the production line is not operating within control limits." These two events are related in the sense that $P(A)$ is not the same as the probability of a defective item, given the prior information that the line is not operating within control limits.

Suppose we restrict our attention to only the subpopulation generated when the event B occurs and look at the fraction of items that are defective. This fraction is the *conditional probability of A, given B*, and in this case would be expected to be larger than $P(A)$. The conditional probability of A, given that B has occurred, is denoted by $P(A \mid B)$, where the vertical bar is read "given" and the event following the bar is the event that has occurred.

The probability that event A will occur, *given that* the event B has occurred, is given by

$$P(A \mid B) = \frac{P(A \cap B)}{P(B)}$$

for $P(B) > 0$.

Use a Venn diagram with events A and B to see that by knowing that the event B has occurred, you effectively exclude any simple events lying outside the event B from further consideration. Since $P(B)$ and $P(A \cap B)$ represent the amounts of probability associated with events B and $A \cap B$, we have

$$P(A \mid B) = \frac{P(A \cap B)}{P(B)}$$

which merely represents the proportion of $P(B)$ that will give rise to the event A.

If $P(A)$ and $P(A \mid B)$ differ in value, then the probability of A changes, depending on whether it is known that B has occurred.

Two events are said to be **independent** if either

1. $P(A) = P(A \mid B)$ when $P(B) > 0$, or
2. $P(B) = P(B \mid A)$ when $P(A) > 0$.

Otherwise, the events are said to be **dependent**.

When $P(A \mid B) = P(A)$, the events A and B are said to be (probabilistically) independent because the probability of the occurrence of A is not affected by knowledge of the occurrence of B. If $P(A \mid B) \neq P(A)$, the events A and B are said to be dependent.

When two events A and B are independent, then the probability of the intersection of A and B is

$$P(A \cap B) = P(A)P(B)$$

You may wish to use this as a check for independence if these quantities are available.

Example 4.19
You hold ticket numbers 7 and 8 in an office lottery in which ten tickets numbered 1 through 10 were sold. The winning ticket is drawn at random from those sold. You are told that the winning number is odd. Does this information alter the probability that you have won the lottery or are the two events independent?

Solution
Define the events A and B as follows:

   A: Number 7 or 8 is drawn
   B: An odd number is drawn

1/5
1/5
independent

The unconditional probability of winning is $P(A) = 2/10 =$ _____, while the conditional probability of winning is $P(A \mid B) =$ _____. Your probability of winning remains unchanged; the events A and B are (dependent, independent).

Example 4.20
Five applicants, all equally qualified, are being considered for a research technician's position. There are three males and two females among the applicants. Define the following events:

   A: Female number one is selected
   B: A female is selected

If the selection is done at random, find $P(A)$ and $P(A \mid B)$. Are A and B mutually exclusive? Are A and B independent?

Solution

The unconditional probability $P(A) = $ _____ and $P(A \cap B) = $

$P(A) = $ _____. Hence, A and B (are, are not) mutually exclusive.

We can find $P(A \mid B)$ in one of two ways.

1. *Direct enumeration.* If B has occurred, then we need only consider the two female applicants as making up the new restricted sample space. Hence, $P(A \mid B) = $ _____.

2. *Calculation.* By definition,

$$P(A \mid B) = \frac{P(A \cap B)}{P(B)}$$

$$= \frac{1/5}{\underline{\hspace{1cm}}} = \frac{1}{2}$$

Since $P(A) \neq P(A \mid B)$, A and B are (dependent, independent) events.

Many events can be viewed as the union or intersection, or both, of simpler events whose probabilities may be known or easily calculated. In such cases, the Additive and Multiplicative Laws of Probability can be used to assess the probability of the event.

Notice that $P(A \cap B) = 1/5$ since A (a female is selected) and B (female number one is selected) can occur in only 1 of 5 ways. $P(A) = 2/5$ and $P(B) = 1/5$ and $P(A)P(B) = 2/25 \neq 1/5$. Therefore you would also conclude using this method that the events A and B are (dependent, independent).

	1/5
	1/5; are not
	1/2
	2/5
	dependent
	dependent

The Additive Law of Probability: The probability of the union of events A and B is given by

$$P(A \cup B) = P(A) + P(B) - P(A \cap B)$$

When the events A and B are mutually exclusive, $P(A \cap B) = 0$ and

$$P(A \cup B) = P(A) + P(B)$$

If event A is contained in event B, then $P(A \cup B) = P(B)$.

The Multiplicative Law of Probability: The probability of intersection of events A and B is given by

$$P(A \cap B) = P(A)P(B \mid A)$$

or, equivalently, by

$$P(A \cap B) = P(B)P(A \mid B)$$

If the events are independent, then

$$P(A \cap B) = P(A)P(B)$$

Example 4.21

In the Venn diagram in Figure 4.3, the ten simple events shown are equally likely. Thus, to each simple event is assigned the probability

1/10

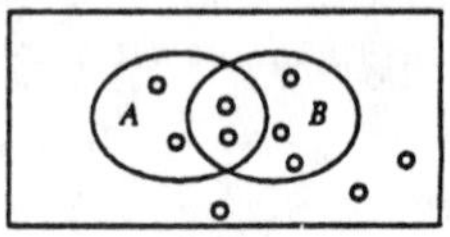

_____ .

Figure 4.3

Find the following probabilities:

7/10
2/10
5/10
2/5

1. $P(A \cup B) =$ _____ .
2. $P(A \cap B) =$ _____ .
3. $P(B) =$ _____ .
4. $P(A \mid B) =$ _____ .

5/10
independent

5. $P(B^C) =$ _____ .
6. A and B are (independent, dependent).

Example 4.22

The personnel files for a large real estate agency list its 150 employees as follows:

	Years Employed with the Agency		
	0–5 (A)	6–10 (B)	11 or More (C)
Not a college graduate (D)	10	20	20
College graduate (E)	40	50	10

If *one* personnel file is drawn at random from the agency's personnel files, calculate the probabilities requested:

1/3
2/3
2/15
1/10
11/15
0
dependent

1. $P(A) =$ _____ .
2. $P(E) =$ _____ .
3. $P(B \cap D) =$ _____ .
4. $P(C \mid E) =$ _____ .
5. $P(A \cup E) =$ _____ .
6. $P(A \mid C) =$ _____ .
7. A and E are (independent, dependent)

Calculating the Probability of an Event: Event Composition Approach

The event composition approach to finding $P(A)$ has four steps:

Step 1. Define the experiment.

Step 2. Clearly visualize the nature of the simple events.
Identify a few to clarify your thinking.

Step 3. Write an equation expressing A as a composition of
two or more events. Make certain that the event expressed by the composition is the same set of simple
events as the event A.

Step 4. Apply the Additive and Multiplicative Laws of Probability as required to the equation found in step 3. It
is assumed that the component probabilities are known
for the particular composition used.

Example 4.23

Player A has entered a tennis tournament but it is not yet certain whether player B will enter. Let A be the event that player A will win the
tournament, and let B be the event that player B will enter the tournament. Suppose that player A has probability 1/6 of winning the tournament if player B enters and probability 3/4 of winning if player B does
not enter the tournament. If $P(B) = 1/3$, find $P(A)$.

Solution

Step 1. The experiment is the observation of whether player B enters
the tournament and whether player A wins the tournament.

Step 2. The experiment can be described using the *tree diagram* in
Figure 4.4. Fill in the missing entries.

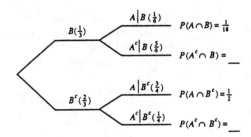

5/18

1/6

Figure 4.4

Step 3. The event A can occur and B can occur, or the event A can
occur and B not occur. Therefore, we can write

$$A = (A \cap B) \cup (A \cap B^C)$$

Step 4. The events $A \cap B$ and $A \cap B^C$ are mutually exclusive.
Therefore,

$$P(A) = P(A \cap B) + P(A \cap B^C)$$
$$= \tfrac{1}{18} + \tfrac{1}{2}$$

$$= \underline{}$$

5/9

Self-Correcting Exercises 4C

1. Refer to Exercise 1, Self-Correcting Exercises 4A.
   a. List the simple events comprising the following events: $A \cup C$, $A \cap D$, $C \cap D$ and $A \cup D$.
   b. Calculate $P(A \cup C)$, $P(A \cap D)$, $P(C \cap D)$, and $P(A \cup D)$.

2. A hospital spokesperson reported that four births had taken place at the hospital during the last twenty-four hours. If we consider only the sex of these four children, recording M for a male child and F for a female child, there are 16 sex combinations possible.
   a. List these 16 outcomes in terms of simple events, beginning with E_1 as the outcome $(FFFF)$.
   b. Define the following events in terms of the simple events $E_1, \ldots, E_{16}$:
      A: two boys and two girls are born.
      B: no boys are born.
      C: at least one boy is born.
   c. List the simple events in the following events
      $A \cap B$, $B \cup C$, $(A \cap C) \cup (B \cap C)$, C^C, $(A \cap C)^C$
   d. If the sex of a newborn baby is just as likely to be male as female, find the probabilities associated with the eight events defined in parts b. and c.
   e. Calculate $P(A \mid C)$. Are A and C mutually exclusive? Are A and C independent?
   f. Calculate $P(B \mid C)$. Are B and C independent? Mutually exclusive?

3. Two hundred corporate executives in the Los Angeles area were interviewed. They were classified according to the size of the corporation they represented and their choice as to the most effective method for reducing air pollution in the Los Angeles basin. (Data are fictitious.)

	Corporation Size		
Option	Small (A)	Medium (B)	Large (C)
Car pooling (D)	20	15	20
Bus expansion (E)	30	25	11
Gas rationing (F)	3	8	4
Conversion to natural gas (G)	10	7	5
Antipollution devices (H)	12	20	10

Suppose one executive is chosen at random to be interviewed on a television broadcast.
   a. Calculate the following probabilities and describe each event involved in terms of the problem: $P(A)$, $P(F)$, $P(A \cap F)$, $P(A \cup G)$, $P(A \cap D)$ and $P(F^C)$.

b. Calculate $P(A \mid F)$ and $P(A \mid D)$. Are A and F independent? Mutually exclusive? Are A and D independent? Mutually exclusive?

4. An investor holds shares in three independent companies which, according to her business analyst, should show an increase in profit per share with probabilities .4, .6, and .7, respectively. Assume that the analyst's estimates for the probabilities of profit increases are correct.
   a. Find the probability that all three companies show profit increases for the coming year.
   b. Find the probability that none of the companies shows a profit increase.
   c. Find the probability that at least one company shows a profit increase.

5. A marksman is able to hit a small target with probability .9. Assume his shots are independent.
   a. What is the probability that he hits the target with the next three shots?
   b. What is the probability that he hits the target with at least one of the next three shots?

6. Suppose that a person who fails an examination is allowed to retake the examination but cannot take the examination more than three times. The probability that a person passes the exam on the first, second, or third try is .7, .8, or .9, respectively.
   a. What is the probability that a person takes the exam twice before passing it?
   b. What is the probability that a person takes the exam three times before passing it?
   c. What is the probability that a person passes this exam?

Solutions to SCE 4C

1. a. $A \cup C$: same as C; $A \cap D$: same as A; $C \cap D$: same as D: $A \cup D$: same as D.
   b. $P(A \cup C) = 7/10$, $P(A \cap D) = 1/10$, $P(C \cap D) = 2/5$, $P(A \cup D) = 2/5$.

2. a.

E_1: $FFFF$	E_5: $FFFM$	E_9: $MFFM$	E_{13}: $MFMM$
E_2: $MFFF$	E_6: $FFMM$	E_{10}: $MFMF$	E_{14}: $MMFM$
E_3: $FMFF$	E_7: $FMFM$	E_{11}: $MMFF$	E_{15}: $MMMF$
E_4: $FFMF$	E_8: $FMMF$	E_{12}: $FMMM$	E_{16}: $MMMM$

   b. A: $\{E_6, E_7, E_8, E_9, E_{10}, E_{11}\}$

   B: $\{E_1\}$

c. $A \cap B$: {no simple events}; $B \cup C$: $\{E_1, E_2, \ldots E_{16}\} = S$;
$(A \cap C) \cup (B \cap C)$: $\{E_6, E_7, E_8, E_9, E_{10}, E_{11}\}$; C^C: $\{E_1\}$;
$(A \cap C)^C$: $\{E_1, E_2, \ldots, E_5, E_{12}, E_{13}, \ldots E_{16}\}$

d. Since each simple event is equally likely,
$P(A) = 6/16 = 3/8$ $P(B \cup C) = 1$
$P(B) = 1/16$ $P[(A \cap C) \cup (B \cap C)] = 3/8$
$P(C) = 15/16$ $P(C^C) = 1/16$
$P(A \cap B) = 0$ $P(A \cap C)^C = 10/16 = 5/8$

e. $P(A \mid C) = \dfrac{P(A \cap C)}{P(C)} = \dfrac{6/16}{15/16} = \dfrac{6}{15}$ while $P(A) = 3/8$.
A and C are dependent but are not mutually exclusive.

f. $P(B \mid C) = \dfrac{P(B \cap C)}{P(C)} = \dfrac{0}{15/16} = 0$ while $P(B) = 1/16$ and
$P(B \cap C) = 0$. B and C are dependent and mutually exclusive.

3. a. $P(A) = P$(the executive represents a small corporation)
$= 75/200 = 3/8$; $P(F) = P$(the executive favors gas
rationing) $= 15/200 = 3/40$; $P(A \cap F) = 3/200$;
$P(A \cup G) = P$(executive favors conversion or represents a
small corporation or both) $= P(A) + P(G) - P(A \cap G)$
$= (75 + 22 - 10)/200 = 87/200$; $P(A \cap D) = P$(executive
represents a small corporation and favors car pooling)
$= 20/200$; $P(F^C) = 1 - P(F) = 1 - (3/40) = 37/40$.

b. $P(A \mid F) = \dfrac{P(A \cap F)}{P(F)} = \dfrac{3/200}{15/200} = \dfrac{3}{15}$, $P(A \mid D) = \dfrac{20/200}{55/200}$

$= \dfrac{20}{55}$. Neither A and F nor A and D are mutually exclusive.

A and F and A and D are both dependent.

4. Define A: company A shows an increase; B: company B shows
an increase; C: company C shows an increase. It is given that
$P(A) = .4$, $P(B) = .6$, $P(C) = .7$, and A, B, and C are independent events.
a. $P(A \cap B \cap C) = P(A)P(B)P(C) = (.4)(.6)(.7) = .168$.
b. $P(A^C \cap B^C \cap C^C) = P(A^C)P(B^C)P(C^C)$
$= [1 - P(A)][1 - P(B)][1 - P(C)] = (.6)(.4)(.3) = .072$
c. P(at least one shows profit)
$= 1 - P$(none show profit)
$= 1 - P(A^C \cap B^C \cap C^C) = 1 - .072 = .928$.

5. Define H: marksman hits the target; H^C: marksman misses the
target. It is given that $P(H) = .9$. Therefore, $P(H^C) = 1 - .9$
$= .1$.
a. Since his shots are independent, $P(H \cap H \cap H)$
$= P(H)P(H)P(H) = (.9)^3 = .729$.

b. The event of interest is A: marksman hits with at least one of the next three shows. Then A^C: marksman misses all three shots.
$$P(A) = 1 - P(A^C) = 1 - P(H^C \cap H^C \cap H^C) = 1 - (.1)^3 = .999.$$

6. Define A_1: person passes on first try; A_2: person passes on second try; A_3: person passes on third try. It is given that $P(A_1) = .7$,
$$P(A_2 \mid A_1^C) = .8, \ P(A_3 \mid A_1^C \cap A_2^C) = .9.$$

a. P(person passes on second try) $= P(A_1^C \cap A_2)$
$$= P(A_1^C)P(A_2 \mid A_1^C) = .3(.8) = .24$$

b. P(person passes on third try) $= P(A_1^C \cap A_2^C \cap A_3)$
$$= P(A_1^C)P(A_2^C \mid A_1^C)P(A_3 \mid A_1^C \cap A_2^C) = .3(.2)(.9) = .054.$$

c. P(person passes) $= P$(passes on first, second, or on third tries) $= P(A_1) + .24 + .054 = .7 + .24 + .054 = .994.$

4.7 BAYES' RULE (Optional)

A very interesting application of conditional probability is found in Bayes' Rule. This rule assumes that the sample space can be partitioned into k mutually exclusive events (subpopulations) $S_1, S_2, \ldots, S_k$ so that $S = S_1 \cup S_2 \cup S_3 \cup \cdots \cup S_k$. If a single repetition of an experiment results in the event A, with $P(A) > 0$, we may be interested in making an inference as to which event (subpopulation) most probably gave rise to the event A. The probability that the subpopulation S_i was sampled, given that event A occurred, is

$$P(S_i \mid A) = \frac{P(S_i \cap A)}{P(A)}$$

$$= \frac{P(S_i)\,P(A \mid S_i)}{\sum\limits_{j=1}^{k} P(S_j)\,P(A \mid S_j)}$$

The Venn diagram in Figure 4.5 with $k = 3$ subpopulations, demonstrates that

$$P(A) = P[(A \cap S_1) \cup (A \cap S_2) \cup (A \cap S_3)]$$
$$= P(A \cap S_1) + P(A \cap S_2) + P(A \cap S_3)$$

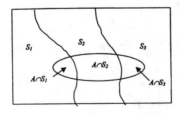

Figure 4.5

Since $P(A \cap S_i) = P(S_i) P(A \mid S_i)$, we can write

$$P(A) = P(S_1) P(A \mid S_1) + P(S_2) P(A \mid S_2) + P(S_3) P(A \mid S_3)$$

which is known as the Law of Total Probability. Hence, for k subpopulations, the Law of Total Probability is

$$P(A) = \sum_{j=1}^{k} P(S_j) P(A \mid S_j)$$

prior

posterior

In order to apply Bayes' Rule, it is necessary to know the probabilities $P(S_1), P(S_2), \ldots, P(S_k)$. These probabilities are called the _____ probabilities. When the prior probabilities are unknown, it is possible to assume that all subpopulations are equally likely, so that $P(S_1) = \cdots = P(S_k) = 1/k$. The conditional probabilities $P(S_i \mid A)$ are called the _____ probabilities because these are the probabilities that result after taking account of the sample information contained in the event A.

Example 4.24
A manufacturer of air-conditioning units purchases 70% of its thermostats from company A, 20% from company B, and the rest from company C. Past experience shows that .5% of company A's thermostats, 1% of company B's thermostats, and 1.5% of company C's thermostats are likely to be defective. An air-conditioning unit randomly selected from this manufacturer's production line was found to have a defective thermostat.

a. Find the probability that the defective thermostat was supplied by company A.
b. Find the probability that the defective thermostat was supplied by company B.

Solution
Let A, B, and C be the events that a thermostat selected at random was supplied by company A, B, or C, respectively. Let D be the event that a defective thermostat is observed. (Notice that $A \cup B \cup C = S$ and A, B, and C are mutually exclusive.) The following information is available:

.2

.1

$P(A) = .7$	$P(D \mid A) = .005$
$P(B) = $ _____	$P(D \mid B) = .010$
$P(C) = $ _____	$P(D \mid C) = .015$

a. Using Bayes' Rule to find $P(A \mid D)$, we have

$$P(A \mid D) = \frac{P(A) P(D \mid A)}{P(A) P(D \mid A) + P(B) P(D \mid B) + P(C) P(D \mid C)}$$

$$= \frac{(.7)(.005)}{(.7)(.005) + (.2)(.010) + (.1)(.015)}$$

$$= \frac{.0035}{.0035 + .0020 + .0015}$$

$$= \underline{\hspace{2cm}}$$

1/2

b. Using the results of part a to find $P(B \mid D)$, we have

$$P(B \mid D) = \frac{P(B)\,P(D \mid B)}{P(A)\,P(D \mid A) + P(B)\,P(D \mid B) + P(C)\,P(D \mid C)}$$

$$= \frac{\overline{\hspace{2cm}}}{.0070}$$

.0020

$$= \underline{\hspace{2cm}}$$

2/7

Furthermore, $P(C \mid D) = 1 - (1/2) - (2/7) = \underline{\hspace{2cm}}$. If one had to make a decision as to which company most probably supplied the defective part, company \underline{\hspace{2cm}} would be so named.

3/14

A

It is interesting to notice that Bayes' Rule entails deductive rather than inductive reasoning. Usually we are interested in investigating a problem beginning with the cause and reasoning to its effect. Bayes' Rule, however, reasons from the effect (a defective is observed) to the cause (which population produced the defective?). The next example further illustrates this type of logic.

Example 4.25
Suppose that a transmission system, whose input X is either a 0 or a 1 and whose output Y is either a 0 or a 1, mixes up the input according to the following scheme:

$$P(Y = 0 \mid X = 0) = .90$$

$$P(Y = 1 \mid X = 0) = .10$$

$$P(Y = 0 \mid X = 1) = .15$$

$$P(Y = 1 \mid X = 1) = .85$$

If $P(X = 0) = .3$, find the following:

1. $P(X = 1 \mid Y = 1)$
2. $P(X = 0 \mid Y = 0)$

Solution
Since $P(X = 0) = .3$, $P(X = 1) = 1 - .3 = \underline{\hspace{2cm}}$

.7

1. To find $P(X = 1 \mid Y = 1)$, write

$$P(X = 1 \mid Y = 1) = \frac{P[(X = 1) \cap (Y = 1)]}{P(Y = 1)}$$

The numerator is

$$P[(X = 1) \cap (Y = 1)] = P(X = 1)\,P(Y = 1 \mid X = 1)$$

$$= (.7)(.85)$$

.595

$$= \underline{\hspace{2cm}}$$

The denominator is

$$P(Y = 1) = P[(X = 0) \cap (Y = 1)] + P[(X = 1) \cap (Y = 1)]$$
$$= P(X = 0)\, P(Y = 1 \mid X = 0)$$
$$\quad + P(X = 1)\, P(Y = 1 \mid X = 1)$$
$$= (.3)(.10) + (.7)(.85)$$

.625

$$= \underline{\hspace{2cm}}$$

Therefore,

.952

$$P(X = 1 \mid Y = 1) = \frac{.595}{.625} = \underline{\hspace{2cm}}$$

2. To find $P(X = 0 \mid Y = 0)$, write

$$P(X = 0 \mid Y = 0) = \frac{P(X = 0 \cap Y = 0)}{P(Y = 0)}$$

The numerator is

$$P[(X = 0) \cap (Y = 0)] = P(X = 0)\, P(Y = 0 \mid X = 0)$$

.90

$$= (.3)(\underline{\hspace{2cm}})$$

.270

$$= \underline{\hspace{2cm}}$$

The denominator is

$$P(Y = 0) = P[(X = 0) \cap (Y = 0)] + P[(X = 1) \cap (Y = 0)]$$
$$= P(X = 0)\, P(Y = 0 \mid X = 0)$$
$$\quad + P(X = 1)\, P(Y = 0 \mid X = 1)$$

.7; .15

$$= (.3)(.90) + (\underline{\hspace{2cm}})(\underline{\hspace{2cm}})$$

.375

$$= \underline{\hspace{2cm}}$$

Therefore,

.720

$$P(X = 0 \mid Y = 0) = \frac{.270}{.375} = \underline{\hspace{2cm}}$$

Self-Correcting Exercises 4D

1. A manufacturer of air-conditioning units purchases 70% of its thermostats from company A, 20% from company B, and the rest from company C. Past experience shows that .5% of company A's thermostats, 1% of company B's thermostats and 1.5% of company C's thermostats are likely to be defective. An air-conditioning unit randomly selected from this manufacturer's

air-conditioning unit randomly selected from this manufacturer's production line was found to have a defective thermostat. Refer to Example 4.24.

a. Find the probability that company A supplied the defective thermostat.

b. Find the probability that company B supplied the defective thermostat.

2. Suppose that, based on past experience, it is known that a lie detector test will indicate that an innocent person is guilty with probability .08, while the test will indicate that a guilty person is innocent with probability .15. Suppose further that 10% of the population under study has committed a traffic violation. If a lie detector test indicates that a randomly chosen individual from this population has committed a traffic violation, what is the probability that this person is innocent of committing a traffic violation? What is the probability that this person is guilty of committing a traffic violation?

3. An oil wildcatter must decide whether or not to hire a seismic survey before deciding whether to drill for oil on a plot of land. Given that oil is present, the survey will indicate a favorable result with probability .8; if oil is not present, a favorable result will occur with probability .3. The wildcatter figures that oil is present on the plot of land with probability equal to .5. Determine the effectiveness of the survey by computing the probability that oil is present on the land, given a favorable seismic survey outcome and comparing this posterior probability with the prior probability of finding oil on the plot of land.

Solutions to SCE 4D

1. Define D: item is defective

   A: item supplied by company A

   B: item supplied by company B

   C: item supplied by company C

It is given that $P(D\,|\,A) = .005$, $P(D\,|\,B) = .010$, $P(D\,|\,C) = .015$.

a. Using Bayes' Law extended to three alternatives,

$$P(A\,|\,D) = \frac{P(A)P(D\,|\,A)}{P(A)P(D\,|\,A) + P(B)P(D\,|\,B) + P(C)P(D\,|\,C)}$$

$$= \frac{(.7)(.005)}{(.7)(.005) + (.2)(.010) + (.1)(.015)} = \frac{1}{2}$$

b. Using the results of part a. to find $P(B\,|\,D)$, we have

$$P(B\,|\,D) = \frac{P(B)P(D\,|\,B)}{P(A)P(D\,|\,A) + P(B)P(D\,|\,B) + P(C)P(D\,|\,C)}$$

$$= \frac{.002}{.007} = \frac{2}{7}$$

2. Define P: lie detector test is positive (guilty)
       I: a person is innocent
       G: a person is guilty
   It is given that $P(I) = .90$, $P(G) = .10$; $P(P \mid I) = .08$,
   $P(P^C \mid I) = .92$; and $P(P \mid G) = .85$, $P(P^C \mid G) = .15$. Find:

$$P(I \mid P) = \frac{P(I \cap P)}{P(I \cap P) + P(G \cap P)}$$

$$= \frac{P(I)P(P \mid I)}{P(I)P(P \mid I) + P(G)P(P \mid G)}$$

$$= \frac{(.90)(.08)}{(.90)(.08) + (.10)(.85)}$$

$$= \frac{.072}{.157} = .459$$

3. Define the events F: a favorable seismic outcome results;
   O: oil is actually present; N: oil is not present. $P(F \mid O) = .8$;
   $P(F \mid N) = .3$; $P(O) = .5$; $P(N) = 1 - P(O) = .5$;

$$P(O \mid F) = \frac{P(F \mid O)P(O)}{P(F \mid O)P(O) + P(F \mid N)P(N)}$$

$$= \frac{(.8)(.5)}{(.8)(.5) + (.3)(.5)} = \frac{8}{11}$$

4.8   DISCRETE RANDOM VARIABLES AND THEIR PROBABILITY DISTRIBUTIONS

In Chapter 1 we defined variables as being either *qualitative* or *quantitative*. We further classified a quantitative variable as being either *discrete*, if its possible values were countable, or *continuous*, if its possible values corresponded to the uncountable points on a line interval.

Example 4.26
The selling price of 50 homes represents a quantitative set of data because each measurement is numerical.

Example 4.27
A particular brand of microwave oven is rated by 25 consumers according to its overall performance as either excellent, very good, good, fair, or poor. The set of 25 measurements represents a qualitative set of data because each measurement is one of the five "qualities" given.

Example 4.28
Identify each of the following sets of data as either quantitative or qualitative.

1. The cost of identical models of a Toyota station wagon was recorded at each of 12 Toyota dealers in southern California. (quantitative, qualitative)

   quantitative

2. In the process of applying for credit, 25 applicants are asked whether or not they currently have an outstanding bank loan. (quantitative, qualitative)

   qualitative

3. In 1999, 150 cars were purchased by a local taxi company, and the make of car was recorded for each. There were 45 Fords, 30 Chevrolets, 60 Plymouths, and 15 Dodges. (quantitative, qualitative)

   qualitative

4. The total sales and the domestic sales were recorded for an electronics firm over the five years 1990 to 1999. (quantitative, qualitative)

   quantitative

Random Variables

Recall that an experiment is the process by which an observation (or measurement) is obtained. Most experiments result in numerical outcomes or events. The outcome itself may be a numerical quantity such as height, weight, time, or some rank ordering of a response; that is, the data are (quantitative, qualitative). If the data are qualitative, many times the observations will fall into one of several categories. When categorical observations are made—such as good or defective, color of eyes, income bracket, and so on, we are usually concerned with the number of observations that fall into a specified category. Again, the experiment results in a numerical outcome. Each time we observe the outcome of an experiment and assign a numerical value to the event that occurs, we are observing one particular value of a variable of interest. Since the value of this variable is determined by the outcome of a random experiment, we call the variable a *random variable*.

quantitative

A **random variable** is a variable that takes values according to a random or chance outcome of an experiment that has an underlying probabilistic mechanism.

If x is a random variable, then one and only one value of x is associated with each simple event in the sample space. The value that x assumes depends on the simple event that occurs according to the chance mechanism underlying the experiment.

Suppose that a sample of 100 people was randomly drawn from a population of voters and the number who favored candidate Jones was recorded. This process defines an experiment. The number of voters in the sample who favored candidate Jones is an example of a _____.

random variable

Random variables are designated as *discrete* or *continuous* according to the values they may assume in an experiment.

A **discrete random variable** is one that can assume a countable* number of values.

The following are examples of discrete random variables:

1. The number of voters who favor a political candidate in a given precinct
2. The number of defective bulbs in a package of 20 bulbs
3. The number of errors in an income tax return

Notice that discrete random variables are basically counts, and the phrase "the number of" can be used to identify a discrete random variable.

A **continuous random variable** can take on the infinitely large number of values associated with the points on a line interval.

The following are examples of continuous random variables:

1. The time required to complete a medical operation
2. The height of an experimental strain of corn
3. The amount of ore produced by a given mining operation

Classify the following random variables as discrete or continuous:

discrete

discrete

discrete
continuous
continuous

continuous

discrete

1. The number of psychological subjects who respond to stimuli in a group of 30. _____
2. The number of building permits issued in a community during a given month. _____
3. The number of amoebae in 1 cubic centimeter of water. _____
4. The juice content of six Valencia oranges. _____
5. The time to failure for an electronic system. _____
6. The amount of radioactive iodine excreted by rats in a medical experiment. _____
7. The number of defects in 1 square yard of carpeting. _____

It is necessary to make the preceding distinction between the discrete and continuous cases because the probability distributions require different mathematical treatment. In fact, calculus is a

---

*_Countable_ means that the values the random variable can assume can be associated with the counting integers, $0, 1, 2, \ldots$, so that the values can be counted.

prerequisite to any complete discussion of continuous random variables. Arithmetic and elementary algebra are all we need to develop discrete probability distributions.

Probability Distributions

A probability distribution is the theoretical equivalent of a frequency distribution for an entire population.

The **probability distribution** for a discrete random variable x consists of the pairs $(x, p(x))$, where x is one of the possible values of the random variable and $p(x)$ is its corresponding probability.

This probability distribution must satisfy two requirements:

1. $\sum_{x} p(x) =$ _____   1

2. _____ $\leq p(x) \leq$ _____   0; 1

You can express the probability distribution for a discrete random variable x in any one of three ways:

1. By listing, opposite each possible value of x, its probability $p(x)$ in a table
2. Graphically as a probability histogram
3. By supplying a formula together with a list of the possible values of x

Example 4.29
A businessman has decided to invest \$10,000 in each of three common stocks. Four stocks, call them A, B, C, and D, have been recommended to him by a broker, and he plans to select three of the four to form an investment portfolio. Unknown to the businessman, stocks A, B, and C will rise in the near future but D will suffer a severe drop in price. If the selection is made at random, find the probability distribution for x, the number of good stocks in the investment portfolio.

Solution
The experiment consists of selecting three of the four stocks. Each of the _____ distinctly different combinations has the same   four
chance for selection, and these _____ simple events form the   four
sample space. The simple events associated with the four combinations along with their probabilities are shown in the table. The value $x = 3$ is assigned to E_1 because this combination includes all three of the good stocks. Assign values of x to the other three simple events.

Simple Events	$P(E_i)$	x
E_1: ABC	1/4	3
E_2: ABD	1/4	_____
E_3: ACD	1/4	_____
E_4: BCD	1/4	_____

2

2

2

1. The probability distribution presented as a table is shown below. When $x = 2$, $p(x)$ is $p(2) = P(E_2) + P(E_3) + P(E_4) =$ _____. Similarly, $p(3) = P(E_1) =$ _____.

3/4; 1/4

x	$p(x)$
2	3/4
3	1/4

Note that the two requirements for a discrete probability distribution are satisfied in this example. These are:

$\sum_x p(x) = 1$

$0 \le p(x) \le 1$

_____

_____

2. The probability distribution can also be presented graphically as a probability histogram.

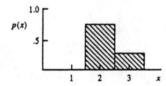

3. If this experiment were repeated many times, approximately what fraction of the outcomes would result in $x = 2$?

3/4

_____ What fraction of the total area under the probability histogram lies over the interval associated with $x = 2$?

3/4

_____

Note that the probability distribution provides a theoretical frequency distribution for the hypothetical population associated with the businessman's experiment and thereby relates directly to the content of Chapter 2 in your text.

Example 4.30

A psychological recognition experiment required a subject to classify a set of objects according to whether or not they had been previously observed. Suppose that a subject can correctly classify each object with probability $p = .7$, that sequential classifications are independent events, and that she is presented with $n = 3$ objects to classify. We are interested in x, the number of correct classifications for the three objects.

Solution

1. This experiment is analogous to tossing three unbalanced coins, where correctly classifying an object corresponds to the observation of a head in the toss of a single coin. Each classification results in one of two outcomes: correct or incorrect. The total number of simple events in the sample space (similar to Example 4.3) is

   _____.

   8

2. Let (I, I, C) represent the simple event for which the classification of the first and second objects is incorrect and the third is correct. Complete the listing of all the simple events in the sample space.

E_1: (I,I,I)	E_2: (I,I,C)
E_3: (I,C,I)	E_4: (I,C,C)
E_5: (C,I,I)	E_6: _____
E_7: _____	E_8: _____

   (C,C,I)

   (C,I,C); (C,C,C)

3. The simple event E_2 is an *intersection* of three independent events. Applying the Multiplicative Law of Probability, we get

   $$P(E_2) = P(I \cap I \cap C) = P(I)P(I)P(C) = (.3)(.3)(.7) = .063$$

   Similarly, $P(E_1) = .027$ and $P(E_3) = .063$. Calculate the probabilities for all the simple events in the sample space.

$P(E_1) = .027$	$P(E_2) = .063$
$P(E_3) = .063$	$P(E_4) = .147$
$P(E_5) =$ _____	$P(E_6) =$ _____
$P(E_7) =$ _____	$P(E_8) =$ _____

   .063; .147

   .147; .343

4. The random variable x, the number of correct classifications for the set of three objects, takes the value $x = 1$ for simple event E_2. Similarly, we would assign the value $x = 0$ to E_1. Assign a value of x to each simple event in the sample space.

Simple Events	x
E_1	0
E_2, E_3, E_5	1
E_4, E_6, E_7	_____
E_8	_____

   2

   3

5. The numerical event $x = 0$ contains only the simple event E_1. Summing the probabilities of the simple events in the event $x = 0$, we have $P(x = 0) = P(E_1) = .027$. Similarly, the numerical event $x = 1$ contains three simple events. Summing the probabilities of these simple events, we have $P(x = 1) = p(1) = .189$.

6. The probability distribution $p(x)$ is presented in tabular form below. Calculate the probabilities $p(2)$ and $p(3)$ and complete the table.

x	$p(x)$
	.027
1	_____
2	_____
3	_____

.189
.441
.343

7. Present $p(x)$ graphically in the form of a probability histogram.

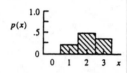

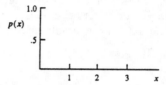

8. After studying Chapter 5, you will be able to express this probability distribution as a formula.

The Mean and Standard Deviation for a Discrete Random Variable

When we develop a probability distribution for a random variable,
we are actually proposing a model that will describe the behavior of
the random variable in repeated trials of an experiment. For example, when we propose the model for describing the distribution of x,
the number of heads in the toss of two fair coins, given by

x	0	1	2
$p(x)$	1/4	1/2	1/4

we mean that if the two coins were tossed a large number of times,
about one fourth of the outcomes would result in the outcome "zero
heads," one half would result in the outcome "one head," and the remaining fourth would result in "two heads." A probability distribution not only is a measure of the belief that a specific outcome will
occur on a single trial but, more important, it actually describes a
population of observations on the random variable x. It is reasonable
then to talk about and calculate the mean and the standard deviation of a random variable by using the probability distribution as a
population model.

The expected value of a random quantity is its average value in
the population. In particular, the expected value of x is simply the
population mean. The expected value of $(x - \mu)^2$ describes the population variance.

The **mean** or **expected value**, $E(x)$, of a discrete random
variable x is

$$\mu = E(x) = \sum_x x\, p(x)$$

The **variance**, σ^2, is given by

$$\sigma^2 = E(x - \mu)^2 = \sum_x (x - \mu)^2 p(x)$$

The **standard deviation**, σ, is given by

$$\sigma = \sqrt{\sigma^2}$$

Example 4.31

Suppose that a psychological experiment is designed in such a way that the patient has two choices for each of three experimental situations into which he is placed. One of the choices is designated as the "correct" choice. The probability distribution for x, the number of correct choices, is:

x	0	1	2	3
$p(x)$	.15	.35	.20	.30

Find the mean and the standard deviation of the number of correct choices.

Solution

Before calculating the mean and variance of x, we see that this (is, is not) a valid probability distribution because

is

$$\sum_x p(x) = \underline{\hspace{1.5cm}} \text{ and } \underline{\hspace{1.5cm}} \leq p(x) \leq \underline{\hspace{1.5cm}}$$

1; 0; 1

1. The expected number of correct choices is calculated as

$$\mu = E(x) = \sum_x x \, p(x)$$
$$= 0(.15) + 1(.35) + 2(.20) + 3(.30)$$

$$= \underline{\hspace{1.5cm}}$$

1.65

2. From part 1, $\mu = E(x) = \underline{\hspace{1.5cm}}$. Then the variance of x is

1.65

$$\sigma^2 = E(x - \mu)^2$$
$$= \sum_x (x - \mu)^2 p(x)$$

$$= (0 - 1.65)^2(.15) + (1 - 1.65)^2(.35)$$
$$\quad + (2 - 1.65)^2(.20) + (3 - 1.65)^2(.30)$$

$$= .408375 + \underline{\hspace{1.5cm}} + .0245 + \underline{\hspace{1.5cm}}$$

.147875; .546750

$$= \underline{\hspace{1.5cm}}$$

1.1275

and $\sigma = \underline{\hspace{1.5cm}}$

1.062

If x is a random variable, either continuous or discrete, it can be shown that

$$E(x - \mu)^2 = E(x^2) - \mu^2$$

This result can be used as a computing formula to calculate the variance, σ^2.

Example 4.32
Use the computing formula to calculate σ^2 for Example 4.31.

Solution
For a discrete random variable x,

$$E(x^2) = \sum_x x^2\, p(x)$$

which is the average value of x^2 over all its possible values. For this example,

$$E(x^2) = 0^2(.15) + 1^2(.35) + 2^2(.20) + 3^2(.30)$$

3.85

$$= \underline{\hspace{2cm}}$$

Then

1.1275

$$\sigma^2 = 3.85 - (.165)^2 = \underline{\hspace{2cm}}$$

and

1.062

$$\sigma = \underline{\hspace{2cm}}$$

Notice that the numerical results are identical to those found in Example 4.31. Moreover, the computational formula involves fewer steps and in general results in less rounding error than does the definition formula.

Example 4.33
Construct the probability histogram for the distribution of the number of correct choices given in Example 4.31. Visually locate the mean and compare it with the computed value, $\mu = 1.65$.

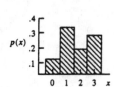
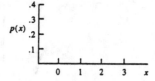

1.5

The approximate value of the mean is $\underline{\hspace{2cm}}$. Its value is close to that calculated and provides an easy check on the calculated value of $E(x)$.

Example 4.34
A corporation has four investment possibilities: A, B, and C with respective gains of \$10, \$20, and \$50 million and investment D with a loss of \$30 million. If one investment will be made and the probabilities of choosing A, B, C, or D are .1, .4, .2, and .3, respectively, find the expected gain for the corporation.

Solution

The random variable is x, the corporation's gain, with possible values of $10, $20, $50, and −$30 million. The probability distribution for x is given as:

x (millions)	$10	20	50	−30
$p(x)$	.1	.4	.2	.3

The expected gain is $E(x) =$ _____ million. $10

Example 4.35

A parcel post service that insures packages against loss up to $200 wishes to reevaluate its insurance rates. If 1 in every 1000 packages was reported lost during the last several years, what rate should be charged on a package insured for $200 so that the postal service's expected gain is zero? Administrative costs will be added to this rate.

Solution

Let x be the gain to the parcel post service and let r be the rate charged for insuring a package for $200.

1. Complete the probability distribution for x:

                                                                                            $(r - 200)$

x	r	_____
$p(x)$	.999	.001

2. If $E(x)$ is to be zero, we need to solve the equation

$$\sum_x x\, p(x) = 0$$

Hence, for our problem,

$$r(.999) + (r - 200)(.001) = 0$$

$$r = \underline{\hspace{2cm}}$$ $.20

Self-Correcting Exercises 4E

1. A subject is shown four photographs, A, B, C, D, of crimes that have been committed and asked to select what she considers the two worst crimes. Although all four crimes shown are robberies involving about the same amount of money, A and B are pictures of robberies committed against private citizens while C and D are robberies committed against corporations. Assuming that the subject does not show discrimination in her selection, find the probability distribution for x, the number of crimes chosen involving private citizens.

2. Suppose that the unemployment rate in a given community is 7%. Four households are randomly selected to be interviewed. In each household, it is determined whether or not the primary wage earner is unemployed. If the 7% rate is correct, find the probability

distribution for x, the number of primary wage earners who are unemployed.

3. Let x be a discrete random variable with a probability distribution given as

x	-2	-1	0	1
$p(x)$	1/9	1/9	4/9	–

a. Find $p(1)$.
b. Find $\mu = E(x)$.
c. Find σ, the standard deviation of x.

4. A police car visits a given neighborhood a random number of times x per evening. $p(x)$ is given by

x	$p(x)$
0	.1
1	.6
2	.2
3	.1

a. Find $E(x)$.
b. Find σ^2 using both the definition and the computational formulas. Verify that the results are identical.
c. Calculate the interval $\mu \pm 2\sigma$ and find the probability that the random variable x lies within this interval. Does this agree with the results given in Tchebysheff's Theorem?

5. Refer to Exercise 4. What is the probability that the patrol car will visit the neighborhood at least twice in a given evening?

6. You are given the following information. An insurance company wants to insure a $80,000 home against fire. One in every 100 such homes is likely to have a fire; 75% of the homes having fires will suffer damages amounting to $40,000, while the remaining 25% will suffer total loss. Ignoring all other partial losses, what premium should the company charge in order to break even?

Solutions to SCE 4E

1. The simple events are (A, B), (A, C), (A, D), (B, C), (B, D), (C, D). If there is no discrimination in the selection, each simple event has probability 1/6. Collecting information,

Simple Events	x	$p(x)$
(C, D)	0	1/6
(A, C), (A, D), (B, C), (B, D)	1	4/6
(A, B)	2	1/6

2. Let U be the event that the primary wage earner is unemployed and let E be the event that the is employed. There are 16 simple events with unequal probabilities.

$$p(0) = P[E \cap E \cap E \cap E] = (.93)^4 = .7481$$

$$p(1) = 4P(E)^3 P(U) = 4(.93)^3(.07)^2 = .2252$$

$$p(2) = 6P(E)^2 P(U)^2 = 6(.93)^2(.07)^2 = .0254$$

$$p(3) = 4P(E)P(U)^3 = 4(.93)(.07)^3 = .0013$$

$$p(4) = P((UUUU)) = (.07)^4 = .000024$$

3. a Since $\sum_x p(x) = 1$, $p(0) = 1 - (6/9) = 3/9$.

   b. $\mu = \sum_x x\, p(x) = -2(1/9) + (-1)(1/9) + 0(4/9) + 1(3/9)$

   $= (-2/9) - (1/9) + (3/9) = 0.$

   c. $\sigma^2 = \sum x^2 p(x) - \mu^2 = (-2)^2(1/9) + (-1)^2(1/9) + 0^2(4/9)$

   $+ 1^2(3/9) - 0^2 = (4/9) + (1/9) + (3/9) = 8/9 = .8889$

   $\sigma = \sqrt{.8889} = .94.$

4. a. $\mu = E(x) = \sum_x x p(x) = 0(.1) + 1(.6) + 2(.2) + 3(.1)$

   $= .6 + .4 + .3 = 1.3$

   b. $\sigma^2 = E((x - \mu)^2) = (0 - 1.3)^2(.1) + (1 - 1.3)^2(.6)$

   $+ (2 - 1.3)^2(.2) + (3 - 1.3)^2(.1) = .61$, or

   $\sigma^2 = \sum x^2 p(x) - \mu^2 = 0^2(.1) + 1^2(.6) + 2^2(.2) + 3^2(.1)$

   $- (1.3)^2 = .6 + .8 + .9 - 1.69 = .61.$

   c. $\mu \pm 2\sigma = 1.3 \pm 2(.61) = 1.3 \pm 1.22$, or .08 to 2.52. Referring
   to the initial probability distribution, $P(.08 < x < 2.52)$
   $= .6 + .2 = .8$, which agrees with Tchebysheff's Theorem.

5. $P(x \geq 2) = p(2) + p(3) = .2 + .1 = .3.$

6. Let x be the gain to the insurance company and let r be the premium
   charged by the company.

x	$p(x)$
r	.9900
$-40,000 + r$	.0075
$-80,000 + r$	.0025

In order to break even, $E(x) = 0$, or

$E(x) = \sum_x x\, p(x) = .99r + .0075(-40,000 + r)$

$+ (-80,000 + r)(.0025) = 0$

$r - 300 - 200 = 0$

$r = \$500.$

KEY CONCEPTS AND FORMULAS

I. *Experiments and the Sample Space*
  1. Experiments, events, mutually exclusive events, simple events
  2. The sample space
  3. Venn diagrams, tree diagrams, probability tables

II. *Probabilities*
  1. Relative frequency definition of probability
  2. Properties of probabilities
      a. Each probability lies between 0 and 1.
      b. Sum of all simple-event probabilities equals 1.
  3. $P(A)$, the sum of the probabilities for all simple events in A

III. *Counting Rules*
  1. mn Rule; extended mn Rule
  2. Permutations: $P^n_r = \dfrac{n!}{(n-r)!}$
  3. Combinations: $C^n_r = \dfrac{n!}{r!(n-r)!}$

IV. *Event Relations*
  1. Unions and intersections
  2. Events
      a. Disjoint or mutually exclusive: $P(A \cap B) = 0$
      b. Complementary: $P(A) = 1 - P(A^C)$
  3. Conditional probability: $P(A \mid B) = \dfrac{P(A \cap B)}{P(B)}$
  4. Independent and dependent events
  5. Additive Rule of Probability:
      $P(A \cup B) = P(A) + P(B) - P(A \cap B)$
  6. Multiplicative Rule of Probability: $P(A \cap B) = P(A)P(B \mid A)$
  7. Law of Total Probability
  8. Bayes' Rule

V. *Discrete Random Variables and Probability Distributions*
  1. Random variables, discrete and continuous
  2. Properties of probability distributions
      a. $0 \le p(x) \le 1$
      b. $\sum p(x) = 1$
  3. Mean or expected value of a discrete random variable:
      $\mu = \sum x\, p(x)$
  4. Variance and standard deviation of a discrete random
      variable: $\sigma^2 = \sum (x - \mu)^2 p(x)$ and $\sigma = \sqrt{\sigma^2}$

Exercises

1. Suppose that an experiment requires the ranking of three applicants, A, B, and C, in order of their abilities to do a certain job.
   The simple events could then be symbolized by the ordered triples (ABC), (BAC), and so on.

a. The event A, that applicant A will be ranked first, comprises which of the simple events?

b. The event B, that applicant B will be ranked third, comprises which of the simple events?

c. List the events in $A \cup B$.

d. List the events in $A \cap B$.

e. If equal probabilities are assigned to the simple events, show whether or not events A and B are independent.

2. An antique dealer had accumulated a number of small items including a valuable stamp collection and a solid gold vase. To make room for new stock he distributed these small items among four boxes. Without revealing which items were placed in which box, the dealer stated that the stamp collection was included in one box and the gold vase in another. The four boxes were sealed and placed on sale, each at the same price. A certain customer purchased two boxes selected at random from the four boxes. What is the probability that the customer acquired

   a. the stamp collection?

   b. the vase?

   c. at least one of these bonus items?

3. If the probability that an egg laid by an insect hatches is $p = .4$, what is the probability that at least three out of four eggs will hatch?

4. Suppose that on the basis of past experience it is known that a lie detector test will indicate that an innocent person is guilty with probability .08, while the test will indicate that a guilty person is innocent with probability .15. Suppose further that 10% of the population under study has committed a traffic violation. If a lie detector test indicates that a randomly chosen individual from this population has committed a traffic violation, what is the probability that this person is innocent of committing a traffic violation?

5. Eight employees have been found equally qualified for promotion to a particular job. It has been decided to choose five of the employees at random for immediate promotion. How many different groups of five employees are possible?

6. Refer to Exercise 5. Suppose that only one vacancy will occur at a time, and that the five employees must be chosen for assignment sequentially. These five will then be promoted as vacancies occur in the order they are listed. How many different promotional lists are possible?

7. In the past history of a certain serious disease it has been found that about 1/2 of its victims recover.

   a. Find the probability that exactly one of the next five patients suffering from this disease will recover.

   b. Find the probability that at least one of the next five patients suffering from this disease will recover.

8. The sample space for a given experiment comprises the simple events E_1, E_2, E_3, and E_4. Let the compound events A, B, and C be defined by these equations:

$$A = E_1 \cup E_2 \qquad B = E_1 \cup E_4 \qquad C = E_2 \cup E_3.$$

Construct a Venn diagram showing the events, E_1, E_2, E_3, E_4, A, B, and C.

9. Refer to Exercise 8. Probabilities are assigned to the simple events as indicated in the following table:

Simple event	E_1	E_2	E_3	E_4
Assigned probability	1/3	1/3	1/6	–

   a. Supply the missing entry in the table.
   b. Find $P(A)$ and $P(A \cap B)$.
   c. Find $P(A \mid B)$ and $P(A \mid C)$.
   d. Find $P(A \cup B)$ and $P(A \cup C)$.
   e. Are A and B mutually exclusive? Independent?

10. An envelope of seeds contains three nonviable seeds and five viable ones. Consider the eight seeds to be distinguishable.
    a. How many different samples of size three can be formed?
    b. How many of these samples of size three comprise two viable seeds and one nonviable seed?
    c. If a sample of size three is selected at random from this envelope, what is the probability that two of these seeds will be viable and the other nonviable?

11. A random sample of size five is drawn from a large production lot with a fraction defective of 10%. The probability that this sample will contain no defectives is .59. What is the probability that this sample will contain at least one defective?

12. A factory operates an 8-hour day shift. Five machines of a certain type are used. If one of these machines breaks down, it is set aside and repaired by a crew operating at night. Suppose the probability that a given machine suffers a breakdown during a day's operation is 1/5.
    a. What is the probability that no machine breakdowns will occur on a given day?
    b. What is the probability that two or more machine breakdowns will occur on a given day?

13. A certain virus disease afflicted the families in 3 adjacent houses in a row of 12 houses. If 3 houses were randomly chosen from a row of 12 houses, what is the probability that the 3 houses would be adjacent? Is there reason to conclude that this virus disease is contagious?

14. A geologist, assessing a given tract of land for its oil content, initially rates the land as having

i.   no oil, with probability 0.7,

ii.   500,000 barrels, with probability 0.2,

iii. 1,000,000 barrels, with probability 0.1.

However, the potential buyer ordered that seismic drillings be performed and found the readings to be "high" based on a "low, medium, high" rating scale. The conditional probabilities $P(E \mid S)$, are given in the following table.

	Seismic Readings, E_i		
i	E_1, Low	E_2, Medium	E_3, High
S_1: no oil	.50	.30	.20
S_2: 500,000 bbl	.40	.40	.20
S_3: 1,000,000 bbl	.10	.50	.40

a. Find $P(S_1 \mid E_3)$, $P(S_2 \mid E_3)$, and $P(S_3 \mid E_3)$.

b. Suppose the seismic readings had been low. Now find $P(S_1 \mid E_1)$, $P(S_2 \mid E_1)$, and $P(S_3 \mid E_1)$.

15. A manufacturer buys parts from a supplier in lots of 10,000 items. The fraction defective in a lot is usually about .1%. Occasionally a malfunction in the supplier's machinery causes the fraction defective to jump to 3%. Records indicate that the probability of receiving a lot with 3% defective is .1. To check the quality of the supplier's lot, the manufacturer selects a random sample of 200 parts from the lot and observes 3 defectives.

   The probability of observing 3 defectives when the fraction defective is .1% is approximated as .0011, and the probability of observing 3 defectives when the fraction defective is 3% is approximated as .0892.

   a. Find the probability that the percentage defective is .1%, given that 3 defectives are observed in the sample.

   b. Find the probability that the percentage defective is 3%, given that 3 defectives are observed in the sample.

   c. Based on your answers to parts a and b, what would you conclude to be the fraction defective in the lot?

16. A car rental agency has three Fords and two Chevrolets left in its car pool. If two cars are needed and the keys are randomly selected from the keyboard, find the probability distribution for x, the number of Fords in the selection.

17. Five equally qualified applicants for a teaching position were ranked in order of preference by the superintendent of schools. If two of the applicants hold master's degrees in education, find the probability distribution for x, the number of applicants holding master's in education ranked first or second.

18. An experiment is run in the following manner: The colors red, yellow, and blue are each flashed on a screen for a short period of time. A subject views the colors and is asked to choose the one he

feels was flashed for the longest amount of time. The experiment is repeated three times with the same subject.

a. If all the colors were flashed for the same length of time, give the probability distribution for x, the number of times the subject chose the color red. Assume that his three choices are independent.

b. Construct a probability histogram for $p(x)$ found in part a.

19. A publishing company is considering the introduction of a monthly gardening magazine. Advance surveys show the initial market for the magazine will be approximated by the following distribution for x, the number of subscribers:

x	$p(x)$
5,000	.30
10,000	.35
15,000	.20
20,000	.10
25,000	.05

Find the expected number of subscribers and the standard deviation of the number of subscribers.

20. Refer to Exercise 19. Suppose the company expects to charge $20 for an annual subscription. Find the mean and standard deviation of the revenue the company can expect from the annual subscriptions of the initial subscribers.

21. Refer to Exercise 20. Production and distribution costs for the gardening magazine are expected to amount to slightly over $200,000. What is the probability that revenue from initial subscriptions will fail to cover these costs?

22. The following is the probability function for a discrete random variable x:

$$p(x) = (.1)(x+1), \quad x = 0, 1, 2, 3.$$

a. Find $\mu = E(x)$ and σ^2.

b. Construct a probability histogram for $p(x)$.

23. The probability of hitting oil in a single drilling operation is 1/4. If drillings represent independent events, find the probability distribution for x, the number of drillings until the first success ($x = 1, 2, 3, \ldots$). Proceed as follows:

a. Find $p(1)$.

b. Find $p(2)$.

c. Find $p(3)$.

d. Give a formula for $p(x)$.

   Note that x can become infinitely large.

e. Will $\sum p(x) = 1$?

24. The board of directors of a major symphony orchestra has voted to create an employee council for the purpose of handling employee complaints. The council will consist of the president and vice-president of the symphony board and two orchestra representatives. The two orchestra representatives will be randomly selected from a list of 6 volunteers, consisting of 4 men and two women.

   a. Find the probability distribution for x, the number of women chosen to be orchestra representatives.

   b. Find the mean and variance for x.

   c. What is the probability that both orchestra representatives will be women?

25. Given a random variable x with the probability distribution

x	$p(x)$
1	1/8
2	5/8
3	1/4

   graph $p(x)$ and make a visual approximation to the mean and standard deviation. (Use your knowledge of Tchebysheff's Theorem to assist in approximating σ.)

26. Refer to Exercise 25 and find the expected value and standard deviation of x. Compare with the answers to Exercise 25.

27. Given the following probability distribution, find the expected value and variance of x.

x	$p(x)$
0	1/2
3	1/3
6	1/6

28. History hs shown that buildings of a certain type of construction suffer fire damage during a given year with probability of .01. If a building suffers fire damage, it will result in either a 50% or a 100% loss with probabilities .7 and .3, respectively. Find the premium required per $1000 coverage in order that the expected gain for the insurance company will equal zero (break-even point).

29. Experience has shown that a rare disease will cause partial disability with probability .6, complete disability with probability .3, and no disability with probability .1. Only 1 in 10,000 will become afflicted with the disease in a given year. If an insurance policy pays $20,000 for partial disability and $50,000 for complete disability, what premium should be charged in order that the insurance company break even (that is, in order that the expected loss to the insurance company will be zero)?

Chapter 5
Several Useful Discrete Distributions

5.1 INTRODUCTION

A random variable that takes a countable number of values corresponding to a countable number of simple events is called a *discrete random variable*. Of the many discrete random variables and their probability distributions found in the sciences and in business and economics, three discrete distributions can be used as a model in many of these situations.

In this chapter, we will learn about the *binomial, Poisson, hypergeometric, and several other distributions* used as models for discrete random variables in different settings and contexts.

5.2 THE BINOMIAL PROBABILITY DISTRIBUTION

Many experiments in the social, biological, and physical sciences can be reduced to a series of trials resembling the toss of a coin in which the outcome on each toss will be either a head or a tail. Consider the following analogies:

1. A student answers a multiple-choice question correctly (head) or incorrectly (tail).
2. A voter casts her ballot either for candidate *A* (head) or against him (tail).
3. A patient treated with a particular drug either improves (head) or does not improve (tail).
4. A subject makes either a correct identification (head) or an incorrect one (tail).
5. A licensed driver either has an accident (head) or does not have an accident (tail) during the period his license is valid.
6. An item from a production line is inspected and classified as either defective (head) or not defective (tail).

If any of the above situations were repeated *n* times and we counted the number of "heads" that occurred in the *n* trials, the resulting random variable would be a _____ random variable. Let us examine what characteristics these experiments have in common. We will call a head a

binomial

success (S) and a tail a failure (F). A success does not necessarily denote a desirable outcome but rather identifies the event of interest.

The five defining characteristics of a binomial experiment are as follows:

---

1. The experiment consists of n identical trials.
2. Each trial results in one of two outcomes: success (S) or failure (F).
3. The probability of success on a single trial is equal to p and remains constant from trial to trial. The probability of failure is $q = 1 - p$.
4. The trials are independent.
5. Attention is directed to the random variable x, the total number of successes observed during the n trials.

Although very few real-life situations perfectly satisfy all five characteristics, this model can be used with fairly good results if the violations are "moderate." The next several examples will illustrate binomial experiments.

Example 5.1
A procedure (the "triangle test") often used to control the quality of name brand food products utilizes a panel of n "tasters." Each member of the panel is presented three specimens, two of which are from batches of the product known to possess the desired taste and the other is a specimen from the latest batch. Each panelist is asked to select the specimen that is different from the other two. If the latest batch does possess the desired taste, then the probability that a given taster will be "successful" in selecting the specimen from the latest batch is _____. If there is no communication among the panelists, their responses will make up n independent _____, with a probability of success on a given trial equal to _____.

1/3

trials
1/3

Example 5.2
Almost all accounts are audited on a sampling basis. Thus, an auditor might check a random sample of n items from a ledger or inventory list comprising a large number of items. If 1% of the items in the ledger are erroneous, then the number of erroneous items in the sample is essentially a _____ random variable with n trials and probability of success (finding an erroneous item) on a given trial equal to _____.

binomial
.01

Example 5.3
No treatment has been known for a certain serious disease for which the mortality rate in the United States is 70%. If a random selection is made of 100 past victims of this disease in the United States, the number x_1 of those in the sample who died of the disease is essentially a binomial random variable with $n =$ _____ and $p =$ _____. More

100; .70

important, if the next 100 persons in the United States who will in the future become victims of this disease are observed, the number x_2 of these who will die from the disease has a distribution approximately the same as that of x_1 if the conditions that affect this disease remain essentially constant for the time period considered.

Example 5.4

The continued operation (reliability) of a complex assembly often depends on the joint survival of all or nearly all of a number of similar components. Thus, a radio may give at least 100 hours of continuous service if no more than two of its ten transistors fail during the first 100 hours of operation. If the ten transistors in a given radio were selected at random from a large lot of transistors, then each of these (ten) transistors would have the same probability p of failing within 100 hours, and the number of transistors in the radio that will fail within 100 hours is a _____ random variable with _____ trials and the probability of success on each trial equal to _____. (*Success* is a word that denotes one of the two outcomes of a single trial and does not necessarily represent a desirable outcome.)

  Three experiments are described below. In each case, state whether or not the experiment is a binomial experiment. If the experiment is binomial, specify the number n of trials and the probability p of success on a given trial. If the experiment is not binomial, state which characteristics of a binomial experiment are not met.

1. A fair coin is tossed until a head appears. The number of tosses x is observed. If binomial, $n =$ _____ and $p =$ _____. If not binomial, list characteristic(s) (1, 2, 3, 4, and 5) violated. _____

2. The probability that an applicant scores above the 90th percentile on a qualifying examination is .10. The examiner is interested in x, the number of applicants (of the 25 taking the examination) who score above the 90th percentile. If binomial, $n =$ _____ and $p =$ _____. If not binomial, list characteristics(s) (1, 2, 3, 4, and 5) violated. _____

3. A sample of 5 transistors will be selected at random from a box of 20 transistors of which 10 are defective. The experimenter will observe the number x of defective transistors appearing in the sample. If binomial, $n =$ _____ and $p =$ _____. If not binomial, list characteristic(s) (1, 2, 3, 4, and 5) violated. _____

As a rule of thumb, if n, the sample size, is large relative to N, the population size, with $n/N \geq .05$, the resulting experiment will not be binomial.

---

binomial

10

p

not binomial

1, 5

25; .10

not binomial

3, 4

The probability distribution for a binomial random variable can be derived by considering the toss of n coins, with the probability of a head $P(\text{head}) = p$, and the probability of a tail $P(\text{tail}) = 1 - p = q$. However, rather than derive the binomial distribution in general, we present the *binomial probability distribution* together with its *mean, variance,* and *standard deviation* in the following display.

The Binomial Probability Distribution

The probability distribution of x, the number of successes in n trials is given by

$$p(x) = C_x^n\, p^x\, q^{n-x}$$

for $x = 0, 1, 2, \ldots, n$; p is the probability of success on a single trial and

$$C_x^n = \frac{n!}{x!\,(n-x)!}$$

where $n! = n(n-1)(n-2)\cdots(2)(1)$ and $0! = 1$.

Mean: $\mu = np$
Variance: $\sigma^2 = npq$
Standard deviation: $\sigma = \sqrt{npq}$

In the formula for $p(x)$, the quantity $p^x q^{n-x}$ represents the probability associated with a simple event having exactly x successes and $(n-x)$ _____. The combinatorial term defined as $n!/[x!(n-x)!]$ counts the number of simple events with exactly x successes. The term for $p(x)$ is just one of the terms in the series expansion of $(p+q)^n$, a binomial raised to power n, and hence the name: the binomial distribution.

failures

Example 5.5

The president of an agency specializing in public opinion surveys claims that approximately 70% of all people to whom the agency sends questionnaires respond by filling out and returning the questionnaire. Four such questionnaires are sent out. Let x be the number of questionnaires that are filled out and returned. Then x is a binomial random variable with $n = $ _____ and $p = $ _____.

4; .70

1. The probability that no questionnaires are filled out and returned is

$$p(0) = C_0^4\,(.7)^0(.3)^4 = \frac{4!}{0!\,4!}(.7)^0(.3)^4$$

$$= (.3)^4 = \underline{\qquad\qquad}$$

.0081

2. The probability that exactly three questionnaires are filled out and returned is

.4116

$$p(3) = C_3^4 (.7)^3 (.3)^1$$
$$= 4(.343)(.3) = \underline{\hspace{2cm}}$$

3. The probability that at least three questionnaires are filled out and returned is

$$P(x \geq 3) = p(3) + p(4) = p(3) + C_4^4 (.7)^4 (.3)^0$$

.2401; .6517

$$= .4116 + \underline{\hspace{2cm}} = \underline{\hspace{2cm}}$$

Example 5.6
A marketing research survey shows that approximately 80% of the car owners surveyed indicated that their next car purchase will be either a compact or an economy car. Assume the 80% figure is correct and five prospective buyers are interviewed.

a. Find the probability that all five indicate that their next car purchase will be either a compact or an economy car.
b. Find the probability that at most one indicates that her next purchase will be either a compact or an economy car.

Solution
Let x be the number of car owners who indicate that their next purchase will be a compact or an economy car. Then $n = \underline{\hspace{2cm}}$

5
.8

and $p = \underline{\hspace{2cm}}$, and the distribution for x is given by

$$p(x) = C_x^5 (.8)^x (.2)^{5-x}, \quad x = 0, 1, \ldots, 5$$

a. The required probability is $p(5)$, which is given by

$$p(5) = C_5^5 (.8)^5 (.2)^0 = (.8)^5 = .32768$$

b. The probability that at most one car owner indicates that her next purchase will be either a compact or an economy car is

$$P(x \leq 1) = p(0) + p(1)$$

For $x = 0$,

.00032

$$p(0) = C_0^5 (.8)^0 (.2)^5 = (.2)^5 = \underline{\hspace{2cm}}$$

For $x = 1$,

.0064

$$p(1) = C_1^5 (.8)^1 (.2)^4 = 5(.8)(.0016) = \underline{\hspace{2cm}}$$

Hence,

$$P(x \leq 1) = .0064 + .00032$$

.00672

$$= \underline{\hspace{2cm}}$$

As you might expect, the calculation of binomial probabilities becomes quite tiresome as the number of trials increases. Table 1 of binomial probabilities in Appendix I in your text can be used to find the binomial probabilities for values of $p = .01, .05, .10, .20, \ldots, .90, .95, .99$ when $n = 2, 3, \ldots, 12, 15, 20, 25$.

1. The table entries are not the individual terms for binomial probabilities but rather cumulative sums of probabilities, beginning with $x = 0$ up to and including the value $x = k$. By formula, the entries for n, p, and k are

$$P(x \leq k) = p(0) + p(1) + \cdots + p(k)$$

2. By using a table entry, which is $p(0) + p(1) + \ldots + p(k)$, we can use these tables to find the following:

a. Left-tail cumulative sums:

$$P(x \leq k) = p(0) + p(1) + \ldots + p(k)$$

b. Right-tail cumulative sums:

$$P(x \geq k) = 1 - [p(0) + p(1) + \ldots + p(k-1)]$$

c. Individual terms such as

$$P(x = k) = [p(0) + p(1) + \ldots + p(k)] \\ - [p(0) + p(1) + \ldots + p(k-1)]$$

Example 5.7
Refer to Example 5.6. Find the probabilities asked for by using Table 1 in Appendix I.

Solution
For this problem, we will use the table for $n = 5$ and $p = .8$.

1. To find the probability that $x = 5$, we proceed as follows:

$$p(5) = [p(0) + p(1) + p(2) + p(3) + p(4) + p(5)] \\ - [p(0) + p(1) + p(2) + p(3) + p(4)]$$

$$= 1 - .672$$

$$= .328$$

2. To find the probability that $x \leq 1$, we need

$$P(x \leq 1) = p(0) + p(1)$$

$$= \underline{\hspace{2cm}}$$

.007

3. Let us extend the problem and find the probabilities associated with the terms $x = 2$ and $x = 3$. For $x = 2$,

$$P(2) = [p(0) + p(1) + p(2)] - [p(0) + p(1)]$$

$$= .058 - .007$$

$$= \underline{\hspace{2cm}}$$

.051

For $x = 3$,

$$p(3) = [p(0) + p(1) + p(2) + p(3)]$$
$$- [p(0) + p(1) + p(2)]$$
$$= .263 - .058$$

.205

$$= \underline{\hspace{2cm}}$$

4. Complete the following table:

x	$p(x)$
0	_____
1	_____
2	_____
3	_____
4	_____
5	_____

.000
.007
.051
.205
.409
.328

1

with $p(0) + p(1) + \ldots + p(5) = \underline{\hspace{2cm}}$.

5. Graph this distribution as a probability histogram.

Example 5.8
Using Table 1 in Appendix I, find the probability distribution for x if $n = 5$ and $p = 1/2$. Graph the resulting probability histogram.

Solution
1. To find the individual probabilities for $x = 0, 1, 2, \ldots, 5$, we need to subtract successive entries in the table for $n = 5$, $p = .5$:

.031

$$p(0) = \underline{\hspace{2cm}}$$

.157

$$p(1) = [p(0) + p(1)] - [p(0)] = .188 - .031 = \underline{\hspace{2cm}}$$

$$p(2) = [p(0) + p(1) + p(2)] - [p(0) + p(1)]$$

.312

$$= .500 - .188 = \underline{\hspace{2cm}}$$

.812; .312

$$p(3) = \underline{\ominus \oslash \pm \doteq} - .500 = \underline{\hspace{2cm}}$$

.969; .157

$$p(4) = \underline{\hspace{2cm}} - .812 = \underline{\hspace{2cm}}$$

.969; .031

$$p(5) = 1.000 - \underline{\hspace{2cm}} = \underline{\hspace{2cm}}$$

2. Using the results of step 1, we find the probability histogram to be symmetric about the value $x = $ _____.

2.5

Example 5.9

Find the probability distribution for x if $n = 5$ and $p = .3$. Graph the probability histogram in this case.

Solution

1. Again, subtracting successive entries for $n = 5$, $p = .3$, we have

$$p(0) = \underline{\hspace{2cm}}$$ .168

$$p(1) = .528 - .168 = \underline{\hspace{2cm}}$$ .360

$$p(2) = .837 - .528 = \underline{\hspace{2cm}}$$ .309

$$p(3) = .969 - .837 = \underline{\hspace{2cm}}$$ .132

$$p(4) = .998 - .969 = \underline{\hspace{2cm}}$$ .029

$$p(5) = 1 - .998 = \underline{\hspace{2cm}}$$ .002

2. Graph the resulting histogram.

In comparing the histograms in Examples 5.7, 5.8, and 5.9, notice that when $p = 1/2$, the histogram is _____. However, if $p = .8$, which is greater than 1/2, the mass of the probability moves to the _____ with p. For $p = .3$, which is less than 1/2, the mass of the probability distribution moves to the _____ with p. Locating the center of the distribution by eye, we see that the mean of the binomial distribution varies directly as _____, the probability of success.

symmetric

right
left

p

Let us consider two more examples. You are now free either to calculate the probabilities by hand or to use the tables when appropriate.

Example 5.10

A preliminary investigation reported that approximately 30% of locally grown poultry were infected with an intestinal parasite that, though not harmful to those consuming the poultry, decrease the usual weight growth rates in the birds and thereby caused a loss in revenue to the growers. A diet supplement believed to be effective against this parasite was added to the birds' rations. During the preparation of poultry that had been fed the supplemental rations for at least two weeks, of 25 birds examined, 3 birds were still found to be infected with the intestinal parasite.

a. If the diet supplement is ineffective, what is the probability of observing 3 or fewer birds infected with the intestinal parasite?

b. If in fact the diet supplement was effective and reduced the infection rate to 10%, what is the probability of observing 3 or fewer infected birds?

Solution

a. With $n = 25$ and $p = .3$, we can use the binomial tables in the text to find $P(x \leq 3)$:

.033

$$P(x \leq 3) = \ p(0) + p(1) + p(2) + p(3) = \underline{\hspace{2cm}}$$

.1

b. We can use the same tables with $n = 25$ and $p = \underline{\hspace{2cm}}$. Hence,

.764

$$P(x \leq 3) = \ p(0) + p(1) + p(2) + p(3) = \underline{\hspace{2cm}}$$

was

Notice that the sample results are much more probable if the diet supplement (was, was not) effective in reducing the infection rate below 30%.

Tchebysheff's Theorem can be used in conjunction with the distribution of a binomial random variable because *at least* $(1 - (1/k^2))$ of *any* distribution lies within k standard deviations of the mean. However, when the number of trials n becomes large and p is not too close to 0 or 1, the Empirical Rule can be used with fairly accurate results. The interval $np \pm 2\sqrt{npq}$ should contain approximately 95% of the distribution, and the interval $np \pm 3\sqrt{npq}$ should contain almost all (approximately 99.7%) of the distribution.

Example 5.11

Suppose it is known that 10% of the citizens of the United States are in favor of increased foreign aid. A random sample of 100 U.S. citizens is questioned on this issue.

a. Find the mean and standard deviation of x, the number of citizens who favor increased foreign aid.

b. Within what limits would we expect to find the number who favor increased foreign aid?

Solution

a. With $n = 100$ and $p = .1$,

10

$$\mu = np = 100(.1) = \underline{\hspace{2cm}}$$

9

$$\sigma^2 = npq = 1000(.1)(.9) = \underline{\hspace{2cm}}$$

$$\sigma = \sqrt{npq} = \sqrt{\underline{\hspace{3cm}}} = \underline{\hspace{2cm}}$$

9; 3

b. From part a, $\mu = 10$ and $\sigma = 3$. Using two standard deviations, we find the interval $\mu \pm 2\sigma$ to be $10 \pm 2(3)$, or 10 ± 6. Since approximately 95% of the distribution lies within this interval, we would expect the number of citizens who favor increased foreign aid to lie between _____ and _____ if, in fact, $p = .1$.

4; 16

Example 5.12
Each person in a random sample of 64 people was asked to state a preference for candidate A or candidate B. If there is no underlying preference for either candidate, then the probability that an individual chooses candidate A will be $p = \underline{\hspace{2cm}}$.

.5

a. What will be the expected number and standard deviation of preferences for candidate A?
b. Within what limits would you expect the number of stated preferences for candidate A to lie?

Solution
Let x be the number of people who state a preference for candidate A. If there really is no preference for either candidate (that is, the voter selects a candidate at random), then x has a binomial distribution with $n = 64$ and $p = \underline{\hspace{2cm}}$.

.5

a. $\mu = np = 64(.5) = \underline{\hspace{2cm}}$

32

$\sigma^2 = npq = 64(.5)(.5) = \underline{\hspace{2cm}}$

16

$\sigma = \sqrt{npq} = \underline{\hspace{2cm}}$

4

b. From part a, $\mu = 32$ and $\sigma = 4$. Hence, $\mu \pm 2\sigma = 32 \pm 8$. We would expect the number of preferences for candidate A to lie between _____ and _____ if, in fact, $p = .5$.

24; 40

In addition to direct calculation and the use of Table 1 in Appendix I, individual and cumulative binomial probabilities are available in many statistical packages. The printouts in Figure 5.1 were produced using MINITAB for Windows with the binomial probability distributions option under the **Calc** (calculate) pull-down menu. The first used the "Probability" option, and the second the "Cumulative probability" option.

Probability Density Function

Binomial with n = 10 and p = 0.500000

x	P(X = x)
0.00	0.0010
1.00	0.0098
2.00	0.0439
3.00	0.1172
4.00	0.2051
5.00	0.2461
6.00	0.2051
7.00	0.1172
8.00	0.0439
9.00	0.0098
10.00	0.0010

Cumulative Distribution Function

Binomial with n = 10 and p = 0.500000

x	P(X <= x)
0.00	0.0010
1.00	0.0107
2.00	0.0547
3.00	0.1719
4.00	0.3770
5.00	0.6230
6.00	0.8281
7.00	0.9453
8.00	0.9893
9.00	0.9990
10.00	1.0000

Figure 5.1

Self-Correcting Exercises 5A

1. A city planner claims that 20% of all apartment dwellers move from their apartments within a year from the time they first moved in. In a particular city, 7 apartment dwellers who had given notice of termination to their landlords are to be interviewed.
   a. If the city planner is correct, what is the probability that 2 of the 7 had lived in the apartment for less than one year?
   b. What is the probability that at least 6 had lived in their apartment for one year or more?

2. Suppose that 70% of the first-class mail from New York to California is delivered within 4 days after being mailed. If 20 pieces of first-class mail are mailed from New York to California:
   a. Find the probability that at least 15 pieces of mail arrive within 4 days of the mailing date.
   b. Find the probability that 10 or fewer pieces of mail arrive later than 4 days after the mailing date.

3. In the past history of a certain serious disease it has been found that about 1/2 of its victims recover.
   a. Find the probability that exactly 4 of the next 15 patients suffering from this disease will recover.

b. Find the probability that at least 4 of the next 15 patients suffering with this disease will recover.

4. Suppose that 20% of the registered voters in a given city belong to a minority group and that voter registration lists are used in selecting potential jurors. If 80 persons were randomly selected from the voter registration lists as potential jurors, within what limits would you expect the number of minority members on this list to lie?

5. A television network claims that its Wednesday evening prime time program attracts 40% of the television audience. If the 40% figure is correct and if each person in a random sample of 400 television viewers was asked whether he or she had seen the previous show, within what limits would you expect the number of viewers who had seen the previous show to lie? What would you conclude if the interviews revealed that 96 of the 400 had actually seen the previous show?

Solutions to SCE 5A

1. Let x be the number of apartment dwellers who move within a year. Then $p = P(\text{move within a year}) = .2$ and $n = 7$.

   a. $P(x = 2) = C_2^7(.2)^2(.8)^5 = .27525$.

   b. $P(x \leq 1) = C_0^7(.2)^0(.8)^7 + C_1^7(.2)^1(.8)^6 = .209715 + .367002$
   $$= .57617.$$

2. Let x be the number of letters delivered within 4 days. Then $p = P(\text{letter delivered within 4 days}) = .7$ and $n = 20$.
   a. $P(x \geq 15) = 1 - P(x \leq 14) = 1 - .584 = .416$.
   b. $P(x \geq 10) = 1 - P(x \leq 9) = 1 - .017 = .983$. Notice that if 10 or fewer letters arrive later than 4 days, then $20 - 10 = 10$ or more will arrive within 4 days.

3. Use Table 1 in Appendix I, indexing $n = 15$, $p = 1/2$.
   a. $P(x = 4) = P(x \leq 4) - P(x \leq 3) = .059 - .018 = .041$
   b. $P(x \geq 4) = 1 - P(x \leq 3) = 1 - .018 = .982$.

4. Let x be the number of minority members on a list of 80, so that $p = .20$ and $n = 80$. Then $\mu = 80(.2) = 16$ and $\sigma = \sqrt{npq}$ $= \sqrt{12.8} = 3.58$. The limits are $\mu \pm 2\sigma = 16 \pm 7.16$ or 8.84 to 23.16. We expect to see between 9 and 23 minority group members on the jury lists.

5. $x =$ number watching the TV program; $p = .4$; $n = 400$; $\mu = 400(.4) = 160$; $\sigma^2 = 400(.4)(.6) = 96$; $\sigma = \sqrt{96} = 9.80$. Calculate $\mu \pm 2\sigma = 160 \pm 2(9.8) = 160 \pm 19.6$, or 140.4 to 179.6. Since we would expect the number watching the show to be between 141 and 179 with probability .95, it is highly unlikely that only 96 people would have watched the show *if* the 40% claim is correct.

It is more likely that the percentage of viewers for this particular show is less than 40%.

5.3 THE POISSON PROBABILITY DISTRIBUTION

The Poisson random variable provides a good model for the number of times a specified event occurs in either time or space. The number of weeds in a wheat field, the number of ships entering a harbor on a given day, and the number of telephone calls arriving at a switchboard during a one-minute interval are random variables that can be modeled using the Poisson probability distribution. In these applications, x represents the number of events in a period of time during which an average of μ events can be expected to occur.

Poisson Probability Distribution

$$P(x = k) = \frac{\mu^k e^{-\mu}}{k!}, \quad k = 0, 1, 2, \dots$$

where

$\mu = $ mean of the random variable x

$\sigma^2 = \mu$

$\sigma = \sqrt{\mu}$

$e = 2.71828\dots$ (e is the base of natural logarithms)

The Poisson model is developed using the assumption that the events occur randomly and independently; hence, its use is appropriate when these conditions are met.

Example 5.13
In a food processing and packaging plant, there are, on the average, two packaging machine breakdowns per week. Assume the weekly machine breakdowns follow a Poisson distribution.

a. What is the probability that there are no machine breakdowns in a given week?
b. Calculate the probability that there are no more than two machine breakdowns in a given week.

Solution
Machine breakdowns occur at the average rate of $\mu = $ _____ breakdowns per week. If the number of breakdowns follows a Poisson distribution, then

$$P(x = k) = \frac{2^k e^{-2}}{k!} \qquad \text{for } k = 0, 1, 2, \dots$$

2

a. $P(x = 0) = p(0) = \dfrac{2^0 e^{-2}}{0!}$

$\qquad = \dfrac{e^{-2}}{1} = \underline{\hspace{2cm}}$

.135335

b. The probability that no more than two machine breakdowns occur in a given week is

$\qquad P(x \le 2) = p(0) + p(1) + p(2)$

From part a, we know that $p(0) = \underline{\hspace{2cm}}$. We need to evaluate $p(1)$ and $p(2)$:

.135335

$\qquad p(1) = \dfrac{2^1 e^{-2}}{1!} = 2(.135335) = \underline{\hspace{2cm}}$

.270670

$\qquad p(2) = \dfrac{2^2 e^{-2}}{2!} = 2(.135335) = \underline{\hspace{2cm}}$

.270670

Hence,

$\qquad P(x \le 2) = \underline{\hspace{2cm}}$

.676675

As in the case of the binomial probability distribution, a statistical package such as Minitab can be used to obtain required probabilities. The printouts in Figure 5.2 were produced using MINITAB for Windows with the Poisson probability option under the **Calc** pull-down menu.

Probability Density Function

Poisson with mu = 2.00000

x	P(X = x)
0.00	0.1353
1.00	0.2707
2.00	0.2707
3.00	0.1804
4.00	0.0902
5.00	0.0361
6.00	0.0120
7.00	0.0034
8.00	0.0009
9.00	0.0002
10.00	0.0000

Cumulative Distribution Function

Poisson with mu = 2.00000

x	P(X <= x)
0.00	0.1353
1.00	0.4060
2.00	0.6767
3.00	0.8571
4.00	0.9473
5.00	0.9834
6.00	0.9955
7.00	0.9989
8.00	0.9998
9.00	1.0000
10.00	1.0000

Figure 5.2

Using the printout in Figure 5.2, index $k = 0$, 1 and 2 to find

$$p(0) = .1353$$

.2707

$$p(1) = \underline{\hspace{2cm}}$$

$$p(2) = .2707$$

so that

.6767

$$P(x \leq 2) = \underline{\hspace{2cm}}$$

which confirms the results of Example 5.13. Alternately, using the printout for the cumulative distribution function in Figure 5.2, index $k = 2$ to find the answer directly.

.6767

$$P(x \leq 2) = \underline{\hspace{2cm}}$$

This result could also have been found using Table 2 in Appendix I, where the Poisson cumulative probabilities, $P(x \leq k) = p(0) + p(1) + \cdots + p(k)$, are given for various values of μ. This table can be used to find right-tailed, left-tailed, or individual Poisson probabilities.

It is important to keep in mind that the Poisson distribution is fixed in time or space. In the last example, the mean number of breakdowns per week was two. The mean number of breakdowns in a three-week period would be 6. The parameter μ in a Poisson distribution is always equal to the *mean* number of rare events observed that occur in a *given unit* of time or space.

The Poisson probability distribution is often used to approxi-

large

small

mate binomial probabilities in cases where n is \underline{\hspace{2cm}} and p or q is \underline{\hspace{2cm}}. Generally, the Poisson approximation to binomial probabilities is adequate when the binomial mean, $\mu = np$,

seven

is less than \underline{\hspace{2cm}}.

Example 5.14
Evidence shows that the probability that a driver will be involved in a serious automobile accident during a given year is .01. A particular corporation employs 100 full-time traveling salesmen. Based on this evidence, what is the probability that exactly two of the salesmen will be involved in a serious automobile accident during the coming year?

Solution

100

.01

This is an example of a binomial experiment with $n = $ \underline{\hspace{2cm}} trials and $p = $ \underline{\hspace{2cm}}. The exact probability distribution for the number of serious automobile accidents in $n = 100$ trials is

$$P(x = k) = \frac{100!}{k!\,(100 - k)!}(.01)^k(.99)^{100-k}, \quad k = 0, 1, 2, \ldots, 100$$

Since we do not have binomial tables for $n = 100$, we note that the binomial mean $\mu = np = 1$. The Poisson approximation to binomial probabilities can be used in this case, with the Poisson mean taken to be $\mu = $ _____ . Therefore,

1

$$p(2) \approx \frac{(1)^2 e^{-1}}{2!}$$

$$= \frac{.367879}{2}$$

$$= \underline{\hspace{2cm}}$$

.1839

Example 5.15

Suppose past records show that the probability of default on an FHA loan is about .01. Assume that 25 homes in a given area are financed by FHA, and use the Poisson approximation to binomial probabilities.

a. Find the probability that there will be no defaults among these 25 loans.
b. Find the probability that there will be two or more defaults.

Compare the values found in parts a and b with the actual binomial probabilities found using Table 1 of Appendix I.

Solution

Although a sample of size $n = 25$ is not usually considered to be large, the value of $p = .01$ is small and $\mu = np = .25$ is less than 7. We will in any case assess the accuracy of the Poisson approximation compared with the actual binomial probabilities. We use

$$P(x = k) = \frac{(.25)^k e^{-.25}}{k!}$$

with $e^{-.25} = .778801$.

a. The probability of $x = 0$ defaults is approximated to be

$$p(0) \approx \frac{(.25)^0 e^{-.25}}{0!} = \underline{\hspace{2cm}}$$

.778801

The actual probability from Table 1 is _____ .

.778

b. The probability of two or more defaults can be found by using

$$P(x \geq 2) = 1 - P(x \leq 1)$$

$$= 1 - [p(\underline{\hspace{2cm}}) + p(\underline{\hspace{2cm}})]$$

0; 1

We need

$$p(1) \approx \frac{(.25)^1 e^{-.25}}{1!}$$

$$= (.25)(\underline{\hspace{2cm}})$$

.778801

$$= \underline{\hspace{2cm}}$$

.194700

Hence,

$$P(x \geq 2) \approx 1 - (.778801 + .194700)$$

.973501

$$= 1 - \underline{\hspace{2cm}}$$

.026499

$$= \underline{\hspace{2cm}}$$

The actual value from Table 1 is

.974

$$P(x \geq 2) = 1 - \underline{\hspace{2cm}}$$

.026

$$= \underline{\hspace{2cm}}$$

Notice that for this problem there is fairly good agreement between the Poisson approximations and the actual binomial probabilities even though n is not large. This is mainly because of the small value

.25

of $\mu = np = \underline{\hspace{2cm}}$.

5.4 THE HYPERGEOMETRIC PROBABILITY DISTRIBUTION

Suppose that we are selecting a sample of n elements from a population that contains N elements, some of which are of one type and the rest are of another type. If we designate one type of element as a "success" and the other as a "failure," the situation is similar to the binomial experiment described in Section 5.2. However, one of the assumptions required for the application of the binomial probability

constant

distribution is that the probability of a success remains $\underline{\hspace{2cm}}$ from trial to trial. This assumption is violated whenever the sam-

without

pling is done (with, without) replacement (that is, once an element has been chosen, it cannot be chosen again).

This departure from the conditions required of the ideal binomial experiment is not important when the population is (small,

large; sample
constant

large) relative to the $\underline{\hspace{2cm}}$ size. In such circumstances, the probability p of a success is approximately $\underline{\hspace{2cm}}$ for each trial or selection. However, if the number of elements in the population is small in relation to the number of elements in the sample, the probability of a success for a given trial is

dependent on

(dependent on, independent of) the outcomes of preceding trials. In this case, the number x of successes follows the hypergeometric probability distribution.

The probability distribution of a random variable x that has the **hypergeometric distribution** is given by the formula

$$P(x = k) = \frac{C_k^M \, C_{n-k}^{N-M}}{C_n^N}$$

where

$$N = \text{number of elements in the population}$$

M = number of elements in the population that are successes

n = number of elements in the sample that are selected from the population

k = number of successes in the sample

The mean and variance of a hypergeometric random variable are very similar to the mean and variance of a binomial random variable, with $p = M/N$ and $q = (N - M)/N$. The quantity $(N - n)/(N - 1)$ is a correction for the finite population size. We have

$$\mu = n\left(\frac{M}{N}\right)$$

$$\sigma^2 = n\left(\frac{M}{N}\right)\left(\frac{N - M}{N}\right)\left(\frac{N - n}{N - 1}\right)$$

C_b^a defined in Section 5.2 is taken to be 0 if $b > a$.

The hypergeometric probability distribution is applicable when you select a sample of elements from a population without _____ and record whether or not each element possesses a certain characteristic.

replacement

Example 5.16

An auditor is checking the records of an accountant who is responsible for ten clients. The accounts of two of the clients contain major errors, and the accountant will fail the inspection if the auditor finds even a single erroneous account. What is the probability that the accountant will fail the inspection if the auditor inspects the records of three clients chosen at random?

Solution

Let x be the number of erroneous accounts found in the (population, sample). Then

sample

$$N = \text{_____}$$ 10

$$M = \text{_____}$$ 2

$$N - M = \text{_____}$$ 8

$$n = \text{_____}$$ 3

The accountant will fail the inspection if $x =$ _____ or   1
_____. So $P(\text{accountant fails}) = P(x \geq \text{_____})$ 2; 1
$= p(\text{_____}) + p(\text{_____})$. We have   1; 2

$\dfrac{8!}{2!\,6!};\ \dfrac{10!}{3!\,7!}$

$$P(x = 1) = \frac{\left(\dfrac{2!}{1!\,1!}\right)\left(\underline{\qquad}\right)}{\left(\underline{\qquad}\right)}$$

.467

$$= \underline{\qquad}$$

$\dfrac{2!}{2!\,0!};\ \dfrac{8!}{1!\,7!}$

$$P(x = 2) = \frac{\left(\underline{\qquad}\right)\left(\underline{\qquad}\right)}{\left(\dfrac{10!}{3!\,7!}\right)}$$

.067

$$= \underline{\qquad}$$

.467; .067; .534

Therefore, the probability that the accountant will fail the inspection is _____ + _____ = _____.

Self-Correcting Exercises 5B

1. The probability of a serious fire during a given year to any one house in a particular city is believed to be .005. A particular insurance company holds fire insurance policies on 1000 homes in this city.
   a. Find the probability that the company will not have any serious fire damage claims by the owners of these homes during the next year.
   b. Find the probability they will have no more than three claims.

2. In a certain manufacturing plant, wood-grain printed 4' by 8' wall board panels are mass produced and packaged in lots of 100. Past evidence indicates that the number of damaged or imperfect panels per bundle follows a Poisson distribution with mean $\mu = 2$.
   a. Find the probability that there are exactly three damaged or imperfect panels in a bundle of 100.
   b. Find the probability that there are at least two damaged or imperfect panels in a bundle of 100.

3. A home improvement store has purchased two bundles (2 bundles of 100 each) of panels from the manufacturer described in Exercise 2. Find the probability his lot contains no more than four damaged or imperfect panels.

4. The board of directors of a company has voted to create an employee council for the purpose of handling employee complaints. The council will consist of the company president, the vice-president for personnel, and four employee representatives. The four employees will be randomly selected from a list of 15 volunteers. This list consists of nine men and six women.

a. What is the probability that two or more men will be selected from the list of volunteers?

b. What is the probability that exactly three women will be selected from the list of volunteers?

5. A bin of 50 parts contains three defective units. A sample of five units is drawn randomly from the bin. What is the probability that no defective units will be selected?

Solutions to SCE 5B

1. Let x be the number of fires observed so that $p = P[\text{fire}] = .005$ and $n = 1000$. The random variable is binomial; however, since n is large and p is small with $\mu = np = 5$, the Poisson approximation is appropriate.

a. $P[x = 0] = \dfrac{\mu^0 e^{-\mu}}{0!} = e^{-5} = .006738$

b. $P[x \le 3] = \dfrac{5^0 e^{-5}}{0!} + \dfrac{5^1 e^{-5}}{1!} + \dfrac{5^2 e^{-5}}{2!} + \dfrac{5^3 e^{-5}}{3!}$

$= .006738(1 + 5 + 12.5 + 20.833) = .2650$

2. Let x be the number of defective panels with $\mu = 2$.

a. $P[x = 3] = \dfrac{2^3 e^{-2}}{3!} = 1.33(.135335) = .180$

b. $P[x \ge 2] = 1 - P[x \le 1] = 1 - \dfrac{2^0 e^{-2}}{0!} - \dfrac{2^1 e^{-2}}{1!}$

$= 1 - e^{-2}(1 + 2) = 1 - 3(.135335) = .594$

3. We are now concerned with the random variable x, the number of defective panels in a bundle of 200, with $\mu = 2(2) = 4$. Then

$$P[x \le 4] = \dfrac{4^0 e^{-4}}{0!} + \dfrac{4^1 e^{-4}}{1!} + \dfrac{4^2 e^{-4}}{2!} + \dfrac{4^3 e^{-4}}{3!} + \dfrac{4^4 e^{-4}}{4!}$$

$$= .018316(1 + 4 + 8 + 10.67 + 10.67)$$
$$= .629$$

4. Four employees will be chosen from 15, nine of whom are men and six of whom are women. Hence, $N = 15$, $M = 9$, $n = 4$, $N - M = 6$.

a. $P[\text{two or more men}] = P[x \ge 2]$

$= \dfrac{C_2^9 C_2^6}{C_4^{15}} + \dfrac{C_3^9 C_1^6}{C_4^{15}} + \dfrac{C_4^9 C_0^6}{C_4^{15}} = \dfrac{36}{1365} + \dfrac{84(6)}{1365} + \dfrac{126}{1365} = \dfrac{666}{1365}$

$= .488$

b. $P[\text{exactly three women}] = P[\text{exactly one man}] = P[x = 1]$

$= \dfrac{C_1^9 C_3^6}{C_4^{15}} = \dfrac{9(20)}{1365} = .132$

5. Define x to be the number of defective units chosen. Then $N = 50$, $M = 3$, $N - M = 47$, $n = 5$.

$$P[x = 0] = \frac{C_0^3 C_5^{47}}{C_5^{50}} = \frac{47!\,5!\,45!}{5!\,42!\,50!} = \frac{47(46)(45)(44)(43)}{50(49)(48)(47)(46)} = .724$$

KEY CONCEPTS AND FORMULAS

I.   *The Binomial Random Variable*
  1. Five characteristics: n identical independent trials, each resulting in either *success S* or *failure F*; probability of success is p and remains constant from trial to trial; and x is the number of successes in n trials
  2. Calculating binomial probabilities
      a. Formula: $P(x = k) = C_k^n p^k q^{n-k}$
      b. Cumulative binomial tables
      c. Individual and cumulative probabilities using Minitab
  3. Mean of the binomial random variable: $\mu = np$
  4. Variance and standard deviation: $\sigma^2 = npq$ and $\sigma = \sqrt{npq}$

II. *The Poisson Random Variable*
  1. The number of events that occur in a period of time or space, during which an average of μ such events are expected to occur
  2. Calculating Poisson probabilities
      a. Formula: $P(x = k) = \dfrac{\mu^k e^{-\mu}}{k!}$
      b. Cumulative Poisson tables
      c. Individual and cumulative probabilities using Minitab
  3. Mean of the Poisson random variable: $E(x) = \mu$
  4. Variance and standard deviation: $\sigma^2 = \mu$ and $\sigma = \sqrt{\mu}$
  5. Binomial probabilities can be approximated with Poisson probabilities when $np < 7$, using $\mu = np$.

III. *The Hypergeometric Random Variable*
  1. The number of successes in a sample of size n from a finite population containing M successes and $N - M$ failures
  2. Formula for the probability of k successes in n trials:

$$P(x = k) = \frac{C_k^M C_{n-k}^{N-M}}{C_n^N}$$

  3. Mean of the hypergeometric random variable: $\mu = n\left(\dfrac{M}{N}\right)$

  4. Variance and standard deviation: $\sigma^2 = n\left(\dfrac{M}{N}\right)\left(\dfrac{N-M}{N}\right)\left(\dfrac{N-n}{N-1}\right)$
     and $\sigma = \sqrt{\sigma^2}$

Exercises

1. A subject is taught to do a task in two different ways. Studies have shown that when subjected to mental strain and asked to perform the task, the subject most often reverts to the method first learned, regardless of whether it was easier or more diffi cult than the second. If the probability that a subject returns to the first method learned is .8 and six subjects are tested, what is the probability that at least five of the subjects revert to their first learned method when asked to perform their task under mental strain?

2. The taste test for PTC (phenylthiourea) is a favorite exercise for every human genetics class. It has been established that a single gene determines the characteristic, and 70% of the Ameri can population are "tasters," while 30% are "nontasters." Suppose 20 people are randomly chosen and administered the test.
   a. Give the probability distribution of x, the number of "non-tasters" out of the 20 chosen.
   b. Using appropriate tables, find $P(x \leq 7)$.
   c. Find $P(3 < x \leq 8)$.

3. A multiple-choice test offers four alternative answers to each of 100 questions. In every case there is but one correct answer. Bill responded correctly to each of the first 76 questions when he noted that just 20 seconds remained in the test period. He quickly checked an answer at random for each of the remaining 24 questions without reading them.
   a. What is Bill's expected number of of correct answers?
   b. If the instructor assigns a grade by taking 1/3 of the wrong from the number marked correctly, what is Bill's expected grade?

4. On a certain university campus a student is fined $10 for the first parking violation of the academic year. The fine is doubled for each subsequent offense, so that the second violation costs $20, the third $40, and so on. The probability that a parking violation on a given day is detected is .10. Suppose that a cer tain student will park illegally on each of 20 days during a given academic year.
   a. What is the probability he will not be fined?
   b. What is the probability that his fines will total no more than $150?

5. Four experiments are described below. Identify which of these might reasonably be treated as a binomial experiment. If a given experiment is clearly not binomial, state what feature disqualifies

it. If it is a binomial experiment, write down the probability function for x.

a. Five percent of the stamps in a large collection are extremely valuable. The stamps are withdrawn one at a time until ten extremely valuable stamps are located. The observed random variable is x, the total number of stamps withdrawn.

b. There are 15 students in a particular class. The names of these students are written on tags placed in a box. Periodically, a tag is drawn at random from the box and the student with that name is asked to recite. The tag is returned to the box and the proceedings continued. Let x denote the number of times a particular student will be called upon to recite when the teacher draws from the box five times.

c. This example is conducted in the manner prescribed for part b except that a tag drawn from the box is not returned. Let x denote the number of times the particular student will be called upon to recite when the teacher draws from the box five times.

d. Sixty percent of the homes in a given country carry fire insurance. A sample of five homes is drawn at random from this country. Let x denote the number of insured homes among the five selected.

6. Eastern University has found that about 90% of its accepted applicants for enrollment in the freshman class will actually take a place in that class. In 1999, 1360 applicants to Eastern were accepted. Within what limits would you expect to find the size of the freshman class at Eastern in the fall of 1999?

7. Suppose that the national unemployment rate is 7.1%. A sample of $n = 100$ persons is taken in the Los Angeles area and the number of unemployed persons is recorded.

a. If the unemployment rate for the Los Angeles area is the same as the national rate, within what limits would you expect the number of unemployed to fall?

b. If 15 of the 100 persons interviewed said that they were unemployed, what would you conclude about the unemployment rate in the Los Angeles area?

8. Suppose that 1 out of 10 homeowners in the state of California have invested in earthquake insurance. If 15 homeowners are randomly chosen to be interviewed,

a. What is the probability that at least 1 had earthquake insurance?

b. What is the probability that 4 or more have earthquake insurance?

c. Within what limits would you expect the number of homeowners insured against earthquakes to fall?

9. Consider 10 management trainees in a firm's rotation program where three of the 10 are members of minority groups. If five of the trainees are randomly assigned to the marketing division, what is the probability that there will be three minority trainees in the group assigned to marketing?

10. Improperly wired control panels were mistakenly installed on two of eight large automated machine tools. It is uncertain which of the machine tools have the defective panels, and a sample of four tools is randomly chosen for inspection.
    a. What is the probability that the sample will include no defective panels? Both defective panels?
    b. Find the binomial approximations for part a. (Optional)

11. A delicatessen has found that the weekly demand for caviar follows a Poisson distribution with a mean of four tins (each tin contains 8 ounces of caviar).
    a. Find the probability that no more than 4 tins are requested during a given week.
    b. As caviar spoils with time, it must be replenished weekly by the delicatessen's owner. How many tins should he buy if it is desired that the probability not exceed .10 that demand cannot be met during a given week?

12. A manufacturer of small mini-computers has found that the average number of service calls per computer each year is 2.2. Assume the number of service calls follows a Poisson distribution.
    a. Find the probability that a particular mini-computer requires no service during a given year.
    b. A small firm has purchased two mini-computers from the manufacturer. Find the probability that the firm requires no service calls during a given year.
    c. Find the probability that the firm requires exactly two service calls during a given year.

13. Refer to Exercise 12. Suppose service calls cost the computer manufacturer an average of $20 each.
    a. What is the expected cost per computer each year?
    b. If the manufacturer has sold 50 mini-computers in the Seattle area, what is the expected annual cost of service in this area?

14. Customers arrive at a certain gasoline filling station at the average rate of one every ten minutes. Assume the arrivals follow a Poisson distribution. The station has only one attendant and he takes an

average of five minutes to service each arrival. What is the probability that two customers arrive while the attendant is servicing an earlier arrival?

Chapter 6
The Normal Probability Distribution

6.1 PROBABILITY DISTRIBUTIONS FOR CONTINUOUS RANDOM VARIABLES

Not all experiments have sample spaces that contain a countable number of simple events. *Continuous random variables*, such as heights, weights, response times, and waiting times, can assume the infinitely many values corresponding to points on a line interval. Since the mathematical treatment of continuous random variables requires the use of calculus, we will merely state some basic concepts. The probability distribution for a continuous random variable can be thought of as the limiting histogram for a very large set of measurements utilizing the smallest possible interval width. In such a case, the outline of the histogram would appear as a continuous curve.

Let us illustrate what happens if we begin with a histogram and allow the interval width to get smaller and smaller while the number of measurements gets larger and larger, as in Figure 6.1. The mathematical

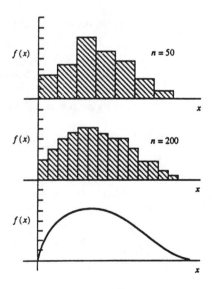

Figure 6.1

probability

1

one

modeling

approximations

function $f(x)$ that traces this curve with varying values of x is called the _____ distribution, or the probability density for the random variable x.

When working with discrete probabilities, the area under the relative frequency histogram was equal to _____, and the probability that x would fall in an interval from a to b was found by summing over the probabilities from a to b. Continuous distributions have similar properties.

- The area under the curve of the probability distribution is equal to _____.

- The probability that x will fall in the interval from a to b is equal to the area under the curve over the interval from a to b as shown in Figure 6.2.

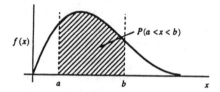

Figure 6.2

In addition to these properties,

- The probability that $x = a$ is equal to zero, that is, $P(x = a) = 0$.

- Therefore, $P(x \geq a) = P(x > a)$ and $P(x \leq a) = P(x < a)$.

- This is not true in general for discrete random variables.

When deciding on a model to describe a population of measurements, we must choose $f(x)$ appropriate to our data. Any inferences we may make will be only as valid as the model we are using. It is therefore very important to know as much as possible about the phenomenon under study that will give rise to the measurements that we record.

The idea of modeling the responses that we will record for an experiment may seem strange to you, but you have probably seen models before, though in a different context. For example, when a physicist says, "The distance (s) traversed by a free-falling body is equal to one-half the force of gravity (g) multiplied by the time (t) squared" and writes $s = gt^2/2$, he is merely _____ a physical phenomenon with a mathematical formula. These mathematical models merely provide _____ to reality, which further need to be verified by experimental techniques.

6.2 THE NORMAL PROBABILITY DISTRIBUTION

Many types of continuous curves are available as models for continuous random variables. However, many continuous random variables that are biological measurements, such as heights and weights or reaction and response times, or measurement errors themselves have mound-shaped distributions that are also bell-shaped. These and other continuous random variables can be modeled by a normal probability distribution whose density function produces the bell-shaped curve shown in Figure 6.3.

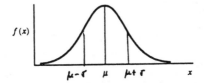

Figure 6.3

The Normal Probability Distribution

$$f(x) = \frac{1}{\sigma\sqrt{2\pi}} e^{-1/2\left(\frac{x-\mu}{\sigma}\right)^2}, \quad -\infty < x < \infty$$

The symbols found in $f(x)$ are the following:

1. The symbols π and e are mathematical constants whose values are approximately 3.14159 and 2.7183.
2. The mean μ is locates the center of the distribution. Notice that the distribution is *symmetric* about the mean μ, so that 1/2 of the area lies to the left of μ and 1/2 lies to the right of μ.
3. The standard deviation σ measures the spread or dispersion of the distribution. One standard deviation is the distance from μ to the two points of inflection where the density curve changes from concave down to concave up. These two points are located at $\mu \pm \sigma$ as indicated in Figure 6.3.

Encountering a random variable whose values can be extremely small (a large negative value) or extremely large might at first be disconcerting to the student who has heard that heights, weights, response times, and errors of measurements are approximately normally distributed. Surely we do not have heights, weights, or times that are less than zero! Certainly not, but almost all of the distribution of a normally distributed random variable lies within the interval $\mu \pm 3\sigma$. In the case of heights or weights, this interval almost always encompasses positive values. Keep in mind this curve is merely a *model* that approximates an actual distribution of measurements. Its great utility lies in the fact that it *can* be used effectively as a model for so many types of measurements.

Figure 6.4 shows two normal probability distributions, one with a mean of $\mu = 25$ and a standard deviation of $\sigma = 5$, and a second with a mean of $\mu = 40$ and a standard deviation of $\sigma = 2$.

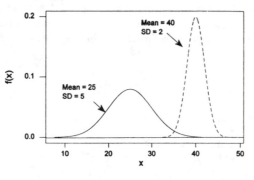

Figure 6.4

Notice again that the distributions are centered at μ. When σ is small, the distribution is peaked and tightly dispersed about its mean, and when σ is large, the distribution is less peaked and more loosely dispersed about its mean.

6.3   TABULATED AREAS OF THE NORMAL PROBABILITY DISTRIBUTION

The probability that a continuous random variable lies between a and b is equal to the area under the probability density function between the values a and b. The areas under a continuous curve such as a probability density function are generally computed using integral calculus. The areas under the normal curve can be found using numerical methods, or using a table that provides these areas. However, there is a second problem, in that you would need different tables for each normal distribution with a given mean and standard deviation. Since this is impractical, using a standardizing procedure that allows you to use the same table for all normal distribu-tions can solve the problem.

The Standard Normal Random Variable

The standardizing procedure expresses the value of x as the number of standard deviations to the left or the right of the mean μ. The standardized normal random variable z is defined as

$$z = \frac{x - \mu}{\sigma}.$$

This can be restated to depict x as

$$x = \mu + z\sigma$$

for which the actual value of z may be positive or negative.

- When x lies below the mean μ, the value of z is _____.   negative

- When x lies above the mean μ, the value of z is _____.   positive

- When $x = \mu$, $z =$ _____.   0

Since the normal distribution is symmetric about its mean, the standard normal random variable, z, has a mean equal to 0. Also, since z represents the distance that the value of a particular value of x lies from the mean in units equal to standard deviations, the standard deviation of z is equal to 1.

The value of the **standard normal random** variable z represents the distance that x lies from its mean in units of standard deviations. Or more simply, z is the number of standard deviations from the mean.

The areas under the standard normal density function are given in Table 3, Appendix 1. The tabulated areas as those less than a given number, say z_0. A convenient notation used to designate the area to the left of z_0 is $A(z_0)$. Refer to Figure 6.5. Since you will need to work with areas to the left and right of a given value of z, you need to remember that the total area under the curve is equal to 1. You will be able to use the relationship $P(z > z_0) = 1 - P(z < z_0)$.

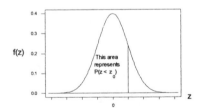

Figure 6.5

1. For $z = 1$, the area left of $z = 1$ is $A(z = 1) = A(1) = .8413$.
2. For $z = 2$. $A(z = 2) = A(2) =$ _____.   .9772
3. For $z = 1.6$, $A(1.6) =$ _____.   .9452
4. For $z = 2.4$, $A(2.4) =$ _____.   .9918

Now try reading the table for values of z given to two decimal places.

.9951
.7734
.9545
.9979

5. For $z = 2.58$, $A(2.58) =$ _____.
6. For $z = .75$, $A(.75) =$ _____.
7. For $z = 1.69$, $A(1.69) =$ _____.
8. For $z = 2.87$, $A(2.87) =$ _____.

We will now find probabilities associated with the standard normal random variable z by using Table 3.

Example 6.1
Find the probability that z is greater than 1.86 – that is, $P(z > 1.86)$.

Solution
Illustrate the problem with a diagram as follows.

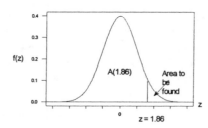

1. The total area under the curve is one.
2. From Table 3 $A(1.86) =$ _____.
3. Therefore, the area to be found is gotten by subtracting $A(1.86)$ from _____.
4. Hence,

.9686

1

$$P(z > 1.86) = 1 - A(1.86)$$
$$= 1 - .9686$$
$$= \underline{\hspace{2cm}}.$$

.0314

Example 6.2
Find $P(z < -2.22)$.

Solution
Illustrate the problem with a diagram.

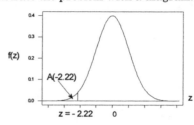

1. The negative value of z indicates that you are to the (left, right) of the mean $z = 0$ and the answer should be less than .5.

left

.0132

2. Therefore, $P(z < -2.22) = A(-2.22) =$ _____.

Example 6.3
Find $P(-1.21 < z < 2.43)$.

Solution
Illustrate the problem with a diagram.

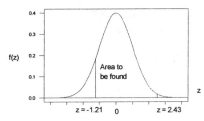

The required area is found by subtracting the area to the left of $z = -1.21$ from the area to the left of $z = 2.43$. Therefore

$$P(-1.21 < z < 2.43) = P(z < 2.43) - P(z < -1.21)$$
$$= A(2.43) - A(-1.21)$$
$$= \underline{\hspace{2cm}} - \underline{\hspace{2cm}}$$ .9925; .1131
$$= \underline{\hspace{2cm}}.$$ .8794

Another type of problem that arises is finding a value of z, say z_0, such that a probability statement about z will be true. The next two examples illustrate this type of problem.

Example 6.4
Find the value of z_0 such that

$$P(z < z_0) = .8925.$$

Solution
Once again, illustrate the problem with a diagram.

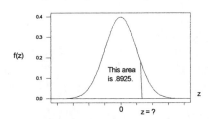

1. Search Table 3 entries until the area .8925 is found. The value of z
    for which $A(z_0) = .8925$ is $z_0 = \underline{\hspace{2cm}}$.   1.24
2. Therefore we know that $P(z < 1.24) = .8925$.

Example 6.5
Find the value of z_0 such that $P(z > z_0) = .2643$.

Solution
The last problem was stated in terms of the area to the left of z_0. This problem is stated in terms of the area to the *right* of z_0. We know that

$$P(z > z_0) = 1 - P(z < z_0) = .2643.$$

.7357
.7357

1. Then $A(z_0) = 1 - .2643 = $ _____.
2. Reading directly from Table 3, we find that $A(.63) = $ _____.
3. Therefore $z_0 = .63$ and $P(z > .63) = .2643$.

USE OF THE TABLE OF NORMAL CURVE AREAS FOR THE NORMAL RANDOM VARIABLE x

We can now proceed to find probabilities associated with any normal random variable x that has mean μ and standard deviation σ. This is accomplished by converting the random variable x to the standard normal random variable z, and then working the problem in terms of z.

Since probability statements are written in the form of inequalities, you are reminded of two facts. A statement of inequality is maintained if (1) the same number is subtracted from each member of the inequality and/or (2) each member of the inequality is divided by the same *positive number*.

Example 6.6
The following are equivalent statements about x:

1. $70 < x < 95$

2. $(70 - 15) < (x - 15) < (95 - 15)$

3. $\dfrac{70 - 15}{5} < \dfrac{x - 15}{5} < \dfrac{95 - 15}{5}$

4. $11 < \dfrac{x - 15}{5} < 16$

Example 6.7
Let x be a normal random variable with mean $\mu = 100$ and standard deviation $\sigma = 4$. Find $P(92 < x < 104)$.

Solution
Recalling that $z = (x - \mu)/\sigma$, we can apply rules 1 and 2 to convert the probability statement about x to one about the standard normal random variable z:

1. $P(92 < x < 104) = P(92 - 100 < x - 100 < 104 - 100)$

$$= P\left(\frac{92 - 100}{4} < \frac{x - 100}{4} < \frac{104 - 100}{4}\right)$$

$$= P(-2 < z < 1)$$

2. The problem now stated in terms of z is readily solved by using the areas of the standard normal random variable z.

$$P(92 < x < 104) = P(-2 < z < 1)$$
$$= A(1) - A(-2)$$

$$= \underline{\hspace{2cm}} - \underline{\hspace{2cm}}$$

.8413; .0228

$$= \underline{\hspace{2cm}}.$$

.8185

Example 6.8

Let x be a normal random variable with mean 100 and standard deviation 4. Find $P(93.5 < x < 105.2)$.

Solution

$$P(93.5 < x < 105.2) = P(\frac{93.5 - 100}{4} < z < \frac{105.2 - 100}{4})$$

$$= P(-1.63 < z < \underline{\hspace{2cm}})$$

1.30

$$= \underline{\hspace{2cm}} - \underline{\hspace{2cm}}$$

.9032; .0516

$$= \underline{\hspace{2cm}}.$$

.8516

Self-Correcting Exercises 6A

1. Find the following probabilities associated with the standard normal random variable z:
   a. $P(z > 2.1)$.
   b. $P(z < -1.2)$.
   c. $P(.5 < z < 1.5)$.
   d. $P(-2.75 < z < -1.70)$.
   e. $P(-1.96 < z < 1.96)$.
   f. $P(z > 1.645)$.

2. Find the value of z, say z_0, such that the following probability statements are true:
   a. $P(z > z_0) = .10$.
   b. $P(z < z_0) = .01$.
   c. $P(-z_0 < z < z_0) = .95$.
   d. $P(-z_0 < z < z_0) = .99$.

3. An auditor has reviewed the financial records of a hardware store and has found that its billing errors follow a normal distribution with mean and standard deviation equal to $0 and $1, respectively.
   a. What proportion of the store's billings are in error by more than $1?
   b. What is the probability that a billing represents an overcharge of at least $1.50?
   c. What is the probability that a customer has been undercharged from $.50 to $1.00?
   d. Within what range would 95% of the billing errors be?
   e. Of the extreme undercharges, 5% would be at least what amount?

4. If x is normally distributed with mean 10 and variance 2.25, evaluate the following probabilities:

 a. $P(x > 8.5)$.
 b. $P(x < 12)$.
 c. $P(9.25 < x < 11.25)$.
 d. $P(7.5 < x < 9.2)$.
 e. $P(12.25 < x < 13.25)$.

5. An industrial engineer has found that the standard household light bulbs produced by a certain manufacturer have a useful life that is normally distributed with a mean of 250 hours and a variance of 2500.
 a. What is the probability that a randomly selected bulb from this production process will have a useful life in excess of 300 hours?
 b. What is the probability that a randomly selected bulb from this production process will have a useful life between 190 and 270 hours?
 c. What is the probability that a randomly selected bulb from this production process will have a useful life not exceeding 260 hours?
 d. Ninety percent of the bulbs have a useful life in excess of how many hours?
 e. The probability if .95 that a bulb does not have a useful life in excess of how many hours?

6. Scores on a trade school entrance examination exhibit the characteristics of a normal distribution with mean and standard deviation of 50 and 5, respectively.
 a. What proportion of the scores on this examination would be greater than 60?
 b. What proportion of the scores on this examination would be less than 45?
 c. What proportion of the scores on this examination would be between 35 and 65?

Solutions to SCE 6A

Note: The student should illustrate each problem with a diagram and list all pertinent information before attempting the solution. Diagrams are omitted here in order to conserve space.

1. a. $P(z > 2.1) = 1 - A(2.1) = 1 - .9821 = .0179$.
   b. $P(z < -1.2) = A(-1.2) = .1151$.
   c. $P(.5 < z < 1.5) = A(1.5) - A(.5) = .9332 - .6915 = .2417$.
   d. $P(-2.75 < z < -1.75) = A(-1.75) - A(-2.75) = .0446 - .0030 = .0416$.
   e. $P(-1.96 < z < 1.96) = A(1.96) - A(-1.96) = .9750 - .0250 = .9500$.
   f. $P(z > 1.645) = 1 - P(z < 1.645) = 1 - A(1.645) = 1 - .9500 = .0500$.

   Linear interpolation was used in part f. That is, since the value $z = 1.6445$ is halfway between the two table values.

$z = 1.64$ and $z = 1.65$, the appropriate area is taken to be half-way between the two table values, $A(1.64) = .9495$ and $A(1.65) = .9505$. As a general rule, values of z will be rounded to two decimal places, except for this particular example, which will occur frequently in our calculations.

2. a.   We know that $P(z > z_0) = .10$ or $1 - A(z_0) = .10$, which implies that $A(z_0) = .9000$. The value of z_0 that satisfies this equation is $z_0 = 1.28$. Therefore, $P(z > 1.28) = .10$.

　　b. $P(z < z_0) = .01$ so that $A(z_0) = .01$. The value of z_0 that satisfies this equation is $z_0 = -2.33$, so that $P(z < -2.33) = .01$. Notice that the value of z is negative since the area of .01 indicates that this value is in the left tail of the distribution.

　　c. $P(-z_0 < z < z_0) = A(z_0) - A(-z_0) = .95$ so that

　　　　$A(-z_0) = 1 - A(z_0) = .0250$. Therefore $z_0 = 1.96$, and

　　　　$P(-1.96 < z < 1.96) = .95$.

　　d. $P(-z_0 < z < z_0) = A(z_0) - A(-z_0) = .99$ so that $A(-z_0) = .005$. The required value of z is $z_0 = 2.58$ and $P(-2.58 < z < 2.58) = .99$.

3.   We will denote by z the random variable of interest that has a standard normal distribution.

　　a. $P(z > 1) + P(z < -1) = 1 - A(1) + A(-1) = 1 - .8413 + .1587 = .3174$.

　　b. $P(z > 1.50) = 1 - P(z < 1.50) = 1 - .9332 = .0668$.

　　c. $P(-1 < z < -.5) = A(-.5) - A(-1) = .3085 - .1587 = .1498$.

　　d.   The problem is to find a value z_0 such that $P(-z_0 < z < z_0) = .95$. This was done in 2c. and the value of $z_0 = 1.96$. Therefore 95% of the billing errors will be between $-\$1.96$ and $\$1.96$.

　　e.   Undercharges imply negative errors. Hence the problem is to find z_0 such that $P(z < z_0) = .05$. The probability of .05 indicates that the value of z is in the left tail of the distribution and hence $A(z_0) = .05$. Referring to 1f, $z_0 = -1.645$ so that 5% of the undercharges will be at lest $\$1.65$.

4.   We have $\mu = 10, \sigma = \sqrt{2.25} = 1.5$.

　　a. $P(x > 8.5) = P(\dfrac{x-\mu}{\sigma} > \dfrac{8.5-10}{1.5}) = P(z > -1) = 1 - P(z < -1) =$

　　　　$1 - .1587 = .8413$.

　　b. $P(x < 12) = P(z < \dfrac{12-10}{1.5}) = P(z < 1.33) = A(1.33) = .9082$.

　　c. $P(9.25 < x < 11.25) = P(\dfrac{9.25-10}{1.5} < z < \dfrac{11.25-10}{1.5}) =$

　　　　$P(-.5 < z < .83) = .7967 - .3085 = .4882$.

　　d. $P(7.5 < x < 9.2) = P(-1.67 < z < -.53) = .2981 - .0475 = .2506$.

　　e. $P(12.25 < x < 13.25) = P(1.5 < z < 2.17) = .9850 - .9332 = .0518$.

5. The random variable of interest is x, the length of life for a standard household light bulb. It is normally distributed with $\mu = 250$, $\sigma^2 = 2500$ and $\sigma = 50$.

    a. $P(x > 300) = P(z > \dfrac{300 - 250}{50}) = P(z > 1) = 1 - A(1) = 1 - .8413 = .1587$.

    b. $P(190 < x < 270) = P(-1.2 < z < .4) = A(.4) - A(-1.2) =$
        $.6554 - .1151 = .5403$.

    c. $P(x < 260) = P(z < .2) = A(.2) = .5793$.

    d. We need to find a value of x, say x_0, such that $P(x > x_0) = .90$.

        Now $P(x > x_0) = P(\dfrac{x - \mu}{\sigma} > \dfrac{x_0 - 250}{50}) = .90$, so that

        $P(z > \dfrac{x_0 - 250}{50}) = 1 - A(\dfrac{x_0 - 250}{50}) = .90$. Hence $A(\dfrac{x_0 - 250}{50}) = .10$.

        From Table 3 we find $A(-1.28) = .10$, and therefore

        $\dfrac{x_0 - 250}{50} = -1.28$ and $x_0 = 250 - 1.28(50) = 186$. Ninety percent of

        the bulbs have a useful life in excess of 186 hours.

    e. Similar to part d. We need to find a value x_0 so that $P(x < x_0) =$
        $.95$. Now $P(x < x_0) = P(z < \dfrac{x_0 - 250}{50}) = .95$. From Table 3 we find

        that $A(1.645) = .95$, so that $\dfrac{x_0 - 250}{50} = 1.645$, and $x_0 = 250 +$

        $1.645(50) = 332.25$. Therefore 95% of all bulbs will burn out before 332.25 hours.

6. The random variable is x, scores on a trade school entrance examination, and has a normal distribution with $\mu = 50$ and $\sigma = 5$.

    a. $P(x > 60) = P(z > \dfrac{60 - 50}{5}) = P(z > 2) = 1 - A(2) = 1 - .9772 = .0228$.

    b. $P(x < 45) = P(z < \dfrac{45 - 50}{5}) = P(z < -1) = A(-1) = .1587$.

    c. $P(35 < x < 65) = P(-3 < z < 3) = A(3) - A(-3) = .9987 - .0013 = .9974$.

6.4 THE NORMAL APPROXIMATION TO THE BINOMIAL PROBABILITY DISTRIBUTION (Optional)

The binomial random variable x was defined in Chapter 5 as the number of successes in n independent trials that make up the binomial experiment. The probabilities associated with this random variable were calculated as

$$P(x = k) = C_k^n \, p^k \, q^{(n-k)} \ \text{ for } \ k = 0, 1, 2, \ldots, n$$

For large values of n (in fact, if n gets much larger than 10), the binomial probabilities are very tedious to compute. Fortunately, several options are available to us in an effort to avoid lengthy calculations.

1. Table 1 of Appendix I contains the tabulated values $P(x \leq k)$ for $n = 1, 2, \ldots, 12, 15, 20, 25$ and for $p = .01, .05, .10, .20, \ldots, .90, .95,$ .99. However, if you require values of n and p for which tables have not been given, Table 1 will not be useful.

2. In Section 5.3, we considered examples in which the Poisson distribution was used to approximate binomial probabilities. This approximation is appropriate when n is large and p (or q) is small, so that $np < 7$. However, there are a great number of situations in which n is large, but $np \geq 7$. In this case, the Poisson approximation (will, will not) be accurate.

will not

3. When n is large and p (or q) is not too small, the binomial probability histogram is fairly symmetric and mound-shaped. This symmetry increases as p gets closer to $p = .5$. Hence, it would seem reasonable to approximate the distribution of a binomial random variable with the distribution of a normal random variable whose mean and variance are identical to those for the binomial random variable.

When n is *sufficiently large* and p is not too close to 0 or 1, the random variable x, the number of successes in n trials, is approximately normally distributed, with mean np and variance npq.

When can we reasonably apply the normal approximation? For small values of n and values of p close to 0 or 1, the binomial distribution will exhibit a "pile-up" around $x = $ _____ or $x = $ _____. The data will not be _____-shaped and the normal approximation will be poor. For a normal random variable, _____% of the measurements will be within the interval $\mu \pm 2\sigma$. For $\mu = np$ and $\sigma = \sqrt{npq}$, the interval $np \pm 2\sqrt{npq}$ should be within the bounds of the binomial random variable x, or within the interval $(0, n)$, to obtain reasonably good approximations to the binomial probabilities. This will occur when $np > 5$ and $nq > 5$.

0; n
bell
95

To show how the normal approximation is used, let us consider a binomial random variable x with $n = 8$ and $p = 1/2$ and attempt to approximate some binomial probabilities with a normal random variable that has the same mean, $\mu = np$, and variance, $\sigma^2 = npq$, as the binomial x. In this case,

$$\mu = np = 8\left(\tfrac{1}{2}\right) = \text{_____}$$

4

2

$$\sigma^2 = npq = 8\left(\tfrac{1}{2}\right)\left(\tfrac{1}{2}\right) = \underline{\hspace{1in}}$$

Note that the interval

$$\mu \pm 2\sigma = 4 \pm 2\sqrt{2} = (1.2, 6.8)$$

is contained within the interval $(0, 8)$; therefore, our approximations should be adequate. Consider the diagrammatic representation of the approximation in Figure 6.6, where $p(x)$ is the frequency distribution for the binomial random variable and $f(x)$ is the frequency distribution for the corresponding normal random variable x.

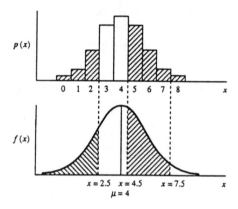

Figure 6.6

Example 6.9
Find $P(x < 3)$ using the normal approximation.

Solution
$P(x < 3)$ for the binomial random variable with mean $\mu = 4$ and $\sigma = \sqrt{2}$ corresponds to the shaded bars in the histogram over $x = 0$, 1, and 2 in Figure 6.6. The approximating probability corresponds to the shaded area to the left of $x = 2.5$ in the normal distribution with mean 4 and standard deviation $\sqrt{2}$.

We proceed as follows:

$$P(x < 3) \approx P(x < 2.5)$$

$$= P\left(\frac{x-4}{\sqrt{2}} < \frac{2.5-4}{\sqrt{2}}\right)$$

$$= P(z < -1.06)$$

$$= A(-1.06)$$

.1446

$$= \underline{\hspace{1in}}.$$

Example 6.10
Find $P(5 \leq x \leq 7)$.

Solution
For the binomial random variable with mean 4 and standard deviation $\sqrt{2}$, $P(5 \leq x \leq 7)$ corresponds to the shaded bars over $x =$ _____, _____, and _____. This corresponds in turn to the shaded area for the approximating normal distribution with mean 4 and standard deviation $\sqrt{2}$ between $x =$ _____ and $x =$ _____. Therefore,

	5
	6; 7
	4.5; 7.5

$$P(5 \leq x \leq 7) \approx P(4.5 < x < 7.5)$$

$$= P\left(\frac{4.5 - 4}{\sqrt{2}} < \frac{x - 4}{\sqrt{2}} < \frac{7.5 - 4}{\sqrt{2}}\right)$$

$$= P(.35 < z < 2.47)$$

$$= A(2.47) - A(.35)$$

$$= \underline{\hspace{2cm}} - \underline{\hspace{2cm}}$$

$$= \underline{\hspace{2cm}}$$

.9932; .6368

.3564

Notice that we used $P(x < 2.5)$ to approximate the binomial probability $P(x < 3)$. In like manner we used $P(4.5 < x < 7.5)$ to approximate the binomial probability $P(5 \leq x \leq 7)$. The addition or subtraction of .5 is called the *correction for continuity* because we are approximating a discrete probability distribution with a probability distribution that is continuous. You may become confused about whether .5 should be added or subtracted in the process of approximating binomial probabilities. A commonsense rule that always works is to examine the binomial probability statement carefully and determine which values of the binomial random variable are included in the statement. (Draw a picture if necessary.) The probabilities associated with these values correspond to the bars in the histogram centered over them. Locating the endpoints of the bars to be included determines the values needed for the approximating normal random variable.

Example 6.11
Suppose x is a binomial random variable with $n = 400$ and $p = .1$. Use the normal approximation to binomial probabilities to find the following:

1. $P(x > 45)$

2. $P(x \leq 32)$

3. $P(34 \leq x \leq 46)$

Solution
If x is binomial, then its mean, variance, and standard deviation are

$$\mu = np = 400(.1) = \underline{\hspace{2cm}}$$

40

36

6

$$\sigma^2 = npq = 400(.1)(.9) = \underline{\hspace{2cm}}$$

$$\sigma = \sqrt{36} = \underline{\hspace{2cm}}$$

Notice that $np = 400(.1) = 40$ and $nq = 400(.9) = 360$, both of which are greater than 5, so that the normal approximation to the binomial probabilities should be reasonably accurate.

1. To find $P(x > 45)$, we need the probabilities associated with the values 46, 47, 48, ..., 400. This corresponds to the bars in the binomial histogram beginning at _____. Hence,

45.5

$$P(x > 45) \approx P(x > 45.5)$$

$$= P\left(z > \frac{45.5 - 40}{6}\right)$$

$$= P(z > .92)$$

.1788

$$= \underline{\hspace{2cm}}$$

2. To find $P(x \leq 32)$, we need the probabilities associated with the values 0, 1, 2, ..., up to and including $x = 32$. This corresponds to finding the area in the binomial histogram to the left of _____. Hence,

32.5

$$P(x \leq 32) \approx P(x < 32.5)$$

$$= P\left(z < \frac{32.5 - 40}{6}\right)$$

−1.25

$$= P(z < \underline{\hspace{1.5cm}})$$

.1056

$$= \underline{\hspace{2cm}}$$

3. To find $P(34 \leq x \leq 46)$, we need the probabilities associated with the values beginning at 34 up to and including 46. This corresponds to finding the area under the histogram between _____ and _____. Hence,

33.5; 465.

$$P(34 \leq x \leq 46) \approx P(33.5 < x < 46.5)$$

$$= P\left(\frac{33.5 - 40}{6} < z < \frac{46.5 - 40}{6}\right)$$

−1.08; 1.08

$$= P(\underline{\hspace{1.5cm}} < z < \underline{\hspace{1.5cm}})$$

.7198

$$= \underline{\hspace{2cm}}$$

Example 6.12

Past records show that at a given college 20% of the students who began as psychology majors either changed their major or dropped out of school. An incoming class has 110 beginning psychology majors. What is the probability that as many as 30 of these students leave the psychology program?

Solution

1. If x represents the number of students who leave the psychology program, with the probability of losing a student given as $p = .2$, then the required probability for $n = 110$ is

   $$P(x \le 30) = p(0) + p(1) + \ldots + p(30)$$

2. To use the normal approximation, we need

   $$\mu = np = 110(.2) = 22$$

   $$\sigma = \sqrt{npq} = \sqrt{110(.2)(.8)} = \sqrt{17.6} = 4.195$$

   Both $np = 22$ and $nq = 88$ exceed 5, so the approximation should be adequate.

   We now proceed to approximate the probability required in part 1 by using a normal probability distribution with a mean of _____ and a standard deviation of _____.   22; 4.195

3. $P(x \le 30)$ corresponds to the area to the left of 30.5 in the approximating normal distribution. Hence,

   $$P(x \le 30) \approx P(x < 30.5)$$

   $$= P\left(\frac{x - 22}{4.195} < \frac{30.5 - 22}{4.195}\right)$$

   $$= P(z < \underline{\hspace{1.5cm}})$$ 2.03

   $$= \underline{\hspace{1.5cm}}$$ .9788

4. Hence, we approximate that $P(x \le 30) = $ _____.   .9788

Self-Correcting Exercises 6B

1. For a binomial experiment with $n = 20$ and $p = .7$, calculate $P(10 \le x \le 16)$ by
   a. using the binomial tables.
   b. using the normal approximation.

2. If the median income in a certain area is claimed to be $22,000, what is the probability that 37 or fewer of 100 randomly chosen wage earners from this area have incomes less than $22,000? Would the $22,000 figure seem reasonable if your sample actually contained 37 wage earners whose income was less than $22,000?

3. A large number of seeds from a certain species of flower are collected and mixed together in the following proportions according to the color of the flowers they will produce: 2 red, 2 white, 1 blue. If these seeds are mixed and then randomly packaged in bags containing about 100 seeds, what is the probability that a bag will contain the following:

    a.   At most 50 "white" seeds?
    b.   At least 65 seed that are not "white"?
    c.   At least 25 but at most 45 "white" seeds?

4.   Refer to Exercise 3. Within what limits would you expect the number of white seeds to lie with probability .95?

Solutions to SCE 6B

1. a. Using Table 1 with $n = 20$ and $p = .7$,
$$P(10 \leq x \leq 16) = P(x \leq 16) - Px \leq 9) = .893 - .017 = .876.$$

    b. The probabilities associated with the values of $x = 10, 11, \ldots, 16$ are needed: hence the area of interest is the area between 9.5 on the left and 16.5 on the right. Further, $\mu = np = 14$, $\sigma^2 = npq \approx 4.2$.

$$P(10 \leq x \leq 16) \approx P(9.5 < x < 16.5)$$
$$= P(\frac{9.5 - 14}{\sqrt{2}} < z < \frac{16.5 - 14}{\sqrt{2}}) = P(-2.20 < z < 1.22)$$
$$= .8888 - .0139 = .8749.$$

2. Let $p = P(\text{income is less than } 22,000) = \frac{1}{2}$, since 22,000 is the median income. Also, $n = 100$, $\mu = np = 50$, $\sigma^2 = npq = 25$.
$$P(x \leq 37) \approx P(x < 37.5) = P(z < \frac{37.5 - 50}{5}) = P(z < -2.5)$$
$$= A(-2.5) = .0062.$$
The observed event is highly unlikely under the assumption that $22,000 is the median income. The $22,000 figure does not seem reasonable.

3. $x = $ number of white seeds, $p = P(\text{white seed}) = .4$, $n = 100$;
$\mu = np = 40$, $\sigma^2 = npq = 24$, $\sigma = 4.90$.

    a. $P(x \leq 50) \approx P(x < 50.5) = P(z < \frac{50.5 - 40}{4.90}) = P(z < 2.14)$
        $= A(2.14) = .9838.$

    b. $P(z \leq 35) \approx P(x < 35.5) = P(z < -.92) = A(-.92) = .1788.$

    c. $P(25 \leq x \leq 45) \approx P(24.5 < x < 45.5) = P(-3.16 < z < 1.12)$
        $= .8686 - .0008 = .8678$

4. $\mu \pm 2\sigma = 40 \pm 2(4.90) = 40 \pm 9.8$, or from 31 to 49.

KEY CONCEPTS AND FORMULAS

I. *Continuous Probability Distributions*
    1. Continuous random variables
    2. Probability distributions or probability density functions
      a. Curves are smooth.

   b. The area under the curve between a and b represents the
      probability that x falls between a and b.
   c. $P(x = a) = 0$ for continuous random variables.
II. *The Normal Probability Distribution*
   1. Symmetric about its mean μ
   2. Shape determined by its standard deviation σ
III. *The Standard Normal Distribution*
   1. The normal random variable z has mean 0 and standard deviation
      1.
   2. Any normal random variable x can be transformed to a standard
      normal random variable using

$$z = \frac{x - \mu}{\sigma}$$

   3. Convert necessary values of x to z.
   4. Use Table 3 in Appendix I to compute standard normal
      probabilities.
   5. Several important z-values have tail areas as follows:

Tail Area	.005	.01	.025	.05	.10
z-Value	2.58	2.33	1.96	1.645	1.28

Exercises

1. Find the following probabilities for the standard normal variable z:
   a. $P(z < 1.9)$.
   b. $P(1.21 < z < 2.25)$.
   c. $P(z > -.6)$.
   d. $P(-2.8 < z < 1.93)$.
   e. $P(-1.3 < z < 2.3)$.
   f. $P(-1.62 < z < .37)$.

2. Find the value of z, say z_0, such that the following probability
   statements are true:
   a. $P(z > z_0) = .2420$.
   b. $P(-z_0 < z < z_0) = .9668$.
   c. $P(-z_0 < z < z_0) = .90$.
   d. $P(z < z_0) = .9394$.

3. If x is distributed normally with mean 25 and standard deviation
   4, find the following:
   a. $P(x > 21)$.
   b. $P(x < 30)$.
   c. $P(15 < x < 35)$.
   d. $P(x < 18)$.

4. A psychological introvert-extrovert test produced scores that had
   a normal distribution with mean and standard deviation 75 and 12,
   respectively. If we wish to designate the *highest* 15% as extrovert,
   what would be the proper score to choose as the cutoff point?

5. A manufacturer's process for producing steel rods can be regulated so as to produce rods with an average length μ. If these lengths are normally distributed with standard deviation of .2 inch, what should be the setting for μ if one wants at most 5% of the steel rods to have a length greater than 10.4 inches?

6. For a given type of cannon and a fixed range setting, the distance that a shell fired from this cannon will travel is normally distributed with a mean and standard deviation of 1.5 and .1 miles, respectively.
   a. What is the probability that a shell will travel farther than 1.72 miles?
   b. What is the probability that a shell will travel less than 1.35 miles?
   c. What is the probability that a shell will travel at least 1.45 miles but at most 1.62 miles?
   d. If three shells are fired, what is the probability that all three will travel farther than 1.72 miles?

7. In investigating the marital status of adults (persons 14 years and older) in the United States, the U.S. Census Bureau reports that in a recent year, 63.4 percent of all males were married, while 29.3 percent were single, 2.5 percent were widowed, and 4.8 percent were divorced. Suppose for simplicity, that 30% of all males were single in this year. If a random sample of $n = 20$ men were taken in this year, what is the probability that 10 or more are single?
   a. Use the binomial tables, Table I, the Appendix.
   b. Use the normal approximation to the binomial.

8. The survival rate for a certain type of cancer is 30%. A random sample of $n = 50$ individuals with this type of cancer were included in a study of a new drug that is hoped will provide a cure for this disease. If the new drug is ineffective, what is the probability that more than 50% of the individuals in the study are cancer-free survivors after five years? Based upon this probability, if 26 or more of the individuals in this study survived cancer-free for five years, what would you conclude about the survival rate for individuals who are treated for this type of cancer with this drug?

9. A preelection poll taken in a given city indicated that 40% of the voting public favored candidate A, 40% favored candidate B, and 20% were as yet undecided. If these percentages are true, in a random sample of 100 voters, what are the following probabilities:
   a. That at most 50 voters in the sample prefer candidate A?
   b. That at least 65 voters in the sample prefer candidate B?
   c. That at least 25 but at most 45 voters in the sample prefer candidate B?

10. On a well-known college campus, the student automobile registration revealed that the ratio of small to large cars (as measured by engine displacement) is 2 to 1. If 72 car owners are chosen at random from the student body, find the probability that this group includes at most 46 owners of small cars.

11. In introducing a new breakfast sausage to the public, an advertising campaign claimed that 7 out of 10 shoppers would prefer these new sausages over other brands. Suppose 100 people are randomly chosen, and the advertiser's claim is true.

    a. What is the probability that at most 65 people prefer the new sausages?

    b. What is the probability that at least 80 people prefer the new sausages?

    c. If only 60 people stated a preference for the new sausages, would this be sufficient evidence to indicate that the advertising claim is false and that, in fact, less than 7 out of 10 people would prefer the new sausages?

Chapter 7
Sampling Distributions

7.1 INTRODUCTION

The last several chapters dealt with probability and probability models. In the study of probability we know the values of population **parameters** such as the mean μ and standard deviation σ, or perhaps we know the value of p to be used for calculating binomial probabilities, or the value of the mean for a Poisson distribution. In the application of probability to the problem of making inferences about a population in terms of its parameters, we are usually able to decide which probability model should be used. Biological variables such as heights and weights generally follow a normal distribution, while the results of sample surveys and polls are based upon the binomial distribution. In many cases, however, the values of the parameters that will specify the exact form of the distribution may not be known. When this is the case, you must rely on the sample to provide as much information as possible about these unknown parameters. The sample mean and standard deviation provide approximate values of μ and σ, and the sample proportion provides information about the underlying value of p. The reliability of this information, however, will depend upon how you select your sample.

7.2 SAMPLING PLANS AND EXPERIMENTAL DESIGNS

The way a sample is obtained is called the sampling plan or the experimental design, and determines the quality and quantity of information in the sample. In addition, by knowing the sampling plan, you can then determine the probability of observing specific samples. It is these probabilities that allow us to assess the reliability or goodness of

inferences

_____ based upon these samples.

Several types of random samples are available for us in a particular situation, depending on the scope of the experiment and the objectives of the experimenter. A commonly used and uncomplicated sampling plan is called a *simple random sample*. For example, suppose that we wish to select a sample of size two from a population of six elements, identified by x_1, x_2, x_3, ..., x_6. Then there are 15 possible samples consisting of two elements.

Sample	Observations in Sample	Sample	Observations in Sample
1	x_1, x_2	9	x_2, x_6
2	x_1, x_3	10	x_3, x_4
3	x_1, x_4	11	x_3, x_5
4	x_1, x_5	12	x_3, x_6
5	x_1, x_6	13	x_4, x_5
6	x_2, x_3	14	x_4, x_6
7	x_2, x_4	15	x_5, x_6
8	x_2, x_5		

If each of these 15 samples has the same chance of selection, then the resulting sample would be called a **simple random sample**, or simply a

_____.   random sample

A **simple random sample** of size n is said to have been drawn if each possible sample of size n in the population has the same chance of being selected.

Although perfect random sampling is difficult to achieve in practice, there are several methods available for selecting a sample that will satisfy the conditions of random sampling when N, the population size, is not too large.

1. *Method A*: List all the possible samples and assign them numbers. Write each of these numbers on a chip or piece of paper and place them in a bowl. Drawing one number from the bowl will select the random sample to be used.
2. *Method B*: Number each of the N members of the population. Write each of these numbers on a chip or piece of paper and place them in a bowl. Now draw n numbers from the bowl and use the members of the population that have these numbers as elements to be included in the sample.
3. *Method C*: A useful technique for selecting random samples is one in which a table of random numbers is used to replace the chance device of drawing chips from a bowl.

Example 7.1
A medical technician needs to choose four animals for testing from a cage that contains six animals. How many samples are available to the technician? List these samples.

Solution

The number of ways to choose four animals from a total of six is 15. Designating each animal by a number from 1 to 6, we list the samples as:

1, 2, 5, 6

(1, 2, 3, 4)	(_____)	(2, 3, 4, 5)
(1, 2, 3, 5)	(1, 3, 4, 5)	(2, 3, 4, 6)
(1, 2, 3, 6)	(1, 3, 4, 6)	(2, 3, 5, 6)
(1, 2, 4, 5)	(1, 3, 5, 6)	(2, 4, 5, 6)
(_____)	(1, 4, 5, 6)	(_____)

1, 2, 4, 6; 3, 4, 5, 6

A simple random sampling plan for this experiment would allow each of these 15 possible samples an equal chance of being selected— namely, 1/15.

A table of random numbers contains blocks of numbers consisting of the ten digits 0, 1, 2, ..., 9. Throughout the table the frequency of each of the digits is about 0.1. In addition, the frequency of every two-digit sequence such as 21 or 56 is about 0.01, the frequency of every three-digit number if about 0.001, and so on. A portion of the random number table taken from your text, is reproduced next.

01563	02011	81647
25595	85393	30995
22527	97265	76393
06243	61680	07856
81837	16656	06121

Table 7.1

If you needed random numbers from 1 to 10, you could select one-digit random numbers beginning with 1, 2, ..., 9 and associate 0 with 10. If you are sampling without replacing an individual already selected, then repeated numbers are not used. We will demonstrate how to use random numbers in selecting random samples.

Example 7.3

How would you proceed if you were interested in obtaining a random sample of $n = 10$ tax returns for audit from a list of 100 returns that were received yesterday?

Solution

1. First you must have or produce a numbered list of the returns to be sampled. In this case, each return would have a number from 1 to 100. The first return would be associated with the random number 01, the second with 02, and so on, with the last or 100th tax return associated with the random number _____.

00

2. We can select two-digit random numbers beginning at a random point in the table. Suppose that we decided in advance to begin in row 1 and column 6. In Table 7 this would mean that you would begin with the first digit in row one of the second block of

five digits. Reading to the right, using *pairs of digits*, you find
the numbers

02 01 18 16 47 25 59 58 53 93 30 99 52 25 27 97 26
57 63 93

and so on.

3. Your sample of ten individuals would consist of those on the list
with numbers

> 02, 01, 18, 16, 47, 25, 59, 58, 53, and 94

or in order,

> 1, 2, 16, 18, 25, 47, 53, 58, 59, and _____.   93

Notice that the random number 25 appears twice in the original
list. If we had encountered 25 a second time before reaching ten
numbers, we would have skipped the second 25 and listed the next
number which was 27.

The situation described in Example 7.2 is called an **observational
study** because the data existed before we decided to observe or describe its characteristics. Sample surveys conducted by the Gallop or
Roper Polls are actually observational studies. Observational studies
very often suffer from the following common problems.

- **Nonresponse.** When individuals fail to return questionnaires or
  refuse to participate in telephone interviews, you are confronted
  with the problem that is called nonresponse. Nonresponse may
  cause **bias** in your results if only those who agree to respond
  have very strong feelings about the subject of the survey.

- **Undercoverage.** When the list of possible participants systematically fails to list one or more segments of the population,
  then the resulting sample will also fail to include these segments
  of the population. This is referred to as undercoverage. Voter
  registration lists (do, do not) in general contain a thorough list   do not
  of all potential voters, and lists of drivers obtained from state
  records systematically excludes non-drivers. The purpose of the
  survey will certainly allow you to determine whether your list
  of potential participants suffers from undercoverage.

- **Wording Bias.** There is definitely an art to phrasing questions so
  as to be clear and non-obtrusive to the individual being surveyed.
  However, it is possible that sensitive subjects cannot be avoided
  if you wish to obtain desired information. For example, the question "Have you ever smoked marijuana?" or "Have you ever been
  arrested?" may not be answered truthfully by respondents.

One or more of these problems with a survey will certainly introduce
bias into the results, and your conclusions may not be valid, even
though you began with what was thought to be a random sample.

Other types of research involve experimentation in which certain factors, called **treatments**, such as diet, exercise, or perhaps a regimen of medication are imposed or applied to experimental units. Simple random sampling is more difficult to implement in these situations.

Example 7.3

An experimenter wishes to determine an appropriate temperature at which to store fresh strawberries to minimize the loss of ascorbic acid. There are 20 storage containers, each with controllable temperature, in which strawberries can be stored. If two storage temperatures are to be used, how would the experimenter assign the 20 containers to one of the two storage temperatures?

Solution

1. The experimenter would begin by numbering the storage containers from 1 to _____. The next step requires a set of random numbers, which can be gotten from a random number table, a statistical package or perhaps from your calculator. We will use the following random numbers generated using Minitab's random data option under the **C**alc pull-down menu.

20

| 94 | 45 | 42 | 68 | 58 | 71 | 14 | 3 | 9 | 41 |
| 74 | 76 | 88 | 52 | 29 | 56 | 5 | 30 | 6 | 60 |

2. One way to make use of every two-digit number is to assign each storage container more than one number. A simple assignment could be the following.

Container Number	Random Numbers
1	1, 21, 41, 61, 81
2	2, 22, 42, 62, 82
3	_____
⋮	⋮
19	19, 39, 59, 79, 99
20	20, 40, 60, 80, 00

3, 23, 43, 63, 83

This assignment gives each container the same chance of being drawn, namely $5/100 =$ _____.

1/20

3. The 10 containers for use with the first temperature setting are:

14, 15, 2, 8, 18, 11, 3, 9, 1, 16.

Notice that container 14 is associated with the random numbers 14, 34, 54, 74, and 94. The numbers 94, 14 and 74 all appear in the list, but 14 is selected only once when 94 appears as the first random number in the list. The remaining containers will be assigned to the second temperature setting group.

In many situations the population of interest consists of one or more subpopulations. Workers within a given industry may be grouped into several natural job categories; business firms or cities may be grouped according to size; areas may be classified as urban, suburban, or rural. In such cases, it is desirable to have each subpopulation represented in the sample.

A **stratified random sample** is a sample obtained by dividing a population into nonoverlapping subpopulations called **strata** and then selecting a simple random sample within each stratum.

In order to select a simple random sample or a stratified random sample, an investigator must have available a frame listing all of the elements in the population to be sampled. When an appropriate frame is not available or is very costly to obtain, a _____ sampling design can be used.

cluster

A **cluster sample** is a simple random sample in which the sampling units are collections or clusters of elements in the population. A cluster sample is obtained by randomly selecting m clusters from the population and then conducting a complete census within each cluster.

In sampling households in a given city, an appropriate cluster of households might be a city block or a political ward. In sampling airline passengers, an appropriate cluster would be an arriving or departing plane load. In sampling students within a large school, an appropriate cluster would be a classroom.

In some instances the population to be sampled is ordered in some way. For example, you may have an alphabetized list of patients, a time-ordered list of service calls for a computer facility, or a list of registered voters. A cost-effective sampling plan in situations such as these is to choose one element at random from the first k elements and then systematically select every kth element in the population.

A **1-in-k systematic sample** involves the random selection of one of the first k elements in an ordered population, and then the systematic selection of every kth element thereafter.

In addition to random sampling designs, some nonrandom sampling designs are commonly used. However, with nonrandom sampling designs, only _____ statements can be made. Convenience

descriptive

sampling, judgment sampling, and quota sampling are three forms of nonrandom sampling.

1. A *convenience sample* consists of elements that can easily be obtained. A group of volunteer subjects (<u>would</u>, would not) make up a convenience sample.

would

2. *Judgment sampling* involves the selection of the elements in the sample by "experts" so that the sample is "representative" of the _____ of interest. For example, one city in the United States might be picked as a typical city to represent all the cities in the United States.

population

3. *Quota sampling* involves the selection of a predetermined number of elements from specific portions of the population so as to construct a sample _____ to the population with respect to certain variables. National opinion polls might rely on quota sampling to insure that given ethnic, socioeconomic, religious, political, and other groups are represented in the sample in roughly the same proportions that they appear in the

proportional

_____.

population

Why is it so important that the sample be randomly drawn? From the practical point of view, one would want to keep the experimenter's biases out of the selection and, at the same time, keep the sample as representative of the _____ as possible. From the statistical point of view, we can assess the probability of observing a random sample and hence make valid _____ about the parent population. If the sample is nonrandom, its probability (can, <u>cannot</u>) in general be determined and hence no valid inferences can be made from it.

population

inferences

cannot

Self-Correcting Exercises 7A

1. An auditing firm has been hired by a company to examine its accounts receivable. The company has 100 current accounts and the auditing firm proposes to examine 20 of those accounts. Explain how you would randomly select 20 accounts from the 100 accounts using a random number table or a set of random numbers generated using your calculator or using a statistical package such as Minitab.

2. Wishing to investigate employee satisfaction with regard to company fringe benefits, the management of a company employing 250 workers proposes to survey 30 workers randomly chosen from among its 250 workers. Provide an efficient selection scheme for randomly choosing the 30 workers to be interviewed.

3. A psychology professor has received a grant to examine the association of words occurring within a given distance of each other in everyday prose writing. One way of attracting students to participate in this study would be to advertise for volunteers with

an offer to pay $10 an hour for participants. Would you consider students selected using this mechanism to comprise a random sample? Why or why not?

4. An investigator has sufficient funds to include 50 wholesale dealers of farm implements in a products-liability study. Of the 100 dealers available, 600 have annual sales of less than $5 million, 300 have annual sales between $5 and $25 million, and the remaining 100 have annual sales exceeding $25 million. Would a simple random sampling plan be appropriate in this situation? Why? If not, what type of sampling plan would you suggest using?

Solutions to SCE 7A

1. Assign random numbers 01 through 99 to accounts no. 1 through no. 99 and assign random number 00 to account no. 100. Randomly select 20 random numbers and sample the associated accounts.

2. a. Using three digit numbers 001 through 250 to identify the $N = 250$ workers, we could use the random digits 001 through 250 to identify the workers to be included in the sample, and for any three-digit random number in the range 251 through 999 use its remainder upon division by 250. For example, worker number 250 would be associated with the random digits 250, 500, 750 and 000.

   b. Suppose a random starting point was determined as line 66, column 5 of the random number table in your text. The first three digits of lines 66-70 in columns 5-10 are

294	218	150	345	333	061
173	376	470	420	974	486
058	248	869	603	164	032
844	605	793	934	688	254
379	610	439	152	806	439

   The remainders after division by 250 are

44	218	150	95	83	61
173	126	220	170	224	236
58	248	119	103	164	32
94	105	43	184	188	4
129	110	189	152	56	189

   Since 189 appeared twice, the next 3 digit number, 952, which reduces to 202, is included. Therefore, the 30 workers associated with the above numbers are to be included in the sample.

3. Although the mechanism used by the professor does not constitute a true random sample, the students selected using this method will probably *behave like* a random sample. The fact that they are being paid to participate should not cause their responses to the

experiment to be substantially different from the responses of a randomly selected group of students.

4. Because of the clear stratification in the $N = 100$ dealers according to annual sales, a stratified random sample would be preferred to a simple random sample. By randomly choosing a set number of dealers within each strata, you can guarantee that each annual sales group will be represented in the sample.

7.3 SAMPLING DISTRIBUTIONS OF STATISTICS

In making inferences about a population based on information contained in a sample, we will use sample statistics to estimate and/or make decisions about population _____.

parameters

Notice that each sample drawn from a population of interest to the experimenter will contain different elements. Hence, the value of a statistic (such as $\bar{x}$ or s^2) will (change, remain the same) from sample to sample. If one were to draw repeated samples of a constant sample size, many different values of the sample statistic would be obtained. These values could be used to create a relative frequency distribution to describe the behavior of the statistic in repeated sampling.

change

Since a sample statistic takes on many different numerical values depending on the outcome of a random sample, it is classified as a _____ _____ and, as such, has a probability distribution associated with it. This probability distribution can be approximated by the relative frequency distribution described above.

random variable

The probability distribution for a sample statistic is called its **sampling distribution**. The sampling distribution results when random samples of size n are repeatedly drawn from the population of interest.

Example 7.4
Consider a population of $N = 6$ elements whose values are $x = 3$, 3.5, 3.5, 4, 4, and 6. If each of the population values is equally likely to be selected in a single random selection, construct the probability distribution for x. Find the mean and the variance of x.

Solution
1. Since each of the $N = 6$ elements has an equal chance of being selected, each has probability $1/N = $ _____. However, some of the elements are identical. Thus,

1/6

$$p(3) = \underline{\hspace{2cm}}$$

1/6

but

$p(3.5) = \frac{1}{6} + \underline{\hspace{2cm}} = \underline{\hspace{2cm}}$

1/6; 1/3

The probability distribution for x is shown below. Fill in the missing entries.

x	$p(x)$
3	1/6
3.5	___
4	1/3
6	___

1/3

1/6

2. Graph the probability distribution for x as in Figure 7.1. The distribution (is, is not) symmetrical.

is not

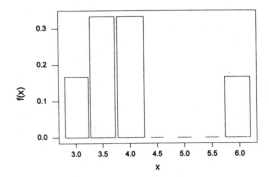

Figure 7.1

3. The mean value of x is calculated as in Chapter 3:

$$\mu = E(x) = \sum_{x} x p(x)$$

$$= 3\left(\frac{1}{6}\right) + 3.5\left(\frac{1}{3}\right) + 4\left(\frac{1}{3}\right) + 6\left(\frac{1}{6}\right)$$

$$= \underline{\hspace{2cm}}$$

4.0

The variance of x is

$$\sigma^2 = \sum (x - \mu)^2 p(x) = \sum x^2 p(x) - \mu^2$$

$$= 3^2\left(\frac{1}{6}\right) + (3.5)^2\left(\frac{1}{3}\right) + 4^2\left(\frac{1}{3}\right) + 6^2\left(\frac{1}{6}\right) - (4.0)^2$$

$$= \underline{\hspace{2cm}} - 16$$

16.9166667

$$= \underline{\hspace{2cm}}$$

.9166667

Example 7.5
Find the sampling distribution for the sample mean $\bar{x}$ when a random sample of size $n = 2$ has been drawn from the population given in Example 7.4.

Solution

When $n = 2$ observations are drawn from $N = 6$, there are $C_2^6 =$ _____ possible samples that can be drawn with equal probability. Since some of the elements in this sample are identical, let us identify the repeated measurements as 3.5 and 3.5*, 4 and 4*. The 15 possible samples are listed below. Fill in the missing entries in column 2.

Sample	Sample Values	$\bar{x}$
1	3, 3.5	3.25
2	3, 3.5*	3.25
3	3, 4	_____
4	3, 4*	3.50
5	3, 6	4.50
6	3.5, 3.5*	3.50
7	3.5, 4	_____
8	_____	3.75
9	3.5, 6	4.75
10	3.5*, 4	3.75
11	_____	3.75
12	3.5*, 6	_____
13	4, 4*	4.00
14	4, 6	_____
15	_____	5.00

For each possible sample, the sample mean can be different. For example, sample 1 has mean

$$\bar{x} = \frac{3 + 3.5}{2} = 3.75$$

whereas for sample 9,

$$\bar{x} = \frac{3.5 + 6}{\_\_\_\_} = _____$$

Fill in the missing entries in column 3 of the sample table above. The value of $\bar{x}$ associated with each sample occurs with probability $1/15$. The sampling distribution for $\bar{x}$ is shown below and graphed in Figure 7.2.

$\bar{x}$	$p(\bar{x})$
3.25	2/15
3.50	3/15
3.75	4/15
4.00	_____
4.50	1/15
_____	2/15
5.00	2/15

Margin answers:

15

3.50

3.75
3.5, 4*

3.5*, 4*
4.75

5.00
4*, 6

2; 4.75

1/15

4.75

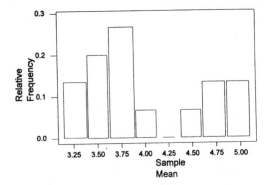

Figure 7.2

Notice several things in comparing the original population distribution (Figure 7.1) to the sampling distribution of $\bar{x}$ (Figure 7.2).

1. The distribution of $\bar{x}$ is (more, <u>less</u>) variable than the original distribution.

      less

2. The distribution of $\bar{x}$ appears (more, <u>less</u>) skewed than the original distribution.

      less

3. The mean of $\bar{x}$ is

$$\mu_{\bar{x}} = 3.25\left(\frac{2}{15}\right) + 3.5\left(\frac{3}{15}\right) + \cdots + 5\left(\frac{2}{15}\right)$$

$$= 4.0$$

which is identical to the mean of the original population.

4. The variance of $\bar{x}$ is

$$\sigma_{\bar{x}}^2 = (3.25)^2\left(\frac{2}{15}\right) + (3.5)^2\,(\underline{\hspace{2cm}}) + (3.75)^2\left(\frac{4}{15}\right) + \cdots$$

      3/15

$$+\, 5^2\left(\frac{2}{15}\right) - (\underline{\hspace{2cm}})^2$$

      4.0

$$= \underline{\hspace{2cm}} - 16 = \underline{\hspace{2cm}}$$

      16.36667; .36667

The variance of $\bar{x}$ is (<u>smaller</u>, larger) than the variance of x, confirming our earlier observation.

      smaller

    Sampling distributions can be constructed for any finite population using this same method. For example, the experimenter might be interested in the sampling distribution of the sample median, the sample variance, s^2, or the sample standard deviation, s. For each possible sample, he calculates the value of the sample statistic and then uses the results to tabulate the appropriate sampling distribution and to describe the behavior of that statistic in repeated sampling. However, as N increases, it becomes extremely difficult to actually enumerate each sample, unless it can be done empirically using a computer. For this reason, we turn to results proven using algebra or higher mathematics to derive the sampling distributions for several commonly used statistics. It can be shown that some of these statistics have desirable properties that will be useful in inference making.

7.4 THE CENTRAL LIMIT THEOREM

Consider a meat processing center that as part of its output prepares 1-pound packages of bacon. If the weights of the packaged bacon were carefully checked, some weights would be slightly heavier than 16 ounces, while others would be slightly lighter than 16 ounces. A frequency histogram of these weights would probably exhibit the mound-shaped distribution characteristic of a normally distributed random variable. Why should this be the case? One can think of the weight of each package as differing from 16 ounces because of an error in the weighing process, a scale that needs adjustment, the thickness of the slices of bacon so that the package contains either one more or less slice than it should, or perhaps the melting of some of the fat. Hence, any one weight would consist of an average weight (preferably 16 ounces) modified by the addition of random errors that might be either positive or negative.

The Central Limit Theorem, loosely stated, says that sums or averages are approximately normally distributed with a mean and standard deviation that depend on the sampled population. If one considers the error in the weight of a 1-pound package of bacon as a *sum* of various effects in which small errors are highly likely and large errors are highly improbable, then the Central Limit Theorem helps explain the apparent normality of the package weights.

Equally as important, the Central Limit Theorem assures us that sample means will be approximately normally distributed with a mean and variance that depend on the population from which the sample has been drawn. This aspect of the Central Limit Theorem will be the focal point for making inferences about populations based on random samples when the sample size is large.

The Central Limit Theorem (1): If random samples of n observations are drawn from a population with finite mean μ and standard deviation σ, then when n is large, the sample mean $\bar{x}$ will be approximately normally distributed with mean μ and standard deviation $\sigma/\sqrt{n}$.

The quantity $\sigma/\sqrt{n}$ is sometimes called the **standard error** of $\bar{x}$. The Central Limit Theorem could also be stated in terms of the sum of the measurements, $\sum x_i$.

The Central Limit Theorem (2): If random samples of n observations are drawn from a population with finite mean μ and standard deviation σ, then when n is large, $\sum x_i$ will be approximately normally distributed with mean $n\mu$ and standard deviation $\sigma\sqrt{n}$.

In both cases, the approximation to normality becomes more and more accurate as n becomes large.

The Central Limit Theorem is important for two reasons:

1. It partially explains *why* certain measurements possess approximately a _____ distribution.

2. Many of the _____ used in making inferences are sums or means of sample measurements and thus possess approximately _____ distributions for large samples. Notice that the Central Limit Theorem (does, does not) specify that the sample measurements come from a normal population. The population (could, could not) have a frequency distribution that is flat or skewed or is nonnormal in some other way. It is the *sample mean* that behaves like a random variable that has an approximately normal distribution.

To clarify a point, we note that the sample mean $\bar{x}$ computed from a random sample of n observations drawn from any infinite population with mean μ and standard deviation σ always has a mean equal to μ and a standard deviation equal to $\sigma/\sqrt{n}$. This result is not due to the Central Limit Theorem. The important contribution of the theorem lies in the fact that when n, the sample size, is *large*, we may approximate the distribution of $\bar{x}$ with a *normal* probability distribution.

The Central Limit Theorem relies heavily on the assumption that the sample size n is large. The question of how large n must be is a difficult one to answer and depends on the characteristics of the underlying population from which we are sampling. Many texts use the rule of thumb that allows application of the Central Limit Theorem if $n > 30$. However, this rule will not always work. If the underlying distribution is (symmetric, skewed), the Central Limit Theorem may be may be appropriate for $n < 30$. However, if the underlying population is (symmetric, heavily skewed), the distribution of $\bar{x}$ may still be skewed even if $n > 30$. The student will need to use judgment to determine the approximate shape of the underlying population from which he or she is sampling in order to determine whether the Central Limit Theorem will be appropriate. For specific types of applications given in the text, we will provide the sample sizes necessary to ensure the applicability of the Central Limit Theorem.

normal

estimators

normal
does not

could

symmetric

heavily skewed

7.5 THE SAMPLING DISTRIBUTION OF THE SAMPLE MEAN

There are several estimators of the population mean μ that are available, each of which under certain circumstances might be preferred over the others. For example, when the population sampled is symmetric, you might expect the sample mean, the sample median and the sample mode to be reasonable estimators of the population mean μ. When selecting an estimator from two or more candidates, there are some questions that you might ask.

- Is this estimator easy to calculate?

- On the average, does the estimator consistently overestimate or underestimate the actual value of μ?

- Is the variability of this estimator greater or smaller than the other candidates?

The sample mean $\bar{x}$ can be shown to have more desirable properties when compared to the sample median or the sample mode. You will find that it is the most commonly used estimator of the population mean μ. What are the properties of the sampling distribution of the sample mean $\bar{x}$?

The Sampling Distribution of the Sample Mean $\bar{x}$

1. If random samples of n measurements are selected from a population with mean μ and standard deviation σ, the sampling distribution of the sample mean $\bar{x}$ will have a mean equal to

$$\mu_{\bar{x}} = \mu$$

   and a standard deviation equal to

$$\sigma_{\bar{x}} = \frac{\sigma}{\sqrt{n}}$$

2. If the sampled population has a *normal* distribution, then the sampling distribution of $\bar{x}$ will be *exactly* normally distributed for *all* sample sizes.

3. If the sampled population is not normal, the sampling distribution of $\bar{x}$ will be approximately normal for large samples according to the Central Limit Theorem. When the sampled population is mound-shaped and approximately normal, the sampling distribution of $\bar{x}$ will be approximately normal for sample sizes as small as $n \geq 25$, or more likely $n \geq 30$.

The standard deviation of a statistic used as an estimator is also called the **standard error of the estimator** since it measures the precision of the estimator. Here and in the text, we will often refer to

$\sigma_{\bar{x}} = \dfrac{\sigma}{\sqrt{n}}$ or its estimate as the _____ of the mean or simply   standard error
as SE or SEM.

Example 7.6
A production line produces items whose mean weight is 50 grams with
a standard deviation of 2 grams. If 25 items are randomly selected
from this production line, what is the probability that the sample mean
$\bar{x}$ exceeds 51 grams?

Solution
The underlying population from which we are sampling is the population of weights of items on the production line. Since weights tend to
have a mound-shaped distribution, the sampled population is
(approximately symmetric, heavily skewed). Thus, the Central Limit   approx. symmetric
Theorem (will, will not) be appropriate.   will
     According to the Central Limit Theorem, the sample mean $\bar{x}$ is ap-
proximately normally distributed with mean μ and standard deviation
$\sigma/\sqrt{n}$. For our problem, $\mu = 50$, $\sigma = 2$, and $n = 25$; hence,

$$\sigma_{\bar{x}} = \frac{\sigma}{\sqrt{n}} = \frac{2}{\sqrt{25}} = \underline{\hspace{2cm}}$$ .4

Therefore,

$$P(\bar{x} > 51) = P(\frac{\bar{x} - 50}{.4} > \frac{51 - 50}{.4}) = P(z > 2.5$$

$$= 1 - A(2.5)$$

$$= 1 - .9938 = \underline{\hspace{2cm}}.$$ .0062

Example 7.7
A bottler of soft drinks packages cans in six-packs.

a. If the fill per can has a mean of 12 fluid ounces and a standard
    deviation of .2 fluid ounce, what is the distribution of the total
    fill for a case of 24 cans?
b. What is the probability that the total fill for a case is less than
    286 fluid ounces?

Solution
The population from which we are sampling is the population of fills per
can, which (should, should not) be an approximately symmetric distribu-   should
tion. Therefore, for $n = 24$, the Central Limit Theorem (will, will not)   will
be applicable.

a. Using the Central Limit Theorem in its second form, we find the
    total fill per case has a mean of $n\mu = 24(12) = \underline{\hspace{1.5cm}}$ fluid   288
    ounces and a standard deviation (or standard error) of $\sigma\sqrt{24}$
    $= .2\sqrt{24} = \underline{\hspace{1.5cm}}$ fluid ounce. The total fill is   .98

normally

approximately _____ distributed with mean 288 and
standard deviation .98.

b. Let T represent the total fill per case. We wish to evaluate
$P(T < 286)$. Since T is approximately normally distributed with

288; .98

$\mu_T = $ _____ and $\sigma_T = $ _____,

$$P(T < 286) = P(\frac{T - 288}{.98} < \frac{286 - 288}{.98}) = P(z < -2.04)$$

$$= A(-2.04)$$

.0207

$$= \underline{\hspace{2cm}}$$

large

The Central Limit Theorem applies when a sufficiently (large,
small) sample is randomly drawn from a very large or infinite
population, regardless of its shape. However, if the population
from which we are sampling is itself normal, the sampling distri-

normal

bution of $\bar{x}$ will be _____, regardless of the size of the
sample.

If a random sample is drawn from a *normal population* with
mean μ and variance σ^2, the sampling distribution of $\bar{x}$ will
be normal with mean μ and variance σ^2/n, *regardless of the
sample size.*

Example 7.8
Suppose that the bottler of soft drinks in Example 7.7 packages cans
whose fill per can is normally distributed with mean 12 fluid ounces
and standard deviation .2 fluid ounce. If a six-pack of soda can be
considered a random sample of size $n = 6$ from the population, what
is the probability that the average fill per can is less than 11.5 fluid
ounces?

Solution

normal
exactly

1. Since we are sampling from a _____ population, the sampling distribution of $\bar{x}$ will be (approximately, exactly) normal

with mean $\mu = 12$ fluid ounces and variance $\sigma^2/n = .2/6 = $

.033333
$\bar{x} < 11.5$

_____.

2. We wish to evaluate $P($_____$)$, or

$$P(\bar{x} < 11.5) = P(\frac{\bar{x} - \mu}{\sigma/\sqrt{n}} < \frac{11.5 - 12}{\sqrt{.0333333}})$$

-2.74

$$= P(z < \underline{\hspace{2cm}})$$

.0031

$$= \underline{\hspace{2cm}}.$$

Self-Correcting Exercises 7B

1. A pharmaceutical company is experimenting with rats to determine if there is a difference in weights for rats fed with and without a vitamin supplement. A frequency distribution of these weights (either with or without the supplement) would probably exhibit the mound-shaped distribution characteristics of a normally distributed random variable. Why should this be the case?

2. An agricultural economist is interested in determining the average diameter of peaches produced by a particular tree. A random sample of $n = 30$ peaches is taken and the sample mean $\bar{x}$ is calculated. Suppose that the average diameter of peaches on this tree is known from previous years' production to be $\mu = 60$ millimeters with $\sigma = 10$ mm. What is the probability that the sample mean, $\bar{x}$, exceeds 65 millimeters?

Solutions to SCE 7B

1. The weights of the rats will result from a combination of random factors, such as initial weight of the rat, genetic make-up of the rat, amount of food consumed, and so on. Hence, the weights behave as a sum of independent random variables, and as such would be normally distributed according to the Central Limit Theorem.

2. Since $\bar{x}$ is normally distributed with mean $\mu = 60$ and with standard deviation $\sigma/\sqrt{n} = 10/\sqrt{30} = 1.8257$, the probability of interest is

$$P(\bar{x} > 65) = P(z > \frac{65 - 60}{1.8257}) = P(z > 2.75)$$

$$= 1 - A(2.75) = 1 - .9970 = .0030$$

7.6   THE SAMPLING DISTRIBUTION OF THE SAMPLE PROPORTION

A statistic that is often used to describe a binomial population is $\hat{p}$, the *proportion* of trials in which a success is observed. In terms of previous notation,

$$\hat{p} = \frac{x}{n} = \frac{\text{number of successes in } n \text{ trials}}{n}$$

Recall that, for a binomial population, the proportion of successes in the population is defined as p. The sample proportion, $\hat{p}$, will be used to estimate or make inferences about p.

For a binomial experiment consisting of n trials, let

$x_1 = 1$ if trial one is a success
$x_1 = 0$ if trial one is a failure

$x_2 = 1$ if trial two is a success
$x_2 = 0$ if trial two is a failure

$\vdots$

$x_n = 1$ if trial n is a success
$x_n = 0$ if trial n is a failure

Then x, the number of successes in n trials, can be thought of as a sum of n independent random variables. From this fact, the statistic

average
Central Limit
Theorem

$\hat{p} = x/n$ is equivalent to an _____ of these n random variables. The _____ _____ _____ assures us that $\hat{p}$ will be approximately normally distributed when n is large. The mean and standard deviation of $\hat{p}$ can be shown to be

$$\mu_{\hat{p}} = p \quad \text{and} \quad \sigma_{\hat{p}} = \sqrt{\frac{p(1-p)}{n}} = \sqrt{\frac{pq}{n}}$$

The Sampling Distribution of the Sample Proportion $\hat{p}$

1. If random samples of n observations are selected from a binomial population with parameter p, the sampling distribution of the sample proportion

   $\hat{p} = x/n$[1]

   has a mean equal to

   $\mu_{\hat{p}} = p$

   and a standard deviation or standard error equal to

   $\sigma_{\hat{p}} = \sqrt{\frac{pq}{n}}$

   for $q = 1 - p$.

2. When the sample size is large, the sampling distribution of $\hat{p}$ can be approximated by a normal distribution with the same mean and standard deviation. The approximation will be adequate if $np > 5$ and $nq > 5$ and will be good if $np > 7$ and $nq > 7$.

---

[1]A carat or "hat" placed over the symbol of a population parameter denotes a statistic used to estimate the population parameter. For example, the symbol $\hat{p}$ denotes the sample proportion.

Example 7.9

Past records show that at a given college, 20% of the students who began as economics majors either changed their major or dropped out of school. An incoming class has 110 beginning economics majors. What is the probability that at most 30% of these students will leave the economics program?

Solution

1. If p represents the proportion of students who leave the economics program, with the probability of losing a student given as $p = .2$, then the required probability for $n = 110$ is $P(\hat{p} \leq .30)$.

2. To use the normal approximation, we need

$$\mu_{\hat{p}} = p = \underline{\hspace{1.5cm}}$$ .2

$$\sigma_{\hat{p}} = \sqrt{\frac{p(1-p)}{n}} = \sqrt{\frac{.2(\underline{\hspace{1.5cm}})}{110}} = \sqrt{\underline{\hspace{2cm}}}$$ .8; .0014545

$$= \underline{\hspace{1.5cm}}$$ .0381

We now proceed to approximate the probability required in part 1 by using a normal probability distribution with a mean of \underline{\hspace{1.5cm}}   .2
and a standard deviation of \underline{\hspace{1.5cm}}.   .0381

3. Using the normal approximation, we find the value $\hat{p} = .30$ corresponds to a z-value of

$$z = \frac{\hat{p} - p}{\sigma_{\hat{p}}} = \frac{.30 - .20}{.0381} = \underline{\hspace{1.5cm}}$$ 2.62

and

$$P(\hat{p} \leq .30) \approx P(z < \underline{\hspace{1.5cm}})$$ 2.62

$$= \underline{\hspace{1.5cm}}.$$ .9956

In Chapter 6, we used the normal approximation to the binomial distribution to calculate probabilities associated with x, the number of successes in n trials. In that situation, it was necessary to have the interval $\mu \pm 2\sigma$, or $np \pm 2\sqrt{npq}$, fall within the binomial limits, 0 to n, or equivalently, when $np > 5$ and $nq > 5$ in order to ensure the accuracy of the approximation. Further, a correction for continuity was used to make the approximation more accurate. Hence, the standard normal random variable used in this earlier approximation was

$$z = \frac{\left(x \pm \frac{1}{2}\right) - np}{\sqrt{npq}} \approx \frac{x - np}{\sqrt{npq}}$$

If we divide both numerator and denominator of this fraction by n, we see the relationship between the sampling distributions of x and $\hat{p} = x/n$:

$$z = \frac{\left(\frac{x}{n} \pm \frac{1}{2n}\right) - p}{\sqrt{\frac{pq}{n}}} = \frac{\left(\hat{p} \pm \frac{1}{2n}\right) - p}{\sqrt{\frac{pq}{n}}} \approx \frac{\hat{p} - p}{\sqrt{\frac{pq}{n}}}$$

are

Hence, the sampling distributions of x and $\hat{p}$ (are, are not) equivalent. *The quantity $\pm 1/2n$ will be ignored for large values of n because the value of z changes very little.* The normal approximation will be appropriate if the interval $p \pm 2\sqrt{pq/n}$ falls within the limits 0 to 1 or when $np > 5$ and $nq > 5$.

Self-Correcting Exercises 7C

1. For a binomial experiment with $n = 20$ and $p = .5$, calculate $P(.8 \le \hat{p} \le .9)$ b
   a. Using the binomial tables, Table 1.
   b. Using the normal approximation to the sampling distribution of $\hat{p}$.

2. A controversial issue in the state of California is the diversion of water from northern to southern regions of the state. Suppose that 30% of the population favor the diversion, while 70% oppose it. If a random sample of $n = 50$ voters is taken, what is the probability that 50% or more favor the diversion? That is, what is the probability that the sample will show a majority in favor of the diversion when in fact only 30% of the population favor it?

Solutions to SCE 7C

1. a. $P(.8 \le \hat{p} \le .9) = P(16 \le x \le 18) = 1 - .994 = .006$ from Table 1.

    b.   Using the fact that $\mu_{\hat{p}} = p = .5$ and $\sigma_{\hat{p}} = \sqrt{\frac{pq}{n}} = \sqrt{\frac{.5(.5)}{20}} = .1118,$

    $$P(.8 \le \hat{p} < .9) = P(\frac{.8 - .5}{.1118} < z < \frac{.9 - .5}{.1118})$$
    $$= P(2.68 < z < 3.58) = 1 - .9963 = .0037.$$

2.   Let $p = P(\text{favor diversion})$. Then $p = .3$ and $n = 50$, while

    $$\mu_{\hat{p}} = p = .3 \text{ and } \sigma_{\hat{p}} = \sqrt{\frac{pq}{n}} = \sqrt{\frac{.3(.7)}{50}} = .0648. \text{ The probability of interest}$$

    is $P(\hat{p} > .50) = P(z > \frac{.50 - .30}{.0648}) = P(z > 3.09) = 1 - .9990 = .0010.$

7.7 A SAMPLING APPLICATION: STATISTICAL PROCESS CONTROL

The field of quality control is concerned with the _____ of products and with techniques for attaining consistently high quality in their production. Statistical quality control techniques are not limited to manufacturing environments but can be used to monitor any business activity that requires reliability and consistency in its goods or its services.

design

Every step in a manufacturing line or in a service-related business is referred to as a _____ with its own inputs and outputs. Producing a good or a service can be viewed as a series of related processes. The use of statistical quality control techniques is called _____ _____ _____.

process

Statistical Process Control (SPC)

Variation exhibited by variables consists of variation that is caused by changes in important process variables and by random variation in the process. Variation that is caused by variables such as machine wear, machine adjustments, and machine operators is referred to as (assignable, random) variation. Control charts are designed to separate assignable variation from random process variation. Using control charts is an example of "on-line" quality control, whereas experimental design procedures are examples of "off-line" quality control techniques.

assignable

If the variation in a process is solely random, the process is said to be *in control*. The first objective in statistical process control is to eliminate assignable causes of variation in the process variable and to get the process in control. The next step is to reduce variation and get the process within specification limits, which are the limits within which measurements on acceptable items must fall.

When a process is in control, process variables are monitored using control charts. Samples of size n are selected from the process at evenly spaced intervals of time and sample statistics are computed and plotted on a control chart so that sudden shifts or slow trends in the process variable can be detected and, it is hoped, corrected.

A Control Chart for the Process Mean: The $\bar{x}$ Chart

A control chart designed to monitor a measurable characteristic of a production process is the _____ chart. The $\bar{x}$ chart plots means of samples drawn from the production process at equally spaced points in time. It provides an ongoing check of average product quality.

$\bar{x}$

The control limits for the $\bar{x}$ chart are constructed from sample data gathered from the production process at a time when the process is known to be (in control, out of control). At least 25 different samples of size $n = 3$, 4, or 5 observations each are recommended for use in constructing the control charts.

in control

The construction of a quality control chart is based on the Central Limit Theorem and the Empirical Rule. When the production process is operating satisfactorily, whether the items are measurable or not, almost all of the sample values will oscillate within _____ standard deviations of the mean of the process values. Hence, it is

three

centerline
three

control limits

very unlikely that we would observe any sample values outside this band. The mean of the process values obtained from the process that is in statistical control forms what is called the _____ on the control chart. The bands located _____ standard deviations on each side of the centerline are called the upper and lower

_____ _____.

In an $\bar{x}$ chart, the centerline is located at $\bar{\bar{x}}$, the average of the sample means, given by

$$\bar{\bar{x}} = \frac{\bar{x}_1 + \bar{x}_2 + \cdots + \bar{x}_k}{k}$$

where $\bar{x}_1, \bar{x}_2, \ldots, \bar{x}_k$ are each based on n observations. From only sample information, the upper and lower control limits would be found using

$$UCL = \bar{\bar{x}} + \frac{3s}{\sqrt{n}} \quad \text{and} \quad LCL = \bar{\bar{x}} - \frac{3s}{\sqrt{n}}$$

where s is the sample standard deviation of all kn observations.

Example 7.9
A manufacturing process is designed to produce drill bits that are precisely .5 inch in diameter. The sample data that follow represent 25 samples selected from the production process, with four measurements in each sample. The sample measurements were obtained during a time when the production process was deemed to be in statistical control. Use the data to construct an $\bar{x}$ chart to monitor the production process. Does the process appear to be in statistical control?

	Measurement Number				Mean
Sample	1	2	3	4	$\bar{x}$
1	.501	.499	.496	.500	.499
2	.499	.497	.494	.498	.497
3	.503	.501	.502	.494	.500
4	.509	.511	.516	.508	.511
5	.507	.494	.505	.510	.504
6	.500	.503	.506	.491	.500
7	.488	.483	.501	.512	.496
8	.497	.499	.495	.501	.498
9	.492	.485	.480	.483	.485
10	.480	.480	.490	.498	.487
11	.496	.493	.493	.489	.490
12	.496	.506	.504	.506	.503
13	.503	.507	.504	.510	.506
14	.508	.502	.509	.509	.507
15	.505	.510	.515	.502	.508
16	.501	.503	.505	.499	.502
17	.503	.500	.497	.500	.500

| | Measurement Number | | | | Mean |
Sample	1	2	3	4	$\bar{x}$
18	.490	.500	.498	.496	.496
19	.502	.509	.523	.502	.506
20	.506	.509	.510	.507	.508
21	.501	.505	.498	.508	.503
22	.500	.497	.497	.502	.499
23	.491	.499	.498	.500	.497
24	.497	.487	.488	.492	.491
25	.489	.493	.503	.507	.498

Solution

The centerline of the $\bar{x}$ chart is located at the average of the sample means, which is given by

$$\bar{\bar{x}} = \frac{.499 + .497 + \cdots + .498}{25}$$

$$= \frac{\rule{4cm}{0.4pt}}{25} = \rule{3cm}{0.4pt}$$

12.491; .49964

To calculate the upper and lower control limits, we also need the value of $s = .008067$. Hence, with $n = 4$,

$$UCL = \bar{\bar{x}} + 3\left(\frac{s}{\sqrt{n}}\right)$$

$$= .49964 + 3\left(\frac{.008067}{\sqrt{4}}\right)$$

$$= .49964 + 3(.004033)$$

$$= .49964 + .012099 = .511739$$

and

$$LCL = \bar{\bar{x}} - 3\left(\frac{s}{\sqrt{n}}\right)$$

$$= .49964 - .012099 = .487541$$

The resulting $\bar{x}$ chart is shown in Figure 7.3

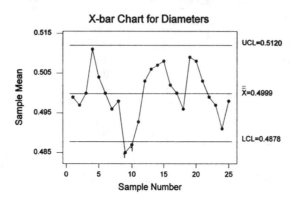

Figure 7.3

A Control Chart for the Proportion Defective: The p Chart

p chart

If the items of a continuous production process can be classified as either "good" or "defective," a _____ is used to monitor the process capabilities. Such items as flash bulbs, transistors, electron tubes, and dry cell batteries when operable can be almost guaranteed to possess uniformity in operating characteristics. However, because of the complexity of design of such items and the rapid pace of the production process, many such items cannot be guaranteed 100% operability in the production process.

statistical control

As in the construction of any control chart, the p chart should be constructed from sample data obtained during periods of time when the process is known to be in _____ _____. At least 100 sample items should be observed when the p chart is constructed but usually many more than 100 are observed.

proportion

binomial

Since the p chart is designed to control the _____ of defective units produced by a production process, the attribute sampling problem can be modeled by the _____ distribution. The centerline of the p chart is the fraction (proportion) of defective items observed from all samples selected when the process is known to be in statistical control and is denoted by $\overline{p}$. Thus,

$$\overline{p} = \frac{\text{total number of defectives in all samples}}{\text{total number of observations in all samples}}$$

$\sqrt{\frac{\overline{p}(1-\overline{p})}{n}}$

Since $\overline{p}$ estimates the true fraction defective of the process, we recall that the standard deviation of $\overline{p}$ is estimated by $\sigma_{\hat{p}} = $ _____, where n is the size of the sample selected at each sampling interval. The upper and lower control limits for the p chart are, as with the previous control charts, the _____ limits around the mean; that is,

3-sigma

$$UCL = \overline{p} + 3\sqrt{\frac{\overline{p}(1-\overline{p})}{n}}$$

and

$$LCL = \bar{p} - 3\sqrt{\frac{\bar{p}(1 - \bar{p})}{n}}$$

Unlike the $\bar{x}$ chart, the _____ control limit for the p chart is rather meaningless as a monitor of the efficiency of the manufacturing process. Observing a sample fraction defective below the lower control limit of a p chart would imply that the fraction defective is much _____ than normal. This is what is desired, not something to be avoided. Such an observation may imply that the efficiency of the manufacturing process has increased and that new control limits should be constructed to reflect this new capability of the process.

Example 7.10

A manufacturing process is designed to produce an electronic component for use in small, portable television sets. The components are all of standard size and need not conform to any measurable characteristic but are sometimes inoperable when emerging from the manufacturing process. Fifteen samples were selected from the process at times when the process was known to be in statistical control. Fifty components were observed within each sample, tests were performed to determine their operability, and the number of defective components was recorded. Construct a p chart to monitor the manufacturing process.

Sample Number	Sample Size	Number of Defectives	$\hat{p}$
1	50	6	.12
2	50	7	.14
3	50	3	.06
4	50	5	.10
5	50	6	.12
6	50	8	.16
7	50	4	.08
8	50	5	.10
9	50	7	.14
10	50	3	.06
11	50	1	.02
12	50	6	.12
13	50	5	.10
14	50	4	.08
15	50	5	.10
Totals	750	75	

Solution

The central line for the p chart is found by computing the average proportion defective using all the sample data. Thus,

lower

lower

75; .10

$\sqrt{\frac{\overline{p}(1-\overline{p})}{n}}$

$$\overline{p} = \frac{(\underline{\hspace{2cm}})}{750} = \underline{\hspace{2cm}}$$

Using the formula _____ for the standard deviation, we find that

$$\hat{\sigma}_{\hat{p}} = \sqrt{\frac{.1(.9)}{50}}$$

.0415

$$= \sqrt{.0018} = \underline{\hspace{2cm}}$$

The control limits for the p chart are then found as follows:

$$UCL = \overline{p} + 3\sqrt{\frac{\overline{p}(1-\overline{p})}{n}}$$

.1; .0424

.2273

$$= \underline{\hspace{2cm}} + 3(\underline{\hspace{2cm}})$$

$$= \underline{\hspace{2cm}}$$

while the lower control limit is

$$LCL = \overline{p} - 3\sqrt{\frac{\overline{p}(1-\overline{p})}{n}}$$

.1; .0424

$$= \underline{\hspace{2cm}} - 3(\underline{\hspace{2cm}})$$

−.0272

$$= \underline{\hspace{2cm}}$$

Since a sample fraction defection, $\hat{p}$, cannot be negative, the computed lower control limit is ignored and we use _____ instead.

zero

The p chart for the electronic component manufacturing process is shown in Figure 6.4.

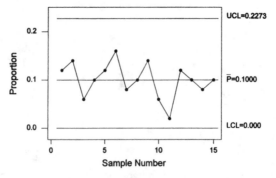

P Chart for Components

Figure 7.4

Self-Correcting Exercises 7D

1. In the production of two-liter bottles of a soda, the fill machine is designed to fill these bottles with 67.6 fluid ounces of soda. Quality control considerations require that all bottles contain between 67.4 and 67.8 fluid ounces. During a production period when the fill process was known to be in statistical control, a series of 30 samples of size $n = 10$ bottles were randomly selected. The average fill was $\bar{\bar{x}} = 67.54$ and the standard deviation was $s = .0123$.

   a. Construct an $\bar{x}$ chart to monitor the fill process.

   b. The next five samples of size $n = 10$ resulted in the following values of the sample mean. 67.77, 67.76, 68.01, 67.62, and 67.65. Plot these points on the chart constructed in part a. Is the process still in control?

2. The side pieces for aluminum picture frames are manufactured in various lengths and packaged in pairs of equal lengths, so that the consumer can construct his own custom frame by purchasing two pairs of side pieces. The machine which cuts the side pieces can be set to cut pieces in varying lengths. During a period in which the process is in control, the cutting machine is set to cut 18 inch side pieces, and the following measurements are obtained.

Sample	Length	Sample	Length
1	18.02, 18.04, 17.95, 17.96	9	17.99, 18.00, 18.01, 17.99
2	17.96, 18.00, 17.99, 17.97	10	18.02, 18.01, 18.03, 17.99
3	18.00, 18.05, 18.01, 18.00	11	17.95, 18.00, 17.98, 18.01
4	17.97, 18.00, 17.99, 18.02	12	17.99, 17.99, 18.00, 17.95
5	18.00, 18.01, 17.98, 17.97	13	17.96, 18.01, 18.02, 17.97
6	17.99, 17.97, 18.01, 18.02	14	17.96, 17.96, 17.99, 18.00
7	17.97, 17.96, 17.99, 18.00	15	18.01, 17.95, 18.02, 18.03
8	17.97, 17.98, 17.97, 18.03	16	18.00, 18.01, 17.98, 18.03

   Construct an $\bar{x}$ chart fo the process. Explain how it will be used.

3. The data that follows summarizes the results of testing permanent magnets used in electric relays. During each of ten weeks, 700 magnets were inspected per week and the number of defective magnets recorded.

Weeks	Number Inspected	Number Defective
1	700	38
2	700	60
3	700	36
4	700	39
5	700	28
6	700	37
7	700	41
8	700	47
(Cont'd)		

(Table Cont'd) Weeks	Number Inspected	Number Defective
9	700	32
10	700	30
Totals	7000	388

Construct a p-chart using these data, under the assumption that the process was under control during this time period.

4. Refer to Exercise 3. The next five weeks produced the following data on the number of defective electromagnets per $n = 700$ weekly samples.

   25, 37, 42, 47, 55

Use the control chart in Exercise 3 to plot the proportion defective per week. Comment on your results.

Solutions to SCE 7D

1. a. Using $\bar{\bar{x}} = 67.54$, $s = .0123$ and $n = 10$,

$$\text{LCL} = \bar{\bar{x}} - 3\left(\frac{s}{\sqrt{n}}\right) = 67.54 - 3\left(\frac{.0123}{\sqrt{10}}\right) = 67.528, \text{ and}$$

$$\text{UCL} = \bar{\bar{x}} + 3\left(\frac{s}{\sqrt{n}}\right) = 67.54 + 3\left(\frac{.0123}{\sqrt{10}}\right) = 67.552.$$

Observations falling outside the interval (LCL, UCL) should cause close examination of the process.

   b. The process means are outside of the control limits on all five days. The process must be examined and adjusted.

2. The means for the 16 samples are shown below. Then

$$\bar{\bar{x}} = \frac{287.891}{16} = 17.9931 \quad \text{and} \quad s = .02436$$

Sample	$\bar{x}$	Sample	$\bar{x}$
1	17.9925	9	17.9975
2	17.9800	10	18.0125
3	18.0150	11	17.9850
4	17.9950	12	17.9825
5	17.9900	13	17.9900
6	17.9975	14	17.9775
7	17.9800	15	18.0025
8	17.9875	16	18.0050

Then $\text{LCL} = \bar{\bar{x}} - 3\left(\frac{s}{\sqrt{4}}\right) = 17.9931 - .03654 = 17.957$

$\text{UCL} = \bar{\bar{x}} + 3\left(\frac{s}{\sqrt{4}}\right) = 17.9931 + .03654 = 18.030$

Sample means falling outside the interval (LCL, UCL) should cause close examination of the process.

3. $\overline{p} = .0554$ and $3\sqrt{\dfrac{\overline{p}(1-\overline{p})}{700}} = 3(.0086) = .02594$. Hence,

$$LCL = .0554 - .02594 = .02948$$

$$UCL = .0554 + .02594 = .08137$$

4. The sample proportions are $25/700 = .0357$, $37/700 = .0529$, $42/700 = .06$, $47/700 = .0671$, $55/700 = .0786$. The proportion defective shows a steady increase, and the process will soon be out of statistical control. The process should be carefully checked for sources contributing to the observed increase.

7.8 SUMMARY

The sampling distribution for a sample statistic is its probability distribution, which results when the behavior of the statistic is examined in repeated sampling. The sampling distribution can be constructed directly for small finite populations. For large or infinite populations, the sampling distribution of a statistic must be found mathematically or approximated empirically.

An important result called the _____ _____ _____ allows us to approximate the sampling distribution for statistics that are sums or averages of random variables. Using this theorem, we examined the sampling distributions of two important sample statistics, $\overline{x}$ and $\hat{p}$. Both could be approximated by a normal distribution (with the appropriate mean and standard deviation) for large sample sizes. The sampling distributions of these two statistics will serve as the basis for making inferences about their corresponding population parameters. This is the subject of Chapter 8.

Central Limit
Theorem

KEY CONCEPTS AND FORMULAS

I. *Sampling Plans and Experimental Designs*
    1. Simple random sampling
        a. Each possible sample is equally likely to occur.
        b. Use a computer or a table of random numbers.
        c. Problems are nonresponse, undercoverage, and wording bias.
    2. Other sampling plans involving randomization
        a. Stratified random sampling
        b. Cluster sampling
        c. Systematic 1-in-k sampling
    3. Nonrandom sampling
        a. Convenience sampling
        b. Judgment sampling
        c. Quota sampling

II. *Statistics and Sampling Distributions*
1. Sampling distributions describe the possible values of a statistic and how often they occur in repeated sampling.
2. Sampling distributions can be derived mathematically, approximated empirically, or found using statistical theorems.
3. The Central Limit Theorem states that sums and averages of measurements from a nonnormal population with finite mean μ and standard deviation σ have approximately normal distributions for large samples of size n.

III. *Sampling Distribution of the Sample Mean*
1. When samples of size n are drawn from a normal population with mean μ and variance σ^2, the sample mean $\bar{x}$ has a normal distribution with mean μ and variance σ^2/n.
2. When samples of size n are drawn from a nonnormal population with mean μ and variance σ^2, the Central Limit Theorem ensures that the sample mean $\bar{x}$ will have an approximately normal distribution with mean μ and variance σ^2/n when n is large ($n \geq 30$).
3. Probabilities involving the sample mean can be calculated by standardizing the value of $\bar{x}$ using z:

$$z = \frac{\bar{x} - \mu}{\sigma/\sqrt{n}}$$

IV. *Sampling Distribution of the Sample Proportion*
1. When samples of size n are drawn from a binomial population with parameter p, the sample proportion $\hat{p}$ will have an approximately normal distribution with mean p and variance pq/n as long as $np > 5$ and $nq > 5$.
2. Probabilities involving the sample proportion can be calculated by standardizing the value $\hat{p}$ using z:

$$z = \frac{\hat{p} - p}{\sqrt{pq/n}}$$

V. *Statistical Process Control*
1. To monitor a *quantitative* process, use an $\bar{x}$ chart. Select k samples of size n and calculate the overall mean $\bar{\bar{x}}$ and the standard deviation s of all nk measurements. Create upper and lower control limits as

$$\bar{\bar{x}} \pm 3\left(\frac{s}{\sqrt{n}}\right)$$

If a sample mean exceeds these limits, the process is out of control.
2. To monitor a *binomial* process, use a p chart. Select k samples of size n and calculate the average of the sample proportions as

$$\bar{p} = \frac{\sum \hat{p}_i}{k}$$

Create upper and lower control limits as

$$\bar{p} \pm 3\sqrt{\frac{\bar{p}(1-\bar{p})}{n}}$$

If a sample procedure exceeds these limits, the process is out of control.

Exercises

1. Suppose that an elevator is designed with a permissible load limit of 3000 pounds with a maximum of 20 passengers. If the weights of people using the elevator are normally distributed with a mean of $\mu = 160$ pounds and a standard deviation of 25 pounds, what is the probability that the weight of a group of 20 persons exceeds the permissible load limit?

2. In a past municipal election, a city bond issue passed with 52% of the vote. If a poll involving $n = 100$ people had been taken just prior to the election, what is the probability that the sample proportion favoring the issue would have been 49% or less?

3. Based on past birth records, the probability of a male livebirth in the United States is .513. Of the first $n = 100$ births in January, what is the probability that the proportion of male livebirths exceeds 60%?

4. Suppose that the average assessed values of single family dwellings in a large municipality is $165,000 with a standard deviation of $20,000.
   a. If a random sample of $n = 100$ dwellings is selected and $\bar{x}$, the average assessed value of these dwellings calculated, what is the mean of $\bar{x}$ in repeated sampling?
   b. What is the standard deviation of $\bar{x}$?

5. A television network allows an average of 5 minutes per half-hour program for advertising with a standard deviation of 3 minutes. What is the probability that the average time devoted to commercials for 50 one-half hour program exceeds 6 minutes?

6. Packages of food whose average weight is 16 ounces with a standard deviation of .6 oz are shipped in boxes of 24 packages. What is the probability that a box of 24 packages will weigh more than 392 ounces (24.5 pounds)?

7. The average number of sick days per year in an electronics industry is 7 days with a standard deviation of 3 days. If a sample of $n = 30$ men is selected from among the employees in this industry, what is the probability that the sample mean number of sick days for men exceeds 8 days?

8. Graduate students applying for entrance to many universities must take a Miller Analogies Test. It is known that the test scores have a mean of 75 and a variance of 16. If, during the past year, 100 students applied for graduate admission to a school requiring the Miller Analogies Test, what is the probability that the average score on these 100 tests exceeds 76?

9. The blood pressures for a population of individuals are normally distributed with a mean of 110 and a standard deviation of 7.
   a. If a sample of $n = 10$ individuals is chosen randomly from this population, within what limits would you expect the sample mean to lie with probability .95?
   b. What is the probability that the sample mean will exceed 115?

10. A hardwoods manufacturing plant has several different production lines to manufacture baseball bats of differing weights. One such production line is designed to produce bats weighing 32 ounces. During a period of time when the production process was known to be in statistical control, the average bat weight was found to be 31.7 ounces. The observed data was gathered from 50 samples each consisting of 5 measurements. The standard deviation of all samples was found to be $s = .2064$ ounces. Construct an $\bar{x}$-chart to monitor the 32-ounce bat production process.

11. Refer to Exercise 10 and suppose that during a day when the state of the 32-ounce bat production process was unknown, the following measurements were obtained at hourly intervals.

Hour	$\bar{x}$	Hour	$\bar{x}$
1	31.6	4	33.1
2	32.5	5	31.6
3	33.4	6	31.8

Each measurement represents a statistic computed from a sample of five bat weights selected from the production process during a certain hour. Use the control chart constructed in Exercise 10 to monitor the process.

12. A manufacturing process, designed to produce flashlight batteries, was examined at 20 different points in time when the process was known to be in control. At each sampling interval, 100 batteries were randomly selected from the process, they were tested, and the number of defective (inoperable) batteries was recorded. The numbers of defectives found in the 20 samples were:

   5, 7, 6, 8, 10, 3, 2, 5, 8, 4, 9, 5, 7, 5, 2, 3, 8, 2, 1, 4

   a. Construct a p-chart to monitor the process.
   b. During a day when the state of the process was unknown, five different samples of size 100 each yielded 7, 9, 11, 12, and 6 defectives, respectively. Comment on the state of the process.

Chapter 8
Large-Sample Estimation

8.1 WHERE WE'VE BEEN

The first seven chapters represent a series of topics and ideas to pre-
pare you for statistical inference. You have learned how to describe
data graphically, and you have also learned to describe data numerically.
Probability and probability distributions were presented as population
models that could be used to describe the behavior of samples that have
been randomly selected from these populations. In the last chapter deal-
ing with sampling distributions, we found that certain descriptive statistics, such as the sample mean or a sample sum have a distribution in
repeated sampling that is either normal or approximately normal when
the sample sizes are large, due to the Central Limit Theorem. You also
found that the normal distribution could be used to approximate binomial probabilities when the binomial mean np and nq are greater than
five. We will build on this information when making inferences about
population parameters based upon information contained in a sample.

8.2 WHERE WE'RE GOING–STATISTICAL INFERENCE

Inference making is actually very common in our everyday lies. Each
of us uses inferential procedures without realizing it.

- You need to make an inference when you decide to either take
  your umbrella with you or leave it at home.

- You may work ahead on the next set of exercises in your text,
  since you are reasonably certain that your next homework set
  will come from these exercises.

Neither of these two situations is earth shaking, but there are situations
in which there is a cost involved in making a decision.

- A stockbroker needs to determine whether market prices will
  continue at their present value, rise, or fall.

- A construction firm must estimate the number of rainy days
  during a construction period so that a reasonable completion
  date can be written into the contract.

- Medical researchers must determine whether a new drug is effective, safe, and has no serious side effects prior to receiving approval to market the drug.

Statistical inferences are made in terms of population parameters. Methods for making inferences fall into one of two categories.

- Estimation or prediction, in which sample information is used to give reasonable bounds for the value of a population parameter.

- Hypothesis testing, in which a decision is made to either accept or reject a value of a population parameter.

Example 8.1

Consider an agricultural experimenter interested in the average yield (in tons per acre) of a variety of alfalfa grown for hay. Let μ be the average yield for this variety of alfalfa. If an inference is to be made about μ, two questions could be asked:

1. What is the most likely value of μ for the population of yields from which we are sampling?
2. Is the mean equal to some specified value, say μ_0, or is it not? For example, the experimenter may know that the average yield of this variety of alfalfa in an adjacent field using chemical pest control is at most 2 tons per acre. Can integrated pest management practices produce yields greater than 2 tons per acre?

The first question is one of predicting or estimating the value of the population parameter μ. The second question concerns a test of a hypothesis about μ. If integrated pest management is no more effective than chemical pest control, then the yield will still be 2 tons per acre. However, if integrated pest management increases the yield of alfalfa, then $\mu > 2$ tons per acre. The objective is to determine which of these two hypotheses is correct.

This chapter will be concerned with making inferences about four parameters:

1. μ, the mean of a population of continuous measurements
2. $\mu_1 - \mu_2$, the difference between the means for two populations of continuous measurements
3. p, the parameter of a dichotomous or binomial population
4. $p_1 - p_2$, the difference in the parameters for two binomial populations

The quantities to be used in making inferences will be the sums or averages of the measurements in a random sample and consequently will possess frequency distributions in repeated sampling that are approximately _____ because of the _____

normal; Central
Limit

_____ Theorem, as explained in Chapter 7.

One of the most important concepts to grasp is that estimation as well as hypothesis testing is a two-step procedure:

1. Making the inference
2. Measuring its goodness

A measure of the goodness of an inference is essential to enable the person using the inference to measure its reliability. For example, we would wonder how close to the population parameter our estimate is expected to lie.

Rather than follow the section numbers exactly as they appear in your text, we have grouped certain topics together and will consider two more general sections: (1) point estimation and (2) interval estimation. The techniques developed in the process of estimating the four parameters mentioned above will also be used to determine how large the sample must be to achieve the accuracy required by the experimenter.

8.3 TYPES OF ESTIMATORS

Using the measurements in a sample to predict the value of one or more parameters of a population is called _____. An _____ is a rule that tells us how to calculate an estimate of a parameter based on the information contained in a sample. We can give many different estimators for a particular population parameter. An estimator is often expressed in terms of a mathematical formula in which the estimate is a function of the sample measurements. For example, $\bar{x}$ is an *estimator* of the population parameter μ. If a sample of $n = 20$ pieces of aluminum cable is tested for strength and the mean of the sample is $\bar{x} = 100.7$, then 100.7 is an *estimate* of the population mean strength μ. The estimator of a parameter is usually designated by placing a "hat" (ˆ) over the parameter to be estimated. Thus, an estimator of μ is $\hat{\mu} = \bar{x}$.

estimation
estimator

Estimates of a population parameter can be made in two ways. The sample can be used to produce an estimate of μ that is a single number or to produce an interval, two points intended to enclose the true value of μ.

A **point estimator** of a population parameter is a rule that tells how to calculate a single number based on the sample measurements. The resulting number is called a **point estimate.**

An **interval estimator** of a population parameter is a rule that tells how to calculate two numbers based on the sample measurements that form an interval within which the true value of the parameter is expected to lie. This pair of numbers is called an **interval estimate.**

Example 8.2

An agronomist is interested in the yield of soybeans as a function of field row spacings. In this situation, the *population* of interest is the theoretical population consisting of yields of soybeans grown under these conditions. The most common measure of yield in this situation is the mean μ. If the agronomist calculates the average yield of $n = 20$ field plots as $\bar{x} = 10$ bushels per quadrate, this represents a *point estimate* for μ. However, he or she may provide an *interval estimate*, such as 8.3 to 11.7 bushels per quadrate.

We will begin by discussing point estimators for population means and proportions, and then show how these point estimates can be used to construct interval estimates.

8.4 POINT ESTIMATION

The properties of an estimator are evaluated by observing its behavior in repeated sampling. Let us discuss the specific case of estimating the population mean μ. An obvious estimator for the population mean is the sample mean $\bar{x}$. You know that the estimator $\bar{x}$ will have a distribution in repeated sampling that will have a mean equal to the mean μ of the population sampled, and a standard deviation equal to $\sigma/\sqrt{n}$ where σ is the variance of the population sampled and n is the sample size. You also know that the sample mean will be normally distributed if the sampled population was normal, and will be approximately normal if the sample size is large. How would $\bar{x}$ compare with other possible estimators? Suppose that there were three other competitors whose distributions are shown in Figure 8.1.

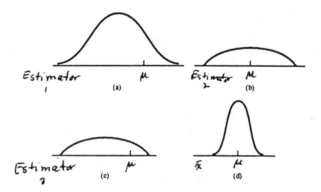

Figure 8.1

An estimator of the population mean would be considered good if μ was in the middle of the distribution of estimates, as in distributions
_____ and _____ in Figure 8.1. These two estimators

b.; d.

would be called *unbiased* because μ is the mean of their sampling distributions. However, we would also like the estimates to cluster tightly about μ as you see in distribution _____ in Figure 8.1. When the variance of an estimator is smaller than any other unbiased estimator, the estimator is said to have *minimum variance* among unbiased estimators. Of the four distributions given in Figure 8.1, the estimator whose distribution is given by _____ satisfies these two conditions. Therefore, a *good estimator* is one that is:

d.

d.

1. Unbiased
2. Minimum variance among unbiased estimators.

In an estimation problem you may known the form of the distribution from which the sample was selected, but you may not know the values of the parameters that describe that distribution. For example, you may know that the population sampled is binomial, but you may not know the value of p. Or, you may know that the sampled distribution is mound-shaped or approximately normal, but you may not know the mean μ or standard deviation σ. Even though we know that $\bar{x}$ is an unbiased, minimum variance estimator, we would like to know how far away from the true value of μ does the sample estimate lie.

The distance between an estimate and the estimated parameter is called the **error of estimation**.

In this chapter we will assume that the sample sizes are large, and that the sampling distribution of the estimator can be approximated by a _____ distribution because of the Central Limit Theorem.

normal

    For an estimator with a normal or approximate normal sampling distribution, the Empirical Rule assures us that 95% of the point estimates will lie within approximately 2 (or exactly 1.96) standard deviations of the mean of its distribution. When the estimator is unbiased, the difference between the value of the point estimate and the true value of the parameter will be less than 1.96 standard errors of the estimator 95% of the time. This distance is called the *margin of error*, and provides a practical upper bound for the error of estimation. See Figure 8.2.

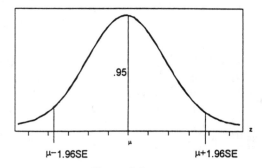

Figure 8.2

To estimate the population mean μ for a quantitative population, the point estimator, $\bar{x}$, is unbiased with standard error equal to

$$SE = \frac{\sigma}{\sqrt{n}}.$$

The margin of error is given as

$$\pm 1.96 \, \frac{\sigma}{\sqrt{n}}$$

If σ is unknown and n is 30 or larger, the sample standard deviation s can be used to approximate σ.

The estimation procedure for the other three cases proceeds using the same approach.

To estimate the population proportion p for a binomial population, the point estimator $\hat{p} = \frac{x}{n}$ is unbiased with standard error

$$SE = \sqrt{\frac{pq}{n}}.$$

The **margin of error** is given by

$$\pm 1.96 \sqrt{\frac{pq}{n}}$$

and estimated by

$$\pm 1.96 \sqrt{\frac{\hat{p}\hat{q}}{n}}$$

Assumption: $np > 5$ and $nq > 5$ (or $n\hat{p} > 5$ and $n\hat{q} > 5$ since p and q are unknown).

Example 8.3
The mean length of stay for patients in a hospital must be known in order to estimate the number of beds required. The length of stay, recorded for a sample of 400 patients at a given hospital, produced mean and a standard deviation of 5.7 and 8.1 days, respectively. Give a point estimate for μ, the mean length of stay for patients entering the hospital, and give the margin of error for this estimate.

Solution
1. The point estimate for μ is $\bar{x} = $ _____ .   5.7
2. Since σ is unknown, the *approximate* margin of error is

$$1.96\left(\frac{s}{\sqrt{n}}\right) = 1.96\left(\frac{8.1}{\sqrt{400}}\right) = \text{_____}$$ .79

Example 8.4
An experimenter is interested in investigating family sizes in an attempt to determine the percentage of families that have more than two children under the age of 18. She randomly samples $n = 100$ families and finds 12 families with more than two children under the age of 18. Estimate the true proportion of families with more than two children under 18, and give the margin of error for estimation.

Solution
1. The point estimate of p is

$$\hat{p} = \frac{x}{n} = \frac{\text{_____}}{100} = \text{_____}$$ 12; .12

That is, _____ % of families have more than two children   12
under 18.
2. The margin of error is $1.96\sigma_{\hat{p}} = 1.96\sqrt{pq/n}$. The quantities p and q are unknown. However, since n is large, $\hat{p}$ and $\hat{q}$ may be substituted for p and q. The *approximate* margin of error is

$$1.96\sqrt{\frac{\hat{p}\hat{q}}{n}} = 1.96\sqrt{\text{_____}} = 1.96\sqrt{\text{_____}}$$ $\frac{(.12)(.88)}{100}$; .001056

$$= 1.96(\text{_____}) = \text{_____}$$ .0325; .064

Self-Correcting Exercises 8A

1. In standardizing an examination, the average score on the exam must be known in order to differentiate among examinees taking the examination. The scores recorded for a sample of 93 examinees yielded a mean and a standard deviation of 67.5 and 8.2, respectively. Estimate the true mean score for this examination and calculate the margin of error in estimation.

2. A new student recreational center for the campus is being planned under the supposition that students will be assessed an annual fee of $135 as part of their college fees over the next 10 years to pay construction costs. It was determined that student support for this issue would be ascertained using a small-scale sample survey. In a random sample of $n = 65$ students, 30 stated that they favored the new center and would be willing to pay the additional fees. Find a point estimate of the percentage of students on campus in favor of

the new recreation center and the associated fees and calculate the margin of error in estimation.

3. A physician wishes to estimate the proportion of accidents on the California freeway system that result in fatal injuries to at least one person. He randomly checks the files on 50 automobile accidents and finds that 8 resulted in fatal injuries. Estimate the true proportion of fatal accidents, and calculate the margin of error.

Solutions to SCE 8A

1. $\bar{x} = 67.5$ with approximate margin of error $1.96s/\sqrt{n} = 1.96(8.2)/\sqrt{93} = \pm 1.67$.

2. $\hat{p} = x/n = 30/65 = .46$ with approximate margin of error $1.96\sqrt{\hat{p}\hat{q}/n} = 1.96\sqrt{.46(.54)/65} = \pm .12$.

3. $\hat{p} = x/n = 8/50 = .16$ with approximate margin of error $1.96\sqrt{\hat{p}\hat{q}/n} = 1.96\sqrt{.16(.84)/50} = 1.96(.0518) = \pm .102$

8.5 INTERVAL ESTIMATION

An interval estimator is a rule that tells us how to calculate two points based on information contained in a sample. The objective is to form a narrow interval that will enclose the parameter. As in the case of point estimation, we can form many interval estimators (rules) for estimating the parameter of interest. Not all intervals generated by an interval estimator will actually enclose the parameter.

The properties of interval estimators are also determined by repeated sampling. Repeated use of an interval estimator generates a large number of _____ _____ for estimating a parameter. If an interval estimator were satisfactory, a large fraction of the interval estimates would enclose the true value of the parameter. The fraction of such intervals enclosing the parameter is known as the _____ coefficient. This is not to be confused with the interval estimate, called the confidence _____. See Figure 8.3.

interval estimates

confidence interval

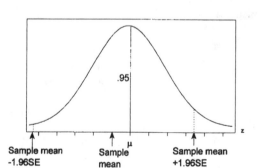

.95

Sample mean
-1.96SE

Sample
mean

μ

Sample mean
+1.96SE

z

Figure 8.3

The probability that a confidence interval will enclose the true value of the estimated parameter is called the **confidence coefficient**.

If the confidence coefficient is .90, and a large number of samples were drawn and each used to calculate a confidence interval, then 90% of these intervals would enclose the true value of the parameter.

To change the *confidence coefficient* for a confidence interval from .95 to a confidence coefficient denoted by the fraction $(1 - \alpha)$, you need to select a value of the standard normal random variable, which we shall call $z_{\alpha/2}$, that has an area to its right equal to $\alpha/2$. The area between $-\alpha/2$ and $\alpha/2$ is equal to $(1 - \alpha)$, the desired confidence coefficient. Some common values for confidence coefficients are given in Table 8.1

$(1 - \alpha)$	$\alpha/2$	$z_{\alpha/2}$
.90	.05	1.645
.95	.025	1.96
.98	.01	2.33
.99	.005	2.58

Table 8.1

A $(1 - \alpha)100\%$ confidence interval

(Point estimator) $\pm z_{\alpha/2}$ (Standard error of the estimator)

Construction of confidence intervals for the cases of interest will require the following information. Complete the following table, filling in the estimator and its standard error (standard deviation) where required.

Parameter	Estimator	Standard Error
μ	$\bar{x} = \sum x_i/n$	$\sigma/\sqrt{n}$
p	$\hat{p} = x/n$	_____ $\sqrt{pq/n}$
$\mu_1 - \mu_2$	_____	_____ $\bar{x}_1 - \bar{x}_2; \sqrt{\dfrac{\sigma_1^2}{n_1} + \dfrac{\sigma_2^2}{n_2}}$
$p_1 - p_2$	$\hat{p}_1 - \hat{p}_2$	$\sqrt{\dfrac{p_1 q_1}{n_1} + \dfrac{p_2 q_2}{n_2}}$

Notice that evaluation of the standard errors given in the table may require values of parameters that are unknown. When the sample sizes are large, the sample estimates can be used to calculate an approximate standard error. As a rule of thumb, we will consider samples of size 30 or greater to be large samples.

Example 8.5
Refer to Example 8.3. To construct a 90% confidence interval for the mean length of hospital stay, μ, based on the sample of $n = 400$ patients ($\bar{x} = 5.7$ and $s = 8.1$), we calculate

$$\bar{x} \pm z_{\alpha/2} \frac{\sigma}{\sqrt{n}}$$

With $z_{.05} = 1.645$ and an estimate for σ given by $s = 8.1$, the interval estimate for the mean length of hospital stay is given as

$$5.7 \pm .67$$

5.03; 6.37

More properly, we estimate that _____ $< \mu <$ _____ with 90% confidence.

$\hat{p} \pm 1.96\sqrt{pq/n}$

The formula for a 95% confidence interval for a binomial parameter p is _____. Note that $\hat{p}$ is used to approximate p in the formula for $\sigma_{\hat{p}}$ because its value is unknown.

Example 8.6
An experimental rehabilitation technique was used on released convicts. It was shown that 79 of a total of 121 men subjected to the technique pursued useful and crime-free lives for a three-year period following prison release. Find a 95% confidence interval for p, the probability that a convict subjected to the rehabilitation technique will follow a crime-free existence for at least three years after prison release.

Solution
The sampling described satisfies the requirements of a binomial experiment consisting of $n = 121$ trials. In estimating the parameter p with a 95% confidence interval, we use the estimator

1.96

$$\hat{p} \pm \underline{\hspace{2cm}} \sqrt{\frac{pq}{n}}$$

Since p is unknown, the sample value $\hat{p}$ will be used in the approximation of $\sqrt{pq/n}$. Collecting pertinent information, we have

.65

1. $\hat{p} = x/n = 79/121 = $ _____

2. $\sqrt{\hat{p}\hat{q}/n} = \sqrt{(.65)(.35)/121} = .04$

3. The interval estimate is given as

.08

$$.65 \pm 1.96(.04) \quad \text{or} \quad .65 \pm \underline{\hspace{2cm}}$$

4. We estimate that _____ $< p <$ _____ with 95% confidence.

.57; .73

The exact values for standard errors of estimators cannot usually be found because they are functions of unknown population parameters. For the following estimators, give the standard error and the best approximation of the standard error for use in confidence intervals.

In order to estimate $\mu_1 - \mu_2$, the difference between two population means or $p_1 - p_2$, the difference between two population proportions, you will need the additional information provided in the next table. Fill in any missing entries.

Point Estimator	Standard Error	Best Approximation of Standard Error
$\bar{x}_1 - \bar{x}_2$	$\sqrt{\dfrac{\sigma_1^2}{n_1} + \dfrac{\sigma_2^2}{n_2}}$	_____
$\hat{p}_1 - \hat{p}_2$	$\sqrt{\dfrac{p_1 q_1}{n_1} + \dfrac{p_2 q_2}{n_2}}$	$\sqrt{\dfrac{\hat{p}_1 \hat{q}_1}{n_1} + \dfrac{\hat{p}_2 \hat{q}_2}{n_2}}$

$\sqrt{\dfrac{s_1^2}{n_1} + \dfrac{s_2^2}{n_2}}$

When the sample sizes are large, the Central Limit Theorem assures us that these estimators will be normally distributed or approximately so. You can now either estimate a population parameter using a point estimate with a margin of error, or you can use a confidence interval with a confidence coefficient of your choice. A point estimate with a margin of error is given by

$$\text{(Point estimator)} \pm 1.96 \text{(Standard Error of the Estimator)}.$$
$$\pm \text{Margin of Error}$$

The interpretation of a point estimate is that we are (95%) confident that the point estimate does not lie further from the true value of the parameter than the value given by the margin of error.

A confidence interval estimate of a parameter when the estimator is normally distributed or approximately so, is given by

$$\text{(Point estimator)} \pm z_{\alpha/2} \text{(Standard error of the estimator)}.$$

In this case, we would be $(1 - \alpha)100\%$ confident that the true value of the parameter estimated lies with the given confidence. Of course we know that the interval either encloses the value of the parameter, or it does not. However, since $(1 - \alpha)$ of these intervals enclose the value of the parameter in repeated sampling, we are confident that the interval we have found encloses the true value of the estimated parameter.

Example 8.7

An experiment was conducted to compare the mean absorptions of drug in specimens of muscle tissue. Seventy-two tissue specimens were randomly divided between two drugs, A and B, with 36 assigned to each drug, and the drug absorption was measured for the 72 specimens. The means and variances for the two samples were $\bar{x}_1 = 7.8$, $s_1^2 = .10$ and $\bar{x}_2 = 8.4$, $s_2^2 = .06$, respectively. Find a 95% confidence interval for the difference in mean absorption rates.

Solution

We are interested in placing a confidence interval about the parameter _____. The confidence interval is

$$(\bar{x}_1 - \bar{x}_2) \pm z_{\alpha/2}\sqrt{\frac{s_1^2}{n_1} + \frac{s_2^2}{n_2}}$$

$$(7.8 - 8.4) \pm \underline{\qquad} \sqrt{\frac{.10}{36} + \frac{.06}{36}}$$

$$\underline{\qquad} \pm .131$$

or

$$\underline{\qquad} < \mu_1 - \mu_2 < \underline{\qquad}$$

Example 8.8

The voting records at two precincts were compared based on samples of 400 voters each. Those voting Democratic numbered 209 and 263, respectively. Estimate the difference in the fraction voting Democratic for the two precincts, using a 90% confidence interval.

Solution

The confidence interval is

$$(\hat{p}_1 - \hat{p}_2) \pm z_{\alpha/2}\sqrt{\frac{\hat{p}_1 \hat{q}_1}{n_1} + \frac{\hat{p}_2 \hat{q}_2}{n_2}}$$

$$(.5225 - .6575) \pm \underline{\qquad} \sqrt{\frac{(.5225)(.4775)}{400} + \frac{(.6575)(.3425)}{400}}$$

$$\underline{\qquad} \pm .057$$

or $\underline{\qquad} < p_1 - p_2 < \underline{\qquad}$

Self-Correcting Exercises 8B

1. In 39 soil samples tested for trace elements, the average amount of copper was found to be 22 milligrams, with a standard deviation of 4 milligrams. Find a 90% confidence interval for the true

[margin annotations:]

$\mu_1 - \mu_2$

1.96

$-.6$

$-.731; -.469$

1.645

$-.135$

$-.192; -.078$

mean copper content in the soils from which these samples were taken.

2. Using the following data, give a point estimate with margin of error for the difference in mortality rates in breast cancers where radical or simple mastectomy was used as a treatment.

	Radical	Simple
Number died	31	41
Number treated	204	191

3. In measuring the tensile strength of two allows, strips of the alloys were subjected to tensile stress and the force (measured in pounds) at which the strip broke recorded for each strip. The data are summarized below.

	Alloy 1	Alloy 2
$\bar{x}$	150.5	160.2
s^2	23.72	36.37
n	35	35

Use these data to estimate the true mean difference in tensile strength by finding a point estimate for $\mu_1 - \mu_2$ and calculate the margin of error in estimation.

4. It is desired to estimate the difference in accident rates between youth and adult drivers. The following data were collected from two random samples, where x is the number of drivers that were involved in one or more accidents:

Youths	Adults
$n_1 = 100$	$n_2 = 200$
$x_1 = 50$	$x_2 = 60$

Estimate the difference $(p_1 - p_2)$ with a 95% confidence interval.

5. The yearly incomes of high school teachers in two cities yielded the following tabulation:

	City 1	City 2
Number of teachers	90	60
Average income	28,520	27,210
Standard deviation	1,510	950

   a. If the teachers from each city are thought of as samples from two populations of high school teachers, use the data to construct a 99% confidence interval for the difference in mean annual incomes.

   b. Using the results of part a, would you be willing to conclude that these two city schools belong to populations having the same mean annual income?

Solutions to SCE 8B

1. $\bar{x} \pm z_{\alpha/2}(s)/\sqrt{n}$; $22 \pm 1.645(4)/\sqrt{39}$;

   $22 \pm (6.58/6.245)$; 22 ± 1.05, or 20.95 to 23.05

2. $\hat{p}_1 - \hat{p}_2 = \dfrac{x_1}{n_1} - \dfrac{x_2}{n_2} = \dfrac{31}{204} - \dfrac{41}{191} = .15 - .21 = -.06$;

   Approximate margin of error:

   $$1.96\sqrt{\dfrac{\hat{p}_1\hat{q}_1}{n_1} + \dfrac{\hat{p}_2\hat{q}_2}{n_2}} = 1.96\sqrt{.000625 + .000869} = 1.96(.039)$$

   $$= .076$$

3. $\bar{x}_1 - \bar{x}_2 = 150.5 - 160.2 = -9.7$ with approximate margin of error

   $$1.96\sqrt{\dfrac{s_1^2}{n_1} + \dfrac{s_2^2}{n_2}} = 1.96\sqrt{\dfrac{23.72}{35} + \dfrac{36.37}{35}} = 1.96\sqrt{1.7169} = 2.57$$

4. $\hat{p}_1 = 50/100 = .50$; $\hat{p}_2 = 60/200 = .30$.

   $$(\hat{p}_1 - \hat{p}_2) \pm 1.96\sqrt{\dfrac{\hat{p}_1\hat{q}_1}{n_1} + \dfrac{\hat{p}_2\hat{q}_2}{n_2}}$$

   $$(.50 - .30) \pm 1.96\sqrt{\dfrac{.5(.5)}{100} + \dfrac{.3(.7)}{200}}$$

   $.2 \pm 1.96(.0596)$ $.2 \pm .12$, or $.08$ to $.32$.

5. a. $(\bar{x}_1 - \bar{x}_2) \pm z_{\alpha/2}\sqrt{\dfrac{s_1^2}{n_1} + \dfrac{s_2^2}{n_2}}$

   $$(28{,}520 - 27{,}210) \pm 2.58\sqrt{\dfrac{(1{,}510)^2}{90} + \dfrac{(950)^2}{60}}$$

   $1{,}310 \pm 2.58(200.938)$

   $1{,}310 \pm 518.42$ or 791.58 to $1{,}828.42$.

   b. If the two schools belonged to populations having the same mean and annual income, then $\mu_1 = \mu_2$, or $\mu_1 - \mu_2 = 0$. This value of $\mu_1 - \mu_2$ does not fall in the confidence interval obtained above. Hence it is unlikely that the two schools belong to populations having the same mean annual income.

8.6   ONE-SIDED CONFIDENCE BOUNDS

The confidence intervals presented in the last section are often called **two-sided** confidence intervals, since they produce upper and lower bounds within which the parameter of interest is expected to lie. There are times when those conducting an experiment or a survey are interested only in either the lower or upper bound for the parameter of interest. For example, a process control person would not be interested in the lower bound for the proportion of defective parts, since the fewer defective parts the more efficient the process, but might rather be very concerned about the _____ bound for the fraction of defective parts. Someone making an investment in property might hedge on buying based upon the estimate of the largest possible interest rate for the next quarter, but would not be concerned if the interest rate were to fall.

upper

When the sampled population is normal or approximately so, we can show that the one-sided $(1 - \alpha)100\%$ confidence bounds for a parameter will enclose the parameter $(1 - \alpha)100\%$ of the time.

A $(1 - \alpha)100\%$ Lower Confidence Bound is given by:

(Point estimate) $- z_\alpha$ (Standard error of the estimate)

A $(1 - \alpha)100\%$ Upper Confidence Bound is given by:
(Point estimate) $+ z_\alpha$ (Standard error of the estimate)

Example 8.9
A chemical manufacturing firm wishes to estimate the maximum percent of impurities in the raw material that it is receiving from its suppliers. Specifically, the firm would like to know if the percent of impurities exceeds 1.5%. A sample of $n = 40$ samples from its suppliers' product was found to have an average of $\bar{x} = 1.1\%$ of impurities with a standard deviation of $s = .50\%$. What is the upper confidence bound for the percent of impurities in the suppliers' material?

Solution
You can assume that the firm is looking for a 95% upper confidence bound for the average percentage of impurities. The upper confidence bound is found using

$$\bar{x} + 1.645 \, \frac{\sigma}{\sqrt{n}} \approx 1.1 + 1.645\left(\frac{}{\sqrt{40}}\right) = 1.1 + \underline{\quad} = 1.18$$

.50; .08

or approximately _____.

1.2%

Therefore, the upper bound on the percent of impurities is less than the 1.5% figure given by the company.

Self-Correcting Exercises 8C

1. Find a 95% upper confidence bound for μ when $\bar{x} = 24.3$, $s = 4.3$ and $n = 40$.

2. Find a 99% lower confidence bound for μ when $\bar{x} = 50.2$, $s = 8.9$ and $n = 30$.

3. A recent survey found that 55% of $n = 200$ potential first-time homebuyers were in favor of adjustable rate mortgages (ARMs). Find a lower 95% confidence bound for the proportion of first-time homebuyers in favor of ARMs.

4. Prior to instituting a new advertising campaign a vintner conducted a product preference study involving $n = 1000$ regular wine buyers in the supermarket chain that serves as his primary distribution outlet and found that 33% were regular purchasers of his wine. After instituting a vigorous new advertising campaign, a survey of $n = 1200$ buyers were surveyed, with 44% indicating that they were regular purchasers of the vintner's product. Find a 95% upper confidence bound on the *increase* in the proportion of buyers following the new advertising campaign.

Solutions to SCE 8C

1. $\bar{x} + z_{.05} \dfrac{s}{\sqrt{n}} \Rightarrow 24.3 + 1.645 \dfrac{4.3}{\sqrt{40}} = 25.42$ or $\mu < 25.42$

2. $\bar{x} - z_{.01} \dfrac{s}{\sqrt{n}} \Rightarrow 50.2 - 2.33 \dfrac{8.9}{\sqrt{30}} = 46.41$ or $\mu > 46.41$

3. $\hat{p} - z_{.05} \sqrt{\dfrac{\widehat{pq}}{n}} \Rightarrow .55 - 1.645 \sqrt{\dfrac{.55(.45)}{200}} = .49$ or $p > .49$

4. $(\hat{p}_2 - \hat{p}_1) + z_{.05} \sqrt{\dfrac{\hat{p}_2 \hat{q}_2}{n_2} + \dfrac{\hat{p}_1 \hat{q}_1}{n_1}}$

   $(.44 - .33) + 1.645 \sqrt{\dfrac{.44(.56)}{1200} + \dfrac{.33(.67)}{1000}}$

   $= .11 + .03 = .14 \quad (p_2 - p_1) < .14$

8.7 CHOOSING THE SAMPLE SIZE

The amount of information contained in a sample depends on two factors:

1. The quantity of information per observation, which depends on the sampling procedure or experimental design
2. The number of measurements or observations taken, which depends on the sample size

In this section, we will concentrate on choosing the sample size required to obtain the desired amount of information.

One of the first steps in planning an experiment is deciding on the quantity of information that we wish to buy. At first glance it would seem difficult to specify a measure of the quantity of information in a sample relevant to a parameter of interest. However, such a practical measure is available in the margin of error in estimation, or, alternatively, we could use the half-width of the confidence interval for the parameter.

The larger the sample size, the greater will be the amount of information contained in the sample. This intuitively appealing fact is evident upon examination of the large-sample confidence intervals. The width of each of the four confidence intervals described in the preceding section is inversely proportional to the square root of the _____ _____. Suppose that an estimator satisfies the conditions for the large-sample estimators previously discussed. Then the bound B on the margin of error will be 1.96SE. This means that the error (in repeated sampling) will be less than 1.96SE with probability _____. If B represents the desired bound on the margin of error, then:

1. For a *point* estimate the restriction is $1.96\text{SE} = B$.
2. In an interval estimation problem with $(1 - \alpha)$ confidence coefficient, the restriction is $z_{\alpha/2}\text{SE} = B$.

Parts 1 and 2 will be equivalent when the confidence coefficient is .95.

Example 8.10
Suppose it is known that $\sigma = 2.25$ and it is desired to estimate μ with a bound on the margin of error less than or equal to .5 unit with probability .95. How large a sample should be taken?

Solution
The estimator for μ is $\bar{x}$ with standard error $\sigma/\sqrt{n}$; $1 - \alpha = .95$, $\alpha/2 = .025$, $z_{.025} = 1.96$. Hence, we solve

$$1.96\left(\frac{\sigma}{\sqrt{n}}\right) = B$$

or

$$1.96\left(\frac{2.25}{\sqrt{n}}\right) = .5$$

$$1.96\left(\frac{2.25}{.5}\right) = \sqrt{n}$$

$$\underline{\hspace{3cm}} = \sqrt{n}$$

$$\underline{\hspace{3cm}} = n$$

The solution is to take a sample of size _____ or greater to ensure that the margin of error is less than or equal to .5 unit. Had we wished to have the same margin of error with probability .99, the value

sample size

.95

8.82

77.79

78

2.58

2.58

134.79

135

1.645

.5

1691.27

1692

$z_{.005} =$ _____ would have been used, resulting in the following solution:

$$\text{\underline{\hspace{2cm}}} \left(\frac{2.25}{\sqrt{n}} \right) = .5$$

$$\sqrt{n} = \frac{2.58(2.25)}{.5}$$

$$n = \text{\underline{\hspace{2cm}}}$$

Hence, a sample of size _____ or greater would be taken to ensure estimation with $B = .5$ unit.

Example 8.11
If an experimenter wished to estimate the fraction of university students who read daily the college newspaper, correct to within .02 with probability .90, how large a sample of students should she take?

Solution
To estimate the binomial parameter with a 90% confidence interval, we would use

$$\hat{p} \pm \text{\underline{\hspace{2cm}}} \sqrt{\frac{pq}{n}}$$

We wish to find a sample size n so that

$$1.645\sqrt{\frac{pq}{n}} = .02$$

Since neither p nor $\hat{p}$ is known, we can solve for n by assuming the worst possible variation, which occurs when $p = q = $ _____. Hence, we solve

$$1.645\sqrt{\frac{.5(.5)}{n}} = .02$$

$$\frac{1.645(.5)}{.02} = \sqrt{n}$$

or

$$\text{\underline{\hspace{2cm}}} = n$$

Therefore, we should take a sample of size _____ or greater to achieve the required margin of error, even if faced with the maximum variation possible.

Example 8.12
An experiment is to be conducted to compare two different sales techniques at a number of sales centers. Suppose that the range of sales for the sales centers is expected to be $4000. How many centers

should be included for each of the sales techniques in order to estimate the difference in mean sales correct to within $500?

Solution
We will assume that the two sample sizes are equal—that is, $n_1 = n_2 = n$—and that the desired confidence coefficient is .95. Then

$$\underline{\hspace{3cm}} = B$$

$$1.96\sqrt{\frac{\sigma_1^2}{n} + \frac{\sigma_2^2}{n}}$$

The quantities σ_1^2 and σ_2^2 are unknown, but we know that the range is expected to be $4000. Then we would take $\sigma_1 = \sigma_2 = \underline{\hspace{2cm}}$ as the best available approximation. Then, substituting into the equation above,

1000

$$1.96\sqrt{\underline{\hspace{2cm}}} = 500$$

$$2(1000)^2/n$$

or

$$n = \underline{\hspace{2cm}}$$

30.73

Thus, $n = \underline{\hspace{2cm}}$ sales centers would be required for each of the two sales techniques.

31

Self-Correcting Exercises 8D

1. A device is known to produce measurements whose errors in measurement are normally distributed with a standard deviation $\sigma = 8$ millimeters. If the average measurement is to be reported, how many repeated measurements should be used so that the error in measurement is no larger than 3 millimeters with probability .95?

2. An experiment is to be conducted to compare the taste threshold levels for each of two food additives as measured by their concentrations in parts per million. How many subjects should be included in each experimental group in order to estimate the mean difference in threshold levels to within 10 units if the range of the measurements is expected to be approximately 80 parts per million for both groups?

3. How many individuals from each of two politically oriented groups should be included in a poll designed to estimate the true difference in proportions favoring a tuition increase at the state university correct to within .01 with probability .95? (In the absence of any prior information regarding the values of p_1 and p_2, solve the problem assuming maximum variation.)

4. If a mental health agency would like to estimate the percentage of local clinic patients that are referred to their counseling center to within 5 percentage points with 90% accuracy, how many patient records should be sampled?

Solutions to SCE 8D

1. The estimator of μ is $\bar{x}$, with standard error $\sigma/\sqrt{n}$. Hence, solve

$$1.96\sigma/\sqrt{n} = B, \ 1.96(8)/\sqrt{n} = 3, \ \sqrt{n} = 5.227, \ n = 27.32$$

   The experimenter should obtain $n = 28$ measurements.

2. For each additive, the range of the measurements is 80, so that $\sigma_1 = \sigma_2 \approx \text{range}/4 = 20$. Assuming equal sample sizes are acceptable, solve

$$1.96\sqrt{\frac{\sigma_1^2}{n} + \frac{\sigma_2^2}{n}} = 10, \ 1.95\sqrt{\frac{2(20)^2}{n}} = 10, \ \sqrt{n} = 5.543,$$

$$n = 30.73$$

   Hence, 31 subjects should be included in each group.

3. Maximum variation occurs when $p_1 = p_2 = .5$. Again assuming equal sample sizes, solve

$$1.96\sqrt{\frac{p_1 q_1}{n} + \frac{p_2 q_2}{n}} = .01, \ 1.96\sqrt{\frac{2(.5)(.5)}{n}} = .01,$$

$$\sqrt{n} = \frac{1.96\sqrt{.5}}{.01}, \ n = 19{,}208$$

4. Maximum variation occurs when $p = .5$. Since 90% accuracy is involved, solve

$$1.645\sqrt{pq/n} = .05, \ 1.645\sqrt{(.5)(.5)/n} = .05,$$

$$\sqrt{n} = 1.645(.5)/.05 = 16.45, \ n = 270.6$$

   271 patient records should be sampled.

8.8 SUMMARY

Estimation as a method of inference making has been discussed in this chapter. Point estimation and interval estimation for four parameters, μ, p, $\mu_1 - \mu_2$, and $p_1 - p_2$, have been presented for large-sample situations.

   To provide a brief summary of the preceding sections, complete the following tables.

1. Give the best estimator for each of the following parameters:

Parameter	Estimator
μ	_____
p	_____
$\mu_1 - \mu_2$	_____
$p_1 - p_2$	_____

$\bar{x}$

$\hat{p}$

$\bar{x}_1 - \bar{x}_2$

$\hat{p}_1 - \hat{p}_2$

2. Give the standard deviations for the following estimators:

Estimator	Standard Error
$\bar{x}$	_____
$\hat{p}$	_____
$\bar{x}_1 - \bar{x}_2$	_____
$\hat{p}_1 - \hat{p}_2$	_____

$\sigma/\sqrt{n}$

$\sqrt{\dfrac{pq}{n}}$

$\sqrt{\dfrac{\sigma_1^2}{n_1} + \dfrac{\sigma_2^2}{n_2}}$

$\sqrt{\dfrac{p_1 q_1}{n_1} + \dfrac{p_2 q_2}{n_2}}$

3. The exact values for standard deviations of estimators cannot usually be found because they are functions of unknown population parameters. Indicate the best approximations of the standard deviations for use in confidence intervals:

Estimator	Best Approximation of Standard Deviation
$\bar{x}$	_____
$\hat{p}$	_____
$\bar{x}_1 - \bar{x}_2$	_____
$\hat{p}_1 - \hat{p}_2$	_____

$s/\sqrt{n}$

$\sqrt{\dfrac{\hat{p}\hat{q}}{n}}$

$\sqrt{\dfrac{s_1^2}{n_1} + \dfrac{s_2^2}{n_2}}$

$\sqrt{\dfrac{\hat{p}_1 \hat{q}_1}{n_1} + \dfrac{\hat{p}_2 \hat{q}_2}{n_2}}$

Margins of error are given as _____ or its approximation, and large-sample confidence intervals are given as _____, using the best approximation for unknown parameters in SE.

1.96SE
Point estimator
± 1.96SE

KEY CONCEPTS AND FORMULAS

I. *Types of Estimators*
   1. Point estimator: a single number is calculated to estimate the population parameter.
   2. Interval estimator: two numbers are calculated to form an interval that contains the parameter.
II. *Properties of Good Estimators*
   1. Unbiased: the average value of the estimator equals the parameter to be estimated.
   2. Minimum variance: of all the unbiased estimators, the best estimator has a sampling distribution with the smallest standard error.

3. The margin of error measures the maximum distance between the estimator and the true value of the parameter.

III. *Large-Sample Point Estimators*

To estimate one of four population parameters when the sample sizes are large, use the following point estimators with the appropriate margin of error.

Parameter	Point Estimator	Margin of Error
μ	$\bar{x}$	$\pm 1.96\left(\dfrac{s}{\sqrt{n}}\right)$
p	$\hat{p} = \dfrac{x}{n}$	$\pm 1.96\sqrt{\dfrac{\hat{p}\hat{q}}{n}}$
$\mu_1 - \mu_2$	$\bar{x}_1 - \bar{x}_2$	$\pm 1.96\sqrt{\dfrac{s_1^2}{n_1} + \dfrac{s_2^2}{n_2}}$
$p_1 - p_2$	$(\hat{p}_1 - \hat{p}_2) = \left(\dfrac{x_1}{n_1} - \dfrac{x_2}{n_2}\right)$	$\pm 1.96\sqrt{\dfrac{\hat{p}_1\hat{q}_1}{n_1} + \dfrac{\hat{p}_2\hat{q}_2}{n_2}}$

IV. *Large-Sample Interval Estimators*

To estimate one of four population parameters when the sample sizes are large, use the following interval estimators.

Parameter	$(1-\alpha)100\%$ Confidence Interval
μ	$\bar{x} \pm z_{\alpha/2}\left(\dfrac{s}{\sqrt{n}}\right)$
p	$\hat{p} \pm z_{\alpha/2}\sqrt{\dfrac{\hat{p}\hat{q}}{n}}$
$\mu_1 - \mu_2$	$(\bar{x}_1 - \bar{x}_2) \pm z_{\alpha/2}\sqrt{\dfrac{s_1^2}{n_1} + \dfrac{s_2^2}{n_2}}$
$p_1 - p_2$	$(\hat{p}_1 - \hat{p}_2) \pm z_{\alpha/2}\sqrt{\dfrac{\hat{p}_1\hat{q}_1}{n_1} + \dfrac{\hat{p}_2\hat{q}_2}{n_2}}$

1. All values in the interval are possible values for the unknown population parameter.
2. Any values outside the interval are unlikely to be the value of the unknown parameter.
3. To compare two population means or proportions, look for the value 0 in the confidence interval. If 0 is in the interval, it is possible that the two population means or proportions are equal, and you should not predict a difference. If 0 is not in

the interval, it is unlikely that the two means or proportions are equal, and you can confidently predict a difference.

V. *One-Sided Confidence Bounds*

Use either the upper $(+)$ or lower $(-)$ two-sided bound, with the critical value of z changed from $z_{\alpha/2}$ to z_α.

VI. *Choosing the Sample Size*

1. Determine the size of the margin of error, B, that you are willing to tolerate.

2. Choose the sample size by solving for n or $n = n_1 = n_2$ in the inequality: $1.96\text{SE} \leq B$, where SE is a function of the sample size n.

3. For quantitative populations, estimate the population standard deviation using a previously calculated value of s or the range approximation $\sigma \approx \text{Range}/4$.

4. For binomial populations, use the conservative approach and approximate p using the value $p = .5$.

Exercises

1. List the two essential elements of any inference-making procedure.

2. What are two desirable properties of a point estimator θ?

3. A bank was interested in estimating the average size of its savings accounts for a particular class of customer. If a random sample of 400 such accounts showed an average amount of \$61.23 and a standard deviation of \$18.20, place 90% confidence limits on the actual average account size.

4. If 36 measurements of the specific gravity of aluminum had a mean of 2.705 and a standard deviation of .028, construct a 98% confidence interval for the actual specific gravity of aluminum.

5. In a study to compare the effects of two pain relievers it was found that of $n_1 = 200$ randomly selected individuals instructed to use the first pain reliever, 93% indicated that it relieved their pain. Of $n_2 = 450$ randomly selected individuals instructed to use the second pain reliever, 96% indicated that it relieved their pain. Find a 98% confidence interval for the difference in proportions experiencing relief from pain for these two pain relievers.

6. In a sample of 400 seeds, 240 germinate. Estimate the true germination percentage with a 95% confidence interval.

7. Refer to Exercise 4. Estimate the specific gravity of aluminum and place bounds on the error of estimation. Compare these results to the results obtained in Exercise 4. Can you explain the difference?

8. In a study to establish the absolute threshold of hearing, 70 male college freshmen were asked to participate. Each subject was seated in a soundproof room and a 150 Hz tone was presented at

a large number of stimulus levels in a randomized order. The subject was instructed to press a button if he detected the tone. The mean for the group was 21.6 db with $s = 2.1$. Estimate the mean absolute threshold of all college freshmen and place bounds on the error of estimation.

9. A researcher classified his subjects as innately right-handed or left-handed by comparing thumb-nail widths. He took a sample of 400 men and found that 80 men could be classified as left-handed according to his criterion. Estimate p for all males with a 95% confidence interval, where p represents the probability a man tests to be left handed.

10. An entomologist wishes to estimate the average development time of the citrus red mite correct to within .5 day. From previous experiments it is known that σ is in the neighborhood of 4 days. How large a sample should the entomologist take to be 95% confident of her estimate?

11. A grower believes that 1 in 5 of his citrus trees are infected with the citrus red mite mentioned in Exercise 10. How large a sample should be taken if the grower wishes to estimate the proportion of his trees that are infected with citrus red mite to within .08.

12. It is desired to estimate $\mu_1 - \mu_2$ from information contained in independent random samples from populations with variances $\sigma_1^2 = 9$ and $\sigma_2^2 = 16$. If the two sample sizes are to be equal $(n_1 = n_2 = n)$, how large should n be in order to estimate $\mu_1 - \mu_2$ with an error less than 1.0 (with probability equal to .95)?

13. To compare the effect of stress in the form of noise upon the ability to perform a simple task, 70 subjects were divided into two groups. The first group of 30 subjects were to act as a control, while the second group of 40 were to be the experimental group. Although each subject performed the task in the same control room, each of the experimental group subjects had to perform the task while loud rock music was being played in the room. The time to finish the task was recorded for each subject and the following summary was obtained:

	Control	Experimental
n	30	40
$\bar{x}$	15 minutes	23 minutes
s	4 minutes	10 minutes

Find a 99% confidence interval for the difference in mean completion times for these two groups.

14. In an experiment to assess the strength of the hunger drive in rats, 30 previously trained animals were deprived of food for 24 hours. At the end of the 24-hour period each animal was put into a cage where food was dispensed if the animal pressed a lever. The length of time the animal continued pressing the bar

(although he was receiving no food) was recorded for each animal. If the data yielded a sample mean of 19.3 minutes with a standard deviation of 5.2 minutes, estimate the true mean time and place bounds on the error of estimation.

15. Last year's records of auto accidents occurring on a given section of highway were classified according to whether the resulting damage was $1000 or more and to whether or nor a physical injury resulted from the accident. The tabulation follows:

	Under $1000	$1000 or more
Number of accidents	32	41
Number involving injuries	10	23

a. Estimate the true proportion of accidents involving injuries and damage of $1000 or more for similar sections of highway and place bounds on the error of estimation.

b. Estimate the true difference in proportion of accidents involving injuries for accidents involving less than $1000 in damage and those involving $1000 or more with a 95% confidence interval.

Chapter 9
Large-Sample Tests
of Hypotheses

9.1 INTRODUCTION

In most situations statistical inference takes the form of

- Estimating population parameters,
- Making a decision about the value of a parameter, or
- Taking actions based upon the value of one or more parameters.

As an administrator of a large health maintenance organization, you may be interested in the average bed occupancy rate for your hospitals in order to determine whether the rates charged are sufficient to cover daily expenses. You may sample n of your hospitals to determine the average number of occupied beds to *estimate* the population mean occupancy rate. However, you may know in advance that there must be at least 75% of the available beds occupied in order to remain solvent at the current prices. In this case, you may wish to test whether your sample result could have occurred if the underlying occupancy rate consisted of 75% of the available hospital beds. In this case you would perform a *test of hypothesis* in which the hypothesis "there is at least 75% average occupancy rate" is tested to determine whether the sample mean value could have reasonably occurred if the occupancy rate was at least 75%. If you then decided to alter the price structure for hospital stays based on the information in the sample, then you would be *taking an action* based upon your assessment of one or more population parameters.

In this chapter, we will concentrate on using the second mode of inference, namely, testing statistical hypotheses about population parameters using the information contained in a sample randomly selected from that population. We will consider testing hypotheses about the parameters, μ, the mean of a population of continuous observations, the binomial parameter p, the difference in two means, $\mu_1 - \mu_2$ and the difference in two binomial proportions, $p_1 - p_2$.

9.2 A STATISTICAL TEST OF HYPOTHESIS

A statistical test of hypothesis consists of five parts.

- The null hypothesis denoted by the symbol H_0.

- The alternative hypothesis denoted by the symbol H_a.

- The test statistic and its p-value.

- The rejection region.

- The conclusion.

A test of hypothesis is very much like a criminal trial under American jurisprudence, in which an accused person is considered _____   innocent
until proven guilty. The prosecutor presents evidence to prove that the
accused is _____. The jury analyzes the evidence, and based   guilty
upon this analysis comes to a conclusion, called the verdict, in which
they find the accused *guilty* or *not guilty.* The not guilty verdict in
general does not infer that the person is innocent, but rather that
there is insufficient evidence to reach a guilty verdict. Our two com-
peting hypotheses are:

- The alternative hypothesis, H_a, is generally the hypothesis
  that the researcher wishes to support.

- The null hypothesis, H_0, is the contradiction of the alternative
  hypothesis, and provides a value of a parameter that the researcher wishes to reject.

The method of proof is to begin by assuming that (H_0, H_a) is true and H_0
presenting evidence to refute H_0 and thereby show that $(\overline{H_0, H_a})$ is H_a
is true. When the statistical evidence has been presented, the re-
searcher must arrive at one of two conclusions:

- Reject H_0 and conclude that H_a is true, or

- Accept (do not reject) H_0 as true.

Example 9.1
The personnel office staff wishes to determine whether the current
average number of yearly sick days for its employees differs from 10,
the average number of sick days for the last five years. Present the
null and alternative hypotheses.

Solution
In this case, the null hypothesis would be

$\quad H_0$: $\mu = 10$

versus the alternative hypothesis

$\quad H_a$: μ _____. $\neq 10$

By examining employee records, we can determine whether the evidence supports rejecting H_0 in favor of H_a.

Example 9.2
The average cost of repairs to a vehicle rear-ended while stopped has been $1350. An insurance company would like to determine if there has been an increase in the average cost of repairs for the current year. What are the null and alternative hypotheses?

Solution
The insurance company is interested in showing that the alternative hypothesis

μ > $1350

H_a: _____

is true, versus the null hypothesis of no change,

H_0: $\mu = \$1350$.

The test set out in Example 9.1 would reject the null hypothesis if the average number of sick days were significantly greater or less than 10. Therefore, the null hypothesis would be rejected if the sample average were either too large or too small. This test would be called a **two-tailed test** of hypothesis. In Example 9.2, we are interested in knowing only if there has been an increase in the cost of repairs. This test would be called a **one-tailed test** of hypothesis.

The decision to reject or accept the null hypothesis is based on the information contained in either the **test statistic** or the **p-value** of the test. When the sample size is large, we can determine how many standard deviations the value of the statistics lies from the mean specified by H_0. If this number is two or greater in magnitude, we would be inclined to reject H_0 and accept H_a. This same information can be summarized by examining the probability of observing the value of the statistic or a value more extreme when H_0 is true. When the number of standard deviations is two or greater in absolute value, the probability of observing such a value is smaller than .05, or less than one chance in 20.

In conducting a statistical test of hypothesis, we must determine which values of the statistic support H_0 and which support H_a. The values of the test statistic that support H_a comprise the

rejection
acceptance

_____ **region**, and those that support H_0 comprise the _____ **region**. The **significance level** denoted by α denotes the amount of confidence that the researcher wishes to attach to the test conclusions.

The **level of significance (significance level)** for a statistical test is defined as

$$\alpha = P(\text{falsely rejecting } H_0) = P(\text{rejecting } H_0 \text{ when } H_a \text{ is true}).$$

In the next section, we will demonstrate the computations required in conducting a test of hypothesis for a population mean μ when the sample size is large, and the Central Limit Theorem assures us that the test statistic $\bar{x}$ is approximately normally distributed.

9.3 A LARGE-SAMPLE TEST OF A POPULATION MEAN

Example 9.3
A large orchard has averaged 140 pounds of apples per tree per year. A new fertilizer is tested to try to increase yield. Forty trees are randomly selected, and the mean and standard deviation of the yield are $\bar{x} = 143.2$ and $s = 9.4$. Do the data indicate a significant increase in yield?

Solution
If there is no increase in the average yield for fertilized trees, then the average value μ of the hypothetical population of fertilized trees from which we have obtained a random sample is $\mu = 140$. If, on the other hand, this is not true and the fertilizer tends to increase yield, then μ, the mean yield for fertilized trees, would be greater than 140. Since the researcher is interested in detecting an increase in yield, the statement $\mu > 140$ is the _____ or _____ hypothesis; the statement $\mu = 140$ is the _____ _____ .

research; alternative
null hypothesis

If we assume that there is no increase in the average yield for fertilized trees, it would be extremely unlikely in a sample of $n = 40$ trees to observe certain values of $\bar{x}$. Since $\bar{x}$ has an approximate normal distribution with $\mu_{\bar{x}} = 140$ and

$$\sigma_{\bar{x}} \approx \frac{s}{\sqrt{n}} = \frac{9.4}{\sqrt{40}} = \underline{\hspace{2cm}}$$

1.486

we can calculate the probability of observing a particular value of $\bar{x}$ or something even more extreme. For example, suppose that $\bar{x} = 150$. This is a highly unlikely event because

$$P(\bar{x} > 150) = P\left(z > \frac{150 - 140}{1.486} \right)$$

$$= P(z > \underline{\hspace{1.5cm}}) \approx \underline{\hspace{2cm}}$$

6.73; 0

The occurrence of $\bar{x} = 141$, however, is not so unlikely because

$$P(\bar{x} > 141) = P(z > \underline{\hspace{1.5cm}})$$

$\dfrac{141 - 140}{1.486}$

$$= P(z > \underline{\hspace{1.5cm}})$$

.67

$$= 1 - \underline{\hspace{1.5cm}} = \underline{\hspace{1.5cm}} .$$

.7486; .2514

In conducting a test of an hypothesis, the possible outcomes for $\bar{x}$ are divided into those for which we agree to reject the null hypothesis [those values of $\bar{x}$ that are much (greater, less) than $\mu = 140$] and those

greater

close to

for which we accept the null hypothesis [those values of $\bar{x}$ that are (close to, far away from) $\mu = 140$]. See Figure 9.1. There are many

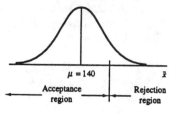

Figure 9.1

possible rejection regions available to the experimenter. One choice that is routinely made is to select a rejection region for which the probability of falsely rejecting H_0 when, in fact, H_0 is true is no larger than α, where $0 \leq \alpha \leq$ _____. Although the value of α in some circumstances might exceed .05, most experimenters prefer to keep α less than .05, thereby being certain of a 1 in 20 chance of error. The error of falsely rejecting H_0 when it is true is referred to as a **Type I Error**.

.05

A **type I Error** for a statistical test is the error made by rejecting the null hypothesis when it is true. The probability of committing a type I error is denoted by α where

$$\alpha = P(\text{type I error}) = P(\text{reject } H_0 \text{ when } H_0 \text{ is true})$$

If we were to formally conduct a test of hypothesis for Example 9.3, we would have the following formal structure.

H_0: $\mu = 140$ versus H_a: _____.

$\mu > 140$

Test Statistic:

$$z = \frac{\bar{x} - \mu}{\sigma/\sqrt{n}} = \frac{143.2 - 140}{9.4/\sqrt{40}} = \underline{\qquad}$$

2.15

Since this value of the sample mean lies more than 2 standard errors from the mean, we would conclude that this value of the sample mean is a suspect outlier from the distribution whose mean is 140.

Rejection Region: If you wish to control a Type I Error to be less than $\alpha = .05$, you would reject H_0 if the observed value of the test statistic is greater than $z_\alpha = z_{.05} =$ _____.

1.645

Conclusion: Since the sample mean lies more than 1.645 standard errors from the hypothesized mean of 140, you can reject H_0 and conclude that the mean yield is greater than 140 lbs. per tree.

Calculating the p-Value

In the previous example, the decision to reject or not reject H_0 was based on a critical value of z found according to the significance level α. There are many different rejection regions that could be utilized depending upon the value of α that you might choose. Another way to determine if the value of a test statistic is significant is to use a variable level of significance called the **p-value** of the test.

> The **p-value** or observed significance level of a statistical test is the smallest value of α for which the null hypothesis can be rejected.

For the right-tailed test in Example 9.3, the smallest value of α for which you could reject H_0 would be

$$P(z \geq 2.15) = 1 - .9842 = \underline{\hspace{2cm}} .$$ .0158

This is the *p-value* for the test in Example 9.3.

Statistical Significance. If the *p-value* of a test is less than a preassigned significance level α, the null hypothesis can be rejected, and the results are reported as significant at level α.

If your value of α is larger than .0158, say $\alpha = .05$, you (would, would not) decide to reject the null hypothesis that the mean   would
is $\mu = 140$ and conclude that the mean yield has increased above the 140 yield for the last harvest. On the other hand, if your value of α happens to be $\alpha = .01$, you (would, would not) reject the null hypothe-   would not
sis since the *p-value* = .0158 is larger than your value of α.

A small *p-value* indicates that the value of the test statistic is far from the mean and would support the rejection of H_0, while a large *p-value* indicates that the value of the test statistic is close to the mean and would not support the rejection of H_0.

Two Types of Errors

When the sample evidence is strong, your action would be to reject H_0, thereby committing a possible Type I Error by rejecting H_0 when H_0 is true with probability α. When the sample evidence does not warrant rejecting H_0, why would you not want to *accept* H_0? If you accept H_0, you may commit an error by accepting H_0 when, in fact, H_a is true. This action would also result in an error designated as a Type II error. These two errors and their probabilities designated by α and β, respectively, are defined in the next table.

The following table is a **decision table**; we look at two possible states of nature (H_0 and H_a) and two possible decisions in a test of hypothesis. Fill in the missing entries as either "correct" or "error":

Null Hypothesis	Decision	
	Reject H_0	Accept H_0
True	_____	Correct
False	Correct	_____

Error
Error

An error of type I is when we reject H_0 when H_0 is (true, false). An error of type II is when we fail to reject H_0 when H_a is (true, false). In considering a statistical test of any hypothesis, it is essential to know the probabilities of committing errors of type I and type II when the test is used in order to assess the goodness of the test.

true
true

A **type I error** for a statistical test is the error made by rejecting the null hypthesis when it is true. The probability of committing a type I error is denoted by α and

$$\alpha = P(\text{type I error}) = P(\text{reject } H_0 \text{ when } H_0 \text{ is true})$$

A **type II error** for a statistical test is the error made by accepting the null hypothesis when it is false and the alternative hythesis is true. The probability of committing a type II error is denoted by β and

$$\beta = P(\text{type II error}) = P(\text{accept } H_0 \text{ when } H_a \text{ is true})$$

For example 9.3, suppose we set the region $\bar{x} > 142$ as the rejection region.

1. If H_0 is true and $\mu = 140$,

$$\alpha = P(\text{reject } H_0 \text{ when } H_0 \text{ is true})$$

$$= P(\bar{x} > 142) = P(z > \frac{142 - 140}{1.486})$$

$$= P(z > \underline{\hspace{1cm}}) = 1 - \underline{\hspace{1cm}}$$

1.35; .9115

$$= \underline{\hspace{1cm}}.$$

.0885

2. If H_0 is false and $\mu = 145$,

$$\beta = P(\text{accept } H_0 \text{ when } H_a \text{ is true})$$

$$= P(\bar{x} < 142 \text{ when } \mu = 145)$$

$$= P(z < \frac{142 - 145}{1.486})$$

$$= P(z < \underline{\hspace{1cm}})$$

-2.02

$$= \underline{\hspace{1cm}}.$$

.0217

Notice that β is a function of H_a because by definition,

$$\beta = P(\text{accept } H_0 \text{ when } H_a \text{ is true})$$

By saying H_a is true, we mean that the true value of the population parameter is that given by H_a and β is computed using that value.

Consider a second rejection region, $\bar{x} > 144$. This rejection region is (larger, smaller) than the first region.

smaller

1. If H_0 is true and $\mu = 140$,

$$\alpha = P(\text{reject } H_0 \text{ when } H_0 \text{ is true})$$

$$= P(\bar{x} > 144 \text{ when } \mu = 140) = P(z > \frac{144 - 140}{1.486})$$

$$= P(z > \underline{\hspace{1cm}}) = 1 - \underline{\hspace{1cm}}$$

2.69; .9964

$$= \underline{\hspace{1cm}}.$$

.0036

2. If H_0 is false and $\mu = 145$,

$$\beta = P(\text{accept } H_0 \text{ when } H_a \text{ is true})$$

$$= P(\bar{x} < 142 \text{ when } \mu = 145)$$

$$= P(z < \frac{144 - 145}{1.486}) = P(z < \underline{\hspace{1cm}})$$

−.67

$$= \underline{\hspace{1cm}}.$$

.2514

Notice that both α and β depend on the rejection region used and that when the sample size n is fixed, α and β are inversely related: as one increases, the other _____. Increasing the sample size provides more information on which to make the decision and reduces the probability of a type II error. Since these two quantities measure the risk of making an incorrect decision, the experimenter chooses reasonable values for α and β and then chooses the rejection region and sample size accordingly. Since experimenters have found that a 1-in-20 chance of a type I error is usually tolerable, common practice is to choose $\alpha \leq$ _____ and a sample size n large enough to provide the desired control of the type II error.

decreases

.05

A graph of β, the probability of a type II error, as a function of the true value of the parameter under test is called the **operating characteristic curve** for the test. The operating characteristic curve associated with using an $\bar{x}$ chart for the process mean or the p chart for the fraction defective described in Chapter 6 would plot the probability of failing to stop and correct the process when in fact it was out of control.

For any statistical test, the value of α is usually selected and fixed in advance, often at values reflecting a 1-in-20 or a 1-in-100 chance of committing a type I error. The value of β, the probability of incorrectly accepting the null hypothesis, depends on the sample size n and the true value of the parameter under test.

For example, if the null hypothesis specifies H_0: $\mu = 140$, which alternative hypothesis would be more easily detected, H_a: $\mu = 141$ or H_a: $\mu = 200$? The answer, of course, is H_a: $\mu = 200$. In fact, a very large sample would be required to detect H_a: $\mu = 141$, that the mean is one unit larger than that specified by H_0. The ability of a test to reject H_0 when H_a is true is called the **power** of the test because a statistical test of an hypothesis is set up with the intention of disproving, and hence rejecting, H_0 when H_a is true.

The **power of a statistical test** is given by

$$1 - \beta = P(\text{reject } H_0 \text{ when } H_a \text{ is true})$$

and serves as a measure of the ability of a test to perform as required.

Let us return to Example 9.3 and determine a rejection region such that the probability of falsely rejecting H_0 is $\alpha = .05$. Since the alternative hypothesis is H_a: $\mu > 140$, the rejection region would be in the right tail of the distribution. Therefore we would reject the null hypothesis if the observed value of z exceeded $z_{.05} = 1.645$, the value of z having an area of .05 to its right. With a fixed rejection region, we can determine how well the test performs in rejecting H_0 when μ is greater than 140. We can find the value of $\bar{x}$ that corresponds to $z = 1.645$.

$$.05 = P\left(\frac{\bar{x} - \mu}{\sigma/\sqrt{n}} > 1.645\right) = P(\bar{x} > 140 + 1.645(\sigma/\sqrt{n}))$$

$$\approx P(\bar{x} > 140 + 1.645(1.486))$$

The value of $\bar{x}$ that corresponds to 1.645 is $\bar{x}_{.05} = 142.44$. Therefore we can evaluate β and the power, $1 - \beta$. The value of β is found by evaluating $P(\bar{x} < 142.44 \mid \mu_a)$ for various values of μ_a. Fill in the missing entries.

μ_a	z	β	$1 - \beta$
140	1.645	.9500	.0500
141	.97	____	.1663
142	.30	.6164	.3836
143	−.38	.3531	____
144	−1.05	.1469	.8531
145	−1.72	____	.9575
146	−2.40	.0083	.9917
147	−3.07	.0011	____
148	−3.74	.0001	.9999

.8337

.6469

.0425

.9989

A plot of the power, $1 - \beta$, against various values of μ is called the power curve as shown in Figure 9.2.

Notice that as the actual value of the mean, μ_a, increases, the probability that H_0 will be rejected increases and approaches 1 as μ_a becomes far removed from the value $\mu_0 = 140$.

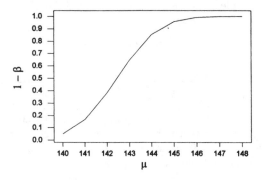

Figure 9.2

There are obvious links between α, β and the sample size n. The following statements are true.

- When α increases and n is held constant, β will (<u>increase, decrease</u>) and vice-versa.

decrease

- When α is held fixed, and n increases, β will (<u>increase, decrease</u>) and vice-versa.

decrease

Some simple sketches will show how these relationships work.

9.4   A LARGE SAMPLE TEST OF HYPOTHESIS FOR THE DIFFERENCE IN TWO POPULATION MEANS

A large number of experiments are routinely conducted to compare the difference in means for two populations. Improved treatments are routinely compared with the standard mode of treatment. In such cases, the object of the experiment is to determine whether the difference in two sample means provides enough information to conclude that there is a significant difference in two population means. The information in the next display provides the mean and standard deviation of the difference in two sample means in which the two samples were _____ drawn from two distinct populations.

independently

The Sampling Distribution of $\bar{x}_1 - \bar{x}_2$

1. The mean of the difference in two sample means is given by $\mu_1 - \mu_2$.

2. The standard deviation of $\bar{x}_1 - \bar{x}_2$ is given by $\sqrt{\dfrac{\sigma_1^2}{n_1} + \dfrac{\sigma_2^2}{n_2}}$.

3. When both sample sizes are ≥ 30, the standard deviation or standard error can be approximated by $\sqrt{\dfrac{s_1^2}{n_1} + \dfrac{s_2^2}{n_2}}$.

Example 9.4

Suppose an educator is interested in testing whether a new teaching technique is superior to an old teaching technique. The criterion will be a test given at the end of a six-week period, and the technique that results in a significantly higher score will be judged superior. Test the hypothesis that the test means for the techniques are the same against the alternative hypothesis that the new technique is superior at the $\alpha = .05$ level, based on the following sample data:

New Technique	Old Technique
$\bar{x}_1 = 69.2$	$\bar{x}_2 = 67.5$
$s_1^2 = 49.3$	$s_2^2 = 64.5$
$n_1 = 50$	$n_2 = 80$

Solution

0

1. H_0: $\mu_1 - \mu_2 = $ _____

2. H_a: $\mu_1 - \mu_2 > 0$

3. Test statistic:

$$z = \frac{(\bar{x}_1 - \bar{x}_2) - 0}{\sqrt{\dfrac{s_1^2}{n_1} + \dfrac{s_2^2}{n_2}}}$$

(since σ_1^2 and σ_2^2 are unknown).

4. Rejection region:

1.645

Reject H_0 if $z > $ _____.

Computing the value of the test statistic, we get

69.2; 67.5

$$z = \frac{(\underline{\quad\quad} - \underline{\quad\quad}) - 0}{\sqrt{\dfrac{49.3}{50} + \dfrac{64.5}{80}}}$$

$$= \frac{}{\sqrt{716.9/400}} = \frac{1.7}{1.34} = \underline{\qquad\qquad}$$

1.7; 1.27

Since the value of z (does, does not) fall in the rejection region, we (will, will not) reject H_0. Before deciding to accept H_0 as true, we may wish to evaluate the probability of a type II error for meaningful values of $\mu_1 - \mu_2$ described by H_a. Until this is done, we will state our decision as "do not reject H_0."

does not
will not

Would your conclusion change if you had used the *p-value* approach? To evaluate the *p-value*, we need to find

$$p\text{-}value = P(z > 1.27) = 1 - .8980 = \underline{\qquad\qquad}.$$

.1020

Since the *p-value* is greater than the value of $\alpha = .05$, you (can, cannot) reject the null hypothesis. Until you are able to evaluate the probability of a Type II error, you should withhold judgment, and not *accept* H_0. This decision agrees with that found using an $\alpha = .05$ fixed rejection region.

cannot

Self-Correcting Exercises 9A

1. To investigate a possible "built-in" sex bias in a graduate school entrance examination, 50 male and 50 female graduate students who were rated as above-average graduate students by their professors were selected to participate in the study by actually taking this test. Their test results on this examination are summarized in the following table.

	Males	Females
$\overline{x}$	720	693
s^2	104	85
n	50	50

Do these data indicate that males will, on the average, score higher than females of the same ability on this exam? Use $\alpha = .05$.

2. A machine shop is interested in determining a measure of the current year's sales revenue in order to compare it with known results from last year. From the 9682 sales invoices for the current year to date, the management randomly selected $n = 400$ invoices and from each recorded x, the sales revenue per invoice. Using the following data summary, test the hypothesis that the mean revenue per invoice is $6.35, the same as last year, versus the alternative hypothesis that the mean revenue per invoice is different from $6.35, with $\alpha = .05$.

Data Summary		
$n = 400$	$\sum\limits_{i=1}^{400} x_i = \$2464.40$	$\sum\limits_{i=1}^{400} x_i^2 = 16{,}156.728$

3. Refer to Exercise 1. Find the *p-value* associated with this test.

4. Refer to Exercise 2. Find the *p-value* associated with this test and interpret your results.

Solutions to SCE 9A

1. $H_0: \mu_1 - \mu_2 = 0$; $H_a: \mu_1 - \mu_2 > 0$; test statistic:

$$z = \frac{(\bar{x}_1 - \bar{x}_2) - 0}{\sqrt{\dfrac{s_1^2}{n_1} + \dfrac{s_2^2}{n_2}}}.$$

With $\alpha = .05$, reject H_0 if $z > 1.645$. Calculate

$$z = \frac{720 - 693}{\sqrt{\dfrac{104}{50} + \dfrac{85}{50}}} = \frac{27}{1.94} = 13.92.$$

Reject H_0. There is a significant difference in the mean scores. Men score higher on the average than women.

2. $H_0: \mu = 6.35$; $H_a: \mu \neq 6.35$; test statistic:

$z = \dfrac{\bar{x} - \mu_0}{s/\sqrt{n}}$. With $\alpha = .05$, reject H_0 if $|z| > 1.96$. Calculate

$\bar{x} = 2464.40/400 = 6.161$;

$s^2 = (16{,}156.728 - 15{,}183.168)/399 = 2.4400$;

$s = \sqrt{2.44} = 1.56$;

$z = \dfrac{\bar{x} - \mu_0}{s/\sqrt{n}} = \dfrac{6.161 - 6.35}{1.56/20} = -2.42.$

Reject H_0. Mean revenue is different from \$6.35.

3. The significance level for a one-tailed test is
p-value $= P(z > 13.92) \approx 0$. These results are considered very highly significant.

4. The level of significance for a two-tailed test is *p-value* $=$
$P(|z| > 2.42) = 2P(z > 2.42) = 2(1 - .9922) = 2(.0078) = .0156$. This small *p-value* indicates that you can reject H_0 at the $\alpha = .05$ level of significance, but not at the $\alpha = .01$ level of significance.

9.5   A LARGE SAMPLE TEST OF HYPOTHESIS FOR A BINOMIAL PROPORTION

When a sample of size n is drawn from a binomial population with parameter p, the distribution of the sample proportion $\hat{p}$ will be approximately normally distributed with mean p and standard deviation, also called the standard error, given by

$$SE = \sqrt{\frac{pq}{n}}.$$

The sampling distribution of $\hat{p}$.

1. The mean of $\hat{p}$ is given by p.

2. The standard deviation or standard error of $\hat{p}$ is given

   by $\sqrt{\frac{pq}{n}}$.

3. When $np > 5$ and $nq < 5$, the distribution of $\hat{p}$ is approximately normally distributed.

In testing the hypothesis

$$H_0: p = p_0$$

against one of the three alternatives

$$H_a: p \neq p_0 \quad \text{or} \quad H_a: p > p_0 \quad \text{or} \quad H_a: p < p_0$$

the test statistic is given by

$$z = \frac{\hat{p} - p_0}{\sqrt{\frac{p_0 q_0}{n}}} = \frac{\hat{p} - p_0}{SE}$$

The rejection region will be two-tailed when $H_a: p \neq p_0$, right-tailed when $H_a: p > p_0$ and left-tailed when $H_a: p < p_0$. Remember that the normal approximation to the binomial distribution is adequate when $np_0 > $ _____ and $nq_0 > $ _____.   5; 5

Example 9.5
To assess the effect of the color of food on taste preferences, 65 subjects were each asked to taste two samples of mashed potatoes, one of which was colored pink and the other was its natural color. Although both samples were identical except for the pink color produced by the addition of a drop of tasteless food coloring, 53 of the subjects preferred the taste of the sample that had its natural color. Does this indicate that subjects tend to be adversely affected by the pink color? Test at the $\alpha = .01$ level of significance.

Solution
If there is no preference according to color of the mashed potatoes, then the probability that an individual chooses the natural or colored potatoes is $p = .5$, or a 1 in 2 chance. Let $x = $ the number who choose the natural color.

.815

Then $\hat{p} = \frac{53}{65} = $ _____ and $p_0 = .5$. The formal test follows.

1. H_a: $p = .5$

2. H_a: $p > .5$

3. Test statistic:

$$z = \frac{\hat{p} - p_0}{\sqrt{p_0 q_0 / n}}$$

4. Rejection region:

2.33

Reject H_0 if $z > $ _____ .

To calculate z, we need $\hat{p} = x/n = 53/65 = .815$. Then

.500

$$z = \frac{.815 - \underline{\hspace{1cm}}}{\sqrt{.5(.5)/65}} = \frac{.315}{.062} = 5.08$$

reject
does

Since $z = 5.08 > 2.33$, we (reject, do not reject) H_0 and conclude that color (does, does not) adversely affect a subject's choice.

How would the *p-value* approach be applied in this situation? Since the test is right-tailed, you would want to assess the probability of observing a value of z as large or larger than the sample value. That is,

$$p\text{-}value = P(z > 5.08) \approx 0.$$

reject H_0

Your decision would agree with that decided upon using a fixed $\alpha = .01$ rejection region, and that is to (reject H_0, do not reject H_0).

9.6   A LARGE-SAMPLE TEST OF HYPOTHESIS FOR THE DIFFERENCE BETWEEN TWO BINOMIAL PROPORTIONS

When random and independent samples are selected from two binomial populations with proportions p_1 and p_2, the investigation is generally centered about the difference in proportions, $p_1 - p_2$. The statistic $\hat{p}_1 - \hat{p}_2$ is an unbiased and minimum variance estimator of

$p_1 - p_2$ with standard error $\sqrt{\frac{p_1 q_1}{n_1} + \frac{p_2 q_2}{n_2}}$. There is a difference in

the standard error of $\hat{p}_1 - \hat{p}_2$ when the null hypothesis is that H_0: $p_1 - p_2 = 0$ or equivalently H_0: $p_1 = p_2 = p$, where p is the common value for both populations. When the null hypothesis of no difference between the two population proportions is true, there are some subtle, but obvious changes in the standard error. When $p_1 = p_2 = p$, the standard error becomes

$$\sqrt{\frac{pq}{n_1} + \frac{pq}{n_2}} = \sqrt{pq\left(\frac{1}{n_1} + \frac{1}{n_2}\right)}$$

where p and q are the common values for both populations. The best estimate of p is the pooled estimate given by

$$\widehat{p} = \frac{x_1 + x_2}{n_1 + n_2}.$$

The sampling distribution of $\widehat{p}_1 - \widehat{p}_2$ when H_0: $p_1 - p_2 = 0$.

1. $\mu_{p_1 - p_2} = p_1 - p_2 = 0.$

2. Standard Error $= \sqrt{\widehat{p}\widehat{q}\left(\frac{1}{n_1} + \frac{1}{n_2}\right)}$

3. Pooled estimate of p is $\widehat{p} = \dfrac{x_1 + x_2}{n_1 + n_2}$

4. The test statistic is $z = \dfrac{(\widehat{p}_1 - \widehat{p}_2) - 0}{\sqrt{\widehat{p}\widehat{q}\left(\frac{1}{n_1} + \frac{1}{n_2}\right)}}$

Sample 9.6
To investigate possible differences in attitude about a current political problem, 100 randomly selected voters between the ages of 18 and 25 were polled and 100 randomly selected voters over age 25 were polled. Each was asked whether he or she agreed with the government's position on the problem. Forty-five of the first group agreed, and 63% of the second group agreed. Do these data represent a significant difference in attitude for these two groups?

Solution
This problem involves a test of the difference between two binomial proportions, $p_1 - p_2$. The relevant data are given in the table:

	Group 1	Group 2
n	100	100
$\widehat{p}$	.45	.63

1. H_0: $p_1 - p_2$ _____ $= 0$

2. H_a: $p_1 - p_2$ _____ $\neq 0$

3. For testing the hypothesis of *no difference* between proportions, the test statistic is

$$z = \frac{(\hat{p}_1 - \hat{p}_2) - 0}{\sqrt{\hat{p}\hat{q}\left(\frac{1}{n_1} + \frac{1}{n_2}\right)}}$$

with

$$\hat{p} = \frac{x_1 + x_2}{n_1 + n_2}$$

4. Rejection region:

1.96

1.96

.54

For a two-tailed test with $\alpha = .05$, we will reject H_0 if $|z| >$ _____. To calculate the test statistic, we need

$$\hat{p} = \frac{x_1 + x_2}{n_1 + n_2} = \frac{45 + 63}{200} = \text{_____}$$

Then

$$z = \frac{(.45 - .63) - 0}{\sqrt{(.54)(.46)(2/100)}}$$

−.18

$$= \frac{\text{_____}}{.0705}$$

−2.55

$$= \text{_____}$$

reject
is

Since $|-2.55| = 2.55 > 1.96$, we (reject, do not reject) H_0 and conclude that there (is, is not) a significant difference in opinion between these two age groups with respect to this issue.

Example 9.7
What is the *p-value* of the test in Example 9.6?

Solution
Since the alternative hypothesis was two-tailed, you need to find the probability of observing a value of the test statistic or a value more extreme. Since $z = -2.55$, you need to find

.0054

$$P(z < -2.55) = \text{_____}.$$

double

Since the test is two-tailed, you must _____ this value. The *p-value* is therefore

.0108

$$p\text{-}value = 2(.0054) = \text{_____}.$$

Any researcher whose value of α was greater than or equal to .0108 would reject H_0 and conclude that there is a significant difference in the underlying proportions, p_1 and p_2.

9.7 THE LEVEL OF SIGNIFICANCE OF A STATISTICAL TEST

The structure of a statistical test of an hypothesis can be summarized as follows:

1. State the null and _____ (or _____) hypotheses.   research; alternative
2. Choose a test statistic.
3. Choose a value of α and, depending on the nature of the alternative hypothesis, establish a one- or two-tailed rejection region for which the α level is approximately the level chosen.
4. Perform the experiment, calculate the test statistic, and come to a conclusion based on the observed value of the test statistic. If the value of the test statistic falls in the rejection region, we _____ the null hypothesis in favor of the alternative hypothesis. If the value of the test statistic is not in the rejection region, we (can, cannot) reject the null hypothesis in favor of the alternative hypothesis.

reject

cannot

One difficulty in utilizing the approach outlined above is that the choice of the α level is to some extent subjective. Another researcher may disagree with your conclusions regarding the research hypothesis because he or she does not agree with your choice of the α level. The observed level of significance or *p-value* for an observed value of the test statistic is the smallest α value for which the null hypothesis could be rejected. In using the rare-event philosophy, we reject the null hypothesis in favor of the alternative whenever the total probability of the observed value of the test statistic and any rarer event is _____.

small

Consider the problem discussed in Example 9.3. We tested H_0: $\mu = 140$ versus H_a: $\mu > 140$ based on an observed value of $\bar{x} = 143.2$. Since this is a one-tailed test and "large" values of $\bar{x}$ belong in the rejection region, the smallest rejection region that contains $\bar{x}$ is the region $\bar{x} \geq 143.2$. The p-value associated with this observation is therefore

$$= P(z > \frac{143.2 - 140}{9.4/\sqrt{40}})$$

$$= P(z > 2.15)$$

$$= 1 - .9842 = \underline{\hspace{2cm}}.$$

.0158

Thus, any researcher who would specify an α value less than or equal to _____ would conclude that there (is, is not) sufficient evidence to accept the research hypothesis. Since this is a relatively small α value,

.0158; is

would

it would probably be acceptable, and the researcher (<u>would</u>, <u>would not</u>) reject H_0.

The *p-value* approach to hypothesis testing does not require specifying an α level before the analysis is undertaken. Rather, the degree of disagreement with the null hypothesis is quantified by the

p-value

calculation of the _____ _____, which is used to make decisions regarding the research hypothesis. Thus, for any researcher

$\geq$

specifying an α value ($\geq$, $\leq$) .002, the conclusion would be to accept the alternative hypothesis, $\mu \neq 10$.

In summary, the procedure used to determine the *p-value* (or level of significance) is as follows:

1. Look at the alternative hypothesis. If it is an upper-tail alternative, list all values of the test statistic greater than or equal to the value observed. If it is a lower-tail alternative, list all values

less

of the test statistic _____ than or equal to the value observed. If it is a two-tailed alternative, determine which tail of the distribution contains the observed value and list all values as rare as or rarer than the observed values that are in that tail.

2. Calculate the probability of getting a value of the test statistic in the collection of values given in step 1 assuming the

null; one

_____ hypothesis is true. If the test is _____ - tailed, this is the *p-value*.

twice

3. If the test is two-tailed, the *p-value* is _____ the probability calculated in step 2.

Self-Correcting Exercises 9B

1. A researcher found that 480 men and 420 women among 900 patients admitted to a hospital had high blood pressure. Is this consistent with the hypothesis that this medical problem occurs with equal frequency among men and women? Test using $\alpha = .10$.

2. Find the *p-value* associated with the test in Exercise 1 and interpret its value.

3. Random samples of 100 units from production line 1 and 50 units from production line 2 revealed 16 units with flaws from production line 1 and 6 units with flaws from line 2. Do these data present sufficient evidence to suggest that there is a significant difference in the number of units with flaws produced by these two production lines? Use $\alpha = .05$.

4. Find the *p-value* for the test in Exercise 3 and interpret its value.

Solutions to SCE 9B

1. H_0: $p = .5$; H_a: $p \neq .5$; test statistic: $z = \dfrac{\hat{p} - p_0}{\sqrt{p_0 q_0 / n}}$.

    With $\alpha = .10$, reject H_0 if $|z| > 1.645$. Since $\hat{p} = 480/900 = .533$,

    $$z = \frac{.533 - .5}{\sqrt{.5(.5)/900}} = \frac{.033}{.5/30} = \frac{.033}{.0167} = 1.98 \, .$$

    Reject H_0. Half of the cases are not male.

2. For a two-tailed test the *p-value* $= P(|z| > 1.98) = 2P(z > 1.98)$
    $= 2(1 - .9761) = 2(.0239) = .0478$. These results are significant at
    the $\alpha = .05$ level, but not at the $\alpha = .01$ level of significance.

3. $H_0 : p_1 - p_2 = 0; H_a : p_1 - p_2 \neq 0$; test statistic:

    $$z = \frac{\hat{p}_1 - \hat{p}_2 - 0}{\sqrt{\hat{p}\hat{q}(\dfrac{1}{n_1} + \dfrac{1}{n_2})}} \, .$$

    Notice that if $p_1 = p_2$ as proposed under H_0, the best estimate of this
    common value of p is $\hat{p} = (x_1 + x_2)/(n_1 + n_2) = (16 + 6)/(100 + 50) = .15$
    With $\alpha = .05$, reject H_0 if $|z| > 1.96$. Calculate

    $$z = \frac{(16/100) - (6/50) - 0}{\sqrt{.15(.85)(\dfrac{1}{100} + \dfrac{1}{50})}} = \frac{.04}{.0618} = .65.$$

    Do not reject H_0. There is insufficient evidence to detect a
    difference in the performances of the machines.

4. The *p-value* for this two-tailed test is given by *p-value* $=$
    $P(|z| > .65) = 2P(z > .65) = 2(1 - .7422) = 2(.2578) = .5156$. Since the *p-value* exceeds .10, you would conclude that there is no difference
    between these two production lines with respect to the number of
    flawed items produced.

KEY CONCEPTS AND FORMULAS

I. *Parts of a Statistical Test*
    1. Null hypothesis: a contradiction of the alternative hypothesis
    2. Alternative hypothesis: the hypothesis the researcher wants
       to support
    3. Test statistic and its *p-value*: sample evidence calculated from
       the sample data
    4. Rejection region–critical values and significance levels: values
       that separate rejection and nonrejection of the null hypothesis.
    5. Conclusion: Reject or do not reject the null hypothesis, stating
       the practical significance of your conclusion

II. *Errors and Statistical Significance*
  1. The significance level α is the probability of rejecting H_0 when it is in fact true.
  2. The *p-value* is the probability of observing a test statistic as extreme as or more extreme than the one observed; also, the smallest value of α for which H_0 can be rejected.
  3. When the *p-value* is less than the significance level α, the null hypothesis is rejected. This happens when the test statistic exceeds the critical value.
  4. In a Type II error, β is the probability of accepting H_0 when it is in fact false. The power of the test is $(1 - \beta)$, the probability of rejecting H_0 when it is false.

III. *Large-Sample Test Statistics Using the z Distribution*
  To test one of the four population parameters when the sample sizes are large, use the following test statistics:

Parameter	Test Statistic
μ	$z = \dfrac{\bar{x} - \mu_0}{s/\sqrt{n}}$
p	$z = \dfrac{\hat{p} - p_0}{\sqrt{\dfrac{p_0 q_0}{n}}}$
$\mu_1 - \mu_2$	$z = \dfrac{(\bar{x}_1 - \bar{x}_2) - D_0}{\sqrt{\dfrac{s_1^2}{n_1} + \dfrac{s_2^2}{n_3}}}$
$p_1 - p_2$	$z = \dfrac{\hat{p}_1 - \hat{p}_2}{\sqrt{\hat{p}\hat{q}\left(\dfrac{1}{n_1} + \dfrac{1}{n_2}\right)}}$

Exercises

1. What are the four essential elements of a statistical test of an hypothesis?

2. Assume that a certain set of "early returns" in an election is actually a random sample of size 400 from the voters in that election. If 225 of the voters in the sample voted for candidate A, could we assert with $\alpha = .01$ that candidate A has won?

3. Refer to Exercise 2 and calculate the *p-value* for the test. Interpret your results.

4. A grocery store operator claims that the average waiting time at a checkout counter is 3.75 minutes. To test this claim, a random sample of 30 observations was taken. Test the operator's claim at the 5% level of significance, using the sample data shown below.

Waiting Time in Minutes				
3	4	3	4	1
1	0	5	3	2
4	3	1	2	0
3	2	0	3	4
1	3	2	1	3
2	4	2	5	2

5. Refer to Exercise 4 and calculate the *p-value* for the test. Interpret your results.

6. In a maze running study, a rat is run in a T maze and the result of each run recorded. A reward in the form of food is always placed at the right exit. If learning is taking place, the rat will choose the right exit more often than the left. If no learning is taking place, the rat should randomly choose either exit. Suppose that the rat is given $n = 100$ runs in the maze and that he chooses the right exit $x = 64$ times. Would you conclude that learning is taking place? Give the *p-value* for the test and make a decision based on this *p-value*.

7. To test the effectiveness of a vaccine, 150 experimental animals were given the vaccine and 150 were not. All 300 were then exposed to the disease. Among those vaccinated, 10 contracted the disease. Among the control group (i.e., those not vaccinated), 30 contracted the disease. Can we conclude that the vaccine is effective in reducing the infection rate for this disease? Use a significance level of .025.

8. Two diets were to be compared. Seventy-five individuals were selected at random from a group of individuals whose doctors recommended they lose weight. Forty of this group were assigned diet A and the other 35 were placed on diet B. The weight losses in pounds over a period of 3 weeks were found and the following quantities recorded:

	Sample Size	Sample Mean (pounds)	Sample Variance
Diet A	40	10.3	7
Diet B	35	7.3	3.25

a. Do these data allow the conclusion that the expected weight loss under diet A (μ_A) is greater than the expected weight loss under diet B (μ_B)? Test at the .01 level. Draw the appropriate conclusion.

b. Construct a 90% confidence interval for $\mu_A - \mu_B$.

9. In a manufacturing plant employing a double inspection procedure, the first inspector is expected to miss an average of 25 defective items per day with a standard deviation of 3 items. If the first inspector has missed an average of 29 defectives per day based upon the last 30 working days, is he working up to company standards? (Use $\alpha = .01$.)

10. Refer to Exercise 9 and calculate the *p-value* for the test. Interpret your results.

11. To compare the tensile strengths of two synthetic fibers, samples of 31 of each kind were selected and tested for breaking strength. The results are summarized in the following table. Find a 98% confidence interval estimate for the difference between the true mean breaking strengths.

Fiber 1	*Fiber 2*
$\bar{x}_1 = 370$	$\bar{x}_2 = 332$
$s_1^2 = 1930$	$s_2^2 = 1670$
$n_1 = 31$	$n_2 = 31$

12. An experimenter has prepared a drug-dose level which he claims will induce sleep for at least 80 percent of people suffering from insomnia. After examining the dosage we feel that his claims regarding the effectiveness of his dosage are inflated. In an attempt to disprove his claim we administer his prescribed dosage to 50 insomniacs and observe that $x = 37$ have had sleep induced by the drug dose. Is there enough evidence to refute his claim?

13. During the 1998-99 academic year, approximately 93% of all students enrolled at a state university were residents of that state. Since the total student population in this university system is assumed to be quite large, the number x of state residents in a random sample of size n is approximately a binomial random variable. Suppose that the administrators of the university are interested in determining whether the percentage of state residents has changed for the academic year 1999-2000. A random sample of $n = 150$ students consists of 142 state residents. Does the data provide sufficient evidence to prove that the percentage has changed?
a. Test at the $\alpha = .05$ level of significance.
b. Find the *p-value* for the test.

14. Polychlorinated biphenols (PCBs) have been found to be dangerously high in some game birds found along the marshlands of the southeastern coast of the United States. The Federal Drug Administration (FDA) considers a concentration of PCBs higher than 5 parts per million (ppm) in these game birds to be dangerous for human consumption. A sample of $n = 38$ game birds

produced an average of $\bar{x} = 7.2$ ppm with a standard deviation of $s = 6.2$ ppm.

a. Find a 90% confidence interval estimate of the true average ppm of PCBs in the population of game birds sampled.

b. Use a statistical test of hypothesis to determine whether the mean ppm of PCBs differs from the FDA's recommended limit of $\mu = 5$ ppm.

c. What are β and $1 - \beta$ if the true mean ppm of PCBs is 6 ppm? If the true mean is 7 ppm?

d. Find $1 - \beta$ when $\mu = 8$, 9, 10, and 12. Use these values to construct a power curve for the test under consideration.

e. For what values of μ does this test have power greater than or equal to .90?

Chapter 10
Inference from Small Samples

10.1 INTRODUCTION

Central Limit

Large-sample methods for making inferences about a population were considered in the preceding chapters. When the sample size was large, the _____ _____ Theorem assured the approximate normality of the distribution of the estimator $\bar{x}$ or $\hat{p}$. However, time, cost, or other limitations may prevent an investigator from collecting enough data to feel confident in using large-sample techniques. When the sample size is small, $n < 30$, the Central Limit Theorem may no longer apply. This difficulty can be overcome if the investigator is reasonably sure

normal

that his or her measurements constitute a sample from a _____ population.

randomly

The results presented in this chapter are based on the assumption that the observations being analyzed have been _____ drawn from a normal population. This assumption is not as restrictive as it sounds because the normal distribution can be used as a model in cases where the underlying distribution is mound-shaped and fairly symmetrical.

10.2 STUDENT'S t DISTRIBUTION

When the sample size is large, the statistic

$$\frac{\bar{x} - \mu}{\sigma/\sqrt{n}}$$

is approximately distributed like the standard normal random variable z. What can be said about this statistic when n, the sample size, is small and the sample variance s^2 is used to estimate σ^2?

If the parent population is not normal (or approximately normal), the behavior of the statistic given above is not known in general when n is small. Its distribution could be empirically generated by repeated sampling from the population of interest. If the parent population **is** normal, we can rely on the results of W. S. Gosset, who published under the pen name Student. He drew repeated samples from a normal population and tabulated the distribution of a statistic that he called t, where

$$t = \frac{\bar{x} - \mu}{s/\sqrt{n}}$$

The resulting distribution for t, shown in Figure 10.1, has the following properties:

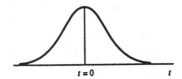

Figure 10.1

1. The distribution is _____-shaped.

   mound

2. The distribution is _____ about the value $t = 0$.

   symmetrical

3. The distribution has more flaring tails than z; hence, t is (more, less) variable than the z statistic.

   more

4. The shape of the distribution changes as the value of _____, the sample size, changes.

   n

5. As the sample size n becomes large, the t distribution becomes identical to the _____ _____ distribution.

   standard normal

These results are based on the following two assumptions:

1. The parent population has a _____ distribution. The t statistic is, however, relatively stable for nonnormal _____-shaped distributions.

   normal
   mound

2. The sample is a _____ sample. When the population is normal, this assures us that $\bar{x}$ and s^2 are independent.

   random

   For a fixed sample size n, the statistic

$$z = \frac{\bar{x} - \mu}{\sigma/\sqrt{n}}$$

contains exactly _____ random quantity, the sample mean _____. However, the statistic

   one
   $\bar{x}$

$$t = \frac{\bar{x} - \mu}{s/\sqrt{n}}$$

contains _____ random quantities, _____ and _____. This accounts for the fact that t is (more, less) variable than z. In fact, $\bar{x}$ may be large while s is small, or $\bar{x}$ may be small while s is large. Hence, it is said that $\bar{x}$ and s are _____, which means that the value assumed by $\bar{x}$ in no way determines the value of s.

   two; $\bar{x}$
   s; more

   independent

As the sample size changes, the corresponding t distribution changes so that each value of n determines a different probability distribution. This is because of the variability of s^2, which appears in the denominator of t. Large sample sizes produce (more, less) stable estimates of σ^2 than do small sample sizes. These different probability

   more

curves are identified by the degrees of freedom associated with the estimator of σ^2.

0

The term **degrees of freedom** can be explained in the following way. The sample estimate s^2 uses the sum of squared deviations in its calculation. Recall that $\sum (x_i - \bar{x}) = $ _____. This means that if we know the values of $n - 1$ deviations, we can determine the last value uniquely because their sum must be 0. Therefore, the sum

$n - 1$

of squared deviations, $\sum (x_i - \bar{x})^2$, contains only _____ independent deviations and not n independent deviations, as one might expect. Degrees of freedom refer to the number of independent deviations that are available for estimating σ^2. When n observations are drawn from one population, we use the estimator

$$s^2 = \frac{\sum (x_i - \bar{x})^2}{n - 1}$$

In this case, the degrees of freedom for estimating σ^2 are

$n - 1$
$n - 1$

_____, and the resulting t distribution is indexed as having _____ degrees of freedom.

The Use of Tables for the t Distribution

right
left

We define t_a as that value of t that has an area equal to a to its _____, and $-t_a$ is that value of t that has an area equal to a to its _____. Consider the diagram in Figure 10.2. The

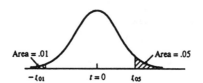

Area = .01   Area = .05

$-t_{.01}$ $t = 0$ $t_{.05}$

Figure 10.2

symmetrical

distribution of t is _____ about the value $t = 0$; hence, only the positive values of t need to be tabulated. Problems involving left-tailed values of t can be solved in terms of right-tailed values, as was done with the z statistic. A negative value of t simply indicates that

left

you are working in the (left, right) tail of the distribution.

Table 4 of the text tabulates *commonly used* critical values, t_a, based on 1, 2, ..., 29, ∞ degrees of freedom for $a = .100, .050, .025$, .010, .005. Along the top margin of the table you will find columns labeled t_a for the various values of a; along the right margin you will find a column marked degrees of freedom, df. By cross-indexing you can find the value t that has an area equal to a to its right and has the proper degrees of freedom.

Example 10.1
To find the critical value of t for $a = .05$ with 5 degrees of freedom, find 5 in the right margin. Now by reading across, you will find $t = 2.015$ in the $t_{.05}$ column. In the same manner, we find that for 12

degrees of freedom, $t_{.025} =$ _____. In using Table 4, think of
your problem in terms of a, the area to the right of the value of t, and
the degrees of freedom used to estimate σ^2.

2.179

Compare the different values of t based on an infinite number of
degrees of freedom with those for a corresponding z. You can perhaps
see the reason for choosing a sample size greater than _____
as the dividing point for using the z distribution when the standard
deviation s is used as an estimate for _____.

30

σ

Example 10.2
Find the critical values for t when t_a is the value of t that has an
area of a to its right, based on the given degrees of freedom:

a	*df*	*t*
.05	2	_____
.005	10	_____
.10	28	_____
.01	16	_____
.025	20	_____

2.920
3.169
1.313
2.583
2.086

Students who are taking their first course in statistics usually
ask the following questions at this point: How will I know whether I
should use z or t? Is sample size the only criterion I should apply?
No, sample size is not the only criterion to be used. When the sample
size is *large*, both

$$T_1 = \frac{\bar{x} - \mu}{\sigma/\sqrt{n}} \quad \text{and} \quad T_2 = \frac{\bar{x} - \mu}{s/\sqrt{n}}$$

behave like a standard normal random variable z, regardless of the
distribution of the parent population. When the sample size is *small*
and the sampled population is *not normal*, then, in general, neither T_1
nor T_2 behaves like z or t. In the special case when the parent population is *normal*, then T_1 behaves like z and T_2 behaves like t.

Use this information to complete the following table when the
sample is drawn from a *normal* distribution:

Statistic	Sample Size	
	$n < 30$	$n \geq 30$
$\dfrac{\bar{x} - \mu}{s/\sqrt{n}}$	_____	t or app. z
$\dfrac{\bar{x} - \mu}{\sigma/\sqrt{n}}$	_____	_____

t

$z; z$

10.3 SMALL-SAMPLE INFERENCES CONCERNING A POPULATION MEAN

Small-Sample Test Concerning a Population Mean μ

A test of an hypothesis concerning the mean μ of a *normal* population when $n < 30$ and σ is unknown can be conducted using either the critical value approach or the *p-value* approach used in large-sample hypothesis testing. These procedures are outlined below.

1. H_0: $\mu = \mu_0$
2. H_a: appropriate one- or two-tailed alternative
3. Test statistic:

$$t = \frac{\bar{x} - \mu_0}{s/\sqrt{n}}$$

4. ***Critical value approach:*** Set the rejection region with $\alpha = P(\text{falsely rejecting } H_0)$:
   a. For H_a: $\mu > \mu_0$, reject H_0 if $t > t_\alpha$ based on $n - 1$ degrees of freedom.
   b. For H_a: $\mu < \mu_0$, reject H_0 if $t < -t_\alpha$ based on $n - 1$ degrees of freedom.
   c. For H_a: $\mu \neq \mu_0$, reject H_0 if $|t| > t_{\alpha/2}$ based on $n - 1$ degrees of freedom.
5. ***p-value approach:*** Use Table 4 to bound the *p-value* based on the tabled areas $a = .005, .01, .025, .05,$ and $.10,$ or use a computer program to compute the exact *p-value* for the test statistic. Reject H_0 if the *p-value* is less than a prespecified value of α.

Example 10.3
A new electronics device that requires 2 hours per item to produce on a production line has been developed by company A. While the new product is being run, profitable production time is used. Hence, the manufacturer decides to produce only six new items for testing purposes. For each of the six items, the time to failure is measured, yielding the measurements 59.2, 68.3, 57.8, 56.5, 63.7, and 57.3 hours. Is there sufficient evidence to indicate that the new device has a mean life longer than 55 hours at the $\alpha = .05$ level? Use the critical value approach.

Solution
To calculate the sample mean and standard deviation, we need

$$\sum x_i = 362.8 \quad \text{and} \quad \sum x_i^2 = 22,043.60$$

60.4667

$$\bar{x} = \frac{\sum x_i}{n} = \frac{362.8}{6} = \underline{\qquad}$$

$$s^2 = \frac{1}{n-1}\left[\sum x_i^2 - \frac{\left(\sum x_i\right)^2}{n}\right]$$

$$= \frac{1}{5}\left[22{,}043.60 - \frac{(362.8)^2}{6}\right]$$

$$= 21.2587$$

$$s = \sqrt{21.2587} = 4.61073$$

The test proceeds as follows:

1. H_0: μ _____ $= 55$

2. H_a: μ _____ > 55

3. Test statistic:

$$t = \frac{\overline{x} - 55}{s/\sqrt{n}}$$

4. Rejection region:

Based on 5 degrees of freedom, reject H_0 if $t > $ _____.   2.015
Now calculate the value of the test statistic:

$$t = \frac{\overline{x} - 55}{s/\sqrt{n}} = \frac{\underline{\hspace{1cm}} - 55}{\underline{\hspace{1cm}}/\sqrt{6}} = \frac{\underline{\hspace{1cm}}}{1.8823}$$ 60.4667; 5.4667
4.61073

$$= \underline{\hspace{2cm}}$$ 2.904

In Table 4, with 5 degrees of freedom, you will find that the observed value of $t = 2.904$ lies between $t = 2.571$ which has a right-tail area of _____, and the value $t = 3.365$, which has a right-tail area   .025
of _____. Therefore the p-value of this test is found using   .010

$$P(t > 2.571) > P(t > 2.904) > P(t > 3.365),$$

or

_____ $> $ p-value $> $ _____.   .025; .010

Since the p-value is less than $\alpha = .05$, you (can, cannot) reject the   can
null hypothesis and conclude that the new device has a mean life
longer than 55 hours at the 5% level of significance.

Confidence Interval for a Population Mean μ

In estimating a population mean, we can use either a point estimator along with its margin of error or an interval estimator that has the required level of confidence.

Small-sample estimation of the mean of a *normal* population with σ *unknown* involves the statistic

$$\frac{\bar{x} - \mu}{s/\sqrt{n}}$$

t; $n-1$

which has a _____ distribution with _____ degrees of freedom. The resulting $(1-\alpha)100\%$ confidence interval estimator is given as

$$\bar{x} \pm t_{\alpha/2} \frac{s}{\sqrt{n}}$$

where $t_{\alpha/2}$ is that value of t based on $n-1$ degrees of freedom that has an area of $\alpha/2$ to its right. The lower confidence limit is

$\bar{x} - t_{\alpha/2}s/\sqrt{n}$; $\bar{x} + t_{\alpha/2}s/\sqrt{n}$; $\bar{x}$

_____, and the upper confidence limit is _____. The point estimator of μ is _____, and the bound on the margin of error can be taken to be $t_{\alpha/2}s/\sqrt{n}$. A proper interpretation of a $(1-\alpha)100\%$ confidence interval for μ would be stated as follows: In repeated sampling, (_____)100% of the _____

$1-\alpha$; confidence intervals

_____ so constructed would enclose the true value of the mean μ.

Example 10.4
Using the data from Example 10.3, find a 95% confidence interval estimate for μ, the mean life in hours for the new device.

Solution
The pertinent information from Example 10.3 follows:

5

$$\bar{x} = 60.4667 \quad df = \text{_____}$$

$$\frac{s}{\sqrt{n}} = 1.8823 \quad \frac{\alpha}{2} = .025$$

2.571

$$t_{.025} = \text{_____}$$

The confidence interval is found by using

$$\bar{x} \pm t_{.025} \frac{s}{\sqrt{n}}$$

Substituting values for $\bar{x}$, $s/\sqrt{n}$, and $t_{.025}$, we have

2.571; 4.84

$$60.4667 \pm \text{_____} (1.8823) = 60.4667 \pm \text{_____}$$

or

55.63; 65.31

$$(\text{_____}, \text{_____}).$$

Example 10.5

In a random sample of ten cans of corn from supplier B, the average weight per can of corn was $\bar{x} = 9.4$ ounces with standard deviation $s = 1.8$ ounces. Does this sample contain sufficient evidence to indicate the mean weight is less than 10 ounces. Use the *p-value* approach.

Solution

The following information is needed:

$$n = \underline{\hspace{2cm}}$$ 10

$$\bar{x} = \underline{\hspace{2cm}}$$ 9.4

$$s = \underline{\hspace{2cm}}$$ 1.8

$$\alpha = \underline{\hspace{2cm}}$$ .01

Set up the test as follows:

1. H_0: $\mu = \underline{\hspace{2cm}}$ 10

2. H_a: $\mu < \underline{\hspace{2cm}}$ 10

3. Test statistic:

$$t = \frac{\bar{x} - \mu}{s/\sqrt{n}}$$

$$= \frac{9.4 - (\underline{\hspace{1.5cm}})}{1.8/\sqrt{10}}$$ 10

$$= \frac{(\underline{\hspace{1.5cm}})}{.57}$$ $-.6$

$$= \underline{\hspace{2cm}}$$ -1.05

The *p-value* for this one-tailed test is defined to be

$$p\text{-}value = P(t \leq -1.05) = P(t \geq 1.05)$$

where t has a Student's t distribution with $n - 1 = 9$ degrees of freedom. From Table 4, with $df = 9$, the critical value of $t = 1.05$ falls to the left of the smallest value given. That value is $t = 1.383$ with area .10 to its right (Figure 10.3). Hence, the area to the right of

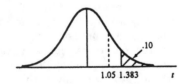

Figure 10.3

$t = 1.05$ is (greater, less) than .10, and the *p-value* is $\underline{\hspace{2cm}}$ than .10.   greater; greater

greater
not reject; are not
do not

Since the *p-value* is (greater, less) than .10, you should (reject, not reject) H_0 and declare that the results (are, are not) statistically significant. The data (do, do not) present sufficient evidence to indicate that the mean weight per can is less than 10 ounces.

Example 10.6
Refer to Example 10.5. Find a 98% confidence interval for μ.

Solution

.02

$\alpha = \underline{\hspace{2cm}}$

9

$df = \underline{\hspace{2cm}}$

9

Based on $\underline{\hspace{2cm}}$ degrees of freedom,

-2.821

$t_{\alpha/2} = \underline{\hspace{2cm}}.$

Calculate:

$$\bar{x} \pm t_{\alpha/2}s/\sqrt{n}$$

9.4; .57

$(\underline{\hspace{2cm}}) \pm 2.821(\underline{\hspace{2cm}})$

9.4; 1.6

$(\underline{\hspace{2cm}}) \pm (\underline{\hspace{2cm}})$

7.8
11.0

Therefore, the 98% confidence interval required is $(\underline{\hspace{2cm}},$ $\underline{\hspace{2cm}}).$

The Minitab package can be used to perform a test of an hypothesis concerning a population mean μ, and to calculate a confidence interval estimate of μ for any specified level of confidence. The commands **Stat → Basic Statistics → 1-Sample t** will generate a dialog box that can be used for both estimation and testing. You must specify the column containing the data, a value of the mean μ_0 to be tested, and whether the test is to be right-tailed, left-tailed or two-tailed. When the test is two-tailed, the upper and lower confidence interval is also provided. If the test is one-tailed, then the appropriate upper or lower confidence bound is provided. The upper and lower confidence bounds are calculated as in Section 8.6 with the tabled value of z replaced by the appropriate tabled value of t. When you click **OK**, Minitab will provide the appropriate printout.

For the displays that follow, the data from Example 10.3 were entered into column 1 and labeled "Failure time," and the estimation and testing command **Stat → Basic Statistics → 1-Sample t** implemented.

One-Sample T: Failure time

```
Test of mu = 55 vs mu > 55

Variable          N      Mean     StDev    SE Mean
Failure time      6     60.47      4.61       1.88

Variable      95.0% Lower Bound        T      P
Failure time               56.67     2.90  0.017
```

One-Sample T: Failure time

```
Test of mu = 55 vs mu not = 55

Variable          N      Mean     StDev    SE Mean
Failure time      6     60.47      4.61       1.88

Variable            95.0% CI            T      P
Failure time  (  55.63,    65.31)    2.90  0.034
```

The printout allows you to verify your calculated values of $\bar{x} =$ _____ and $s =$ _____. In addition, the estimated standard deviation of $\bar{x}$, called the standard error of the mean, (SE Mean), is given by $s/\sqrt{n} = 1.88$. This quantity is used in calculating t and in constructing confidence limits based on t.

60.47; 4.61

In testing H_0: $\mu = 55$ versus H_a: $\mu > 55$ (the alternative is greater than 55), the calculated value of t in the "One-Sample T" printout agrees with our earlier calculations. However, the value of α is not set in advance: instead, the decision to reject or not reject H_0 is made after examining the *p-value* for the test. Since the *p-value* of .017 is (smaller, larger) than the value of $\alpha = .05$ used in Example 10.3, we agree to (reject, not reject) H_0. the lower and upper 95% confidence limits for μ found in the second printout are 55.63 and 65.31, respectively, the same as the values found in Example 10.4.

smaller
reject

Self-Correcting Exercises 10A

1. A school administrator claimed that the average time spent on a school bus by those students in his school district who rode school buses was 35 minutes. A random sample of 20 students who did ride the school buses yielded an average of 42 minutes riding time with a standard deviation of 6.2 minutes. Does this sample of size 20 contain sufficient evidence to indicate that the mean riding time is greater than 35 minutes at the 5% level of significance? What *p-value* would you report?

2. Find a 95% confidence interval for the mean time spent on a school bus using the data from Exercise 1.

3. The telephone company was interested in measuring the average daily usage in minutes for household telephones in a specific area in order to determine whether this rate is different from a state-wide average daily usage for households. Suppose that a random sample of nine households were sampled on random days, producing the following times (in minutes):

$$35, 59, 42, 44, 31, 46, 24, 56, 50$$

a. Estimate the average daily usage using a 90% confidence interval.

b. Suppose that the statewide average for households has been found to be 45 minutes. Does this data provide evidence to indicate that the average in this area differs from the statewide average? Use $\alpha = .05$.

Solutions to SCE 10A

1. H_0: $\mu = 35$; H_a: $\mu > 35$; test statistic: $t = \dfrac{\bar{x} - \mu}{s/\sqrt{n}}$. With $\alpha = .05$, and $n - 1 = 19$ degrees of freedom, reject H_0 if $t > t_{.05} = 1.729$. Calculate

$$t = \frac{42 - 35}{6.2/\sqrt{20}} = \frac{7}{1.386} = 5.05.$$

Reject H_0. The mean riding time is greater than 35 minutes. From Table 4 with 19 degrees of freedom, the observed $t = 5.05$ exceeds $t_{.005} = 2.861$. Hence, p-value $< .005$.

2. $\bar{x} \pm t_{.025} \dfrac{s}{\sqrt{n}}$, $42 \pm 2.093(1.386)$, 42 ± 2.90, or 39.1 to 44.9.

3. a. $\sum x_i = 387$; $\sum x_i^2 = 17,695$; $\bar{x} = 43$; $s^2 = \dfrac{17,695 - 16,641}{8}$

$$= 141.75. \quad \bar{x} \pm t_{.05} \frac{s}{\sqrt{n}}, \quad 43 \pm 1.86 \sqrt{\frac{131.75}{9}}, \quad 43 \pm 7.12 \text{ or}$$

35.88 to 50.12.

b. H_0: $\mu = 45$, H_a: $\mu \neq 45$; test statistic: $t = \dfrac{\bar{x} - \mu}{s/\sqrt{n}} = \dfrac{43 - 45}{\sqrt{\dfrac{131.75}{9}}}$

$= -.52$. With $\alpha = .05$ and 8 degrees of freedom, reject H_0 if $|t| > 2.306$. Do not reject H_0.

10.4 SMALL-SAMPLE INFERENCES CONCERNING THE DIFFERENCE BETWEEN TWO MEANS, $\mu_1 - \mu_2$: INDEPENDENT RANDOM SAMPLES

Inferences concerning $\mu_1 - \mu_2$ based on small samples are founded upon the following assumptions:

normal

variances

1. Each population sampled has a _____ distribution.
2. The population _____ are equal; that is, $\sigma_1^2 = \sigma_2^2$.
3. The samples are independently drawn.

$\bar{x}_1 - \bar{x}_2$

An unbiased estimator for $\mu_1 - \mu_2$, regardless of sample size, is _____. The standard deviation of this estimator is

$$\sqrt{\frac{\sigma_1^2}{n_1} + \frac{\sigma_2^2}{n_2}}$$

When $\sigma_1^2 = \sigma_2^2$, we can replace σ_1^2 and σ_2^2 by a common variance σ^2. Then the standard deviation of $\bar{x}_1 - \bar{x}_2$ becomes

$$\sqrt{\frac{\sigma_1^2}{n_1} + \frac{\sigma_2^2}{n_2}} = \sqrt{(\underline{\quad\quad})\left(\frac{1}{n_1} + \frac{1}{n_2}\right)} \qquad\qquad \sigma^2$$

If σ^2 were known, then in testing an hypothesis concerning $\mu_1 - \mu_2$, we would use the statistic

$$z = \frac{(\bar{x}_1 - \bar{x}_2) - D_0}{\sqrt{\sigma^2\left(\frac{1}{n_1} + \frac{1}{n_2}\right)}}$$

where $D_0 = \mu_1 - \mu_2$. For small samples with σ^2 unknown, we would use

$$t = \frac{(\bar{x}_1 - \bar{x}_2) - D_0}{\sqrt{s^2\left(\frac{1}{n_1} + \frac{1}{n_2}\right)}}$$

where s^2 is the estimate of σ^2, calculated from the sample values. When the data are normally distributed, this statistic has a _____ _____ distribution with degrees of freedom the same as those available for estimating _____.

Student's
t
σ^2

In selecting the best estimate (s^2) for σ^2, we have three immediate choices:

1. s_1^2, the sample _____ from population I   variance

2. s_2^2, the sample _____ from population II   variance

3. A combination of _____ and _____ $s_1^2;\ s_2^2$

The best choice is $(1, 2, 3)$ because it uses the information from both samples. A logical method of combining this information into one estimate, s^2, is

3

4. $s^2 = \dfrac{(n_1 - 1)s_1^2 + (n_2 - 1)s_2^2}{(n_1 - 1) + (n_2 - 1)}$

a weighted average of the sample variances using the degrees of freedom as weights.

The expression in 4 can be written in another form by replacing s_1^2 and s_2^2 by their defining formulas. Then, if x_{1j} and x_{2j} represent the jth observation in samples 1 and 2, respectively,

$$s^2 = \frac{\sum (x_{1j} - \bar{x}_1)^2 + \sum (x_{2j} - \bar{x}_2)^2}{(n_1 - 1) + (n_2 - 1)}$$

$n_1 + n_2 - 2$

In this form we see that we have pooled or added the sums of squared deviations from each sample and divided by the pooled degrees of freedom, $n_1 + n_2 - 2$. Hence, s^2 is a **pooled estimate** of the common variance σ^2 and is based on _____ degrees of freedom. Since our samples were drawn from normal populations, the statistic

$$t = \frac{(\bar{x}_1 - \bar{x}_2) - (\mu_1 - \mu_2)}{\sqrt{s^2\left(\frac{1}{n_1} + \frac{1}{n_2}\right)}}$$

Student's t;
$n_1 + n_2 - 2$

has a _____ _____ distribution with _____ degrees of freedom.

Example 10.7
A medical student conducted a diet study using two groups of 12 rats each as subjects. Group I received diet I, while group II received diet II. After 5 weeks, the student calculated the gain in weight for each rat. The data yielded the following information:

Group I	Group II
$\bar{x}_1 = 6.8$ ounces	$\bar{x}_2 = 5.3$ ounces
$s_1 = 1.5$ ounces	$s_2 = .9$ ounce
$n_1 = 12$	$n_2 = 12$

Do these data present sufficient evidence to indicate, at the $\alpha = .05$ level, that rats on diet I will gain more weight than those on diet II? Find a 90% confidence interval for $\mu_1 - \mu_2$, the mean difference in weight gained.

Solution
We will take the gains in weight to be normally distributed with equal variances and calculate a pooled estimate for σ^2:

$$s^2 = \frac{(n_1 - 1)s_1^2 + (n_2 - 1)s_2^2}{n_1 + n_2 - 2}$$

1.530

$$= \frac{11(1.5)^2 + 11(.9)^2}{12 + 12 - 2} = \frac{33.66}{22} = \underline{\qquad}$$

The test is as follows:

0

1. H_0: $\mu_1 - \mu_2 =$ _____

0

2. H_a: $\mu_1 - \mu_2 >$ _____

3. Test statistic:

$$t = \frac{(\bar{x}_1 - \bar{x}_2) - D_0}{\sqrt{s^2\left(\frac{1}{n_1} + \frac{1}{n_2}\right)}}$$

4.   Rejection region:

With $n_1 + n_2 - 2 = $ _____ degrees of freedom, we would reject H_0 if $t > $ _____ . Now we calculate the test statistic:

$$t = \frac{(\bar{x}_1 - \bar{x}_2) - D_0}{\sqrt{s^2 \left(\frac{1}{n_1} + \frac{1}{n_2} \right)}}$$

$$= \frac{(\underline{\hspace{1.5cm}}) - (\underline{\hspace{1.5cm}})}{\sqrt{1.530(.1667)}}$$

$$= \frac{(\underline{\hspace{1.5cm}})}{.505} = \underline{\hspace{1.5cm}}$$

Decision: _____ .
    To find a 90% confidence interval for $\mu_1 - \mu_2$, we need $t_{.05}$
based on 22 degrees of freedom: $t_{.05} = $ _____ . Hence, we
would use

$$(\bar{x}_1 - \bar{x}_2) \pm 1.717 \sqrt{s^2 \left(\frac{1}{n_1} + \frac{1}{n_2} \right)}$$

_____ $\pm 1.717($ _____ $)$

$($ _____ $) \pm ($ _____ $)$

Therefore, a 90% confidence interval for $\mu_1 - \mu_2$ would be

$($ _____ , _____ $)$

22	
1.717	
1.5; 0	
1.5; 2.97	
Reject H_0	
1.717	
1.5; .505	
1.5; .867	
.63; 2.37	

Example 10.8
In an effort to compare the average swimming times for two swimmers, each swimmer was asked to swim freestyle for a distance of
100 yards at randomly selected times. The swimmers were thoroughly
rested between laps and did not race against each other, so that each
sample of times was an independent random sample. The times for
each of ten trials are shown for the two swimmers.

Swimmer 1	Swimmer 2
59.62	59.81
59.48	59.32
59.65	59.76
59.50	59.64
60.01	59.86

Swimmer 1	Swimmer 2
59.74	59.41
59.43	59.63
59.72	59.50
59.63	59.83
59.68	59.51

Suppose that swimmer 2 was last year's winner when the two swimmers raced. Does it appear that the average time for swimmer 2 is still faster than the average time for swimmer 1 in the 100-yard freestyle? Find the approximate *p-value* for the test and interpret the results.

Solution
We will assume that the times for the two swimmers are normally distributed with equal variances.

1. The hypotheses to be tested are:

$$H_0: \ \mu_1 - \mu_2 = 0$$

> 0

$$H_a: \ \mu_1 - \mu_2 \ \underline{\hspace{2cm}}$$

The following calculations are necessary:

Swimmer 1	Swimmer 2
596.27	$\sum x_{1j} = 596.46$	$\sum x_{2j} = \underline{\hspace{1.5cm}}$
35,554.109	$\sum x_{1j}^2 = 35{,}576.698$	$\sum x_{2j}^2 = \underline{\hspace{1.5cm}}$
59.627	$\bar{x}_1 = 59.646$	$\bar{x}_2 = \underline{\hspace{1.5cm}}$

2. Calculate the pooled estimate for σ^2 using the alternate form of s^2:

$$s^2 = \frac{\sum (x_{1j} - \bar{x}_1)^2 + \sum (x_{2j} - \bar{x}_2)^2}{(n_1 - 1) + (n_2 - 1)}$$

$$= \frac{\sum x_{1j}^2 - \dfrac{\left(\sum x_{1j}\right)^2}{n_1} + \sum x_{2j}^2 - \dfrac{\left(\sum x_{2j}\right)^2}{n_2}}{n_1 + n_2 - 2}$$

35,554.109; 596.27; 18

.56255; .0312528

$$= \frac{35{,}576.698 - \dfrac{(596.46)^2}{10} + \underline{\hspace{1.5cm}} - \dfrac{(\underline{\hspace{1cm}})^2}{10}}{\underline{\hspace{1.5cm}}}$$

$$= \frac{\underline{\hspace{1.5cm}}}{18} = \underline{\hspace{1.5cm}}$$

3. The test statistic is

$$t = \frac{(\bar{x}_1 - \bar{x}_2) - D_0}{\sqrt{s^2\left(\frac{1}{n_1} + \frac{1}{n_2}\right)}} = \frac{59.646 - 59.627}{\sqrt{.0312528\left(\frac{1}{10} + \frac{1}{10}\right)}}$$

$$= \frac{\rule{3cm}{0.4pt}}{\rule{3cm}{0.4pt}} = \rule{3cm}{0.4pt}$$

.019; .24
.079

4. The *p-value* for this (one-tailed, two-tailed) test is *p-value* = $P(t > \underline{\hspace{2cm}})$. Notice in Table 4 in the text that this value is not tabulated. Rather, for each different value of the degrees of freedom, the table gives t_a such that $P(t > t_a) = a$. This is the value of t that cuts off an area equal to a to its (right, left). However, since the observed value, $t = .24$, falls to the (left, right) of the smallest tabulated value, $t_{.10} = 1.330$, the area to the right of $t = .24$ must be (greater, less) than .10. Hence, the *p-value* is bounded as *p-value* > .10. This is too large a value to allow rejection of H_0. Hence, we conclude that there (is, is not) sufficient evidence to detect a difference in the two averages.

one-tailed
.24

right
left

greater

is not

The Minitab commands **Stat** → **Basic Statistics** → **2-Sample t** can be used to generate the output necessary for the two-independent sample *t*-test and confidence interval. You must indicate the columns in which you have stored your data, and you must select the necessary confidence coefficient and/or alternate hypothesis. Further, the box marked "Assume equal variances" must be checked so that the sample variances are pooled to estimate the underlying common (equal) variance. The Minitab printout using the **2-Sample t** command on the data for Example 10.8 is given in the following display:

Two-Sample T-Test and CI: Swimmer 1, Swimmer 2

```
Two-sample T for Swimmer 1 vs Swimmer 2

            N     Mean    StDev   SE Mean
Swimmer   10   59.646    0.165     0.052
Swimmer   10   59.627    0.188     0.059

Difference = mu Swimmer 1 - mu Swimmer 2
Estimate for difference:  0.0190
95% lower bound for difference: -0.1181
T-Test of difference = 0 (vs >): T-Value = 0.24  P-Value = 0.406  DF = 18
Both use Pooled StDev = 0.177
```

Notice that the results on the printout agree with the earlier hand calculations.

At times you might suspect that the two underlying population variances are *not equal*. As a **rule of thumb**, you should not assume that the underlying population variances are equal if the ratio of the two sample variances,

$$\frac{\text{larger } s^2}{\text{smaller } s^2} > 3$$

In this case, the statistic used for testing hypotheses about $\mu_1 - \mu_2$ is

$$t^* = \frac{(\bar{x}_1 - \bar{x}_2) - D_0}{\sqrt{\frac{s_1^2}{n_1} + \frac{s_2^2}{n_2}}}$$

When the sample sizes are small, the appropriate critical values for this test statistic can be read from the critical values of t given in Table 4, using degrees of freedom equal to

$$\frac{\left(\frac{s_1^2}{n_1} + \frac{s_2^2}{n_2}\right)^2}{\frac{\left(\frac{s_1^2}{n_1}\right)^2}{n_1 - 1} + \frac{\left(\frac{s_2^2}{n_2}\right)^2}{n_2 - 1}}$$

This quantity must be rounded to the nearest integer to use Table 4. When n_1 and n_2 are large, the distribution of t^* can be approximated by the standard normal distribution, as we did in Section 9.3. Finally, Minitab will produce output for this statistical test if you do not check the "Assume equal variances" box in the 2-Sample t Dialog box.

Self-Correcting Exercises 10B

1. What are the assumptions required for the proper use of the statistic

$$t = \frac{(\bar{x}_1 - \bar{x}_2) - (\mu_1 - \mu_2)}{\sqrt{s^2\left(\frac{1}{n_1} + \frac{1}{n_2}\right)}} \; ?$$

2. In the process of making a decision to either continue operating or close a civic health center, a random sample of 25 people who had visited the center at least once was chosen and each person asked whether he or she felt the center should be closed. In addition, the distance between each person's place of residence and the health center was computed and recorded. Of the 25 people responding, 16 were in favor of continued operation. For these 16 people, the average distance from the center was 5.2 miles with a standard deviation of 2.8 miles. The remaining 9 people who were in favor of closing the center lived at an average of 8.7 miles from the center with a standard deviation of 5.3 miles. Do these data indicate that there is a significant difference in mean distance to the health center for these two groups?

3. Estimate the difference in mean distance to the health center for the two groups in Exercise 2 with a 95% confidence interval.

4. In investigating which of two presentations of subject matter to use in a computer-programmed course, an experimenter

randomly chose two groups of 18 students each, and assigned one group to receive presentation I and the second to receive presentation II. A short quiz on the presentation was given to each group and their grades recorded. Do the following data indicate that a difference in the mean quiz scores (hence a difference in effectiveness of presentation) exists for the two methods? Find the approximate *p-value* and interpret your results.

	$\bar{x}$	s^2
Presentation I	81.7	23.2
Presentation II	77.2	19.8

Solutions to SCE 10B

1. See Section 10.4, paragraph 1.

2. H_0: $\mu_1 - \mu_2 = 0$; H_a: $\mu_1 - \mu_2 \neq 0$; test statistic:

$$t = \frac{(\bar{x}_1 - \bar{x}_2) - D_0}{\sqrt{s^2\left(\frac{1}{n_1} + \frac{1}{n_2}\right)}}.$$

With $\alpha = .05$ and $n_1 + n_2 - 2 = 16 + 9 - 2 = 23$ degrees of freedom, reject H_0 if $|t| > t_{.025} = 2.069$. Calculate

$$s^2 = \frac{(n_1 - 1)s_1^2 + (n_2 - 1)s_2^2}{n_1 + n_2 - 2}$$

$$= \frac{15(2.8)^2 + 8(5.3)^2}{23} = \frac{117.6 + 224.72}{23} = 14.8835$$

$$t = \frac{5.2 - 8.7}{\sqrt{14.8835\left(\frac{1}{16} + \frac{1}{9}\right)}} = -2.16.$$

Reject H_0. The mean distance to the health center differs for the two groups.

3. $(\bar{x}_1 - \bar{x}_2) \pm t_{.025}\sqrt{s^2\left(\frac{1}{n_1} + \frac{1}{n_2}\right)}$,

$-3.5 \pm 2.069(1.62)$, -3.5 ± 3.35, or -6.85 to -0.15 miles.

4. H_0: $\mu_1 - \mu_2 = 0$; H_a: $\mu_1 - \mu_2 \neq 0$; test statistic:

$$t = \frac{(\bar{x}_1 - \bar{x}_2) - D_0}{\sqrt{s^2\left(\frac{1}{n_1} + \frac{1}{n_2}\right)}}.$$

With $\alpha = .05$ and $n_1 + n_2 - 2 = 34$ degrees of freedom, reject H_0 if $|t| > t_{.025} = 1.96$. Calculate

$$s^2 = \frac{(n_1 - 1)s_1^2 + (n_2 - 1)s_2^2}{n_1 + n_2 - 2}$$

$$= \frac{17(23.2) + 17(19.8)}{34} = 21.5.$$

$$t = \frac{81.7 - 77.2}{\sqrt{21.5\left(\frac{2}{18}\right)}} = \frac{4.5}{1.5456} = 2.91.$$

Reject H_0. There is a difference in mean scores for the two methods. From Table 4 with 34 degrees of freedom, $t = 2.91$ exceeds $t_{.005} = 2.576$. Hence, $p\text{-}value < 2(.005) = .01$.

10.5 SMALL-SAMPLE INFERENCE CONCERNING THE DIFFERENCE BETWEEN TWO MEANS: A PAIRED-DIFFERENCE TEST

In many situations an experiment is designed so that a comparison of the effects of two "treatments" is made on the same person, twin offspring, two animals from the same litter, two pieces of fabric from the same loom, or two plants of the same species grown on adjacent plots. Such experiments are designed so that the pairs of experimental units (people, animals, fabrics, plants) are as much alike as possible. Because measurements are taken on the two treatments within the relatively homogeneous pairs of experimental units, the difference in the measurements for the two treatments in a pair will primarily

treatment

reflect the difference between _____ effects rather than the difference between experimental units. This experimental design reduces the error of comparison and increases the quantity of information in the experiment.

To analyze such an experiment using the techniques of the last section would be incorrect. In planning this type of experiment, we *intentionally violate* the assumption that the measurements are *independent* and hope that this violation will work to our advantage by

reducing

(increasing, reducing) the variability of the differences of the paired observations. Consider the situation in which two sets of identical twin calves are selected for a diet experiment. One of each set of twins is randomly chosen to be fed diet A, while the other is given diet B. At the end of a given period of time, the calves are weighed and the data are presented for analysis:

Set	Diet		Difference
	A	B	
1	A_1	B_1	$A_1 - B_1$
2	A_2	B_2	$A_2 - B_2$

Now A_1 and B_1 are *not* independent because the calves are identical twins and as such have the same growth trend, weight gain trend, and so on. Although A_1 could be larger or smaller than B_1, if A_1 were large, we would also expect B_1 to be large. A_2 and B_2 are not independent for the same reason. However, because we are looking at the differences $(A_1 - B_1)$ and $(A_2 - B_2)$, the characteristics of the twin calves no longer cloud the issue, since these differences would represent the difference due to the effects of the two treatments.

In using a paired-difference design, we analyze the differences of the paired measurements and, in so doing, attempt to *reduce* the *variability* that would be present in two *randomly* selected groups without pairing. A test of the hypothesis that the difference in two population means, $\mu_1 - \mu_2$, is equal to a constant, D_0, is equivalent to a test of the hypothesis that the mean of the differences, μ_d, is equal to a constant, D_0. That is, $H_0: \mu_1 - \mu_2 = D_0$ is equivalent to $H_0: \mu_d = D_0$. Usually, we will be interested in the hypothesis that $D_0 = 0$.

Example 10.9

To test the results of a conventional versus a new approach to the teaching of reading, 12 pupils were selected and matched according to IQ, age, present reading ability, and so on. One from each of the pairs was assigned to the conventional reading program and the other to the new reading program. At the end of 6 weeks, their progress was measured by a reading test. Do the following data present sufficient evidence to indicate that the new approach is better than the conventional approach at the $\alpha = .05$ level?

Pair	Conventional	New	$d_i = N - C$
1	78	83	5
2	65	69	4
3	88	87	−1
4	91	93	2
5	72	78	6
6	59	59	0

Find a 95% confidence interval for the difference in mean reading scores.

Solution

We analyze the set of six differences as we would a single set of six measurements. The change in notation required is straightforward:

$$\sum d_i = 16 \qquad \sum d_i^2 = 82$$

The sample mean is

$$\bar{d} = \tfrac{1}{6} \sum d_i = \underline{\hspace{2cm}}$$ 2.6667

The sample variance of the differences is

$$s_d^2 = \frac{\sum d_i^2 - \left(\sum d_i\right)^2/6}{5}$$

$$= \frac{82 - (16)^2/6}{5} = \frac{39.3333}{5} = \underline{\hspace{2cm}}$$ 7.8667

$$s_d = \sqrt{7.8667} = \underline{\hspace{2cm}}$$ 2.8048

The test is conducted as follows. Remember that $\mu_d = \mu_N - \mu_C$.

1. H_0: $\mu_d = 0$

> 0

2. H_a: μ_d _____

3. Test statistic:

$$t = \frac{\bar{d} - 0}{s_d/\sqrt{n}}$$

4. Rejection region: Based on 5 degrees of freedom, we will reject

2.015 H_0 if the observed value of t is greater than $t_{.05} = $ _____.

The sample value of t is

$$t = \frac{\bar{d} - 0}{s_d/\sqrt{n}}$$

2.6667; 2.33 $= \dfrac{\underline{\hspace{1cm}} - 0}{2.80/\sqrt{6}} = \dfrac{2.6667}{1.145} = $ _____

reject   Since the value of the test statistic is greater than 2.015, we (reject, do not reject) H_0. This sample indicates that the new method appears to be superior to the conventional method at the $\alpha = .05$ level, if we assume that the reading test is a valid criterion upon which to base our judgment.

A 95% confidence interval estimate for μ_d is given by

$$\bar{d} \pm t_{.025}\left(\frac{s_d}{\sqrt{n}}\right)$$

2.571   Using sample values and $t_{.025} = $ _____, we have

2.571 $2.6667 \pm$ _____ (1.145)

2.9439 $2.6667 \pm$ _____

With 95% confidence, we estimate that μ_d lies within the interval

−.28; 5.61   _____ to _____.

Notice that in using a paired-difference analysis, the degrees of freedom for the critical value of t drop from $2n - 2$ for an unpaired design to $n - 1$ for the paired, a loss of $(2n - 2) - (n - 1) = $

$n - 1$; larger   _____ degrees of freedom. This results in a (larger, smaller) critical value of t. Therefore, a larger value of the test statistic is needed to reject H_0. Fortunately, *proper* pairing will reduce $\sigma_{\bar{d}}$. Hence, the paired-difference experiment results in both a loss and a gain of information. However, the *loss* of $(n - 1)$ degrees of freedom is usually far overshadowed by the gain in information when $\sigma_{\bar{d}}$ is substantially reduced.

The statistical design of the paired-difference test is a simple example of a randomized block design. In such a design, the pairing must occur when the experiment is planned and not after the data are collected. Once the experimenter has used a paired design for an experiment, he no longer has the choice of using the unpaired design for testing the difference between means. This is because he has violated the assumptions needed for the unpaired design—namely, that the samples are random and independent.

A paired-difference experiment can be analyzed using the commands **Stat → Basic Statistics → Paired-t** in Minitab 12 and 13, or by using the **Stat → Basic Statistics → 1-Sample t** command in earlier versions of Minitab. In order to use the **1-Sample t** command, the paired responses must be stored in two columns (say, C1 and C2) and the **Calc → Calculator** command is used to produce a column of differences in a third column –say, C3. The one sample t test against the hypothesized value of $\mu_1 - \mu_2 = 0$ is performed on the data in C3. In the following display, the data from Example 10.9 has been stored in C1 and C2.

Paired T-Test and CI: New, Conventional

```
Paired T for New - Conventional

                N     Mean    StDev   SE Mean
New             6    78.17    12.43    5.08
Conventional    6    75.50    12.63    5.16
Difference      6     2.67     2.80    1.15

95% lower bound for mean difference: 0.36
T-Test of mean difference = 0 (vs > 0): T-Value = 2.33   P-Value = 0.034
```

The printout shows the observed value of the test statistic, $t = $ _____, with the *p-value* for the test given as _____.
Since the *p-value* is (less then, greater than) α, the null hypothesis (is, is not) rejected. You (can, cannot) conclude that the new method is superior to the conventional method, based on the average scores on the reading test.

-2.33; .067
less than
is; can

Self-Correcting Exercises 10C

1. The owner of a small manufacturing plant is considering a change in salary base by replacing an hourly wage structure with a per-unit rate. She hopes that such a change will increase the output per worker but has reservations about a possible decrease in quality under the per-unit plan. Before arriving at any decision, she forms 10 pairs of workers so that within each pair the two workers have produced about the same number of items per day and their work has been of comparable quality. From each pair, one worker is randomly selected to be paid as usual and the other is to be pair on a per-unit basis. In addition to the number of items produced, a cumulative quality score for the items produced is kept for each

worker. The quality scores follow. (A high score is indicative of high quality.)

	Rate	
Pair	Per Unit	Hourly
1	86	91
2	75	77
3	87	83
4	81	84
5	65	68
6	77	76
7	88	89
8	91	91
9	68	73
10	79	78

Do these data indicate that the average quality for the per-unit production is significantly lower than that based on an hourly wage? What *p-value* would you report?

3. Refer to Exercise 1. The following data represent the average number of items produced per worker, based on one week's production records:

	Rate	
Pair	Per Unit	Hourly
1	35.8	31.2
2	29.4	27.6
3	31.2	32.2
4	28.6	26.4
5	30.0	29.0
6	32.6	31.4
7	36.8	34.2
8	34.4	31.6
9	29.6	27.6
10	32.8	29.8

a. Estimate the mean difference in average daily output for the two pay scales with a 95% confidence interval.

b. Test the hypothesis that a per-unit pay scale increases production at the .05 level of significance.

Solutions to SCE 10C

1. $H_0: \mu_d = \mu_P - \mu_H = 0$; $H_a: \mu_d = \mu_P - \mu_H < 0$; test statistic:

$$t = \frac{\bar{d} - \mu_d}{s_d / \sqrt{n}}$$

With $\alpha = .05$ and $n - 1 = 10 - 1 = 9$ degrees of freedom, reject H_0 if $t < -t_{.05} = -1.833$. The 10 differences and the calculation of the test statistic follow.

d_i	d_i^2
−5	25
−2	4
4	16
−3	9
−3	9
1	1
−1	1
0	0
−5	25
1	1
−13	91

$$\bar{d} = \frac{-13}{10} = -1.3, \quad s_d^2 = \frac{91 - (-13)^2/10}{9} = \frac{74.1}{9} = 8.2333,$$

$$s_d = \sqrt{8.2333} = 2.869, \quad t = \frac{-1.3 - 0}{2.869/\sqrt{10}} = -1.43.$$

Do not reject H_0. With 9 degrees of freedom, $t = -1.43$ falls between $-t_{.10}$ and $-t_{.05}$. Hence, $.05 < p\text{-}value < .10$.

2.

d_i	d_i^2
4.6	21.16
1.8	3.24
−1.0	1.00
2.2	4.84
1.0	1.00
1.2	1.44
2.6	6.76
2.8	7.84
2.0	4.00
3.0	9.00
20.2	60.28

$$\bar{d} = \frac{20.2}{10} = 2.02, \quad s_d^2 = \frac{60.28 - 40.804}{9} = \frac{19.476}{9} = 2.164,$$

$$s_d = 1.47.$$

a. $\bar{d} \pm t_{.025} \dfrac{s_d}{\sqrt{n}}$, $2.02 \pm 2.262 \dfrac{1.47}{\sqrt{10}}$, 2.02 ± 1.05, or

   $.97 < \mu_d < 3.07$.

b. H_0: $\mu_P - \mu_H = 0$; H_a: $\mu_P - \mu_H > 0$; test statistic:

   $$t = \frac{\bar{d} - \mu_d}{s_d/\sqrt{n}}.$$ With $\alpha = .05$ and $n - 1 = 9$ degrees of freedom,

   reject H_0 if $t > t_{.05} = 1.833$. Calculate $t = 2.02/.465 = 4.34$.
   Reject H_0. Per-unit scale increases production.

10.6 INFERENCES CONCERNING A POPULATION VARIANCE

In many cases the measure of variability is more important than that of the central tendency. For example, an educational test consisting of 100 items has a mean score of 75 with a standard deviation of 2.5. The value $\mu = 75$ may sound impressive, but $\sigma = 2.5$ would imply that this test has very poor discriminating ability because approximately 95% of the scores would be between 70 and 80. In like manner, a production line that makes bearings with $\mu = .5$ inch and $\sigma = .2$ inch would produce many defective items. The fact that the bearings have a mean diameter of .5 inch would be of little value when the bearings are fitted together. *The precision of an instrument, whether it be an educational test or a machine, is measured by the standard deviation of the error of measurement.* Hence, we proceed to a test of a population variance σ^2.

unbiased

The sample variance s^2 is an _____ estimator for σ^2. To use s^2 for inference making, we find that in repeated sampling, the distribution of s^2 has the following properties:

σ^2

nonsymmetric

0

$n; \sigma^2$

1. $E(s)^2 =$ _____
2. The distribution of s^2 is (symmetric, nonsymmetric).
3. s^2 can assume any value greater than or equal to _____.
4. The shape of the distribution changes for different values of _____ and _____.
5. When sampling is from a *normal* population, s^2 is independent of the population mean μ and the sample mean $\bar{x}$. As with the z statistic, the distribution for s^2 when the sampling is from a normal population can be standardized by using

$$\chi^2 = \frac{(n-1)s^2}{\sigma^2}$$

which is the chi-square random variable that has the following properties in repeated sampling:

$n - 1$

nonsymmetric

0

a. $E(\chi)^2 = df =$ _____

b. The distribution of χ^2 is (symmetric, nonsymmetric).

c. $\chi^2 \geq$ _____

d. The distribution of χ^2 depends on the degrees of freedom, $n - 1$.

Since χ^2 does not have a symmetric distribution, critical values of χ^2 have been tabulated for both the upper and lower tails of the distribution in Table 5 of the text. The degrees of freedom are listed along both the right and left margins of the table. Across the top margin are values of χ_a^2, indicating a value of χ^2 that has an area equal to a to its right; that is,

$$P(\chi^2 > \chi_a^2) = a$$

Example 9.10
Use Table 5 to find the critical values of χ^2:

a	df	χ_a^2	
.05	2	_____	5.99147
.99	10	_____	2.55821
.01	20	_____	37.5662
.95	30	_____	18.4926
.995	9	_____	1.734926
.025	15	_____	27.4884
.005	24	_____	45.5585
.90	17	_____	10.0852

The statistical test of an hypothesis concerning a population variance σ^2 at the α level of significance is given as follows:

1. H_0: $\sigma^2 = \sigma_0^2$

2. H_a: appropriate one- or two-tailed test

3. Test statistic:

$$\chi^2 = \frac{(n-1)s^2}{\sigma_0^2}$$

4. Rejection region:
   a. For H_a: $\sigma^2 > \sigma_0^2$, reject H_0 if $\chi^2 > \chi_\alpha^2$ based on $n-1$ degrees of freedom.
   b. For H_a: $\sigma^2 < \sigma_0^2$, reject H_0 if $\chi^2 < \chi_{(1-\alpha)}^2$ based on $n-1$ degrees of freedom.
   c. For H_a: $\sigma^2 \neq \sigma_0^2$, reject H_0 if $\chi^2 > \chi_{\alpha/2}^2$ or $\chi^2 < \chi_{(1-\alpha/2)}^2$ based on $n-1$ degrees of freedom.
5. Bound the *p-value* using Table 5. If *p-value* $< \alpha$, reject H_0.

Example 10.11
A producer of machine parts claimed that the diameters of the connector rods produced by his plant had a variance of at most .03 inch^2. A random sample of 15 connector rods from his plant produced a sample mean and variance of .55 inch and .053 inch^2, respectively. Is there sufficient evidence to reject his claim at the $\alpha = .05$ level of significance?

Solution
Collecting pertinent information, we have

$$s^2 = .053 \text{ inch}^2$$

$$df = n - 1 = 14$$

$$\sigma_0^2 = .03 \text{ inch}^2$$

The test of the hypothesis follows:

1. H_0: $\sigma^2 = .03$

> 2. H_a: σ^2 _____ .03

3. Test statistic:

$$\chi^2 = \frac{(n-1)s^2}{\sigma_0^2}$$

4. Rejection region:

23.6848

For 14 degrees of freedom, we will reject H_0 if $\chi^2 \geq$ _____. Calculate

$$\chi^2 = \frac{(n-1)s^2}{\sigma_0^2}$$

.053; .03

$$= \frac{14(\underline{\hspace{2cm}})}{(\underline{\hspace{2cm}})}$$

$$= 24.733$$

Reject

Decision: (Reject, Do not reject) H_0 because

$$24.733 > \chi_{.05}^2 = 23.6848$$

is

The data produced sufficient evidence to reject H_0. Therefore, we can conclude that the variance of the rod diameters (is, is not) greater than .03 inch^2.

Example 10.12
Refer to Example 10.11. Find the approximate *p-value* and interpret your results.

Solution
The *p-value* for this test is

$$p\text{-}value = P(\chi^2 > 24.733)$$

where χ^2 has a chi-square distribution with $n - 1 = 14$ degrees of freedom. From Table 5 in the text, the observed value falls between

23.6848; 26.1190

$\chi_{.05}^2 =$ _____ and $\chi_{.025}^2 =$ _____. Hence,

.025; .05

_____ < *p-value* < _____

The null hypothesis can be rejected for any value of α (greater than, less than) or equal to $\alpha =$ _____. Since α was .05 in Example 10.11, H_0 was rejected.

greater than
.05

The sample variance s^2 is an unbiased point estimator for the population variance σ^2. Utilizing the fact that $(n-1)s^2/\sigma^2$ has a chi-square distribution with $n-1$ degrees of freedom, we can show that a $(1-\alpha)100\%$ confidence interval for σ^2 is

$$\frac{(n-1)s^2}{\chi^2_{\alpha/2}} < \sigma^2 < \frac{(n-1)s^2}{\chi^2_{(1-\alpha/2)}}$$

where $\chi^2_{\alpha/2}$ is the tabulated value of the chi-square random variable based on _____ degrees of freedom that has an area equal to $\alpha/2$ to its right, and $\chi^2_{(1-\alpha/2)}$ is the tabulated value from the same distribution that has an area of $\alpha/2$ to its left or, equivalently, an area of $1-\alpha/2$ to its right.

$n-1$

Example 10.13
Find a 95% confidence interval estimate for the variance of the rod diameters from Example 10.11.

Solution
From Example 10.11, the estimate of σ^2 was $s^2 = .053$ with 14 degrees of freedom. For a confidence coefficient of .95, we need

$$\chi^2_{(1-\alpha/2)} = \chi^2_{.975} = \underline{\hspace{2cm}}$$

5.62872

$$\chi^2_{\alpha/2} = \chi^2_{.025} = \underline{\hspace{2cm}}$$

26.1190

1. Using the confidence interval estimator

$$\frac{(n-1)s^2}{\chi^2_{\alpha/2}} < \sigma^2 < \frac{(n-1)s^2}{\chi^2_{(1-\alpha/2)}}$$

we have

$$\frac{14(.053)}{(\underline{\hspace{1cm}})} < \sigma^2 < \frac{14(.053)}{(\underline{\hspace{1cm}})}$$

26.1190; 5.62872

$$\underline{\hspace{2cm}} < \sigma^2 < \underline{\hspace{2cm}}$$

.028; .132

2. By taking the square roots of the upper and lower confidence limits, we have an equivalent confidence interval for the standard deviation σ. For this problem,

$$\underline{\hspace{2cm}} < \sigma < \underline{\hspace{2cm}}$$

.167; .363

Comment: Although the sample variance is an unbiased point estimator for σ^2, notice that the confidence interval estimator for σ^2 *is* *not symmetrically located about s^2* as was the case with confidence intervals that were based on the z or t distributions. This follows from the fact that a chi-square distribution is not symmetric, whereas the z and t distributions are symmetric.

Self-Correcting Exercises 10D

1. In an attempt to assess the variability in the time until a pain reliever became effective for a patient, a doctor, on five different occasions, prescribed a controlled dosage of the drug for his patient. The five measurements recorded for the time until effective relief were 20.2, 15.7, 19.8, 19.2, and 22.7 minutes. Would these measurements indicate that the standard deviation of the time until effective relief was less than 3 minutes?

2. An educational testing service, in developing a standardized test, would like the test to have a standard deviation of at least 10. The present form of the test has produced a standard deviation of $s = 8.9$ based on $n = 30$ test scores. Should the present form of the test be revised based on these sample data? What *p-value* would you report?

3. A quick technique for determining the concentration of a chemical solution has been proposed to replace the standard technique, which takes much longer. In testing a standardized solution, 30 determinations using the new technique produced a standard deviation of $s = 7.3$ parts per million.
   a. Does it appear that the new technique is less sensitive (has larger variability) than the standard technique whose standard deviation is $\sigma = 5$ parts per million?
   b. Estimate the true standard deviation for the new technique with a 95% confidence interval.

Solutions to SCE 10D

1. H_0: $\sigma = 3$ ($\sigma^2 = 9$); H_a: $\sigma < 3$ ($\sigma^2 < 9$); test statistic:
$$\chi^2 = (n-1)s^2/\sigma_0^2.$$

With $\alpha = .05$ and $n - 1 = 4$ degrees of freedom, reject H_0 if $\chi^2 < \chi_{.95}^2 = .710721$. Calculate
$$s^2 = \frac{\sum x_i^2 - (\sum x_i)^2/n}{n-1} = \frac{1930.5 - 1905.152}{4} = 6.337.$$
$\chi^2 = 25.348/9 = 2.816$. Do not reject H_0.

2. H_0: $\sigma^2 = 100$; H_a: $\sigma^2 < 100$; test statistic:
$$\chi^2 = (n-1)s^2/\sigma_0^2.$$

With $\alpha = .05$ and $n - 1 = 29$ degrees of freedom, reject H_0 if $\chi^2 < \chi_{.95}^2 = 17.7083$. Calculate $\chi^2 = 29(8.9)^2/100 = 22.97$. Do not reject H_0. From Table 5 with 29 degrees of freedom, $\chi^2 = 22.97$ exceeds $\chi^2_{.90}$. Hence, $p\text{-}value > .10$.

3. a. H_0: $\sigma = 5$; H_a: $\sigma > 5$; test statistic: $\chi^2 = (n-1)s^2/\sigma_0^2$.

With $\alpha = .05$, reject H_0 if $\chi^2 > 42.5569$. Since $\chi^2 = 29(7.3)^2/25 = 61.81$, reject H_0. The new technique is less sensitive.

b. $\dfrac{(n-1)s^2}{\chi_U^2} < \sigma^2 < \dfrac{(n-1)s^2}{\chi_L^2}$,

$$\frac{29(7.3)^2}{45.7222} < \sigma^2 < \frac{29(7.3)^2}{16.0471},$$

$33.80 < \sigma^2 < 96.30$, $5.81 < \sigma < 9.81$.

10.7 COMPARING TWO POPULATION VARIANCES

An experimenter may wish to compare the variability of two testing procedures or compare the precision of one manufacturing process with another. One may also wish to compare two population variances prior to using a t test.

To test the hypothesis of equality of two population variances,

$$H_0: \ \sigma_1^2 = \sigma_2^2$$

we need to make the following assumptions:

1. Each population sampled has a _____ distribution.   normal
2. The samples are _____.   independent

The statistic s_1^2/s_2^2 is used to test

$$H_0: \ \sigma_1^2 = \sigma_2^2$$

A _____ value of this statistic implies that $\sigma_1^2 > \sigma_2^2$, a   large
_____ value of this statistic implies that $\sigma_1^2 < \sigma_2^2$, and a value of   small
the statistic close to 1 implies that $\sigma_1^2 = \sigma_2^2$. In repeated sampling this
statistic has an F distribution as shown in Figure 10.4 when $\sigma_1^2 = \sigma_2^2$
with the following properties:

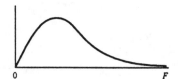

Figure 10.4

1. The distribution of F is (symmetric, nonsymmetric).   nonsymmetric
2. The shape of the distribution depends on the degrees of freedom
   associated with _____ and _____. s_1^2; s_2^2

0

3. F is always greater than or equal to _____.

The tabulation of critical values of F is complicated by the fact that the distribution is nonsymmetric and must be indexed according to the values of df_1 and df_2, the degrees of freedom associated with the numerator and denominator, respectively, of the F statistic. As we will see, however, it will be sufficient to have only right-tailed critical values of F for the various combinations of df_1 and df_2.

Table 6 in the text has tabulated right-tailed critical values for the F statistic, where F_a is that value of F that has an area of a to its right, based on df_1 and df_2, the degrees of freedom associated with the **numerator** and **denominator** of F, respectively. See Figure 10.5.

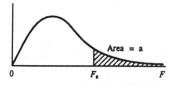

Figure 10.5

F_a satisfies the relationship $P(F > F_a) = a$. Table 6 has values of F_a for $a = .10$, $.05$, $.025$, $.01$, and $.005$ and various values of df_1 and df_2 between 1 and ∞.

Example 10.14
Find the value of F based on $df_1 = 5$ and $df_2 = 7$ degrees of freedom such that

$$P(F > F_{.05}) = .05$$

Solution
1. We wish to find a critical value of F with an area $a = .05$ to its right based on $df_1 = 5$ and $df_2 = 7$ degrees of freedom. Therefore, we will use Table 6.
2. Values of df_1 are found along the top margin of the table, and values of df_2 appear on both the right **and** left margins of the table. Find the value of $df_1 = 5$ along the top margin and cross-index this value with $df_2 = 7$ along the left margin to find $F_{.05(5, 7)} = 3.97$.

Example 10.15
Find the critical right-tailed values of F:

19.30
2.16
3.42
2.40
3.76

df_1	df_2	a	F_a
5	2	.05	_____
7	15	.10	_____
20	10	.025	_____
30	40	.005	_____
17	13	.01	_____

We can always avoid using left-tailed critical values of the F distribution by using the following approach. In testing H_0: $\sigma_1^2 = \sigma_2^2$ against the alternative H_a: $\sigma_1^2 > \sigma_2^2$, we would reject H_0 *only if* s_1^2/s_2^2 *is too large* (larger than a right-tailed critical value of F). In testing H_0: $\sigma_1^2 = \sigma_2^2$ against H_a: $\sigma_1^2 < \sigma_2^2$, we would reject H_0 *only if* s_2^2/s_1^2 *is too large*. In testing H_0: $\sigma_1^2 = \sigma_2^2$ against the two-tailed alternative H_a: $\sigma_1^2 \neq \sigma_2^2$, we will agree to *designate the population that produced the larger sample variance as population 1 and the larger sample variance as s_1^2*. We then agree to reject H_0 if s_1^2/s_2^2 is *too large*.

When we agree to designate the population with the larger sample variance as population 1, the test of H_0: $\sigma_1^2 = \sigma_2^2$ versus H_a: $\sigma_1^2 \neq \sigma_2^2$ using s_1^2/s_2^2 will be right-tailed. However, in so doing we must remember that the tabulated tail area must be doubled to get the actual significance level of the test. For example, if the critical right-tailed value of F has been found using $a = .05$, the actual significance level of the test will be $\alpha = 2(.05) = $ _____. If the critical value uses $a = .01$, the actual level will be $\alpha = 2(.01) = $ _____, and so forth.

.10
.02

Example 10.16
An experimenter has performed a laboratory experiment using two groups of rats. One group was given a standard treatment, while the second received a newly developed treatment. Wishing to test the hypothesis H_0: $\mu_1 = \mu_2$, the experimenter suspects that the population variances are not equal, an assumption necessary for using the t statistic in testing the equality of the means. Use the following data to test whether the experimenter's suspicion is warranted at the $\alpha = .02$ level:

Old Treatment	New Treatment
$s = 2.3$	$s = 5.8$
$n = 10$	$n = 10$

Solution
This problem involves a test of the equality of two population variances. Let population 1 be the population that receives the new treatment.

1. H_0: $\sigma_1^2 = \sigma_2^2$

2. H_a: $\sigma_1^2 \neq \sigma_2^2$

3. Test statistic:

$$F = \frac{s_1^2}{s_2^2}$$

4. Rejection region: With $df_1 = df_2 = 9$ degrees of freedom, reject H_0 if $F > F_{.01} = $ _____.

5.35

Now we calculate the test statistic:

$$F = \frac{s_1^2}{s_2^2} = \frac{(5.8)^2}{(2.3)^2}$$

33.64; 5.29

$$= \frac{(\underline{\hspace{1.5cm}})}{(\underline{\hspace{1.5cm}})}$$

6.36

$$= \underline{\hspace{2cm}}$$

greater

is

Decision: Since F is (greater, less) than $F_{.01} = 5.35$, H_0: $\sigma_1^2 = \sigma_2^2$ (is, is not) rejected.

Since the population variances were judged to be different, the experimenter is not justified in using the t statistic to test H_0: $\mu_1 - \mu_2 = 0$. She must resort to using t^* (see 10.4) or other methods, several of which will be discussed in Chapter 15.

Example 10.17
Refer to Example 10.16. Find the approximate *p-value* for the test and interpret your results.

Solution

two-tailed

6.36

9; 9

The *p-value* for this (one-tailed, two-tailed) test is *p-value* $= 2P(F > \underline{\hspace{2cm}})$, where F has an F distribution with $df_1 = \underline{\hspace{2cm}}$ and $df_2 = \underline{\hspace{2cm}}$ degrees of freedom. In order to find an approximate *p-value*, we must first find critical values, F_a, for various values of a with $df_1 = df_2 = 9$. These values are found in Table 6. Fill in the missing entries in the table that follows:

a	F_a
.10	2.44
.05	
.025	4.03
.01	
.005	

3.18

5.35

6.54

5.35

Since the observed value, $F = 6.36$ falls between $F_{.01} = \underline{\hspace{2cm}}$ and $F_{.005} = 6.54$, the *p-value* will be between $2(.005) = .01$ and $2(.01) = .02$; that is, $.01 < p\text{-}value < .02$. Hence, H_0 can be rejected for any value of α (greater, less) than or equal to $\underline{\hspace{2cm}}$.

greater; .02

Utilizing the fact that $(s_1^2/s_2^2)(\sigma_2^2/\sigma_1^2)$ has an F distribution with $df_1 = (n_1 - 1)$ and $df_2 = (n_2 - 1)$ degrees of freedom, we can show that a $(1 - \alpha)100\%$ confidence interval for σ_1^2/σ_2^2 is

$$\frac{s_1^2}{s_2^2} \frac{1}{F_{df_1 df_2}} < \frac{\sigma_1^2}{\sigma_2^2} < \frac{s_1^2}{s_2^2} F_{df_2 df_1}$$

where $F_{df_1 df_2}$ is the tabulated value of F with df_1 and df_2 degrees of freedom and area $\alpha/2$ to its right, and $F_{df_2 df_1}$ is the tabulated value of F with df_2 and df_1 degrees of freedom and area $\alpha/2$ to its right.

Example 10.18
A comparison of the precisions of two machines developed for extracting juice from oranges is to be made using the following data:

Machine A	Machine B
$s^2 = 3.1$ ounces^2	$s^2 = 1.4$ ounces^2
$n = 25$	$n = 25$

Is there sufficient evidence to indicate that $\sigma_A^2 > \sigma_B^2$ at the $\alpha = .05$ level? Find a 90% confidence interval for σ_A^2/σ_B^2.

Solution
Let population 1 be the population of measurements on machine A. The test would proceed as follows:

1. H_0: $\sigma_1^2 = \sigma_2^2$

2. H_a: σ_1^2 _____ σ_2^2 $>$

3. Test statistic:

   $$F = \frac{s_1^2}{s_2^2}$$

4. Rejection region: Based on $df_1 = df_2 = $ _____ degrees of   24
   freedom, we will reject H_0 if $F > F_{.05}$ with $F_{.05} = $ _____.   1.98

The value of the statistic is

$$F = \frac{s_1^2}{s_2^2} = \frac{3.1}{1.4} = \underline{\hspace{2cm}}$$ 2.21

Decision: We reject H_0 and conclude that the variability of machine A (is, is not) greater than that of machine B.   is

A 90% confidence interval for $\sigma_A^2/\sigma_B^2 = \sigma_1^2/\sigma_2^2$ is

$$\frac{s_1^2}{s_2^2}\frac{1}{F_{24,24}} < \frac{\sigma_A^2}{\sigma_B^2} < \frac{s_1^2}{s_2^2} F_{24,24}$$

$$\frac{2.21}{1.98} < \frac{\sigma_A^2}{\sigma_B^2} < 2.21(1.98)$$

1.12; 4.38

$$\underline{\hspace{3cm}} < \frac{\sigma_A^2}{\sigma_B^2} < \underline{\hspace{3cm}}$$

Self-Correcting Exercises 10E

1. Refer to Exercise 2, Self-Correcting Exercises 10B. In using the t statistic in testing an hypothesis concerning $\mu_1 - \mu_2$, one assumes that $\sigma_1^2 = \sigma_2^2$. Based on the sample information, could you conclude that this assumption had been met for this problem? Use $\alpha = .05$.

2. An experiment to explore the pain thresholds to electrical shock for males and females resulted in the following data summary:

	Males	Females
n	10	13
$\overline{x}$	15.1	12.6
s^2	11.3	26.9

Do these data supply sufficient evidence to indicate a significant difference in variability of thresholds for these two groups at the 10% level of significance? What *p-value* would you report?

Solutions to SCE 10E

1. $H_0: \sigma_1^2 = \sigma_2^2$; $H_a: \sigma_1^2 \neq \sigma_2^2$; test statistic: $F = s_1^2/s_2^2$, where population 1 is the population of distances for people wanting the center closed. With $\alpha = .02$ and $df_1 = 8$, $df_2 = 15$, reject H_0 if $F > 4.00$ from Table 6.

   $F = (5.3)^2/(2.8)^2 = 28.09/7.84 = 3.58$.

   Do not reject H_0. Assumption has been met.

2. $H_0: \sigma_1^2 = \sigma_2^2$; $H_a: \sigma_1^2 \neq \sigma_2^2$; test statistic: $F = s_1^2/s_2^2$, where population 1 is the population of thresholds for females. With $\alpha = .10$, $df_1 = 12$, $df_2 = 9$, reject H_0 if $F > 3.07$. Calculate $F = 26.9/11.3 = 2.38$. Do not reject H_0. The two groups exhibit the same basic variation. From Table 6, the following values for F_a are found:

a	F_a
.10	2.12
.05	2.64
.025	3.20
.01	4.00
.005	4.67

Since the observed value of $F = 2.38$ falls between $F_{.10}$ and $F_{.05}$, $2(.05) < p\text{-}value < 2(.10)$ or $.10 < p\text{-}value < .20$.

10.8 ASSUMPTIONS

The testing and estimation procedures presented in this chapter are based on the t, χ^2, and F statistics. In order that the probability statements associated with these testing and estimation procedures accurately reflect the prescribed probability values, specific assumptions concerning the sampled population(s) and the method of sampling must be satisfied.

The valid use of the t, χ^2, and F statistics requires that all samples be randomly selected from _____ populations. With the exception of the paired-difference experiment, when two samples are drawn, the samples must be drawn _____. In addition, when inferences are made about the difference in two population means, μ_1 and μ_2, using two independent samples, the population variances σ_1^2 and σ_2^2 must be _____.

normal

independently

equal

It would be unusual to have all these assumptions satisfied in practice. However, if the sampled population were not normal, or $\sigma_1^2 \neq \sigma_2^2$, we would like our procedures to produce error probabilities that are approximately equal to the specified values. A statistical procedure that is insensitive to departures from the assumptions upon which it is based is said to be _____.

robust

Procedures based on the t statistic are fairly robust to departures from normality provided that the sampled population(s) is (are) not strongly skewed. This (is, is not) true for procedures based on the χ^2 and F statistics. The t statistic used in comparing two means is moderately robust to departures from the assumption $\sigma_1^2 = \sigma_2^2$ when $n_1 = n_2$. However, when $\sigma_1^2 \neq \sigma_2^2$ and one sample size becomes large relative to the other, the procedure fails to be robust.

is not

When the experimenter is aware of possible violations of assumptions, the usual procedure can be used if it is robust with respect to the assumptions violated. Otherwise, the nonparametric procedures presented in Chapter 15 can be used. Nonparametric methods require few or no assumptions concerning the sampled population(s); however, samples must nonetheless be _____ selected, and when appropriate, the samples must also be independently drawn. When the sample sizes are relatively large, techniques such as those presented in Chapter 9 can be used in place of nonparametric procedures.

randomly

KEY CONCEPTS AND FORMULAS

I. *Experimental Designs for Small Samples*
1. Single random sample: The sampled population must be normal.
2. Two independent random samples: Both sampled populations must be normal.
   a. Populations have a common variance σ^2.
   b. Populations have difference variances: σ_1^2 and σ_2^2.
3. Paired-difference or matched pairs design: The samples are not independent.

II. *Statistical Tests of Significance*
1. Based on the t, F, and χ^2 distributions
2. Use the same procedure as in Chapter 9
3. Rejection region–critical values and significance levels: based on the t, F, or χ^2 distributions with the appropriate degrees of freedom
4. Tests of population parameters: a single mean, the difference between two means, a single variance, and the ratio of two variances.

III. *Small-Sample Test Statistics*
To test one of the population parameters when the sample sizes are small, use the following test statistics:

Parameter	Test Statistic	Degrees of Freedom
μ	$t = \dfrac{\overline{x} - \mu_0}{s/\sqrt{n}}$	$n - 1$
$\mu_1 - \mu_2$ (equal variances)	$t = \dfrac{(\overline{x}_1 - \overline{x}_2) - (\mu_1 - \mu_2)}{\sqrt{s^2\left(\frac{1}{n_1} + \frac{1}{n_2}\right)}}$	$n_1 + n_2 - 2$
$\mu_1 - \mu_2$ (unequal variances)	$t \approx \dfrac{(\overline{x}_1 - \overline{x}_2) - (\mu_1 - \mu_2)}{\sqrt{\frac{s_1^2}{n_1} + \frac{s_2^2}{n_2}}}$	Satterthwaite's approximation
$\mu_1 - \mu_2$ (paired samples)	$t = \dfrac{\overline{d} - \mu_d}{s_d/\sqrt{n}}$	$n - 1$
σ^2	$\chi^2 = \dfrac{(n - 1)s^2}{\sigma_0^2}$	$n - 1$
$\dfrac{\sigma_1^2}{\sigma_2^2}$	$F = \dfrac{s_1^2}{s_2^2}$	$n_1 - 1$ and $n_2 - 1$

Exercises

1. Why can we say that the test statistics employed in Chapter 9 are approximately normally distributed?

2. What assumptions are made when Student's t statistic is used to test an hypothesis concerning a population mean μ?

3. How does one determine the degrees of freedom associated with a t statistic?

4. Ten butterfat determinations for brand G milk were carried out yielding $\bar{x} = 3.7\%$ and $s = 1.7\%$. Do these results produce sufficient evidence to indicate that brand G milk contains, on the average, less than 4.0% butterfat? (Use $\alpha = .05$.)

5. Refer to Exercise 4. Estimate the mean percentage of butterfat for brand G milk with a 95% confidence interval.

6. An experimenter has developed a new fertilizing technique that should increase the production of cabbages. Do the following data produce sufficient evidence to indicate that the mean weight of those cabbages grown by using the new technique is greater than the mean weight of those grown by using the standard technique?

Population I (New Technique)	Population II (Standard Technique)
$n_1 = 16$	$n_2 = 10$
$\bar{x}_1 = 33.4$ ounces	$\bar{x}_2 = 31.8$ ounces
$s_1 = 3$ ounces	$s_2 = 4$ ounces

7. Find a 90% confidence interval for the difference in means $\mu_1 - \mu_2$ for the data given in Exercise 6.

8. To test the comparative brightness of two red dyes, nine samples of cloth were taken from a production line and each sample was divided into two pieces. One of the two pieces in each sample was randomly chosen and red dye 1 applied; red dye 2 was applied to the remaining piece. The following data represent a "brightness score" for each piece. Is there sufficient evidence to indicate a difference in mean brightness scores for the two dyes?

Sample	Dye 1	Dye 2
1	10	8
2	12	11
3	9	10
4	8	6
5	15	12
6	12	13
7	9	9
8	10	8
9	15	13

9. To test the effect of alcohol in increasing the reaction time to respond to a given stimulus, the reaction times of seven persons were measured. After consuming 3 ounces of 40% alcohol, the reaction time for each of the seven persons was measured again. Do the following data indicate that the mean reaction time after consuming alcohol was greater than the mean reaction time before consuming alcohol? (Use $\alpha = .05$.)

Person	Before (time in seconds)	After (time in seconds)
1	4	7
2	5	8
3	5	3
4	4	5
5	3	4
6	6	5
7	2	5

10. A manufacturer of odometers claimed that mileage measurements indicated on his instruments had a variance of at most .53 mile per 10 miles traveled. An experiment, consisting of eight runs over a measured 10-mile stretch, was performed in order to check the manufacturer's claim. The variance obtained for the eight runs was .62. Does this provide sufficient evidence to indicate that $\sigma^2 > .53$? (Use $\alpha = .05$.)

11. Construct a 99% confidence interval estimate for σ^2 in Exercise 10.

12. In a test of heat resistance involving two types of metal paint, two groups of ten metal strips were randomly formed. Group one was painted with type I paint, while group two was painted with type II paint. The metal strips were placed in an oven in random order, heated, and the temperature at which the paint began to crack and peel recorded for each strip. Do the following data indicate that the variability in the critical temperatures differs for the two types of paint? Use $\alpha = .05$.

	$\bar{x}$	s^2	n
Type I	280.1°F	93.2	10
Type II	269.9°F	51.9	10

13. Construct a 98% confidence interval for σ_1^2/σ_2^2 for the data given in Exercise 12.

14. In an attempt to reduce the variability of machine parts produced by process A, a manufacturer has introduced process B (a modification of A). Do the following data, based on two samples of 25 items, indicate that the manufacturer has achieved his goal? Use $\alpha = .01$.

	n	s^2
Process A	25	6.57
Process B	25	3.19

15. Before contracting to have stereo music piped into each of his suites of offices, an executive had his office manager randomly select seven offices in which to have the system installed. The average time spent outside these offices per excursion among the employees involved was recorded before and after the music system was installed with the following results.

	Time in minutes	
Office number	*No music*	*Music*
1	8	5
2	9	6
3	5	7
4	6	5
5	5	6
6	10	7
7	7	8

Would you suggest that the executive proceed with the installation? Find the approximate *p-value* and interpret your results.

16. The weights in grams of 10 male and 10 female juvenile ring-necked pheasants are given below, together with a data summary produced by the Minitab command Stat → Basic Statistics → Display Descriptive Statistics.

Males	*Females*
1384	1073
1286	1058
1503	1053
1627	1038
1450	1018
1672	1146
1370	1123
1659	1089
1725	1034
1394	1281

Descriptive Statistics

Variable	N	Mean	Median	TrMean	StDev	SE Mean
Males	10	1507.0	1676.5	1507.4	153.1	48.4
Females	10	1091.3	1065.5	1076.7	77.7	24.6

Variable	Minimum	Maximum	Q1	Q3
Males	1286.0	1725.0	1380.5	1662.2
Females	1018.0	1281.0	1037.0	1128.8

a. Use a statistical test to determine if the population variance of the weights of the male birds differs from that of the females.

b. Test whether the average weight of juvenile male ring-necked pheasants exceeds that of the females by 300 grams. The procedure that you use should take into account the results of the analysis in part a. (If $\sigma_1^2 \neq \sigma_2^2$, the alternate form of the t statistic must be used and the degrees of freedom estimated.)

The analysis in part b can be implemented using the Minitab command **Stat → Basic Statistics → 2-Sample t** without the "Assume equal variances" option. Since this procedure always tests H_0: $\mu_1 - \mu_2 = 0$, you can subtract 300 from the weights of the males before implementing the analysis using Minitab.

Chapter 11
The Analysis of Variance

11.1 THE MOTIVATION FOR AN ANALYSIS OF VARIANCE

Suppose a national home builder is interested in comparing the prices per 1000 board feet of standard or better grade Douglas fir framing lumber from five major suppliers in each of the four states where the builder is planning to begin construction. The prices are given in the following display:

	State			
	1	2	3	4
	$241	$216	$230	$245
	235	220	225	250
	238	205	235	238
	247	213	228	255
	250	220	240	255
Means	242.2	214.8	231.6	248.6

Do the data provide sufficient evidence to conclude that the average prices of 1000 board feet of Douglas fir differ among the four states?

To answer this question, we might consider testing all possible pairs of means using Student's t statistic. Finding at least one significant difference would certainly provide evidence that the average prices do differ in at least two states. However, we would require $C_2^4 = 6$ tests. Even if all population means were the same, there is a probability α that we would *incorrectly* conclude that there is a significant difference between means. The overall error rate across all six tests could be quite large. Hence, we need *one* test procedure that will simultaneously test H_0, that the means μ_1, μ_2, μ_3, and μ_4 are equal, against the alternative hypothesis H_a, that at least one pair of means differs.

The procedure for comparing more than two population means is known as *analysis of variance*. This procedure compares the variation among the means with a within-sample pooled estimate of variation. This can be seen in the dotplots in Figure 11.1.

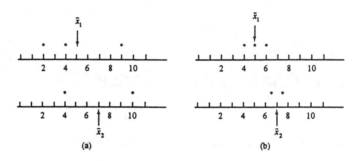

Figure 11.1

In your judgment, which dotplot suggests that μ_1 is not equal to μ_2? In dotplot (a), the difference between the sample means is $7 - 5 = $ _____, but the variation within samples is large, and it appears that both samples may have arisen from the same population. However, in dotplot (b), the difference in the means is still 2, but the variation within the samples is much _____ than in (a). In dotplot (b), the indication of differences in population means is strengthened by the fact that the ranges of the two sets of observations (do, do not) overlap.

Notice that visually you compared the difference between the means with the variation within samples. This is actually what happens in an analysis of variance and its testing procedures.

11.2 THE ASSUMPTIONS FOR AN ANALYSIS OF VARIANCE

Testing for the equality of three or more means is a generalization of the two-sample t test presented in Chapter 10. Although different designs or sampling procedures may be used, we will assume that the observations in each sampled population are _____ distributed with a common variance σ^2.

We will be concerned with the analysis of variance for two specific designs. The first design is based on independent random sampling from the populations being tested. The second design is a generalization of the paired-difference design and involves the random assignment of treatments within matched sets of observations. In addition we will also consider a two-way classification known as a factorial experiment.

2

smaller

do not

normally

Assumptions for an Analysis of Variance:

1. Observations within each population are normally distributed with a common variance σ^2.
2. Assumptions concerning the sampling procedure are specified for each design.

11.3 THE COMPLETELY RANDOMIZED DESIGN: A ONE-WAY CLASSIFICATION

In the completely randomized design, a useful and simply implemented design, random samples are independently selected from each of k populations. In implementing this design, we classify the resulting observations only according to the population from which they were selected; hence, this design is also referred to as a _____ *classification*.

one-way

> A **completely randomized design** involves the selection of randomly and independently drawn samples from each of k populations.

In designing an experiment, we need to make the distinction between populations that exist in fact and those that exist in concept. The voters in a state, the residents of a city, the giant redwoods of northern California, and the Medicare recipients in 1994 are all examples of populations that exist in _____. When populations such as these are sampled, a randomization device, such as a table of random numbers, must be used to identify those individuals to be included in the sample.

fact

When treatments consist of procedures that are to be evaluated, it may be that the underlying populations exist only in concept. For example, in evaluating the performance of three drugs used to control hypertension, individuals with hypertension would be selected as subjects. In this case, those individuals who receive the first drug would be a sample of individuals from the _____ population of all people with hypertension who could or will have received the drug. The same could be said for the second and third drugs. In this case, the drug to be used by each individual would be assigned at random in the following way. If n individuals are to be assigned to each drug group, then each individual in the pool of subjects would be assigned a number. A randomization device would be used to produce n random numbers, and the corresponding individuals would be assigned the

conceptual

first drug. The individuals who correspond to the next set of n random numbers would receive the second drug, and the remaining n individuals the third. At this point, the individuals would be classified according to the drug they were to receive, resulting in a one-way classification of the data.

11.4 THE ANALYSIS OF VARIANCE FOR A COMPLETELY RANDOMIZED DESIGN

Analysis of variance gets its name from the fact that the total variation in a set of data is decomposed into component parts associated with one or more factors (variables) plus a component due to random variation. For a *completely randomized design* or a *one-way classification*, the j-th observation in the i-th group is denoted by x_{ij} with $j = 1, 2, \ldots, n_i$, and $i = 1, 2, \ldots, k$. The total variation of all observations about the grand mean $\bar{x}$ is called the **total sum of squares**, and given by

$$\text{Total SS} = \sum (x_{ij} - \bar{x})^2 = \sum x_{ij}^2 - \frac{\left(\sum x_{ij}\right)^2}{n}.$$

You will recognize the calculational form for the total sum of squares as the formula for calculating the numerator of the sample variance. The second term in the calculational formula is called the **correction for the mean** and denoted as

$$\text{CM} = \frac{\left(\sum x_{ij}\right)^2}{n} = \frac{G^2}{n}.$$

You can see that its effect is to change the sum of squares of observations $\sum x_{ij}^2$ to a sum of squared deviations about the grand mean, hence its name.

The total sum of squares is partitioned into two components. The first component is called the **sum of squares for treatments (SST)**, and measures the variation among the k sample means.

CM

$$\text{SST} = \sum n_i (\bar{x}_i - \bar{x})^2 = \sum \frac{T_i^2}{n_i} - \underline{\hspace{2cm}}$$

where T_i is the sum of the observations for treatment i with n_i the number of observations for treatment i. The second component is called the **sum of squares for error (SSE)** and is the pooled within sample variation for each of the k groups.

$$\text{SSE} = (n_1 - 1)s_1^2 + (n_2 - 1)s_2^2 + \ldots + (n_k - 1)s_k^2.$$

Notice that this is simply an extension of the pooled estimate of σ^2 from Chapter 10. It can be shown that

SSE

$$\text{Total SS} = \text{SST} + \underline{\hspace{2cm}}$$

Because of this relationship, you need only calculate two of the three quantities. Since SSE is the most difficult of the three to calculate directly, it is usually found by subtraction, using the other two quantities.

SSE = Total SS − SST.

Associated with each sum of squares is a quantity called the **degrees of freedom**. The total sum of squares involves n squared deviations about the grand mean $\bar{x}$ and therefore has $df = n - 1$ degrees of freedom. The sum of squares for treatments involves k squared deviations of the sample means about the grand mean, and therefore has $df = k - 1$ degrees of freedom. The sum of squares for error has pooled degrees of freedom equal to

$$df = (n_1 - 1) + (n_2 - 1) + \ldots + (n_k - 1) = \underline{\hspace{2cm}}. \qquad n - k$$

Like the sums of squares, the degrees of freedom also have an additive property with

$$df(\text{Total}) = df(\text{Treatments}) + df(\text{Error}).$$

The sums of squares for treatments and error, when divided by their respective degrees of freedom, produce mean squares. All of this information is displayed in an **analysis of variance table** as shown in the following display.

ANOVA Table for Completely Randomized Design

Source	df	SS	MS	F
Treatments	$k-1$	SST	$MST = SST/(k-1)$	MST/MSE
Error	$n-k$	SSE	$MSE = SSE/(n-k)$	
Total	$n-1$	Total SS		

Quantities used to calculate sums of squares include:

$$\text{CM} = \frac{(G)^2}{n}$$

$$\text{Total SS} = \sum x_{ij}^2 - \text{CM}$$

$$\text{SST} = \sum \frac{T_i^2}{n_i} - \text{CM} \qquad\qquad MST = SST/(k-1)$$

$$\text{SSE} = \text{Total SS} - \text{SST} \qquad\qquad MSE = SSE/(n-k)$$

T_i = total of all observations in sample i.

n_i = number of observations in sample i.

$n = n_1 + n_2 + \ldots + n_k$

Example 11.1
The prices of 1000 board feet of Douglas fir framing lumber for five suppliers in each of four states follows. Construct an analysis of variance table for these data.

	State			
	1	2	3	4
	$241	$216	$230	$245
	235	220	225	250
	238	205	235	238
	247	213	228	255
	250	220	240	255
Means	242.2	214.8	231.6	248.6
Totals	1211	1074	1158	1243

Solution

For this problem, we will refer to the four states as the treatments. The calculational formulas require the $k = 4$ treatment totals as well as the grand total. For this problem the four sample sizes are each equal to _____ and the total number of observations is $n =$ _____.

Some initial computations will show that

$$T_1 = \underline{\qquad} \qquad T_2 = 1074 \qquad T_3 = 1158$$

$$T_4 = 1243 \qquad\qquad G = \underline{\qquad}$$

a. $\quad CM = \dfrac{(4886)^2}{20} = 1{,}097{,}929.8$

b. $\quad$ Total SS $= 241^2 + 235^2 + \ldots + 255^2 - CM$

$$= 1{,}101{,}862 - \underline{\qquad}$$

$$= 3932.2$$

c. $\quad SST = \dfrac{1211^2 + 1074^2 + 1158^2 + 1243^2}{20} - CM$

$$= \dfrac{5{,}506{,}010}{5} - CM$$

$$= 1{,}101{,}202 - 1{,}097{,}929.8$$

$$= \underline{\qquad}$$

d. $\quad$ SSE $=$ SS Total $-$ SST

$$= 3932.2 - 3272.2 = \underline{\qquad}$$

These quantities can be entered into an ANOVA table.

Source	df	SS	MS
Treatments	3	3272.2	1090.73
Error	16	660.0	_____
Total	19	3932.2	

The analysis of variance Minitab program for a one-way analysis of variance (**Stat → ANOVA → One-way**) is shown in Figure 11.2.

Margin answers:

5

20

1211

4686

1,097,929.8

3272.2

660.0

41.25

One-way Analysis of Variance

```
Analysis of Variance for X
Source      DF        SS        MS        F        P
States       3    3272.2    1090.7    26.44    0.000
Error       16     660.0      41.2
Total       19    3932.2
                                    Individual 95% CIs For Mean
                                    Based on Pooled StDev
Level       N      Mean     StDev   -+---------+---------+---------+-----
1           5    242.20      6.22                        (---*----)
2           5    214.80      6.22   (---*---)
3           5    231.60      5.94                (---*---)
4           5    248.60      7.23                          (---*---)
                                    -+---------+---------+---------+-----
Pooled StDev =     6.42            210       225       240       255
```

Figure 11.2

In addition to the basic ANOVA table, the printout provides the sample means and standard deviations, as well as plotted confidence intervals for each of the treatment means. It also provides the results of a test of the equality of all four treatment means in the columns labeled F and p.

Testing the Equality of Treatment Means

A test of the hypothesis H_0: $\mu_1 = \mu_2 = \ldots = \mu_k$, versus the alternative hypothesis H_a: $\mu_i \neq \mu_j$ for at least one pair of means (i, j), $i \neq j = 1, 2, \ldots, k$ involves the ratio of MST and MSE.

- MSE is an unbiased estimator of the underlying variance σ^2 whether H_0 is true of not.

- When the population means are all equal, MST is an unbiased estimator of σ^2, the common variance for all populations and the ratio $F = \text{MST}/\text{MSE}$ will in general be close to one.

- However, when the means **are not** all equal, MST will tend to be _____ than MSE and the ratio $F = \text{MST}/\text{MSE}$ will tend to be _____ than one. Hence, a test of equality of the k treatment means is based on a *right-tailed* F-test with $df_1 = (k - 1)$ numerator degrees of freedom and $df_2 = (n - k)$ denominator degrees of freedom.

larger
larger

The test is summarized in the next table.

F test for Comparing k population means.

- H_0: $\mu_1 = \mu_2 = \ldots = \mu_k$

- H_a: at least one pair of means differ.

- Test Statistic: $F = \text{MST}/\text{MSE}$ with $df_1 = (k - 1)$ and $df_2 = (n - k)$.

- Rejection region: Reject H_0 if $F > F_\alpha$ where F_α is the critical value of F with area α to its right. Alternately, reject H_0 if the *p-value* $< \alpha$.

Assumptions: The samples are randomly and independently drawn from normal populations with equal variances.

Example 11.2
Do the data in Example 11.1 provide sufficient evidence to conclude that average cost of lumber varies from state to state?

Solution
A test of H_0: $\mu_1 = \mu_2 = \mu_3 = \mu_4$ against the alternative hypothesis that at least one pair of means differ is based upon the statistic

26.44

$$F = \frac{\text{MST}}{\text{MSE}} = \frac{1090.73}{41.25} = \underline{\hspace{2cm}}$$

shown in the Minitab printout under the column labeled F. The *p-value* $= .000$ which is given in the column labeled p can be interpreted as *p-value* $< .0005$. The null hypothesis can be rejected

.0005

for any α greater than $\underline{\hspace{2cm}}$. You will arrive at the same conclusion by comparing $F = 26.44$ with the $\alpha = .05$ critical value of

3.29

$F_{.05} = \underline{\hspace{2cm}}$.

Estimating Differences in Treatment Means

Estimating a single mean or the difference between two means follows the procedures in Chapter 10, with the change that MSE, the pooled ANOVA estimate of σ^2 is used rather than the two-sample pooled estimator. In this chapter, we will use

$$s^2 = \text{MSE} \quad \text{with} \quad df = (n - k).$$

For a single population mean the confidence interval estimator is

$$\bar{x}_i \pm t_{\alpha/2} \frac{s}{\sqrt{n_i}}$$

where $\bar{x}_i$ is the sample mean for the i-th treatment. A confidence interval for the difference in two means is

$$(\bar{x}_i - \bar{x}_j) \pm t_{\alpha/2} \sqrt{s^2 \left(\frac{1}{n_i} + \frac{1}{n_j} \right)}$$

The estimator of σ^2 is $s^2 = \text{MSE}$ with $n - k$ degrees of freedom.

Example 11.3
Find a 95% confidence interval estimate for the average cost of 1000 board feet of Douglas fir for suppliers in state 1.

Solution
The treatment mean can be found by calculation or it can also be found on the Minitab printout together with the value of s and s^2.

$$\bar{x}_1 = \frac{T_1}{n_1} = \frac{1211}{5} = \underline{\hspace{2cm}} \quad \text{and} \qquad\qquad 242.20$$

$$s = \sqrt{\text{MSE}} = \sqrt{41.25} = \underline{\hspace{2cm}} \quad \text{with} \qquad\qquad 6.42$$

$$df = n - k = 20 - 4 = 16$$

Then with 16 df and $t_{.025} = 2.120$, the 95% confidence interval is

$$242.20 \pm 2.120 \left(\frac{\overline{\hspace{2cm}}}{\sqrt{5}} \right) = 242.20 \pm 6.09 \qquad\qquad 6.4226163$$

or from \$236.11 to \$248.29.

Example 11.4
Find a 95% confidence interval estimate for the difference in the average cost of 1000 board feet of Douglas fir for states 1 and 2.

Solution
The two sample means are

$$\bar{x}_1 = 242.2, \text{ and } \bar{x}_2 = \underline{\hspace{2cm}}. \qquad\qquad 214.8$$

The critical value of t and s^2 are given in Example 11.3. Therefore, a 95% confidence interval estimate of the difference in prices between states 1 and 2 is found as

$$(\bar{x}_1 - \bar{x}_2) \pm t_{.025} \sqrt{s^2 \left(\frac{1}{n_1} + \frac{1}{n_2} \right)}$$

$$(242.20 - 214.80) \pm 2.120 \sqrt{(\underline{\hspace{2cm}}) \left(\frac{1}{5} + \frac{1}{5} \right)} \qquad\qquad 41.25$$

$$27.40 \pm \underline{\hspace{2cm}} \qquad\qquad 8.61$$

or from \$18.79 to \underline{\hspace{2cm}}. \qquad\qquad \$36.01

Self-Correcting Exercises 11A

1. In the evaluation of three rations fed to chickens grown for market, the dressed weights of five chickens fed from birth on one of the three rations were recorded.

	Rations		
	1	*2*	*3*
	7.1	4.9	6.7
	6.2	6.6	6.0
	7.0	6.8	7.3
	5.6	4.6	6.2
	6.4	5.3	7.1
Total	32.3	28.2	33.3
Average	6.46	5.64	6.66

a. Do the data present sufficient evidence to indicate a difference in the mean growth for the three rations as measured by the dressed weights?

b. Estimate the difference in mean weight for rations 2 and 3 with a 95% confidence interval.

2. In the investigation of a citizens' committee complaint about the availability of fire protection within the county, the distance in miles to the nearest fire station was measured for each of five randomly selected residences in each of four areas.

	Areas				
	1	*2*	*3*	*4*	
	7	1	7	4	
	5	4	9	6	
	5	3	8	3	
	6	4	7	7	
	8	5	8	5	
T_i	31	17	39	25	Total $= 112$
n_i	5	5	5	5	$n = 20$
$\bar{x}_i$	6.2	3.4	7.8	5.0	

a. Do these data provide sufficient evidence to indicate a difference in mean distance for the four areas at the $\alpha = .01$ level of significance?

b. Estimate the mean distance to the nearest fire station for those residents in area 1 with a 95% confidence interval.

c. Construct a 95% confidence interval for $\mu_1 - \mu_3$.

Solutions to SCE 11A

1. $CM = (93.8)^2/15 = 586.5627$; Total SS $= 596.26 - CM = 9.6973$;

$$SST = \frac{32.3^2 + 28.2^2 + 33.3^2}{5} - CM = 589.484 - CM = 2.9213;$$

$SSE = 9.6973 - 2.9213 = 6.7760.$

		ANOVA		
Source	*df*	*SS*	*MS*	*F*
Treatments	2	2.9213	1.4607	2.5867
Error	12	6.7760	.5647	
Total	14	9.6973		

a. H_0: $\mu_1 = \mu_2 = \mu_3$; H_a: at least one of the equalities is incorrect. Test statistic: $F = MST/MSE$. With $df_1 = 2$ and $df_2 = 12$, reject H_0 if $F > F_{.05} = 3.89$. Since $F = 2.5867$, do not reject H_0. We cannot find a significant difference.

b. $(\bar{x}_2 - \bar{x}_3) \pm t_{.025}\sqrt{MSE\left(\frac{1}{n_2} + \frac{1}{n_3}\right)}$

$= (5.64 - 6.66) \pm 2.179\sqrt{\frac{2(.5647)}{5}}$

$= -1.02 \pm 2.179(.4753)$

$= -1.02 \pm 1.04$

$-2.06 < \mu_2 - \mu_3 < .02$ with 95% confidence.

2. $CM = (112)^2/20 = 627.2$; Total SS $= 708 - CM = 80.80$;

$$SST = \frac{31^2 + 17^2 + 39^2 + 25^2}{5} - CM = 679.2 - CM = 52.00;$$

$SSE = 80.80 - 52.00 = 28.80.$

		ANOVA		
Source	*df*	*SS*	*MS*	*F*
Areas	3	52.00	17.33	9.63
Error	16	28.80	1.80	
Total	19	80.80		

a. H_0: $\mu_1 = \mu_2 = \mu_3 = \mu_4$; H_a: at least one equality does not hold. $F = 9.63 > F_{.01} = 5.29$. Reject H_0. There is a difference for the four areas.

b. $\bar{x}_1 \pm t_{.025}\sqrt{MSE/n_1} = 6.2 \pm 2.120\sqrt{1.80/5}$
   $= 6.2 \pm 2.12(.6) = 6.2 \pm 1.272.$

c. $(\bar{T}_1 - \bar{T}_3) \pm t_{.025}\sqrt{MSE\left(\frac{1}{n_1} + \frac{1}{n_3}\right)}$

$= (6.2 - 7.8) \pm 2.120\sqrt{1.80(2/5)}$

$= -1.6 \pm 2.12(.85) = -1.6 \pm 1.802.$

11.5 RANKING POPULATION MEANS

The analysis of variance is a method of analysis that allows the user
to determine whether the variability of the observed sample means is
small relative to error variability (thereby indicating no differences
among the associated population means) or the variability of the
sampled means is large relative to error variability (thereby indicating that one or more of the population means differs from at least
one other mean). If we conclude that there are no significant
differences among the means, then our analysis is finished. On the
other hand, if the F test is significant and we are led to the
conclusion that at least one pair of means differ significantly, then it
is natural to ask, Which means are different from others? Furthermore, what procedure should we use to find out? From what we
learned in Chapters 9 and 10, we could perform a series of t tests,
one for each pair of means. However, the overall Type I error rate
associated with the $[k(k-1)]/2$ possible pairwise tests climbs very
rapidly with t and quickly becomes unacceptably large.

One way to avoid a high risk of declaring false significant differences is to use the percentiles of the Studentized range statistic,

$$\frac{\overline{x}_{max} - \overline{x}_{min}}{s}$$

rather than those for the Student's t statistic. Tukey's *honestly significant differences* (HSD) procedure, which uses the Studentized range
statistic, has an overall probability equal to α of declaring at least
one false difference when no differences exist.

When the samples are independent and the same size, the critical difference used in determining whether two sample means are
significantly different is given by

$$\omega = q_\alpha(k, df)\left(\frac{s}{\sqrt{n_t}}\right)$$

where $q_\alpha(k, df)$ is an entry from the Studentized range table based
on k treatment means with df degrees of freedom for error, and n_t is
the common sample size for each mean.

Tukey's HSD Procedure:

The difference between two sample means, say $\overline{x}_i$ and $\overline{x}_j$,
is declared to be significantly different if the absolute
difference, $\left|\overline{x}_i - \overline{x}_j\right|$ exceeds

$$\omega = q_\alpha(k, df)\left(\frac{s}{\sqrt{n_t}}\right)$$

for

k = the number of treatment means
s = the pooled estimate of σ given by $\sqrt{\text{MSE}}$
df = the degrees of freedom associated with MSE
n_t = the number of observations in each mean
α = the overall protection level covering all possible comparisons.
$q_\alpha(k, df)$ = tabulated value from Table 11, Appendix I for $\alpha = .05$ and $.01$ and for various values of k and df.

Rule: Two means are declared significantly different if they differ by ω or more.

Example 11.5
Refer to Example 11.1. Use Tukey's HSD procedure to determine which means differ significantly at the $\alpha = .05$ protection level.

Solution
1. From the Minitab printout in Example 11.1, we can find

$\bar{x}_1 = 242.20$	$k = $ _____	4
$\bar{x}_2 = $ _____	$n_t = 5$	214.80
$\bar{x}_3 = 231.60$	$df = 16$	
$\bar{x}_4 = $ _____	$\alpha = .05$	248.60
$s = \sqrt{41.25} = $ _____		6.42

2. The tabled value of $q_{.05}(4, 16)$ is 5.19. Therefore, the critical difference, ω, is

$$\omega = q_{.05}(4, 16)\left(\frac{s}{\sqrt{n_t}}\right) = 4.05\left(\frac{6.42}{\sqrt{5}}\right) = 4.05(\underline{\hspace{1.5cm}})$$ 2.87

$$= \underline{\hspace{1.5cm}}$$ 11.63

3. The next step is to arrange the means in order of magnitude, beginning with the largest:

$$\bar{x}_4 = 248.60 \quad \bar{x}_1 = 242.20 \quad \bar{x}_3 = 231.60 \quad \bar{x}_2 = 214.80$$

4. Now compare the means two at a time, first finding the differences between mean 4 and all the means to the right of mean 4; then finding the differences between mean 1 and all the means to its right; and finally, the difference between means 3 and 2.

$\bar{x}_4 - \bar{x}_2 = 248.60 - 214.80 = $ _____ > 11.63		33.80
$\bar{x}_4 - \bar{x}_3 = 248.60 - 231.60 = $ _____ > 11.63		17.00
$\bar{x}_4 - \bar{x}_1 = 248.60 - 242.20 = $ _____ < 11.63		6.40
$\bar{x}_1 - \bar{x}_2 = 242.20 - 214.80 = $ _____ > 11.63		27.40

10.60

16.80

3
2

4.05

positive

are

4

$$\bar{x}_1 - \bar{x}_3 = 242.20 - 231.60 = \underline{\hspace{2cm}} < 11.63$$

$$\bar{x}_3 - \bar{x}_2 = 231.60 - 214.80 = \underline{\hspace{2cm}} > 11.63$$

5. In summary, we see that mean 4 differs significantly from means 2 and \underline{\hspace{2cm}}, mean 1 differs significantly from mean \underline{\hspace{2cm}}, mean 3 differs from mean 2, and all other differences are nonsignificant. Another way to report the results of paired comparisons is to display the means from left to right, and use an underline to indicate which of the means do not differ from the others, as shown here:

$$\underline{\bar{x}_4 \qquad\qquad \bar{x}_1} \qquad\qquad \bar{x}_3 \qquad\qquad \bar{x}_2$$

Most computer programs have the option or capability to perform pairwise tests of treatment means. The Minitab printout of the paired comparison option for Example 11.1 is given in Figure 11.3.

```
Tukey's pairwise comparisons

   Family error rate = 0.0500
Individual error rate = 0.0113

Critical value = 4.05

Intervals for (column level mean) - (row level mean)

                1          2          3

      2       15.77
              39.03

      3       -1.03     -28.43
              22.23      -5.17

      4      -18.03     -45.43     -28.63
               5.23     -22.17      -5.37
```

Figure 11.3

The critical value of \underline{\hspace{2cm}} is the tabled value of $q_{.05}(4, 16)$ used to find the critical difference, ω. The intervals displayed are found as $(\bar{x}_i - \bar{x}_j) \pm \omega$. If an interval spans zero, that is, if the lower limit is negative and the upper limit is \underline{\hspace{2cm}}, the means are not declared to be significantly different. If an interval has both lower and upper limits positive or both negative, then the means (are, art not) declared to be significantly different. The results in Figure 11.3 indicate that the pairs 1 and 2, 2 and 3, 2 and 4, and 3 and \underline{\hspace{2cm}} are significantly different, a result that agrees with our calculations in Example 11.2.

Tukey's procedure can be used to determine which means differ from others when an analysis of variance indicates that there are significant differences among the treatment means for a completely randomized design and for the designs that will be introduced in the next section.

11.6 THE RANDOMIZED BLOCK DESIGN

The randomized block design is a natural extension of the _____ _____ experiment. The purpose is to increase the _____ in the design by making comparisons between treatments within relatively homogeneous blocks of experimental material. The randomized block design for k treatments and b blocks assumes blocks of relatively homogeneous material, with each block containing _____ experimental units. Each treatment is applied to one experimental unit in each block. Consequently, the number of observations for a given treatment for the entire experiment will equal _____. Thus, for the randomized block design, $n_1 = n_2 = \cdots = n_k = b$. A randomized block design for $k = 3$ treatments and $b = 4$ blocks is shown in Figure 11.4.

paired-
difference;
information

k

b

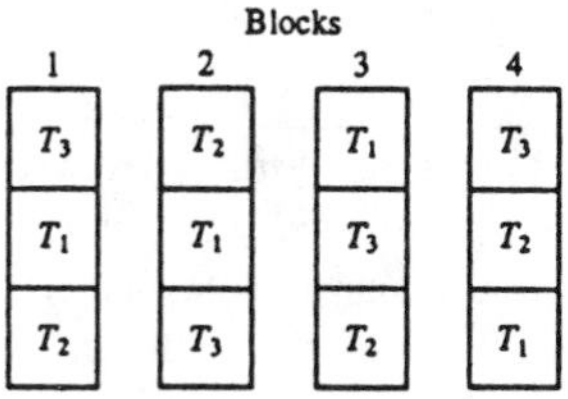

Figure 11.4

We denote the treatments as T_1, T_2, and T_3. The total number of observations for a randomized block design with b blocks and k treatments is $n =$ _____.

bk

> A **randomized block design** containing k treatments consists of b blocks of _____ experimental units each. The treatments are randomly assigned to the units in each _____, with each treatment appearing exactly _____ in each block.

k

block
once

Let us look at some situations where a block design can be used to reduce uncontrolled variation.

1. The potencies of several drugs are to be compared by three analysts. If each analyst makes one determination for each drug, the variability of these determinations should be homogeneous for a given analyst. Hence, we an consider the _____ as blocks.

analysts

2. An experiment is to be conducted to assess the relative merits of five different gasolines. Since vehicle-to-vehicle variation is inevitable in such experiments, four vehicles are chosen and each of the five gasolines is used in each vehicle. In this case, we would take _____ as blocks.

vehicles

3. An experiment to assess the effects of three raw material suppliers and four different mixtures on the crushing strength of concrete

blocks is to be run. To eliminate the variability from supplier to supplier, each of the four mixtures is prepared using the material from each of the three suppliers. Thus, we have reduced the variability by measuring the crushing strengths of the four mixtures in each of three relatively homogeneous blocks, which are the

suppliers
treatments

_____.

The word *randomized* means that the _____ are randomly distributed over the experimental units within each block. The randomized block design involves two independent variables:

blocks; treatments

_____ and _____. For an experiment run in a randomized block design, the total variation can now be partitioned into three sources of variation: blocks (B), treatments (T), and error (E).

Randomized block designs prove to be very useful because many investigations involve human (or animal) subjects that exhibit a large subject-to-subject variability.

block

1. By using a subject as a "_____" and having each subject receive all the treatments in a random order, treatment comparisons made within subjects would exhibit less variation than treatment comparisons made between subjects.
2. Since every subject receives each treatment in some random order, a _____ number of subjects is required in a randomized block design than in a completely randomized design.

smaller

11.7 AN ANALYSIS OF VARIANCE FOR A RANDOMIZED BLOCK DESIGN

An analysis of variance for a randomized block design partitions the total sum of squares into three components: the first component is the variation in treatment means, the second is the variation in block means, and the third is random variation. Therefore,

SSE

$$\text{Total SS} = \text{SST} + \text{SSB} + \underline{\qquad\qquad}$$

where

$$\text{Total SS} = \sum (x_{ij} - \bar{x})^2$$

and SST, SSB, and SSE denote the sums of squares for treatments, blocks, and error, respectively.

We will use the following notation when conducting an analysis of variance for a randomized block design:

$k = $ number of treatments

$b = $ number of blocks

$n = bk = $ total number of observations in the experiment

$G = \sum_{ij} x_{ij} = $ total of all observations in the experiment

$\bar{x} = \dfrac{G}{n} = $ mean of all observations in the experiment

T_i = total of all observations receiving treatment i, $i = 1, 2, \ldots, k$

B_j = total of all observations in block j, $j = 1, 2, \ldots, b$

A summary of the computing formulas follows for an analysis of variance for a randomized block design, with k treatments in b blocks follows.

$$CM = \frac{(G)^2}{n}$$

where $n = bk$ and G = sum of all n observations.

$$\text{Total SS} = \sum_{ij} x_{ij}^2 - CM$$

$$= \text{sum of squares of all } x \text{ values} - CM$$

$$SST = \sum \frac{T_i^2}{b} - CM \qquad MST = \frac{SST}{k-1}$$

$$SSB = \sum \frac{B_j^2}{k} - CM \qquad MSB = \frac{SSB}{b-1}$$

The typical ANOVA table for k treatments in b blocks is shown in the following table:

Source	df	SS	MS	F
Treatments	$k-1$	SST	$MST = \frac{SST}{k-1}$	$\frac{MST}{MSE}$
Blocks	$b-1$	SSB	$MSB = \frac{SSB}{b-1}$	$\frac{MSB}{MSE}$
Error	$n-b-k+1$	SSE	$MSE = \frac{SSE}{n-b-k+1}$	
Total	$n-1$			

Notice that the degrees of freedom for error can also be expressed as

$$bk - b - k + 1 = b(k-1) - (k-1)$$

$$= (b-1)(k-1)$$

In short, the error degrees of freedom are found as the product of the _____ and _____ degrees of freedom.   block; treatment

Example 11.6

In a study to increase the productivity of employees whose jobs involve a monotonous assembly procedure, 12 employees selected at random were asked to perform their usual jobs under three different trial conditions. The variable measured was assembly line stoppage time during a four-hour period. The data follow:

| | | Conditions | | |
Employee	1	2	3	Totals
1	31	22	26	79
2	20	15	23	58
3	26	21	18	65
4	21	12	22	55
5	12	16	18	46
6	13	19	23	55
7	18	7	16	41
8	15	9	12	36
9	21	11	26	58
10	15	15	19	49
11	11	14	21	46
12	18	11	21	50
Totals	221	172	245	638

Provide an analysis of variance for this experiment.

Solution

The treatment totals, the block totals and the grand total are given in the data table. In general, when the sample sizes are equal, in finding a sum of squares for one of the sources in the source column, you sum the squares of each total, divide by the number of observations in each total, and from this quantity you subtract the correction for the mean. Remember that $b = 12$, $k = 3$ and $n = bk = 12(3)$

36

$=$ _____.

a.   The correction for the mean is

638

$$\text{CM} = \frac{G^2}{n} = \frac{(\underline{\quad\quad})^2}{36} = 11,306.78$$

b.   The total sum of squares is found as

$$\text{Total SS} = \sum x_{ij}^2 - \text{CM} = 31^2 + 20^2 + \ldots + 21^2 - \text{CM}$$
$$= 12,340 - 11,306.78$$

1033.22

$$=$$ _____

c.   The sum of squares for treatments is

$$\text{SST} = \frac{T_1^2 + T_2^2 + T_3^2}{b} - \text{CM} = \frac{221^2 + 172^2 + 245^2}{12} - \text{CM}$$
$$= 11,537.5 - 11,306.78$$

230.72

$$=$$ _____

d.   The sum of squares for blocks is found as

$$\text{SSB} = \frac{B_1^2 + B_2^2 + \ldots + B_b^2}{k} - \text{CM} = \frac{79^2 + 58^2 + \ldots + 50^2}{3} - \text{CM}$$

$$= 11,784.67 - 11,306.78$$

$$= \underline{\hspace{3cm}}$$

477.89

e. The ANOVA table follows.

Source	df	SS	MS	F
Blocks	11	477.89	43.44	_____
Treatments	2	230.72	_____	7.82
Error	22	324.61	14.76	
Total	35	1033.22		

2.94

115.36

f. The Minitab ANOVA (**Stat** $\rightarrow$ **ANOVA** $\rightarrow$ **Two-way**) printout for this two-way classification is given in Figure 11.5.

Two-way Analysis of Variance

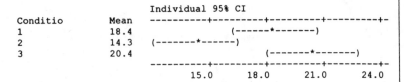

```
Analysis of Variance for X
Source      DF       SS        MS       F        P
Employee    11      477.9     43.4     2.94    0.015
Conditio     2      230.7    115.4     7.82    0.003
Error       22      324.6     14.8
Total       35     1033.2

                      Individual 95% CI
Conditio       Mean  ---------+---------+---------+---------+-
1              18.4                (------*------)
2              14.3   (-------*------)
3              20.4                     (-------*------)
                      ---------+---------+---------+---------+-
                         15.0      18.0      21.0      24.0
```

Figure 11.5

The entries in Figure 11.5 agree with the earlier calculations for these data. Notice that the Minitab printout includes values of F-statistics and their *p-values*.

Testing the Equality of Treatment and Block Means

The mean squares for treatments and blocks can be used to determine if there are significant differences among treatment and block means. When the treatment means differ strongly, MST will be large, and in like manner, when the block means vary strongly, MSB will be large. A formal procedure for testing treatment means is based upon the statistic

$$F = \frac{(\underline{\hspace{2cm}})}{\text{MSE}}$$

MST

which has an F distribution with $df_1 = (k-1)$ and $df_2 = (b-1)(k-1)$ when H_0: the treatment means are all equal, is, in fact, true. When H_a: at least two or more treatment means differ is true, F will be larger than expected, so that a right-tailed rejection region is used.

MSB

In a similar manner, a test of the equality of block means is based on the statistic

$$F = \frac{(\underline{\hspace{1cm}})}{\text{MSE}}$$

which has an F distribution with $df_1 = (b-1)$ and $df_2 = (b-1)(k-1)$ when the null hypothesis of no differences among the block means is true, and will be too large otherwise.

The formal testing procedures for testing treatment and block means follow.

For comparing treatment means:
1. *Null hypothesis:* H_0: the treatment means are equal
2. *Alternative hypothesis:* H_a: at least two of the treatment means differ
3. *Test statistic:* $F = \text{MST}/\text{MSE}$, where F is based on $df_1 = k-1$ and $df_2 = (b-1)(k-1)$ degrees of freedom.
4. *Rejection region:* Reject if $F > F_\alpha$, where F_α lies in the upper tail of the F distribution (see the figure).

For comparing block means:
1. *Null hypothesis:* the block means are equal
2. *Alternative hypothesis:* at least two of the block means differ
3. *Test statistic:* $F = \text{MSB}/\text{MSE}$, where F is based on $df_1 = b-1$ and $df_2 = (b-1)(k-1)$ degrees of freedom.
4. *Rejection region:* Reject if $F > F_\alpha$, where F_α lies in the upper tail of the F distribution (see the figure).

Example 11.7
For the experiment described in Example 11.6, test for significant differences among block and treatment means.

Solution
When the hypothesis of no treatment differences and/or the hypothesis of no blocks differences is true, MST, MSB, and MSE all provide unbiased estimates of the underlying variance σ^2.

a. The statistic for testing the hypothesis H_0: no differences among the k treatment means is

7.82

$$F = \frac{\text{MST}}{\text{MSE}} = \frac{115.36}{14.76} = \underline{\hspace{2cm}}$$

We would reject H_0 if $F > F_\alpha$ based on $(k-1)$ numerator and $(b-1)(k-1)$ denominator degrees of freedom. The computed value of F appears in column 5 of Figure 11.5 and is $F =$ _____. With $\alpha = .05$, the critical value of F with $(k-1) = (3-1) = 2$ numerator degrees of freedom and $(b-1)(k-1) = (12-1)(3-1) = 22$ denominator degrees of freedom is $F_{.05} =$ _____. Since the observed value of F exceeds 3.98, we can conclude that there is sufficient evidence of a difference among at least two of the three means. The *p-value* for this test, found in column 6 in the line corresponding to Conditions, is 0.003. Therefore, we could reject H_0 for any value of α greater than .003.

7.82

3.98

b. To test H_0: no differences among the block means, we use the statistic

$$F = \frac{MSB}{MSE}$$

with $(b-1)$ numerator and $(b-1)(k-1)$ denominator degrees of freedom and reject H_0 when $F > F_\alpha$. The observed value of F is $F =$ _____ found in column 5 of Figure 11.5 with an associated *p-value* = _____. Hence, we could reject H_0 for any value of α greater than .015, and specifically for $\alpha = .05$. Since at least one pair of block means differ, we can conclude that the randomized block design has removed a significant proportion of the variation from error, and therefore that our design was effective.

2.94
.015

Identifying Differences in Treatment and Block Means

When you find that an analysis of variance indicates that there are significant differences among treatment means or block means, you may wish to isolate those means that are different from others. This can be accomplished using confidence intervals for some means, or by using Tukey's paired comparison technique for all pairs of means. Remember that you will use $s = \sqrt{MSE}$ with _____ degrees of freedom since s^2 is the unbiased estimator of σ^2. The appropriate formulas are given in the next display.

$(b-1)(k-1)$

Comparing Treatment and Block Means

Tukey's Yardstick for Comparing Block Means:

$$\omega = q_\alpha(b, df)\,\frac{s}{\sqrt{k}}$$

Tukey's Yardstick for Comparing Treatment Means:

$$\omega = q_\alpha(k, df)\,\frac{s}{\sqrt{b}}$$

$(1 - \alpha)100\%$ *Confidence Interval for the Difference in Two Block Means*[1]:

$$(\overline{B}_i - \overline{B}_j) \pm t_{\alpha/2}\sqrt{s^2\left(\frac{1}{k} + \frac{1}{k}\right)}$$

where $\overline{B}_i$ is the average of all observations in block i.

$(1 - \alpha)100\%$ *Confidence Interval for the Difference in Two Treatment Means:*

$$(\overline{T}_i - \overline{T}_j) \pm t_{\alpha/2}\sqrt{s^2\left(\frac{1}{b} + \frac{1}{b}\right)}$$

where $\overline{T}_i$ is the average of all observations in treatment i.

Note: The values $q_\alpha(*, df)$ from Table 11, $t_{\alpha/2}$ from Table 4 and $s^2 = \text{MSE}$ all depend on $df = (b-1)(k-1)$ degrees of freedom.

Example 11.8

Identify significant pairs of treatment means for the data in Example 11.6 using Tukey's procedure.

Solution

To use Tukey's procedure to isolate significant differences among the three conditions, you need

3.84

a. $s = \sqrt{\text{MSE}} = \sqrt{14.76} = \underline{\hspace{1.5cm}}$ with $df = 22$

b. Using simple linear interpolation in Table 11,

$$q_{.05}(3, 22) = 3.58 - .5(3.58 - 3.53)$$

$$= 3.58 - .025$$

3.56

$$= 3.555 \text{ or } \underline{\hspace{1.5cm}}$$

3.94

c. The critical difference is $\omega = q_{.05}(3, 22)\,\dfrac{s}{\sqrt{b}} = 3.56\left(\dfrac{3.84}{\sqrt{12}}\right)$

$$= \underline{\hspace{1.5cm}}$$

d. The differences in treatment means can be determined from the Minitab ANOVA printout.

$$\overline{x}_1 = 18.4, \qquad \overline{x}_2 = 14.3 \quad \text{and} \quad \overline{x}_3 = 20.4$$

4.1

$$\overline{x}_1 - \overline{x}_2 = 18.4 - 14.3 = \underline{\hspace{1.5cm}}$$

$$\overline{x}_1 - \overline{x}_3 = 18.4 - 20.4 = -2.0, \text{ and}$$

−6.1

$$\overline{x}_2 - \overline{x}_3 = 14.3 - 20.4 = \underline{\hspace{1.5cm}}.$$

[1]You cannot construct a confidence interval for a single mean unless the blocks have been randomly selected from among the population of all blocks. The procedure for constructing intervals for single means is beyond the scope of this text.

We can conclude that the means corresponding to conditions 1 and 2, and 2 and 3 are significantly different.

The result of using the "Tukey's comparisons" option in Minitab's general linear model command (**Stat → ANOVA → General linear model**) is given in Figure 11.6.

```
Tukey 95.0% Simultaneous Confidence Intervals
Response Variable X
All Pairwise Comparisons among Levels of Conditions

Condition= 1 subtracted from:

Condition   Lower    Center    Upper    -------+---------+---------+---------
2           -8.020   -4.083    -0.1469  (-------*-------)
3           -1.936    2.000     5.9365           (-------*-------)
                                        -------+---------+---------+---------
                                           -5.0      0.0       5.0


Condition= 2 subtracted from:

Condition   Lower    Center    Upper    -------+---------+---------+---------
3            2.147    6.083    10.02                     (-------*-------)
                                        -------+---------+---------+---------
                                           -5.0      0.0       5.0
```

Figure 11.6

Although done with greater accuracy than we used, these results agree with ours since the confidence interval estimate for the difference for Condition means 1 and 2 has both confidence limits _____, while the interval estimate for Condition means 2 and 3 have both confidence limits _____, indicating that these two pairs are significantly different from zero.

negative

positive

Self-Correcting Exercises 11B

1. An experiment was conducted to compare four feed additives on the growth of pigs. To eliminate genetic variability, pig litters were used as blocks. Five litters were employed, with four pigs selected randomly from each litter. Each group of four pigs, tending to be more homogeneous than those between litters, was considered a block. The data (growth in pounds) are shown below.

Litter	*Additive*			
	1	*2*	*3*	*4*
1	78	69	78	85
2	66	64	70	70
3	81	78	72	83
4	76	66	77	74
5	61	66	69	70

a. Do the data present sufficient evidence to indicate a difference in mean growth for the four additives?

b. Do the data present sufficient evidence to indicate a difference in mean growth between litters? Was blocking desirable?

c. Find a 95% confidence interval for the difference in mean growth for additives 1 and 2.

2. Example 10.9 was analyzed in Chapter 10 as a paired-difference experiment. The data have been reproduced below.

Pair	Conventional	New
1	78	83
2	65	69
3	88	87
4	91	93
5	72	78
6	59	59

a. Analyze the data to detect a difference in means by using the analysis of variance for a randomized block design. (Use $\alpha = .05$.)
b. What is the relationship between the calculated value of t (Example 10.9) and the calculated value of F?

Solutions to SCE 11B

1. $CM = (1453)^2/20 = 105{,}560.45$; Total SS $= 106{,}399 - CM$
   $= 838.55$;

   $$SST = \frac{362^2 + 343^2 + 366^2 + 382^2}{5} - CM = 105{,}714.6 - CM$$

   $= 154.15$;

   $$SSB = \frac{310^2 + 270^2 + \ldots + 266^2}{4} - CM = 106{,}050.25 - CM$$

   $= 489.80$;
   $SSE = 838.55 - 154.15 - 489.80 = 194.60$.

		ANOVA		
Source	*df*	*SS*	*MS*	*F*
Litters	4	489.80	122.45	7.55
Additive	3	154.15	51.38	3.17
Error	12	194.60	16.22	
Total	19	838.55		

a. $F = 3.17 < 3.49$. Do not reject H_0. Insufficient evidence to detect a difference due to additives.
b. $F = 7.55 > 3.26$. Reject H_0. Thee is a difference due to litters. Blocking is desirable.
c. $(\overline{T}_1 - \overline{T}_2) \pm t_{.025}\sqrt{MSE(2/b)} = (362 - 343)/5 \pm 2.179\sqrt{6.488}$

   $= 3.8 \pm 5.6$.

2. a. $CM = (922)^2/12 = 70,840.3333$; Total $SS = 72,432 - CM$

$= 1,591.6667$;

$$SST = \frac{453^2 + 469^2}{6} - CM = \frac{425,170}{6} - CM = 21.3334;$$

$$SSB = \frac{161^2 + 134^2 + \ldots + 118^2}{2} - CM = 72,391 - CM$$

$= 1,550.6667$;

$SSE = 19.6666.$

		ANOVA		
Source	*df*	*SS*	*MS*	*F*
Treatments	1	21.3334	21.3334	5.42
Blocks	5	1550.6667	310.1333	78.85
Error	5	19.6666	3.9333	
Total	11	1591.6667		

b. $F = 5.42 = t^2 = (2.328)^2$

11.8 SOME CAUTIONARY COMMENTS ON BLOCKING

There are two points you should be aware of when considering the use of a block design. First, a randomized block design should not be used when the block and treatment factors are experimental factors. In general, experimental factors tend to interact, and this violates one of the assumptions associated with a randomized block design—namely, that the effect of each block is the same for all treatments in the experiment. When interaction occurs, the effect of the interaction of the block and treatment factors is to inflate SSE and thereby MSE and s^2. This can lead to false conclusions when testing for differences among treatments. Interaction (inflates, deflates) MSE and therefore causes the observed F to be smaller than it should be. This in turn can cause us to not reject H_0 when we should and thereby lead to a large Type II error.

    The second point you should be aware of is that *blocking is not always beneficial*. Randomized block designs work well when the blocks are chosen so that the variation within each block is _____ and the variation between the blocks is _____. This will result in large gains by reducing SSE and thereby increasing the precision of the ANOVA F test. When the blocks are not very different, there is no great reduction in SSE, but we have lost $b - 1$ degrees of freedom from error, and therefore we have lost the advantage that comes with having larger degrees of freedom for error. As the degrees of freedom with SSE decrease, the value of the associated right-tailed critical F increases, thereby making it more difficult to _____ H_0 when it is false. We should use a randomized block design when *the*

inflates

small
large

reject

elimination of block variation from error outweighs the loss of $b-1$ degrees of freedom.

Unless the experiment involves only a small number of observations, and therefore a small number of degrees of freedom for error, the loss of $b-1$ degrees of freedom will usually be outweighed by the decrease in SSE. Therefore, if you have reason to suspect that the experimental units are not homogeneous, you may wish to consider

randomized block

using a _____ _____ design.

11.9 TWO-WAY CLASSIFICATIONS: THE $a \times b$ FACTORIAL EXPERIMENT

The randomized block design is an example of an experiment that involves a two-way classification, since an observation is classified as receiving the ith treatment in the jth block. The block classification, however, was not a treatment classification and was introduced into

reduce

the experiment to help _____ the error variation. However, in many situations, an experimenter may wish to investigate two or

factors

more independent treatment variables, called _____, with

levels

each factor held at several settings, called _____. An experiment that utilizes every combination of factor levels is called a

factorial

_____ experiment. When two factors are investigated in a factorial experiment, the experiment produces a two-way classification of the data in which each classification corresponds to a "treatment" factor.

Consider a factorial experiment involving factors A and B. By including all factor combinations in the experiment, the investigator can assess the effect of factor A alone, the effect of factor B alone, or their effect in concert. When factors A and B do not behave inde-

interact

pendently, they are said to _____. The nonindependent

interaction

behavior of factors A and B is called _____.

Example 11.9

A researcher at a pharmaceutical company wishes to investigate the effect of hormone H and vitamin V upon the activity of laboratory animals. Two levels of hormone H and two levels of vitamin V are to be investigated. This experiment will involve $(2)(2) =$

4

_____ factor combinations.

If hormone H tends to increase activity, we might observe the following response.

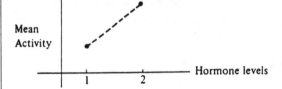

If vitamin V also increases activity, we might observe this response.

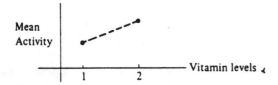

If the factors H and V do not interact, we should observe a similar effect of hormone for each vitamin dose (or vice versa).

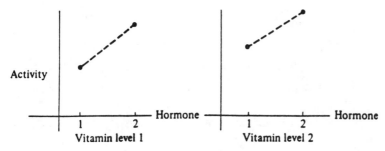

If V and H do interact, we might observe the following situation.

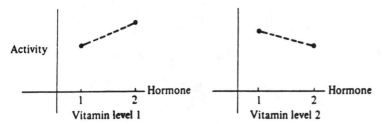

The last two diagrams indicate that at vitamin dosage (level) 1, the effect of increasing the amount of hormone is to _____ activity, while at vitamin dosage (level) 2, the effect of increasing the amount of hormone is to _____ activity. Hence the factors V and H do not behave independently in this situation, and would be said to

_____.

increase

decrease

interact

In the analysis of variance for a factorial experiment involving factor A at a levels and factor B at b levels with r observations of each factor combination,

$$\text{Total SS} = \text{SSA} + \text{SSB} + \text{SS}(AB) + \text{SSE}$$

To compute these sums of squares, define

$A_i =$ the sum of all observations at the ith level of
factor A, $i = 1, \ldots, a$

B_j = the sum of all observations at the jth level of factor B, $j = 1, \ldots, b$

$(AB)_{ij}$ = the sum of all observations at the ith level of factor A and the jth level of factor B, $i = 1, \ldots, a$; $j = 1, \ldots, b$

Each $(AB)_{ij}$ total will contain r observations, each A_i total will contain rb observations and each B_j total will contain ra observations.

The computational formulas for the sums of squares are:

$$\text{Total SS} = \sum x^2 - \text{CM}$$

$$\text{SSA} = \sum \frac{A_i^2}{rb} - \text{CM}$$

$$\text{SSB} = \sum \frac{B_j^2}{ra} - \text{CM}$$

$$\text{SS}(AB) = \sum \ \sum \frac{(AB)_{ij}^2}{r} - \text{CM} - \text{SSA} - \text{SSB}$$

$$\text{SSE} = \text{Total SS} - \text{SSA} - \text{SSB} - \text{SS}(AB)$$

with

$$\text{CM} = \frac{(\text{Grand total})^2}{n} = \frac{G^2}{n}$$

The corresponding mean squares are

$a - 1$

$$\text{MSA} = \text{mean square for factor } A = \frac{\text{SSA}}{(\underline{\hspace{1cm}})}$$

SSB

$$\text{MSB} = \text{mean square for factor } B = \frac{\overline{}}{b - 1}$$

interaction

$$\text{MS}(AB) = \text{mean square for } \underline{\hspace{2cm}} = \frac{\text{SS}(AB)}{(a - 1)(b - 1)}$$

$$\text{MSE} = \text{mean square for error} = \frac{\text{SSE}}{n - ab}$$

Example 11.10

An experiment was conducted to investigate the effect of management training on the decision-making abilities of supervisors in a large corporation. Two factors were considered in the experiment: A, the presence or absence of managerial training, and B, the type of decision-making situation with which the supervisor was confronted. Sixteen supervisors were selected, and eight were randomly chosen to receive managerial training. Four trained and four untrained supervisors were then randomly selected to function in a situation in which a standard problem arose. The other eight supervisors were presented with an emergency situation in which standard procedures could not be used. The response was a management behavior rating for each supervisor as assessed by a rating scheme devised by the experimenter. The basic elements of the experiment are as follows.

Experimental units: _____   supervisors

Factors: The type of training (qualitative) and the type of
    decision-making situation (qualitative,
    quantitative)   qualitative

Levels of factors: Two levels for type of training,
    _____ (give number) levels for type of   two
    decision-making situation.

Treatments: The _____ (give number) training-   4
    situation combinations.

The data for this factorial experiment are shown below.

| Situation | A | | |
(B)	Trained	Not Trained	Totals
	85	53	
	91	49	
Standard	80	38	
	78	45	
	334	185	519
	76	40	
	67	52	
Emergency	82	46	
	71	39	
	296	177	473
Totals	630	362	992

Partition the total sum of squares into SSA, SSB, SS(AB) and SSE.

Solution

Notice that in the data table, the right margin entries are the totals for the two levels of factor B, the bottom margin entries are the totals for factor A, while the entries inside the table are the subtotals for the four factorial combinations.

1. Since there are $n = 16$ observations in the experiment,

$$\text{CM} = \frac{(\underline{})^2}{16} = 61,504$$ 992

and

$$\text{Total SS} = (85^2 + 91^2 + \ldots + 46^2 + 39^2) - \text{CM}$$

$$= 66,640 - \underline{\hspace{1.5in}}$$

61,504

5136

$$= \underline{\hspace{1.5in}}$$

2. Each A total contains eight measurements. Hence,

$$\text{SSA} = \frac{(630)^2 + (362)^2}{(\underline{\hspace{0.4in}})} - \text{CM}$$

8

61,504

$$= 65,993 - \underline{\hspace{1.5in}}$$

4489

$$= \underline{\hspace{1.5in}}$$

3. Each B total also contains eight measurements. Therefore

$$\text{SSB} = \frac{(519)^2 + (473)^2}{(\underline{\hspace{0.4in}})} - \text{CM}$$

8

61,504

$$= 61,636.25 - \underline{\hspace{1.5in}}$$

132.25

$$= \underline{\hspace{1.5in}}$$

4. Each AB total contains four measurements, and SSA and SSB have been computed, so that

$$\text{SS}(AB) = \frac{334^2 + 185^2 + 296^2 + 177^2}{4} - \text{CM} - \text{SSA} - \text{SSB}$$

$$= [66,181.5 - 61,504] - \text{SSA} - \text{SSB}$$

4677.5

$$= \underline{\hspace{1.5in}} - 4489 - 132.25$$

56.25

$$= \underline{\hspace{1.5in}}$$

5. We can find SSE by using the additivity of the sums of squares.

$$\text{SSE} = \text{Total SS} - \text{SSA} - \text{SSB} - \text{SS}(AB)$$

458.5

$$= \underline{\hspace{1.5in}}$$

The analysis-of-variance summary table for a $a \times b$ factorial experiment is given below. Fill in any missing entries.

ANOVA			
Source	*df*	*SS*	*MS*
Factor A	$\underline{\hspace{0.5in}}$	SSA	MSA
Factor B	$\underline{\hspace{0.5in}}$	SSB	MSB
$A \times B$	$(a-1)(b-1)$	SS(AB)	MS(AB)
Error	$n - ab$	$\underline{\hspace{0.5in}}$	MSE
Total	$n - 1$	Total SS	

$a - 1$
$b - 1$

SSE

The entries in the mean-square column are found by dividing each sum of squares by its appropriate degrees of freedom.

There are three tests available for a $a \times b$ factorial experiment.

1. A test of H_0: no interaction among the factors, utilizes

$$F = \frac{MS(AB)}{MSE}$$

with $df_1 = $ _____ and $df_2 = $ _____ degrees of freedom. H_0 is rejected if $F > F_\alpha$ with the appropriate degrees of freedom.

$(a-1)(b-1)$;
$n - ab$

2. To test H_0: no difference in the effect of levels of B, we use

$$F = \frac{MSB}{MSE}$$

with $df_1 = $ _____ and $df_2 = $ _____ degrees of freedom. We reject H_0 if $F > F_\alpha$ with $(b-1)$ and $(n-ab)$ degrees of freedom, respectively.

$b - 1$; $n - ab$

3. In testing H_0: no difference in the effect of levels of A, use

$$F = \frac{MSA}{MSE}$$

with _____ numerator and _____ denominator degrees of freedom, respectively, rejecting H_0 if $F > F_\alpha$ with the specified degrees of freedom.

$(a - 1)$; $(n - ab)$

Example 11.11

a. Present the results of Example 11.10 in an analysis-of-variance table.

b. Is there a significant interaction between A and B at the 5% level of significance?

c. Do the data indicate a significant difference in behavior ratings for the two types of situations at the 5% level of significance?

d. Do behavior ratings differ significantly for the two types of training categories at the 5% level of significance?

Solution

Recall that the experiment involved a total of $n = $ _____ observations with $a = b = $ _____ levels for each factor.

16
2

a. Collecting the results of Example 11.10, we have the following table. Fill in any missing entries.

		ANOVA	
Source	*df*	*SS*	*MS*
Training (A)	1	4489.00	_____
Situation (B)	1	132.25	_____
$A \times B$	1	56.25	_____
Error	12	458.50	_____
Total	15	5136.00	

4489.00
132.25
56.25
38.21

b. Test H_0: no interaction between A and B using

56.25; 1.47

$$F = \frac{MS(AB)}{MSE} = \frac{}{38.21} = \underline{\hspace{1cm}}$$

with $df_1 = 1$ and $df_2 = 12$ degrees of freedom.

4.75; cannot

$F_{.05} = \underline{\hspace{1cm}}$; hence, we $\underline{\hspace{1cm}}$ reject H_0. The data indicate that there is no significant interaction between these two factors.

c. In testing H_0: no difference in behavior ratings for the two situation types, use

132.25; 3.46

$$F = \frac{MSB}{MSE} = \frac{}{38.21} = \underline{\hspace{1cm}}$$

4.75; cannot

Since $3.46 < F_{.05} = \underline{\hspace{1cm}}$, we (can, cannot) conclude that there is a significant difference in behavior ratings for the two types of situations.

d. To test H_0: no difference in behavior ratings for the two training categories, use

4489; 117.48

$$F = \frac{MSA}{MSE} = \frac{}{38.21} = \underline{\hspace{1cm}}$$

4.75

In comparing 117.48 to $F_{.05} = \underline{\hspace{1cm}}$, we find that there is a highly significant difference in behavior ratings for the two training classifications.

e. In summary, there is no significant interaction between the factors and no significant difference in behavior ratings for standard and emergency situations. However, there is a highly significant difference in behavior ratings for the two conditions "Trained" and "Not Trained." Notice that the average behavior rating for

78.75
45.25

trained supervisors was $630/8 = \underline{\hspace{1cm}}$ compared to an average of $362/8 = \underline{\hspace{1cm}}$ for untrained supervisors.

You can use one several Minitab ANOVA options to analyze a two-way classification. In addition to (**Stat** → **ANOVA** → **Two-way...**) and (**Stat** → **ANOVA** → **General linear model**), you can also use (**Stat** → **ANOVA** → **Balanced ANOVA**) for analyzing a factorial experiment. The printout that follows is the analysis of the data from Example 11.11 using the "Balanced ANOVA" option, with which you can also produce the main effect as well as the interaction means.

The numerical results are the same as those you found in Exercise 11.11. However, the *p-values* given in the ANOVA table simplifies testing. You can see immediately that the interaction is not significant (*p-value* = .248) and only the factor training is significant (*p-value* = .000, *p-value* = .087 for Situations).

Figure 11-7

Analysis of Variance (Balanced Designs)

```
Factor      Type Levels Values
Situation fixed      2    1    2
Training  fixed      2    1    2
```

Analysis of Variance for Rating

Source	DF	SS	MS	F	P
Situation	1	132.3	132.3	3.46	0.087
Training	1	4489.0	4489.0	117.49	0.000
Situation*Training	1	56.2	56.2	1.47	0.248
Error	12	458.5	38.2		
Total	15	5136.0			

Means

Situation	N	Rating
1	8	64.875
2	8	59.125

Training	N	Rating
1	8	78.750
2	8	45.250

Situation	Training	N	Rating
1	1	4	83.500
1	2	4	46.250
2	1	4	74.000
2	2	4	44.250

Figure 11.7

Minitab main effect plots and an interaction plot follow.

Main Effect Plots for Example 11.11

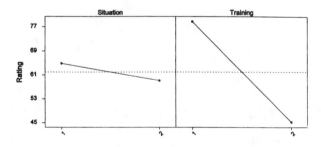

Figure 11.8

The main effect plots show the strong effect of " _____ " and the non-significant difference in the " _____ " means, supporting the results of the analysis of variance F tests.

Training
Situation

Figure 11.9

interaction

Notice that the interaction plots are almost parallel, indicating that there is no significant _____ of the factors "Training" and "Situations." This is exactly what the analysis of variance tests indicated.

Self-Correcting Exercises 11C

1. The effect of three fungicide treatments on the germination rate of three varieties of beans was measured as the percentage of germinating beans out of 100 planted. The data follow.

| | Fungicide Treatment | | |
Variety	1	2	3
A	78	82	92
	62	78	85
	72	70	87
	68	75	90
B	65	72	85
	70	68	79
	75	73	84
	69	76	80
C	81	87	94
	78	83	90
	75	82	89
	85	85	95

a. What type of experimental design has been used?
b. Perform an analysis of variance on the data.
c. Is there a significant interaction between treatments and varieties? Use $\alpha = .05$.
d. Are there significant differences among varieties? Among fungicides? Use $\alpha = .05$.

2. One aspect of a study to compare the quality of medical care delivered by two health maintenance organizations (HMO) was to get a satisfaction rating for patients using the various departments of the HMOs. Five patients were randomly selected from three departments of each HMO. The data which follow consist of a satisfaction rating on a scale from 0 (least satisfaction) to 10 (most satisfied).

	Health Maintenance Organization	
Department	*1*	*2*
Pediatrics	7	8
	6	9
	5	7
	7	9
	4	8
OB/GYN	8	6
	7	5
	5	6
	5	7
	6	5
Family Practice	4	7
	3	6
	5	5
	6	5
	5	7

a. What type of experimental design has been used?
b. Perform an analysis of variance on the data.
c. Is there evidence of a significant interaction between the type of department and HMOs? Use $\alpha = .05$.
d. Is it appropriate to consider main effects in the presence of significant interaction?
e. Estimate the difference in ratings for the two HMOs for each of the three departments using 95% confidence intervals.

Solutions to SCE 11C

1. a. The experiment is a 3×3 factorial, with 4 replicates per treatment combination, in a completely randomized experimental design.

b. $CM = \dfrac{(2859)^2}{36} = 227,052.25$; Total SS $= 229,557 - CM = 2504.75$

$SS(\text{Fungicide}) = \dfrac{2,740,145}{12} - CM = 228,345.4167 - CM$
$= 1293.1667$

$SS(\text{Variety}) = \dfrac{2,733,113}{12} - CM = 277,759.416 - CM = 707.1667$

$SS(V \times F) = \dfrac{916,441}{4} - CM - SSF - SSV = 57.6667$

$SSE = \text{Total SS} - SSF - SSV - SS(VF) = 446.75$

Source	df	SS	MS	F
V	2	707.1667	353.58	21.37
F	2	1293.1667	646.59	39.08
$V \times F$	4	57.6667	14.4175	.8713
Error	27	446.7500	16.5462963	
Total	35	2504.7500		

c. The test for interaction is $F = \dfrac{\text{MS(VF)}}{\text{MSE}} = .8713$, which is not significant ($F_{.05} = 2.73$). Hence, we can investigate the differences between treatments or between varieties separately.

d. The test for differences between the three varieties is

$F = \dfrac{\text{MSV}}{\text{MSE}} = 21.37$, while the test for differences between fungicide treatments is $F = \dfrac{\text{MSF}}{\text{MSE}} = 39.08$. Both are significant at $\alpha = .05$ ($F_{.05} = 3.35$). Hence, the germination rate is affected both by the type of bean and the type of fungicide treatment.

2. a. The experiment is a 3×2 factorial, with 5 replicates per treatment combination, in a completely randomized experimental design.

b. $\text{CM} = \dfrac{(183)^2}{30} = 1116.3$; Total SS $= 1179 - \text{CM} = 62.7$

$\text{SS(HMO)} = \dfrac{16{,}889}{15} - \text{CM} = 1125.9333 - \text{CM} = 9.6333$

$\text{SS(Dept)} = \dfrac{11{,}309}{10} - \text{CM} = 1130.9 - \text{CM} = 14.6000$

$\text{SS}(H \times D) = \dfrac{5753}{5} - \text{CM} - \text{SSH} - \text{SSD} = 10.0667$

$\text{SSE} = \text{Total SS} - \text{SSH} - \text{SSD} - \text{SS(HD)} = 28.4000$

Source	df	SS	MS	F
H	1	9.6333	9.6333	8.138
D	2	14.6000	7.3000	6.17
$H \times D$	2	10.0667	5.0333	4.25
Error	24	28.4000	1.1833	
Total	29	62.7000		

c. The test for interaction is $F = \dfrac{\text{MS}(HD)}{\text{MSE}} = 4.25$, which is significant at $\alpha = .05$ ($F_{.05} = 3.40$).

d. Since the interaction is significant, the ratings for the three departments behave differently in one HMO than the other. Hence, main effects need not be considered. Attention should be focused on the mean ratings for the six treatment combinations.

e. *Pediatrics:* $(\bar{x}_1 - \bar{x}_2) \pm t_{.025}\sqrt{\text{MSE}\left(\frac{1}{5} + \frac{1}{5}\right)}$

$$(5.8 - 8.2) \pm 2.064\sqrt{1.1833\left(\frac{1}{5} + \frac{1}{5}\right)} = -2.4 \pm 1.42$$

OB/GYN: $(6.2 - 5.8) \pm 2.064\sqrt{1.1833\left(\frac{1}{5} + \frac{1}{5}\right)} = 0.4 \pm 1.42$

Family Practice: $(4.6 - 6.0) \pm 2.064\sqrt{1.1833\left(\frac{1}{5} + \frac{1}{5}\right)}$
$$= -1.4 \pm 1.42$$

11.10 ASSUMPTIONS FOR THE ANALYSIS OF VARIANCE

To validly apply the testing and estimation procedures in an analysis of variance, the following assumptions concerning the probability distribution of the response x must be satisfied:

---

1. For any treatment or block combination, the response x is *normally distributed* with a *common variance* σ^2.
2. The observations are selected *randomly* and *independently* so that any two observations are *independent*.

Although it is never known in practice whether these assumptions are satisfied, we should be reasonably sure that violations of these assumptions are moderate. If the random variable under investigation is discrete, it is possible that the distribution for x is mound-shaped and, hence, approximately normal. This would not be the case if the discrete random variable assumed only three or four values.

Although binomial and Poisson random variables have probability distributions that may be approximately normal, these random variables usually violate the assumption of equal variances. Recall that for a binomial random variable, $\mu = \underline{\hspace{1.5cm}}$ and $\sigma^2 = \underline{\hspace{1.5cm}}$ $= \mu(1 - p)$, whereas for a Poisson random variable, $\sigma^2 = \underline{\hspace{1.5cm}}$. In either case, if the treatments are effective in changing the means, they will also cause the variances to change. Hence, the homogeneity assumption will be violated.

Even when the response is normally distributed, the variances for the treatment groups may still be unequal. In agricultural variety trials, it is not unusual for the variability in the height of a plant to increase with the average height. Similarly, in economic studies the variability of income within economic groupings is known to increase as the average income increases. Relationships of this sort can be detected by plotting the treatment or group means against the group variances or standard deviations. The coefficient of variation, defined as the ratio of the standard deviation to the $\underline{\hspace{1.5cm}}$ expressed as a percentage,

$np;\ npq$

μ

mean

can also be used to identify situations in which the standard deviations, and hence the variances, are unequal through their dependence on the mean.

When the data fail to meet the assumptions of normality and equal variances, an appropriate transformation of the data, such as their square roots, logarithms, or some other function of the data values, may be used in order that the transformed values approximately satisfy these assumptions.

When the data consist of only rankings or ordered preferences, appropriate nonparametric testing and estimation procedures can be used. These procedures can also be used when the data fail to satisfy the assumptions of normality and equal variances because the only requirement needed to use nonparametric techniques is that the

independent

observations be _____ within the constraints of the design used.

Residual Plots

It is possible to check some of the assumptions underlying the analysis of variance using residual plots. To examine the equality of variance assumption, you can plot the residuals from the model against the treatment factor, or if the experiment used a randomized block design, you can also plot the residuals against blocks. For factorial experiments, you can plot residuals against each of the main effect factors. In each case, for each level of the factor you should see a relatively constant spread in the residuals if there is no inequality of variance. To ascertain whether the variance is increasing with increasing values of x, a plot of the residuals versus x should not display an increasing pattern, but rather a random scatter of points.

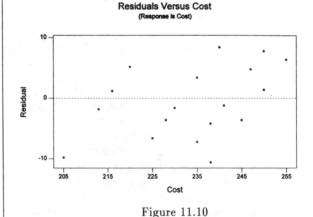

Residuals Versus Cost
(Response is Cost)

Figure 11.10

Figure 11.10 displays the residuals versus the actual value of the response variable, cost. The variance (does, does not) appear to vary

does not

with cost. In Figure 11.11, the residuals are plotted against the variable, time.

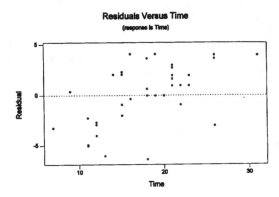

Figure 11.11

There is no apparent pattern and the points would not be judged to be (random, nonrandom). In the same way you could look at the dispersion of the residuals for one or more treatment factors. In the next two plots, you will see that the residuals appear to have equal spreads for each of the factor settings.

nonrandom

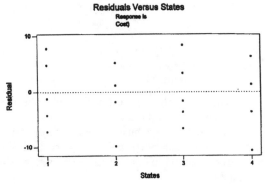

Figure 11.12

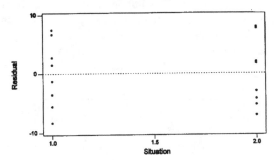

Figure 11.13

constant

In both Figures 11.12 and 11.13, the spread of the residuals appears to be _____, and therefore not violating the equality of variance assumption.

The normality assumptions can also be checked with a normal probability plot that you have used in earlier chapters. We will examine only two plots corresponding to Examples 11.1 and 11.6.

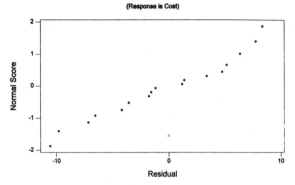

Figure 11.14

This normal probability plot does not show any strong violations of the normality assumption, although there do appear to be one fairly small and one fairly large observation indicating some curvature in the tails of the distribution.

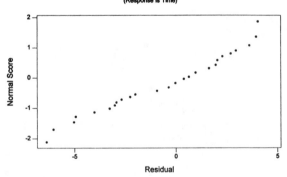

Figure 11.15

Figure 11.15 indicates that the smallest and largest observations may be outliers. The remaining points appear to be linear. Remember that the analysis of variance is robust to moderate departures from normality. Further, you must not expect every residual plot to be "perfect," but rather that the plots show no strong deviations from the ideal straight line.

KEY CONCEPTS AND FORMULAS

I. *Experimental Designs*

1. Experimental units, factors, levels, treatments, response variables
2. Assumptions: Observations within each treatment group must be normally distributed with a common variance σ^2.
3. One-way classification—completely randomized design: Independent random samples are selected from each of k populations.
4. Two-way classification—randomized block design: k treatments are compared within b relatively homogeneous groups of experimental units called blocks.
5. Two-way classification—$a \times b$ factorial experiment: Two factors A and B, are compared at several levels. Each factor-level combination is replicated r times to allow for the investigation of an interaction between the two factors.

II. *Analysis of Variance*

1. The total variation in the experiment is divided into variation (sums of squares) explained by the various experimental factors and variation due to experimental error (unexplained).
2. If there is an effect due to a particular factor, its mean square ($MS = SS/df$) is large and $F = MS(\text{factor})/MSE$ is large.
3. Test statistics for the various experimental factors are based on F statistics, with appropriate degrees of freedom ($df_2 =$ Error degrees of freedom).

III. *Interpreting an Analysis of Variance*

1. For the completely randomized and randomized block design, each factor is tested for significance.
2. For the factorial experiment, first test for a significant interaction. If the interaction is significant, main effects need not be tested. The nature of the differences in the factor-level combinations should be further examined.
3. If a significant difference in the population means is found, Tukey's method of pairwise comparisons or a similar method can be used to further identify the nature of the differences.
4. If you have a special interest in one population mean or the difference between two population means, you can use a confidence interval estimate. (For a randomized block design, confidence intervals do not provide unbiased estimates for single population means.)

IV. *Checking the Analysis of Variance Assumptions*

1. To check for normality, use the normal probability plot for the residuals. The residuals should exhibit a straight-line pattern, increasing at a 45° angle.
2. To check for equality of variance, use the residuals versus fit plot. The plot should exhibit a random scatter, with the same vertical spread around the horizontal "zero error line."

Exercises

1. A large piece of cotton fabric was cut into 12 pieces and randomly partitioned into three groups of four. Three different chemicals designed to produce resistance to stain were applied to the units, one chemical for each group. A stain was applied (as uniformly as possible) over all $n = 12$ units and the intensity of the stain measured in terms of light reflection.

   a. What type of experimental design was employed?

   b. Perform an analysis of variance and construct the ANOVA table for the following data:

Chemical		
1	*2*	*3*
12	14	9
8	9	7
9	11	9
6	10	5

   c. Do the data present sufficient evidence to indicate a difference in mean resistance to stain for the three chemicals?

   d. Give a 95% confidence interval for the difference in means for chemicals 1 and 2.

   e. Give a 90% confidence interval for the mean stain intensity for chemical 2.

   f. Approximately how many observations per treatment would be required to estimate the difference in mean response for two chemicals, correct to within 1.0?

   g. Obtain SSE directly for the data of Exercise 1 by calculating the sums of squares of deviations within each of the three treatments and pooling. Compare with the value found by using SSE = Total SS − SST.

2. A substantial amount of variation was expected in the amount of stain applied to the experimental units of Exercise 1. It was decided that greater uniformity could be obtained by applying the stain three units at a time. A repetition of the experiment produced the following results:

Application	Chemical			Total
	1	*2*	*3*	
1	12	15	9	36
2	9	13	9	31
3	7	12	7	26
4	10	15	9	34
Total	38	55	34	127

   a. Give the type of design.

   b. Conduct an analysis of variance for the data.

   c. Do the data provide sufficient evidence to indicate a difference between chemicals?

d. Give the formula for a $(1 - \alpha)100\%$ confidence interval for the difference in a pair of chemical means. Calculate a 95% confidence interval for $(\mu_2 - \mu_3)$.

e. Approximately how many blocks (applications) would be required to estimate $(\mu_1 - \mu_2)$ correct to within .5?

f. We noted that the chemist suspected an uneven distribution of stain when simultaneously distributed over the 12 pieces of cloth. Do the data support this view? (That is, do the data present sufficient evidence to indicate a difference in mean response for applications?)

3. Twenty maladjusted children were randomly separated into four equal groups and subjected to 3 months of psychological treatment. A slightly different technique was employed for each group. At the end of the 3-month period, progress was measured by a psychologist test. The scores are shown below (one child in group 3 dropped out of the experiment).

	Group				
	1	*2*	*3*	*4*	
	112	111	140	101	
	92	129	121	116	
	124	102	130	105	
	89	136	106	126	
	97	99		119	
Total	514	577	497	567	2155

a. Give the type of design that appears to. be appropriate.

b. Conduct an analysis of variance for the data.

c. Do the data present sufficient evidence to indicate a difference in mean response on the test for the four techniques?

d. Find a 95% confidence interval for the difference in mean response on the test for groups 1 and 2.

e. How could one employ blocking to increase the information in this problem? Under what circumstances might a blocking design applied to this problem fail to achieve the objective of the experimenter?

4. The Graduate Record Examination scores were recorded for students admitted to three different graduate programs in a university.

Graduate Programs		
1	*2*	*3*
532	670	502
601	590	607
548	640	549
619	710	524
509		542
627		
690		

a. Do these data provide sufficient evidence to indicate a difference in mean level of achievement on the GRE for applicants admitted to the three programs?

b. Find a 90% confidence interval for the difference in mean GRE scores for programs 1 and 2.

5. In a study where the objective was to investigate methods of reducing fatigue among employees whose job involved a monotonous assembly procedure, 12 randomly selected employees were asked to perform their usual job under each of three trial conditions. As a measure of fatigue, the experimenter used the total length of time in minutes of assembly line stoppages during a 4-hour period for each trial condition. The data follow.

Employee	Condition		
	1	*2*	*3*
1	31	22	26
2	20	15	23
3	26	21	18
4	21	12	22
5	12	16	18
6	13	19	23
7	18	7	16
8	15	9	12
9	21	11	26
10	15	15	19
11	11	14	21
12	18	11	21

a. Perform an analysis of variance for these data, testing for whether there is a significant difference among the mean stoppage times for the three conditions.

b. Is there a significant difference in mean stoppage times for the 12 employees? Was the blocking effective?

c. Estimate the difference in mean stoppage time for conditions 2 and 3 with 95% confidence.

Chapter 12
Linear Regression and Correlation

12.1 INTRODUCTION

We have investigated the problem of making inferences about population parameters in the cases of large and small sample sizes. We will now consider another aspect of this problem. Suppose that the average value of a random variable y, depends on the values assigned to other variables, $x_1, x_2, \ldots, x_k$. Then we say that a functional relationship exists between the average value of y and $x_1, x_2, \ldots, x_k$. Since the values of y depend on the values assumed by $x_1, x_2, \ldots, x_k$, y is called the **dependent variable** and $x_1, x_2, \ldots, x_k$ are called the **independent variables**. The variables $x_1, x_2, \ldots, x_k$ can also be called **predictor variables** because they are used for predicting the value of y that will be observed for given values of $x_1, x_2, \ldots, x_k$. We restrict our investigation to the case where the average value of y is a **linear** function of one variable x. By linear, we mean that the relationship between y and x can be described by a straight line. This problem was originally addressed in Chapter 3, and we revisit it now.

Review: The Algebraic Representation of a Straight Line

To understand the development of the following linear models, you must be familiar with the algebraic representation of a straight line and its properties.

The mathematical equation for a straight line is

$$y = \alpha + \beta x$$

where x is the independent variable, y is the dependent variable, and α and β are fixed constants. When values of x are substituted into this equation, pairs of numbers, (x_i, y_i), are generated that, when plotted or graphed on a rectangular coordinate system, form a straight line as in Figure 12.1.

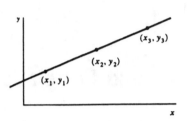

Figure 12.1

Consider the graph of a linear equation $y = \alpha + \beta x$, shown in Figure 12.2.

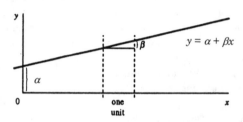

Figure 12.2

1. By setting $x = 0$, we have $y = \alpha + \beta(0) = \alpha$. Because the line intercepts or cuts the y-axis at the value $y = \alpha$, α is called the y _____.

intercept

2. The constant β represents the increase in y for a one-unit increase in x and is called the _____ of the line.

slope

Example 12.1
Plot the equation $y = 1 + .5x$ on a rectangular coordinate system.

Solution
Two points are needed to uniquely determine a straight line, and therefore a minimum of two points must be found. A third point is usually found as a check on calculations.

1. Using 0, 2, and 4 as values of x, find the corresponding values of y.

1

    When $x = 0$, $y = 1 + .5(0) =$ _____.

2

    When $x = 2$, $y = 1 + .5(2) =$ _____.

3

    When $x = 4$, $y = 1 + .5(4) =$ _____.

2. Plot these points on a rectangular coordinate system and join them by using a straightedge.

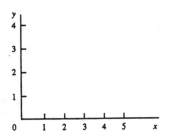

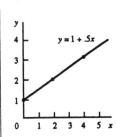

Practice plotting the following linear equations on a rectangular coordinate system:

1. $y = -1 + 3x$

2. $y = 2 - x$

3. $y = -.5 - .5x$

4. $y = x$

5. $y = .5 + 2x$

12.2 A SIMPLE LINEAR PROBABILISTIC MODEL

Suppose we are given a set consisting of n pairs of values for x and y, with each pair representing the value of a response y for a given value of x. Plotting these points might result in the scatter diagram shown in Figure 12.3. Someone might say that these points appear to lie on a

Figure 12.3

straight line. This person would be hypothesizing that a *model* for the relationship between x and y is of the form

$$y = \alpha + \beta x$$

According to this model, for a given value of x, the value of y is *uniquely determined*. Therefore, this is called a _____ model.   deterministic

Another person might say that these points appear to be *deviations* about a straight line, hypothesizing the model

$$y = \alpha + \beta x + \epsilon$$

where ϵ represents the deviation of a particular point (x, y) from the straight line $y = \alpha + \beta x$.

Suppose that we were able to make four observations on y at each of the values x_1, x_2, and x_3. We might observe the 12 pairs of values shown in Figure 12.4. To account for what appear to be random deviations about the deterministic line $y = \alpha + \beta x$, we will

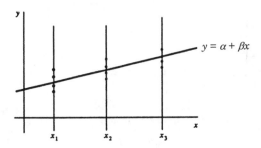

Figure 12.4

consider the deviations to be *random errors* with the following properties:

1. For any fixed value of x, in repeated sampling, the random errors have a mean of 0 and a variance equal to σ^2.
2. Any two random errors are independent in the probabilistic sense.
3. Regardless of the value of x, the random errors have the same *normal* distribution with mean 0 and variance σ^2.

probabilistic

Since this model uses a random error component that has a probability distribution, it is referred to as a _____ model.

The probabilistic model assumes that the average value of y is linearly related to x and the observed values of y will deviate above and below the **line of means**

$$E(y) = \alpha + \beta x$$

by a random amount. The random components all have the same normal distribution and are independent of each other. According to the properties given above, repeated observations on y at the values x_1, x_2, and x_3 would result in the visual representation of the random errors in Figure 12.5. The probabilistic model appears to be the model that best describes the data.

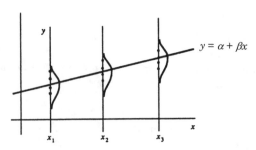

Figure 12.5

Example 12.2

Consider the following set of data for which x is the square footage (in thousands of feet) of heated floor space and y is the sales price (in thousands of dollars) of six houses randomly selected from those sold in a small city during a given week.

x	1.5	2.1	1.7	1.5	1.9	2.4
y	89	109	101	91	102	113

A **scatterplot** of the data is shown in Figure 12.6. The data (satisfy, do not satisfy) the requirements for a simple linear probabilistic model. | satisfy

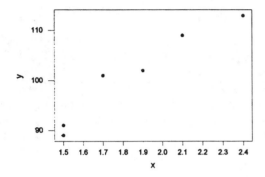

Figure 12.6

In the next section, we will consider the problem of finding the best-fitting straight line for a given set of data. This line is also known as a *regression line*.

12.3 THE METHOD OF LEAST SQUARES

The criterion used for estimating α and β in the model

$$y = \alpha + \beta x + \epsilon$$

is to find an estimated line

$$\hat{y} = a + bx$$

that in some sense minimizes the deviations of the observed values of y from the fitted line. If the deviation of the ith observed value from the fitted value is $(y_i - \hat{y}_i)$, we define the best estimated line as one that minimizes the sum of squares of the deviations of the observed values of y from the fitted values of y. The quantity

$$\text{SSE} = \sum (y_i - \hat{y}_i)^2$$

represents the sum of squares of deviations of the observed values of y from the fitted values and is called the **sum of squares for error** (SSE).

The process of minimization is called the **method of least squares** and produces **least squares estimates** of α and β. The resulting estimates can be found using the formulas in Chapter 3, or using the equivalent formulas below. These formulas are based on the **sum of squares** for the variables x and y, and are similar in form to the sum of squares used in Chapter 11.

Least Squares Estimators of α and β

$$b = \frac{S_{xy}}{S_{xx}} \quad \text{and} \quad a = \bar{y} - b\bar{x}$$

with

$$S_{xy} = \sum(x_i - \bar{x})(y_i - \bar{y}) = \sum x_i y_i - \frac{(\sum x_i)(\sum y_i)}{n}$$

$$S_{xx} = \sum(x_i - \bar{x})^2 = \sum x_i^2 - \frac{(\sum x_i)^2}{n}$$

These estimators can be generated using a calculator with a statistics mode or an appropriate computer software package. Since personal computers and pocket calculators are so readily available, it is likely that you will choose to use one of them to analyze a linear regression problem.

Example 12.3
Obtain the least squares prediction line for the data in Example 12.2. Plot the data and draw the least squares line. Does the line appear to fit the data? What is the predicted selling price of a house with 2000 square feet of heated floor space?

Solution
1. Fill in the blanks in the summary table shown below:

x_i	y_i	x_i^2	y_i^2	$x_i y_i$
1.5	89	____	7921	133.5
2.1	109	4.41	11,881	228.9
1.7	101	2.89	____	171.7
1.5	91	2.25	8281	136.5
1.9	102	3.61	10,404	193.8
2.4	113	5.76	12,769	____
____	____	21.17	61,457	1135.6

2. Calculate

$$S_{xy} = \sum x_i y_i - \frac{(\sum x_i)(\sum y_i)}{n} = 1135.6 - \frac{(11.1)(605)}{6}$$

$$= \underline{\qquad}$$

Margin notes:

2.25

10,201

271.2
11.1; 605

16.35

$$S_{xx} = \sum x_i^2 - \frac{(\sum x_i)^2}{n} = 21.17 - \frac{(\underline{\qquad})^2}{6} = .635.$$

11.1

Then

$$b = \frac{S_{xy}}{S_{xx}} = \frac{16.35}{.635} = 25.748$$

and

$$a = \bar{y} - b\bar{x} = \left(\frac{\overline{\qquad}}{6}\right) - 25.748\left(\frac{\overline{\qquad}}{6}\right) = 53.199.$$

605; 11.1

3. The least squares line is

$$\hat{y} = a + bx = \underline{\qquad\qquad} + 25.748x.$$

53.199

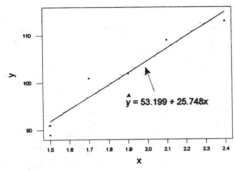

Figure 12.7

The plot of the six data points and the regression line are shown in
Figure 12.7. The line (does, does not) appear to provide a good fit to
the data.

does

4. To predict the value of y for a house that has 2000 square feet of
heated floor space, use $x = 2.0$ in the fitted regression line. Then,

$$\hat{y} = 53.199 + 25.748(2) = 104.695$$

We predict that the house will sell for $104,695.

12.4 AN ANALYSIS OF VARIANCE FOR LINEAR REGRESSION

Just as you did in Chapter 11, you can use an analysis of variance procedure to divide the total variation in the response variable y into portions explained by two different sources. If you define the total
variation as

$$\text{Total SS} = S_{yy} = \sum (y_i - \bar{y})^2 = \sum y_i^2 - \frac{(\sum y_i)^2}{n}$$

you can divide it into

$$\text{SSR} = \frac{(S_{xy})^2}{S_{xx}} \text{ which is the portion of the total variation}$$

explained by the linear regression model, and

$$\text{SSE} = S_{yy} - \frac{(S_{xy})^2}{S_{xx}} \text{ which is the portion of the total variation}$$

representing "residual" or random variation not explained by the model. Remember from Chapter 11 that the sums of squares, when divided by the appropriate degrees of freedom, estimate the various sources of variation in the experiment.

1. Since there are a total of n pairs (x, y) in the experiment, the total number of degrees of freedom are _____. *[answer: $n-1$]*
2. Since the equation for the estimated regression line is
$$\widehat{y} = a + bx = \overline{y} - b\overline{x} - bx$$
involves estimating _____ (give number) additional parameter β, there is _____ (give number) degree of freedom associated with SSR. *[answers: 1; 1]*
3. Since the degrees of freedom are additive, the degrees of freedom for error are

$$(n-1) - \underline{\hspace{1.5cm}} = \underline{\hspace{1.5cm}}$$ *[answers: 1; $n-2$]*

Summarize the partitioning of the total variation using the ANOVA table shown below. Fill in the blanks where necessary.

Source	df	SS	MS
Regression	1	_____	SSR$/1 = $ MSR
Error	_____	$S_{yy} - \dfrac{(\quad)^2}{S_{xx}}$	$\dfrac{\text{SSE}}{} = $ MSE
Total	$n-1$	S_{yy}	

[margin answers: $(S_{xy})^2/S_{xx}$; $n-2$; S_{xy}; $n-2$]

Example 12.4
Construct the ANOVA table for the linear regression of Example 12.2.

Solution
1. In Example 12.3, you found $S_{xy} = 16.35$ and $S_{xx} = .635$.
2. Use the table in Example 12.3 to calculate

$$S_{yy} = \sum y_i^2 - \frac{(\sum y_i)^2}{n} = 61{,}457 - \frac{(\quad)}{6} = 452.8333$$ *[margin answer: 605]*

3. Partition the total sum of squares as

$$\text{SSR} = \frac{(S_{xy})^2}{S_{xx}} = \frac{(\quad)^2}{.635} = 420.9803$$ *[margin answer: 16.35]*

$$\text{SSE} = S_{yy} - \frac{(S_{xy})^2}{S_{xx}} = 452.8333 - 420.9803 = \underline{\hspace{2cm}}$$

31.8530

4. The total degrees of freedom are $n - 1 = 6 - 1 = 5$ and there is
   \underline{\hspace{2cm}} degrees of freedom for regression. Use this information to fill in the blanks in the analysis of variance table shown
   below.

1

Source	df	SS	MS
Regression	___	420.9803	420.9803
Error	4	31.8530	___
Total	___	___	

1
7.9633
5; 452.8333

You can use Minitab (**Stat → Regression → Regression**) to generate the
equation of the least squares line and the analysis of variance table
shown below.

Regression Analysis
```
The regression equation is
y = 53.2 + 25.7 x

Predictor      Coef      StDev        T        P
Constant     53.199     6.652      8.00    0.001
x            25.748     3.541      7.27    0.002

S = 2.822      R-Sq = 93.0%      R-Sq(adj) = 91.2%

Analysis of Variance
Source          DF        SS        MS        F        P
Regression       1     420.98    420.98    52.87    0.002
Residual Error   4      31.85      7.96
Total            5     452.83
```

Minitab reports the equation of the regression line in the second line
of the printout as \underline{\hspace{2cm}}. The analysis of variance table is in
the bottom section of the printout, and matches your hand calculations
(rounded to two decimals).

$y = 53.2 + 25.7x$

   Look for a value in the center portion of the table marked $s =$
\underline{\hspace{2cm}}. This is the best estimate of the population standard
deviation σ and is calculated as

2.822

$$s = \sqrt{\text{MSE}} = \sqrt{7.9633} = 2.822$$

You will probably not be surprised to know that this estimate of σ
will be used to form statistics which can be used for estimating parameters and testing statistical hypotheses in the next section. These tests
result in the t and F statistics and their *p-values* which you see in
the Minitab printout.

Self-Correcting Exercises 12A

1. The registrar at a small university noted that the preenrollment
   figures and the actual enrollment figures for the past 6 years (in
   hundreds of students) were as shown here:

x, preenrollment	30	35	42	48	50	51
y, actual enrollment	33	41	46	52	59	55

a. Plot these data. Does it appear that a linear relationship exists between x and y?

b. Find the least-squares line $\hat{y} = a + bx$.

c. Using the least-squares line, predict the actual number of students enrolled if the preenrollment figure is 5000 students.

d. Construct the ANOVA table for the linear regression.

2. An entomologist, interested in predicting cotton harvest using the number of cotton bolls per quadrate counted during the middle of the growing season, collected the following data, where y is the yield in bales of cotton per field quadrate and x is hundreds of cotton bolls per quadrate counted during midseason.

y	21	17	20	19	15	23	20
x	5.5	2.8	4.7	4.3	3.7	6.1	4.5

a. Fit the least-squares line $\hat{y} = a + bx$ using these data.

b. Plot the least-squares line and the actual data on the same graph. Comment on the adequacy of the least-squares predictor to describe these data.

3. Refer to Exercise 2. The same entomologist also had available a measure of the number of damaging insects present per quadrate during a critical time in the development of the cotton plants. The data follow.

y, yield	21	17	20	19	15	23	20
x, insects	11	20	13	12	18	10	12

a. Fit the least-squares line to these data.

b. Plot the least-squares line and the actual data points on the same graph.

c. Construct the ANOVA table for the linear regression.

Solutions to SCE 12A

1. a.

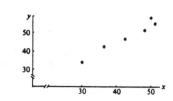

$$\sum x_i = 256 \qquad\qquad \sum x_i y_i = 12{,}608$$

$$\sum y_i = 286 \qquad\qquad n = 6$$

$$\sum x_i^2 = 11{,}294 \qquad\qquad \sum y_i^2 = 14{,}096$$

The trend appears to be linear.

b. $S_{xy} = 12,608 - (256)(286)/6 = 405.3333$

   $S_{xx} = 11,294 - (256)^2/6 = 371.3333$

   $b = S_{xy}/S_{xx} = 405.3333/371.3333 = 1.0916$

   $a = (286/6) - 1.0916(256/6) = 47.6667 - 46.573 = 1.093$

c. $\hat{y} = 1.093 + 1.0916(50) = 55.67$ or 5567 students.

d. Total SS $= S_{yy} = 14,096 - \dfrac{(286)^2}{6} = 463.33$

   SSR $= (S_{xy})^2/S_{xx} = 442.446$

Source	df	SS	MS
Regression	1	442.446	442.446
Error	4	20.887	5.22
Total	5	463.333	

2. a. $\sum x_i = 31.6$ $\sum x_i y_i = 624.6$

   $\sum y_i = 135$ $n = 7$

   $\sum x_i^2 = 149.82$ $\sum y_i^2 = 2645$

   $S_{xy} = 624.6 - (31.6)(135)/7 = 15.1714$

   $S_{xx} = 149.82 - (31)^2/7 = 7.1686$

   $b = S_{xy}/S_{xx} = 15.1714/7.1686 = 2.116$

   $a = 19.286 - (2.116)(4.514) = 9.73$

b.

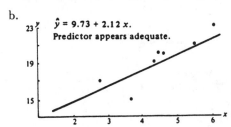

$\hat{y} = 9.73 + 2.12\,x.$
Predictor appears adequate.

3. a. $\sum x_i = 96$ $\sum x_i y_i = 1799$

   $\sum y_i = 135$ $n = 7$

   $\sum x_i^2 = 1402$ $\sum y_i^2 = 2645$

   $S_{xy} = 1799 - (96)(135)/7 = -52.4286$

   $S_{xx} = 1402 - (96)^2/7 = 85.4286$

   $b = S_{xy}/S_{xx} = -52.4286/85.4286 = -0.6137$

$$a = 19.286 - (-.6137)(13.714) = 27.702$$

$$\widehat{y} = 27.702 - .614x$$

b.

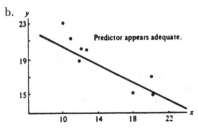

Predictor appears adequate.

c. Total SS $= S_{yy} = 2645 - \dfrac{(135)^2}{7} = 41.4286$

$\text{SSR} = (S_{xy})^2/S_{xx} = 32.176$

Source	df	SS	MS
Regression	1	32.176	32.176
Error	5	9.2526	
Total	6	41.4286	

12.5 TESTING THE USEFULNESS OF THE LINEAR REGRESSION MODEL

There are several statistical tests and measures that you can use to determine whether the linear model you have used for a particular application is actually useful in predicting the response y as a function of x. Once you have determined that the model is *actually useful*, you can use the model for prediction.

Inferences Concerning β, the Slope of the Line of Means

The slope β is the average increase in _____ for a one-unit increase in _____. The question of the existence of a linear relationship between x and y must be phrased in terms of the slope β. If $\beta = 0$, then x contributes nothing to the estimation of y, and the points should appear like a random scatter in a plot of the (x, y) pairs. However, if $\beta \neq 0$, then y should increase (or decrease) linearly as x increases over the region of observation. Therefore, given the data, is it probable that we would observe the given con-figuration of points if x and y were unrelated?

You can answer this question by either testing a hypothesis about or creating a confidence interval for the slope β, using a statistic which is a function of the best estimate of β, given by $b = S_{xy}/S_{xx}$. Remember that the probabilistic model

$$y = \alpha + \beta x + \epsilon$$

y

x

requires that the random error component ϵ have a normal distribution with variance σ^2. When this is true, then estimated slope b is also normally distributed with mean β and standard error given as

$$\text{SE} = \sqrt{\frac{\sigma^2}{S_{xx}}}$$

The following test statistics can be constructed using the fact that b is a *normally* distributed, *unbiased* estimator of β:

1. $z = \dfrac{b - \beta}{\sigma/\sqrt{S_{xx}}}$ if σ^2 is known.

2. $t = \dfrac{b - \beta}{s/\sqrt{S_{xx}}}$ if s^2 is used to estimate σ^2 and hence to estimate the

standard error of b.

Since σ^2 is rarely known, you will have to estimate its value using

$$s^2 = \text{MSE} = \frac{\text{SSE}}{n-2}$$

which can be found in the analysis of variance portion of the Minitab printout or can be calculated using

$$\text{SSE} = S_{yy} - \frac{(S_{xy})^2}{S_{xx}}$$

You can then test for a significant linear relationship between x and y using the statistic given in 2, which has a Student's t distribution with _____ degrees of freedom. A test of the hypothesis H_0: $\beta = 0$ versus H_a: $\beta \neq 0$ is given as follows:

$n - 2$

1. H_0: $\beta = 0$

2. H_a: $\beta \neq 0$

3. Test statistic:

$$t = \frac{b - (0)}{\sqrt{\text{MSE}/S_{xx}}}$$

4. Rejection region: Reject H_0 if $|t| > t_{\alpha/2}$ based on $n-2$ degrees of freedom.

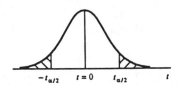

or

5. Calculate the approximate *p-value* for the test.

Example 12.5
Use the data in Example 12.2 to determine whether the value of b provides sufficient evidence to indicate that β differs from 0; that is, does a linear relationship exist between the selling price of a house and the square feet of heated floor space?

Solution
1. We wish to test the hypothesis

$$H_0: \ \beta = 0$$

versus

$$H_a: \ \beta \neq 0$$

for the data in Example 12.2.
2. A portion of the Minitab printout for the data in Example 12.2 is shown below. You should be able to find the equation of the fitted line, the estimated slope and y-intercept, and their standard errors, together with the value of the t-statistic for the test of β.

Regression Analysis
The regression equation is
y = 53.2 + 25.7 x

Predictor	Coef	StDev	T	P
Constant	53.199	6.652	8.00	0.001
x	25.748	3.541	7.27	0.002

S = 2.822   R-Sq = 93.0%   R-Sq(adj) = 91.2%

Since MSE $= 7.9633$ and $S_{xx} = .635$, you can verify that the standard error of b is

3.541

$$\text{SE} = \sqrt{\frac{\text{MSE}}{S_{xx}}} = \sqrt{\frac{7.9633}{.635}} = \underline{\qquad}$$

and the observed value of the test statistic is

25.748

$$t = \frac{b - \beta}{\sqrt{\text{MSE}/S_{xx}}} = \frac{}{3.541} = 7.271$$

3. The Minitab printout gives the *p-value* associated with this value of t with $n - 2 = 6 - 2 = 4$ degrees of freedom in the column labeled P as $P = .002$. Therefore, if you were using a value of α

reject

greater than .002, your decision would be to (reject, not reject) $H_0: \beta = 0$.

Once we have decided that a linear relationship between two variables does exist, our next step is to provide an estimate of the slope in the form of a confidence interval or a point estimate with a given margin of error. Since the slope is the change in y for a one-unit increase in x, it is the linear rate of change in the response variable, and this may be the most important parameter in the investigation. The confidence interval for β is constructed following the procedures of Chapter 10, where we used estimate $\pm t_{\alpha/2}$ (standard error of estimate)

A $(1 - \alpha)100\%$ **confidence interval** for β is

$b \pm t_{\alpha/2}$ (standard error of b)

where t is based on $n - 2$ degrees of freedom.

Example 12.6
Find a 95% confidence interval estimate for β for the data in Example 12.2.

Solution
1. The estimates of the slope (25.748) and its standard error (3.541) were calculated earlier, and can be found in the Minitab printout as well.
2. For 95% confidence, $1 - \alpha = .95$, $\alpha = .05$, and $\alpha/2 = .025$.
3. The degrees of freedom are $n - 2 = 6 - 2 = 4$, and $t_{.025} = 2.776$.
4. Therefore, the 95% confidence interval is found to be

$$25.748 \pm (2.776)(3.541) = 25.748 \pm 9.830$$

or from 15.918 to 35.578.

The Analysis of Variance F Test

There is an equivalent test statistic that can be used to test for significant linear regression. If you write $b = S_{xy}/S_{xx}$, it is not hard to see that

$$t^2 = \frac{b^2}{\text{MSE}/S_{xx}} = \frac{(S_{xy})^2/S_{xx}}{\text{MSE}} = \frac{\text{SSR}/1}{\text{MSE}} = \frac{\text{MSR}}{\text{MSE}} = F$$

This statistic is the F statistic given in the analysis of variance portion of the Minitab printout. It has an F distribution with $df_1 = 1$ and $df_2 = n - 2$ and can be used to test

$$H_0: \beta = 0$$

instead of the t statistic. You will always reach identical conclusions with these two tests. For the data in Example 12.2, you found

$$\text{MSE} = \underline{\hspace{2cm}} \quad \text{and MSR} = \text{SSR} = 420.9803 \qquad 7.9633$$

so that

$$F = \frac{\text{MSR}}{\text{MSE}} = \frac{420.9803}{7.9633} = \underline{\hspace{2cm}} \qquad 52.87$$

Notice that

52.87

$$\sqrt{F} = \sqrt{52.87} = 7.27 = t \text{ and } t^2 = (7.271)^2 = \underline{\hspace{2cm}}$$

You will also notice that the *p-values* for the F and t tests in the printout are identical ($P = .002$), as would be necessary for equivalent test procedures. In Chapter 13, we will use this F test in its more general form to test for the overall usefulness of a more general type of regression model.

Measuring the Strength of the Relationship: The Coefficient of Determination

In Chapter 3, we used a measure called the **correlation coefficient r** to measure the strength of the relationship between two variables x and y. Although the formulas given in that chapter looked somewhat different, you will find that

$$r = \frac{s_{xy}}{s_x s_y} = \frac{S_{xy}}{\sqrt{S_{xx}S_{yy}}}$$

where r can take values between -1 and 1. Another measure of the strength of the relationship between x and y can be found using the square of the correlation coefficient, known as the **coefficient of determination** and calculated as

$$r^2 = \frac{(S_{xy})^2}{S_{xx}S_{yy}} = \frac{\text{SSR}}{\text{Total SS}}$$

If you remember that SSR measures the portion of the total variation (Total SS) that can be explained by the linear regression model, you can see that r^2 measures the *proportion* of the total variation that can be explained by using the regression line $\widehat{y} = a + bx$, rather than ignoring x and using the sample mean $\overline{y}$ to predict the response y. For the data in Example 12.2, the value of r^2 is calculated as

.930

$$r^2 = \frac{\text{SSR}}{\text{Total SS}} = \frac{420.9803}{452.8333} = \underline{\hspace{2cm}}$$

This value is sometimes reported as a percentage. That is, you could conclude that 93.0% of the total variation in the response y can be explained by using the linear regression model. Since this value is quite close to 100%, you would conclude that the linear model (<u>does, does not</u>) fit the data quite well.

does

Points Concerning the Interpretation of Results

If the test H_0: $\beta = 0$ is performed and H_0 is *not rejected*, this (<u>does, does not</u>) mean that x and y are *not related* because:

does not

II
linearly

1. A type _____ error may have been committed, or
2. x and y may be related, but not _____. For example, the true relationship may be of the form $y = \beta_0 + \beta_1 x + \beta_2 x^2$.

If the test H_0: $\beta = 0$ is performed and H_0 *is rejected:*

1. We (can, cannot) say that x and y are solely linearly related because there may be other terms (x^2 or x^3) that have not been included in our model.

cannot

2. We should not conclude that a *causal* relationship exists between x and y because the related changes we observe in x and y may actually be *caused* by an unmeasured third variable—say z.

Consider the problem where the true relationship between x and y is a "curve" rather than a straight line. Suppose we fitted a straight line to the data for values of x between c and d as in Figure 12.8.

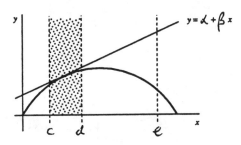

Figure 12.8

Using $\hat{y} = a + bx$ to predict values of y for $c \leq x \leq d$ would result in quite an accurate prediction. However, if the prediction line were used to predict y for the value $x = e$, the prediction would be highly _____. Although the line adequately describes the indicated trend in the region $c \leq x \leq d$, there is no justification for assuming that the line will fit equally well for values of x outside the region $c \leq x \leq d$. The process of predicting outside the region of experimentation is called _____. As our example shows, an experimenter should *not* extrapolate unless he or she is willing to assume the consequences.

inaccurate

extrapolation

Self-Correcting Exercises 12B

1. Refer to Self-Correcting Exercises 12A, Exercise 1. Calculate SSE, s^2 and s for these data.
   a. Test the hypothesis that there is no linear relationship between actual and preenrollment figures at the $\alpha = .05$ level of significance.
   b. Estimate the average increase in actual enrollment for an increase of 100 in preenrolled students with a 95% confidence interval.
   c. Use the analysis of variance F test to test the hypothesis in part a. Confirm that the observed test statistic in part a is related to F by $t^2 = F$.

2. Refer to Self-Correcting Exercises 12A, Exercise 2.
   a. Test for a significant linear relationship between yield and number of bolls.
   b. Calculate the coefficient of determination. What does this measure tell you?

3. Refer to Self-Correcting Exercises 12A, Exercise 3. Test for a significant linear relationship between yield and the number of insects present at the $\alpha = .05$ level of significance.

Solutions to SCE 12B

1. $SSE = S_{yy} - (S_{xy})^2/S_{xx}$

   $= 14{,}096 - [(286)^2/6] - [(405.3333)^2/371.3333]$
   $= 463.333 - 442.446 = 20.887$
   $s^2 = 20.887/4 = 5.22175;\ s = \sqrt{5.22175} = 2.285$
   a. $H_0: \beta = 0;\ H_a: \beta \neq 0$.
   Reject H_0 if $|t| > t_{.025} = 2.776$.

   Test statistic: $t = \dfrac{b - \beta}{s/\sqrt{S_{xx}}} = \dfrac{1.0916}{\sqrt{5.22175/371.333}} = 9.20$

   Reject H_0.
   b. $b \pm t_{.025}\ s/\sqrt{S_{xx}} = 1.09 \pm 2.776(.119) = 1.09 \pm .33$, or $.76 < \beta < 1.42$.
   c. From SCE 12A, MSR $= 442.446$; MSE $= 5.2218$;
   $F = 442.446/5.2218 = 84.73$
   With $df_1 = 1$ and $df_2 = 4$, reject H_0 if $F > 7.71$ and H_0 is rejected. $\sqrt{F} = \sqrt{84.73} = 9.20 = t$.

2. $SSE = 2645 - (1/7)(135)^2 - (15.1714)^2/7.1686$
   $= 41.42857 - 32.10827 = 9.3203$
   $s^2 = 9.3203/5 = 1.86406;\ s = \sqrt{1.86406} = 1.365$
   a. $H_0: \beta = 0;\ H_a: \beta \neq 0$.
   Rejection region: With 5 degrees of freedom and $\alpha = .05$, reject H_0 if $|t| > t_{.025} = 2.571$.

   Test statistic: $t = \dfrac{b - 0}{s/\sqrt{S_{xx}}} = \dfrac{2.116}{\sqrt{1.864/7.1686}} = \dfrac{2.116}{.5099} = 4.15$,

   Reject H_0.
   b. $r^2 = \dfrac{S_{xy}^2}{S_{xx}S_{yy}} = \dfrac{(15.1714)^2}{7.1686(41.42857)} = .775$

   77.5% of the total variation in y is explained by using the linear model $\hat{y} = a + bx$ to predict y.

3. $SSE = 2645 - (135)^2/7 - (-52.4286)^2/85.4286$
   $= 41.42857 - 32.17609 = 9.25248$
   $MSE = s^2 = 9.25248/5 = 1.8505$

$H_0: \beta = 0; \; H_a: \beta \neq 0.$ Reject H_0 if $|t| > t_{.025} = 2.571.$

$$t = \frac{-.6137}{\sqrt{1.8505/85.4286}} = -4.17. \text{ Reject } H_0.$$

12.6 ESTIMATION AND PREDICTION USING THE FITTED LINE

Assume that x and y are related according to the model

$$y = \alpha + \beta x + \epsilon$$

We have found an estimator for this line, which is

$$\hat{y} = \underline{\hspace{3cm}} \qquad\qquad a + bx$$

Suppose we are interested in estimating $E(y \mid x)$ for a given value of x —say x_0. See Figure 12.9. Notice that:

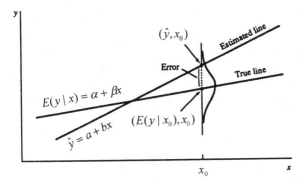

Figure 12.9

1. The error in estimating the expected or average value of y when $x = x_0$ is the difference between the two lines at the point $x = x_0$.

2. The error increases as we move toward the endpoints of the interval over which x has been measured.

   Suppose now that we consider the problem of predicting the actual single value of y when $x = x_0$ rather than $E(y \mid x_0)$, the average value of y. The predictor for this actual value of y is given by

$$\hat{y} = a + bx$$

By examining Figure 12.10, we see that the difference between y and $\hat{y}$ consists of two parts:

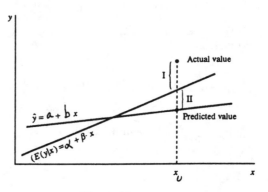

Figure 12.10

1. The difference between the actual point y and $E(y \mid x_0)$, labeled I in the figure.
2. The difference between $E(y \mid x_0)$ and $\hat{y}$, labeled II in the figure.

Therefore, the variability of the error in predicting an individual value of y exceeds the variability for estimating $E(y \mid x_0)$.

A definition of the $100(1 - \alpha)\%$ confidence interval for the expected value of y given $x = x_0$ follows.

A **$100(-\alpha)\%$ confidence interval** for $E(y \mid x)$ when $x = x_0$ is

$$\hat{y} \pm t_{\alpha/2}\text{SE}(\hat{y})$$

where $t_{\alpha/2}$ is based on $n - 2$ degrees of freedom and

$$\text{SE}(\hat{y}) = \sqrt{\text{MSE}\left(\frac{1}{n} + \frac{(x_0 - \bar{x})^2}{S_{xx}}\right)} \quad \text{is the standard error in}$$

estimating $E(y \mid x)$.

Example 12.7
Find a 95% confidence interval for the expected value of y, the selling price of a house, given that there are 2000 square feet of heated floor space.

Solution
To find the point estimate of the average selling price of a house with 2000 square feet of heated floor space, we need to evaluate the fitted regression line for $x = 2$ (since x was measured in 1000 square feet).

1. The point estimate for $E(y \mid x = 2)$ is

$$\hat{y} = 53.1995 + 25.748x$$
$$= 53.1995 + 25.748(2)$$
$$= 53.1995 + 51.496$$
$$= 104.6955$$

2. The value of $t_{.025}$ based on 4 degrees of freedom is 2.776.

3. The standard error of $\hat{y}$ is

$$SE(\hat{y}) = \sqrt{MSE\left(\frac{1}{n} + \frac{(x_0 - \bar{x})^2}{S_{xx}}\right)}$$

$$= \sqrt{7.9633\left(\frac{1}{6} + \frac{(2 - 1.85)^2}{.635}\right)} = \underline{\hspace{2cm}}$$

1.2686

4. The 95% confidence interval estimate of $E(y \mid x = 2)$ is

$$104.6955 \pm 2.776(1.2686) = 104.6955 \pm 3.522$$

or $(101.173, 108.217)$. Since the value of y was given in \$1000 units, the confidence interval estimate is from \$101,173 to \$108,217.

Remember that when you are predicting a particular value of y for a given value of x, the error in estimation consists of two components: (1) the difference between y and $E(y \mid x)$ and (2) the difference between $E(y \mid x)$ and $\hat{y}$. We will denote the error in prediction by $SE(y - \hat{y})$ where

$$SE(y - \hat{y}) = \sqrt{MSE\left(1 + \frac{1}{n} + \frac{(x_0 - \bar{x})^2}{S_{xx}}\right)}$$

A **100$(-\alpha)$% prediction interval** for y when $x = x_0$ is

$$\hat{y} \pm t_{\alpha/2} \, SE(y - \hat{y})$$

where $t_{\alpha/2}$ is based on $n - 2$ degrees of freedom and

$$SE(y - \hat{y}) = \sqrt{MSE\left(1 + \frac{1}{n} + \frac{(x_0 - \bar{x})^2}{S_{xx}}\right)}$$

Example 12.8

Find a 95% prediction interval for the selling price of a house that has 2000 square feet of heated floor space.

Solution

1. Refer to Example 12.7. The predicted value of y when $x = 2$ is also found using the regression line to be

$$\hat{y} = 53.1995 + 25.748(2) = 104.695$$

2.776

2. With 4 degrees of freedom, $t_{.025} = $ _____.

3. The 95% prediction interval would be

2.776; 7.9633

$$104.695 \pm \underline{} \sqrt{(\underline{})\left(1 + \frac{1}{6} + \frac{(2 - 1.85)^2}{.635}\right)}$$

8.589

$$104.645 \pm (\underline{})$$

or from \$96,106 to \$113,284.

Recall that the 95% confidence interval for our estimate of $E(y \mid x = 2)$ was 104.695 ± 3.522. Consequently, the prediction

wider

interval is _____ for the actual value of y at $x = 2$.

Most statistical software packages provide an option within the regression procedure for estimation and prediction. In the Minitab program, clicking on **Options** in the Regression Dialog box will allow you to enter either a single value of $x = x_0$ or a column of values. When you implement the regression command, both the confidence interval and the prediction intervals will appear at the end of the printout, as shown below for the data in Example 12.2, with $x = 2$.

```
Predicted Values

  Fit   StDev Fit        95.0% CI            95.0% PI
104.70     1.27    ( 101.17,  108.22)  (  96.11,  113.29)
```

1. The first column ("Fit") is the value of the fitted line:

104.70

$$\hat{y} = a + bx_0 = 53.119 + 25.748(2) = 104.695 \approx \underline{}$$

2. The second column ("StDev Fit) is the calculated value of $\text{SE}(\hat{y})$ —that is:

7.9633

$$\text{SE}(\hat{y}) = \sqrt{\text{MSE}\left(\frac{1}{n} + \frac{(x_0 - \bar{x})^2}{S_{xx}}\right)} = \sqrt{\underline{}\left(\frac{1}{6} + \frac{(2 - 1.85)^2}{.635}\right)}$$

$$= 1.269 \approx 1.27$$

Notice that the standard error of $(y - \hat{y})$, used in the prediction interval, can be calculated from $\text{SE}(\hat{y})$ using the fact that

$$SE(y - \widehat{y}) = \sqrt{MSE + [SE(\widehat{y})]^2}$$

3. The third and fourth columns provide the confidence interval for $E(y \mid x = 2)$ and the prediction interval for a particular value of y when $x = 2$.

4. The Minitab intervals should match your hand-calculated intervals. Fill in the following table as a check on your calculations.

	95% Confidence Interval	95% Prediction Interval
Minitab	$(101.17, 108.22)$	$(96.11, 113.29)$
Hand Calculation	$(101.173, \underline{\quad})$	$(\underline{\quad}, 113.284)$

108.217; 96.106

The hand calculations (<u>do</u>, do not) match the Minitab printout.

do

It is important to keep the distinction between estimating the average value of y for a given value of x and predicting a new observation. *In the first case we are estimating a parameter, and in the second case we are predicting a new value of y.* This difference shows up in the width of the respective intervals; a prediction interval is always (narrower, <u>wider</u>) than a confidence interval. The graph in Figure 12.11 displays a plot of the regression line together with a plot of the confidence interval bands and the prediction interval bands. Notice that the confidence interval bands are inside the prediction intervals bands and that *both* get (narrower, <u>wider</u>) as we move toward the ends of the interval over which x was measured.

wider

wider

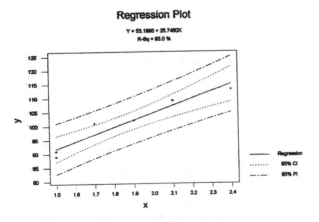

Regression Plot

Y = 53.1995 + 25.7460X
R-Sq = 93.0 %

Figure 12.11

Self-Correcting Exercises 12C

1. For Self-Correcting Exercises 12A, Exercise 2, estimate the expected yield in cotton when the mid-season boll count is 450 with

a 90% confidence interval. Could you use the prediction line to predict the cotton yield if the mid-season boll count was 250?

2. Refer to Self-Correcting Exercises 12A, Exercise 3. Using the least-squares line for these data, estimate the expected cotton yield if the insect count is 12 with a 90% confidence interval. Compare this interval with that found in Exercise 1 and comment on these two predictors of cotton yield.

3. Use the least-squares line from Exercise 1, Self-Correcting Exercises 12A to predict the enrollment with a 95% prediction interval if the preenrollment figure is 4000 students.

Solutions to SCE 12C

1. $(\hat{y} \mid x = 4.5) = 9.73 + 2.116(4.5) = 19.25$

$$[SE(\hat{y})]^2 = 1.86406\left[\frac{1}{7} + \frac{(4.50 - 4.5143)^2}{7.1686}\right] = 1.86406(.14289)$$
$$= .26636$$

The 90% confidence interval is $\hat{y} \pm t_{.05}\, SE(\hat{y})$

$= 19.25 \pm 2.015\sqrt{.26636} = 19.25 \pm 1.04$ or $18.21 < E(y \mid x = 4.5)$
< 20.29. Since $x = 250$ is outside the limit for the observed x, one should not predict for that value.

2. $(\hat{y} \mid x = 12) = 27.702 - .614(12) = 20.33$

$$[SE(\hat{y})]^2 = 1.8505\left[\frac{1}{7} + \frac{(12 - 13.71429)^2}{85.4286}\right] = .328015$$

The 90% confidence interval is $20.33 \pm 2.015\sqrt{.328015}$
$= 20.33 \pm 2.015(.57) = 20.33 \pm 1.15$, or $19.18 < E(y \mid x = 12)$
< 21.48. Note that this interval predicts a slightly higher expected yield.

3. $(\hat{y} \mid x = 40) = 1.093 + 1.0916(40) = 44.76$

$$SE(y - \hat{y}) = 5.22175\left[1 + \frac{1}{6} + \frac{(40 - 42.6667)^2}{371.3333}\right] = 6.19204.$$

The 95% prediction interval is $44.76 \pm 2.776\sqrt{6.19204}$
$= 44.76 \pm 6.91$ or $37.85 < y < 51.67$. Enrollment will be between 3785 and 5167 with 95% confidence.

12.7 REVISITING THE REGRESSION ASSUMPTIONS

For the techniques in this chapter to be valid, the pairs of data points must satisfy the following assumptions:

1. The response y can be modeled as

$$y = \beta_0 + \beta_1 x + \epsilon$$

2. The independent variable x is measured without error.
3. The quantity ϵ is a random variable such that for any fixed value of x,

$$E(\epsilon) = 0 \quad \sigma_\epsilon^2 = \sigma^2$$

   and all pairs ϵ_i, ϵ_j are independent.
4. The random variable ϵ has a _____ distribution.   normal

   The first assumption requires that the response y be
_____ related to the independent variable x. If the response is   linearly
not just a simple linear function of x, but the fit to the data as evi-
denced by a high value of r^2 is good, then the least squares equation
will produce adequate predictions within the range of the experimental
values of x. Extrapolation, however, can cause problems in prediction
because the fitted linear relationship may fail to adequately describe
the data outside the range of values used in the analysis. This is espe-
cially true when the independent variable is _____.   time
   The assumption of a constant variance for the ϵ's may not be valid
in all situations. For example, the variability in the amount of impurities
in a chemical mixture as well as the amount of impurities itself may
increase with increasing temperature of the mixture. However, if repeated values of the independent variable x are included within the experiment, a plot of the data points will usually reveal whether the
variance of the ϵ's depends on x. When this is the case, weights are assigned to each value of x and a regression analysis is performed on y
and the weighed values of x.
   It is not unusual to have the error terms correlated when the data
are collected over time. For example, a high inflation rate during the
first 3 months of the year is likely to be followed by a high inflation
rate during the next quarter as well. Ordinary regression techniques
applied to such data produce underestimates of the true variance and
hence cause distortion in significance levels and confidence coefficients.
Time series analysis should be used in the analysis of such data.
   Departure from the normality assumption will not distort results
too strongly provided the distributions of the ϵ's are not strongly
skewed.

Residual Plots

The assumptions of normality, equal variances, and the adequacy of
the linear model can all be checked using **residual plots** which can be
easily generated using the **Graphs** option in the Minitab regression
procedure.

1. The *plot of residuals versus fit* should be free of any patterns
   and should appear as a random scatter of points about _____.   zero
   The vertical spread of the points should remain _____ for all   constant

of the fitted values. If a pattern appears, this may indicate that a different model should have been fit. If the vertical spread does not remain constant, there may be a problem with the assumption of equal variances. The residuals versus fit plot for the data in Example 12.2 shows (no, some) unusual patterns, indicating that assumptions 1 and 3 (are, are not) reasonable.

no

are

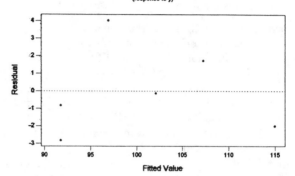

Figure 12.12

2. To check the normality assumption, use the *normal probability plot*. Remember from Chapter 11 that the plot should appear as a straight line, sloping upwards at a _____ angle. The normal probability plot for the data in Example 12.2 (does, does not) appear to satisfy the normality assumption.

45°

does

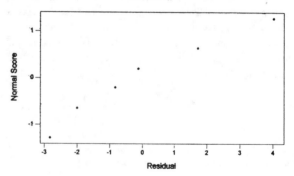

Figure 12.13

12.8 CORRELATION ANALYSIS

A common measure of the strength of the _____ relationship
between two variables is the Pearson product-moment _____
_____ _____, symbolized by _____. This correla-
tion coefficient is (dependent on, independent of) the scales of measurement of the two variables. The Pearson product-moment coefficient
of correlation was presented in Chapter 3 and is calculated as

$r =$ _____

where S_{xx}, S_{yy}, and S_{xy} are as defined earlier in this chapter, and
where

_____ $\leq r \leq$ _____

Most scientific calculators will calculate the correlation coefficient r
when the data pairs (x, y) have been properly entered into the calculator. You can also find either r or r^2 on most regression printouts.
   Examine the formula for r above and notice the following:

1. The denominator of r is the square root of the product of two positive quantities and will always be _____.
2. The numerator of r is identical to the numerator used to calculate
   _____, whose denominator is also always positive.
3. Hence, _____ and r will always have the same algebraic sign.

   a. When $b > 0$, then r _____.

   b. When $b = 0$, then r _____.

   c. When $b < 0$, then r _____.

When $r > 0$, there is a _____ linear correlation; when $r < 0$,
there is a _____ linear correlation; when $r = 0$, there is
_____ linear correlation. See the examples in Figure 12.14.

linear		
coefficient		
of correlation; r		
independent of		

$$\dfrac{S_{xy}}{\sqrt{S_{xx}S_{yy}}}$$

$-1; 1$

positive

b

b

> 0

$= 0$

< 0

positive

negative

no

$>; <; =$

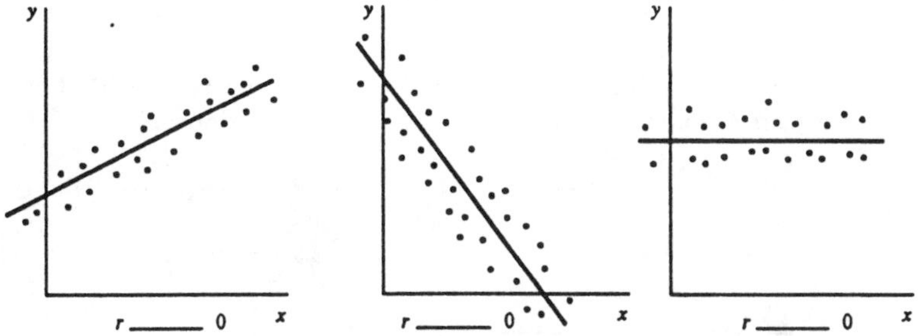

Figure 12.14

   Notice that both r and b measure the linear relationship between
x and y. Whereas r is independent of the scale of measurement, b re-
tains the measurement units for both x and y because b is the number

y

x

of units increase in _____ for a one-unit increase in _____. When should the investigator use r, and when should a least squares estimate of β be used? Although the situation is not always clear-cut, r is used when either x or y can be considered the random variable of interest (i.e., when x and y are both random), whereas regression estimates and confidence intervals are appropriate when one variable, say x, is not random and the other (y) is random.

Example 12.9
Two personnel evaluation techniques are available. The first requires a 2-hour test–interview session; the second can be completed in less than an hour. A high correlation between test scores would indicate that the second test, which is shorter to use, could replace the 2-hour test and hence save time and money. The following data give the scores on test I (x) and test II (y) for $n = 15$ job applicants. Find the coefficient of correlation for the pairs of scores.

Applicant	Test I (x)	Test II (y)
1	75	38
2	89	56
3	60	35
4	71	45
5	92	59
6	105	70
7	55	31
8	87	52
9	73	48
10	77	41
11	84	51
12	91	58
13	75	45
14	82	49
15	76	47

Solution
1. Use a summary table similar to the one used in Example 12.3 or use your calculator to find the following summations:

1192; 725

$$\sum x_i = \underline{\hspace{2cm}} \qquad \sum y_i = \underline{\hspace{2cm}}$$

$$\sum x_i^2 = 96{,}990 \qquad \sum y_i^2 = 36{,}461$$

59,324

$$\sum x_i y_i = \underline{\hspace{2cm}}$$

2. To use the formula for r, we need the following:

$$S_{xy} = \sum x_i y_i - \frac{\left(\sum x_i\right)\left(\sum y_i\right)}{n}$$

$$= 59,324 - \frac{(\underline{\hspace{2cm}})(\underline{\hspace{2cm}})}{15}$$

1192; 725

$$= \underline{\hspace{3cm}}$$

1710.6667

$$S_{xx} = \sum x_i^2 - \frac{\left(\sum x_i\right)^2}{n}$$

$$= 96,990 - \frac{(\underline{\hspace{1.5cm}})^2}{15}$$

1192

$$= 2265.7333$$

$$S_{yy} = \sum y_i^2 - \frac{\left(\sum y_i\right)^2}{n}$$

$$= 36,461 - \frac{(\underline{\hspace{1.5cm}})^2}{15}$$

725

$$= 1419.3333$$

Then

$$r = \frac{S_{xy}}{\sqrt{S_{xx}S_{yy}}}$$

$$= \frac{1710.6667}{\sqrt{(2265.7333)(1419.3333)}}$$

$$= \frac{1710.6667}{1793.1154} = \underline{\hspace{2.5cm}}$$

.9540

Since the correlation coefficient has a maximum value of 1 and a minimum value of -1, it would appear that the correlation between these two test scores is quite strong, with the relationship being a _____ linear one, as indicated by the sign of the correlation coefficient. In other words, a high score on test I would predict a _____ score on test II, or a low score on test I would predict a _____ score on test II. It should be noted that a dependent random variable, y, usually depends on _____ predictor variables, rather than just one. Consequently, the correlation between y and a single predictor variable is of doubtful value. It is even more important to bear in mind that r measures only the _____ relationship between two variables—say x and y. So even when $r = 0$, x and y could be *perfectly* related by a _____ function.

positive

high
low
several

linear

nonlinear

What more can be gleaned from knowing the value of r? How can one assess the strength of the linear relationship between two variables? If $r = 0$, it is fairly obvious that there appears to be no linear relationship between x and y. To evaluate nonzero values of r, let us consider two possible predictors of y.

If the x variate were not measured, we would be forced to use the model

$$y = \alpha + \epsilon$$

with $\alpha = \bar{y}$, so that we would use

$$\hat{y} = \bar{y}$$

as the predictor of the response y. The sum of squares for error for this predictor would be

$$S_{yy} = \sum (y_i - \bar{y})^2$$

Knowing the value of x as well as y, we would use the model

$$y = \alpha + \beta x + \epsilon$$

in which case the resulting sums of squares for error would be

$$SSE = \sum (y_i - \hat{y}_i)^2$$

If a linear relationship between x and y does exist, then S_{yy} would be larger than SSE.

In Section 12.5, we shoed that

SSE

$$r^2 = \frac{SSR}{Total\ SS} = \frac{S_{yy} - SSE}{S_{yy}} = 1 - \frac{}{S_{yy}}$$

You can see that r^2 lies in the interval

$$0 \leq r^2 \leq 1$$

so that r will equal $+1$ or -1 only when all the points lie exactly on the fitted line and SSE equals 0.

Since the difference $S_{yy} - SSE$ represents a reduction in the sum of squares accomplished by using a linear relationship,

r^2 = ratio of the reduction in the sum of squares achieved by using the linear model to the total sum of squares about the sample mean that would be used as a predictor of y if x were ignored

A more understandable way of saying the same thing is to note that r^2 represents the amount of variability in y that is accounted for by knowing x. Thus, to evaluate a correlation coefficient r, you can examine r^2 to interpret the strength of the linear relationship between x and y. The quantity r^2 is called the **coefficient of determination**.

In our example, the value of r was found to be $r = .9540$; there-

.9101
91%

fore, $r^2 = $ _____. Hence, we have reduced the variability of our predictor by _____ by knowing the value of x.

The coefficient r, which is calculated from sample data, is actu-

population
ρ; -1; 1
β

ally an estimator of the _____ coefficient of correlation, symbolized by _____, where _____ $\leq \rho \leq$ _____.
Since ρ and _____ both measure the linear relationship between x and y, the test of H_0: $\beta = 0$ is equivalent to testing H_0:

$\rho = 0$ and is based on a similar set of assumptions. The test is given here:

1. H_0: $\rho = 0$
2. Appropriate one- or two-tailed alternative hypothesis
3. Test statistic:

$$t = \frac{r\sqrt{n-2}}{\sqrt{1-r^2}}$$

4. For a specified value of α and $n-2$ degrees of freedom, the appropriate one- or two-tailed rejection region is found using Table 3 of Appendix I. Or, calculate the approximate *p-value* and compare to α.

Example 12.10
For the data given in Example 12.2, find r and test for significant correlation at the $\alpha = .05$ level of significance.

Solution
In Examples 12.3 and 12.4, you found the necessary sums of squares, so that

$$r = \frac{S_{xy}}{\sqrt{S_{xx}S_{yy}}} = \frac{\rule{2cm}{0.4pt}}{\sqrt{.635(452.8333)}} = \rule{1.5cm}{0.4pt}$$

16.35; .96419

You can use Minitab to find the value of r by using the command **Stat → Basic Statistics → Correlation** which produces the following printout:

Correlations (Pearson)

Correlation of x and $y = 0.964$, *p-value* $= 0.002$

The test of the hypothesis is as follows:

1. H_0: $\rho = 0$
2. H_a: $\rho \neq 0$
3. Test statistic:

$$t = \frac{r\sqrt{n-2}}{\sqrt{1-r^2}} = \frac{(\rule{1.5cm}{0.4pt})\sqrt{4}}{\sqrt{1-(\rule{1.5cm}{0.4pt})}} = \rule{1.5cm}{0.4pt}$$

.96419; 7.271; .96419

Note that this is exactly the same value for t as was given in Example 12.5, in which we tested H_0: $\beta = 0$ against H_a: $\beta \neq 0$. The Minitab printout also gives identical *p-values* $(P = .002)$ for the two equivalent tests.

4. Rejection region: With 4 degrees of freedom, we will reject H_0 if

$$|t| > t_{.025} = \rule{1.5cm}{0.4pt}$$

2.776

falls
reject; is

5. Since $t = 7.25$ (falls, does not fall) in the rejection region, we (reject, do not reject) H_0. There (is, is not) a significant correlation between x and y.

Self-Correcting Exercises 12D

1. Refer to Exercise 2, Self-Correcting Exercises 12A.
   a. Find the correlation between the number of bolls and the yield of cotton.
   b. Find the coefficient of determination r^2, and explain its significance in using the number of cotton bolls to predict the yield of cotton.
   c. Test to see if there is a significant positive correlation between x and y. Use $\alpha = .05$.

2. Refer to Exercise 3, Self-Correcting Exercises 12A.
   a. Find the value of r^2 and r for these data and explain the value of using the number of damaging insects present to predict cotton yield.
   b. Compare the values of r^2 using these two predictors of cotton yield. Which predictor would you prefer?

3. The data in Exercises 2 and 3, Self-Correcting Exercises 12A, are related in that for each field quadrate, the yield, the number of bolls, and the number of damaging insects were simultaneously recorded. Using this fact, calculate the correlation between the number of cotton bolls and the number of insects present for the seven field quadrates. Does this value of r explain in any way the similarity of results when using the predictor in Exercise 2 and that in Exercise 3?

Solutions to SCE 12D

1. a. $r = S_{xy}/\sqrt{S_{xx}S_{yy}} = 15.1714/\sqrt{(7.1686)(41.42857)}$
      $= 15.1714/17.2332 = .88036$

   b. $r^2 = (.88036)^2 = .775$. Total variation is reduced by 77.5% by using number of cotton bolls to aid in prediction.
   c. $H_0: \rho = 0$; $H_a: \rho > 0$
      Test statistic:
      $$t = \frac{r\sqrt{n-2}}{\sqrt{1-r^2}} = \frac{.88036\sqrt{5}}{\sqrt{1-.77503}} = 4.15$$
      Rejection region: With 5 degrees of freedom and $\alpha = .05$, reject H_0 if $t > t_{.05} = 2.015$. Reject H_0.

2. a. $r = -52.4286/\sqrt{85.4286(41.42857)} = -52.4286/59.49105$
      $= -.88129$

Total variation is reduced by 77.7% by using number of damaging insects to aid in prediction.

b. The predictors are equally effective.

3.

x_1 (Bolls)	x_2 (Insects)
5.5	11
2.8	20
4.7	13
4.3	12
3.2	18
6.1	10
4.5	12

$\sum x_1 = 31.6$

$\sum x_1^2 = 149.82$

$n = 7$

$\sum x_2 = 96$

$\sum x_2^2 = 1402$

$\sum x_1 x_2 = 410.80$

$$r = \frac{410.8 - (31.6)(96)/7}{\sqrt{(7.1686)(85.4286)}} = \frac{-22.5714}{24.7468} = -.91.$$

High correlation explains the fact that either variable is equally effective in predicting cotton yield.

KEY CONCEPTS AND FORMULAS

I. *A Linear Probabilistic Model*
1. When the data exhibit a linear relationship, the appropriate model is $y = \alpha + \beta x + \epsilon$.
2. The random error ϵ has a normal distribution with mean 0 and variance σ^2.

II. *Method of Least Squares*
1. Estimates a and b, for α and β, are chosen to minimize SSE, the sum of squared deviations about the regression line $\hat{y} = a + bx$.
2. The least-squares estimates are $b = S_{xy}/S_{xx}$ and $a = \bar{y} - b\bar{x}$.

III. *Analysis of Variance*
1. Total SS $=$ SSR $+$ SSE, where Total SS $= S_{yy}$ and SSR $= (S_{xy})^2/S_{xx}$.
2. The best estimate of σ^2 is MSE $=$ SSE$/(n-2)$.

IV. *Testing, Estimation, and Prediction*
1. A test for the significance of the linear regression—H_0: $\beta = 0$ —can be implemented using one of two test statistics:

$$t = \frac{b}{\sqrt{\text{MSE}/S_{xx}}} \quad \text{or} \quad F = \frac{\text{MSR}}{\text{MSE}}$$

2. The strength of the relationship between x and y can be measured using

$$R^2 = \frac{\text{MSR}}{\text{Total SS}}$$

which gets closer to 1 as the relationship gets stronger.
3. Use residual plots to check for nonnormality, inequality of variances, and an incorrectly fit model.
4. Confidence intervals can be constructed to estimate the intercept α and slope β of the regression line and to estimate the average value of y, $E(y)$, for a given value of x.
5. Prediction intervals can be constructed to predict a particular observation, y, for a given value of x. For a given x, prediction intervals are always wider than confidence intervals.

V. *Correlation Analysis*
1. Use the correlation coefficient to measure the relationship between x and y when both variables are random:

$$r = \frac{S_{xy}}{\sqrt{S_{xx}S_{yy}}}$$

2. The sign of r indicates the direction of the relationship; r near 0 indicates no linear relationship, and r near 1 or -1 indicates a strong linear relationship.
3. A test of the significance of the correlation coefficient is identical to the test of the slope β.

Exercises

1. For the following equations, (i) give the y intercept, (ii) give the slope, and (iii) graph the line corresponding to the equation.
   a. $y = 3x - 2$
   b. $2y = 4x$
   c. $-y = .5 + x$
   d. $3x + 2y = 5$
   e. $y = 2$

2. a. Find the least squares line for the following data:

x	-3	-2	-1	0	1	2	3
y	-1	-1	0	1	2	2	3

   b. As a check on your calculations, plot the data points and graph the least-squares line.
   c. Construct the ANOVA table. Under what conditions could SSE $= 0$?
   d. Do the data present sufficient evidence to indicate that x and y are linearly related at the $\alpha = .05$ level of significance?
   e. Estimate the average change in y for a one-unit change in x with a 95% confidence interval.
   f. Calculate the coefficient of linear correlation for the data and interpret your results.

g. Calculate r^2 and state in words the significance of its magnitude.

h. Construct a 90% confidence interval estimate for a particular value of y when $x = 1$.

3. For the following data,

x	0	2	4	6	8	10
y	9	7	3	1	−2	−3

a. Fit the least-squares line, $\hat{y} = a + bx$.

b. Plot the points, and graph the line to check your calculations.

c. Construct the ANOVA table.

d. Is there a linear relationship between x and y at the $\alpha = .05$ level of significance?

e. Calculate r^2, and explain its significance in predicting the response, y.

f. Predict the particular value of y when $x = 5$ with 95% confidence.

g. Estimate the expected value of y when $x = 5$ with 95% confidence.

4. What happens if the coefficient of linear correlation, r, assumes the value one? The value −1?

5. The following data were obtained in an experiment relating the dependent variable, y (texture of strawberries), with x (coded storage temperature).

x	−2	−2	0	2	2
y	4.0	3.5	2.0	0.5	0.0

a. Find the least-squares line for the data.

b. Plot the data points and graph the least-squares line as a check on your calculations.

c. Construct the ANOVA table.

d. Do the data indicate that texture and storage temperature are linearly related? ($\alpha = .05$)

e. Estimate the expected strawberry texture for a coded storage temperature of $x = -1$ with a 99% confidence interval.

f. Of what value is the *linear* model in increasing the accuracy of prediction as compared to the predictor, $\bar{y}$?

g. Predict the particular value of y when $x = 1$ with a 99% confidence interval.

h. At what value of x will the width of the confidence interval for a particular value of y be a minimum, assuming n remains fixed?

6. In addition to increasingly large bounds on error, why should an experimenter refrain from predicting y for values of x outside the experimental region?

7. If the experimenter stays within the experimental region, when will the error in predicting a particular value of y be maximum?

8. An agricultural experimenter, investigating the effect of the amount of nitrogen (x) applied in 100 pounds per acre on the yield of oats (y) measured in bushels per acre, collected the following data:

x	1	2	3	4
y	22	38	57	68
	19	41	54	65

a. Fit a least-squares line to the data.
b. Construct the ANOVA table.
c. Is there sufficient evidence to indicate that the yield of oats is linearly related to the amount of nitrogen applied? $(\alpha = .05.)$
d. Predict the expected yield of oats with 95% confidence if 250 pounds of nitrogen per acre are applied.
e. Predict the average increase in yield for an increase of 100 pounds of nitrogen with 90% confidence.
f. Calculate r^2 and explain its significance in terms of predicting y, the yield of oats.

9. In an industrial process, the yield, y, is thought to be linearly related to temperature, x. The following coded data is available:

Temperature	0	0.5	1.5	2.0	2.5
Yield	7.2	8.1	9.8	11.3	12.9
	6.9	8.4	10.1	11.7	13.2

a. Find the least-squares line for this data.
b. Plot the points and graph the line. Is your calculated line reasonable?
c. Construct the ANOVA table.
d. Does the data line indicate a linear relationship between yield and temperature at the $\alpha = .01$ level of significance?
e. Calculate r, the coefficient of linear correlation and interpret your results.
f. Calculate r^2, and interpret its significance in predicting the yield, y.
g. Predict the particular value of y for a coded temperature $x = 1$ with 90% confidence.

10. A horticulturist devised a scale to measure the viability of roses that were packaged and stored for varying periods of time before transplanting. y represents the viability measurement and x represents the length of time in days that the plant is packaged and stored before transplanting.

x	5	10	15	20	25
y	15.3	13.6	9.8	5.5	1.8
	16.8	13.8	8.7	4.7	1.0

a. Fit a least-squares line to the data.
b. Construct the ANOVA table.
c. Is there sufficient evidence to indicate that a linear relationship exists between freshness and storage time? (Use $\alpha = .05$.)
d. Estimate the mean rate of change in freshness for a 1-day increase in storage time by using a 98% confidence interval.
e. Estimate the expected freshness measurement for a storage time of 14 days with 95% confidence.
f. Of what value is the linear model in preference to $\bar{y}$ in predicting freshness?

Chapter 13
Multiple Regression Analysis

13.1 INTRODUCTION

We have examined estimation, testing, and prediction techniques for the situation in which y, the response of interest, is linearly related to an independent or predictor variable x in the following way:

$$y = \alpha + \beta x + \epsilon$$

In this chapter, we will extend these techniques to the more general situation in which the response y is linearly related to one or more independent or predictor variables. These extended techniques can be used when the response is linearly related to several different independent variables, or when the response is a polynomial function of just one variable. Modeling, testing, and prediction in these cases belong to an area of statistics called **multiple regression analysis.**

13.2 THE MULTIPLE REGRESSION MODEL AND ASSOCIATED ASSUMPTIONS

The general results given in the remainder of this chapter are applicable and produce standard solutions for a multiple regression problem when the response y is a _____ function of the unknown regression coefficients. We write the extended model as

linear

$$y = \beta_0 + \beta_1 x_1 + \beta_2 x_2 + \cdots + \beta_k x_k + \epsilon$$

where

1. y is the response variable we wish to predict.
2. $\beta_0, \beta_1, \ldots, \beta_k$ are (known, <u>unknown</u>) constants.

unknown
are

3. $x_1, x_2, \ldots, x_k$ are independent variables that (<u>are</u>, are not) measured without error.
4. ϵ is a random error that has a normal distribution with mean 0 and variance σ^2, independent of $x_1, x_2, \ldots, x_k$. Furthermore, the error terms for any two values of y are taken to be _____.

independent

5. Since $E(\epsilon) = 0$, we can write the mean value of y for a given set $x_1, x_2, \ldots, x_k$ as

$$E(y) = \beta_0 + \beta_1 x_1 + \beta_2 x_2 + \cdots + \beta_k x_k$$

Although the actual observed values of y will not exactly equal the values generated by the model, they will deviate from $E(y)$ by a random amount ϵ if the model is in fact true.

The methodology that we use requires only that the intercept β_0 and the β_i's occur in a linear fashion; that is, β_i must be the coefficient of a term that does not involve any *unknown* parameters.

Example 13.1
The following are examples of the general linear model:

1. $y = \beta_0 + \beta_1 t + \beta_2 \sin\left(\frac{2\pi t}{n}\right) + \epsilon$

2. $y = \beta_0 + \beta_1 x + \beta_2 x^2 + \epsilon$

3. $y = \beta_0 + \beta_1 x_1 + \beta_2 x_2 + \beta_3 x_1 x_2 + \epsilon$

Although the model given in part 1 involves the sine function, no un-known parameters occur within the function itself. If we let $x_1 = t$ and $x_2 = \sin(2\pi t/n)$, this model could be written as

$$y = \beta_0 + \beta_1 x_1 + \beta_2 x_2 + \epsilon$$

which is a linear function of the unknown regression parameters β_0, _____, and _____. In a similar fashion, the model given in part 2 can be rewritten by letting $x_1 =$ _____ and $x_2 =$ _____. In the third model, we can achieve the same re-sult by letting $x_1 = x_1$, $x_2 = x_2$, and $x_3 =$ _____.
Terms that involve only $x_1, x_2, \ldots, x_k$ are called *first-order*

$\beta_1; \beta_2$
x
x^2
$x_1 x_2$

terms; terms that involve $x_1^2, x_2^2, \ldots, x_k^2$ or the product $x_i x_j$ are called *second-order terms*.

Example 13.2
The next three models (do, do not) fit the requirements of the general linear model.

do not

1. $y = \beta_0 e^{\beta_1 x} + \epsilon$

2. $y = \beta_0 + \beta_1 \sin\left(\frac{2\pi t}{n} + \beta_2\right) + \epsilon$

3. $y = \beta_0 + \beta_1 x^{\beta_2} + \epsilon$

Notice that models 1 and 3 involve an unknown parameter as a power. Model 2 involves an unknown parameter, β_2, within the sine function itself.

Example 13.3
Determine whether the following models are linear or nonlinear models:

1. $y = \beta_0 + \beta_1 x_1^2 + \beta_2 x_2^2 + \beta_3 x_1 x_2 + \epsilon$

linear

nonlinear

linear

may

2. $y = \beta_0 + \beta_1 \cos\left(\frac{2\pi x}{5}\right) + \beta_2 x^{\beta_3} + \epsilon$

3. $y = \beta_0 e^{-5x} + \epsilon$

A linear statistical model (may, may not) contain nonlinear terms provided all unknown parameters occur in a linear fashion within the model.

Formulation of the linear model to be used in the data analysis is perhaps the most difficult aspect of regression analysis because the results we achieve depend strictly on the model we have chosen to fit. For example, if y is related to x in a quadratic fashion and we include only a linear term in x in our model, our analysis would produce a poor estimator of y in general. In like manner, if we include the independent predictor variables x_1 and x_2 in our model, but fail to include x_3, which has high predictive potential, we may indeed end up with a poor estimator of y. Model formulation will be addressed in Section 13.6.

Self-Correcting Exercises 13A

1. Graph the following equations:
   a. $E(y) = 1 + 2x$;
   b. $E(y) = 1 + .5x$;
   c. $E(y) = 2 - 2x$.

2. Graph the following equations, which graph as parabolas:
   a. $E(y) = x^2$;
   b. $E(y) = 1 + x^2$;
   c. $E(y) = -x^2$.
   d. How does the sign of the coefficient of x^2 affect the graph of the parabola?

3. Graph the following equation:
   a. $E(y) = 1 - 2x + x^2$.
   b. Compare the graph in part a with the graph in Exercise 2, part b. What effect does the term $-2x$ have on the graph?
   c. How would the graph change if $-2x$ were replaced by $+2x$?

4. Suppose $E(y)$ is related to two predictor variables x_1 and x_2 by the equation

   $$E(y) = 2 + 3x_1 - x_2.$$

   a. Graph the relationship between $E(y)$ and x_1 when $x_2 = 0$. Repeat for $x_2 = 1$ and $x_2 = 2$.
   b. How are the graphs of the three lines in part a related?
   c. Graph the relationship between $E(y)$ and x_2 when $x_1 = 0$. Repeat for $x_1 = 1$ and $x_1 = 2$.
   d. How are the graphs of the three lines in part c related?

Solutions to SCE 13A

1.

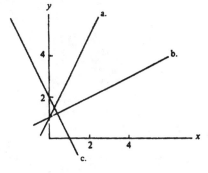

2.

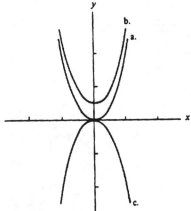

3.

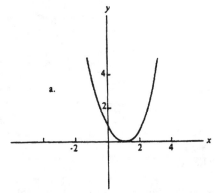

b. The addition of $-2x$ to the equation has the effect of moving the parabola one unit to the right along the x-axis.

c. If the term $2x$ were added to the equation, the parabola would be moved one unit to the left along the x-axis.

4. a. When $x_2 = 0$, $E(y) = 2 + 3x_1$. When $x_2 = 1$, $E(y) = 1 + 3x_1$ and when $x_2 = 2$, $E(y) = 3x_1$.

   b. The three lines are parallel.

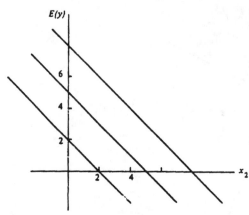

c. When $x_1 = 0$, $E(y) = 2 - x_2$. When $x_1 = 1$, $E(y) = 5 - x_2$ and when $x_1 = 2$, $E(y) = 8 - x_2$.

   d. The lines are again parallel.

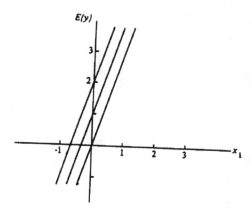

13.3 A MULTIPLE REGRESSION ANALYSIS

In generalizing the results of simple linear regression to multiple linear regression, we have considered the model

$$y = \beta_0 + \beta_1 x_1 + \beta_2 x_2 + \cdots + \beta_k x_k + \epsilon$$

where ϵ is a normally distributed random error component with mean 0 and variance σ^2. In addition, the error terms for any two values of y are taken to be _____. The parameters $\beta_1, \beta_2, \ldots,$ β_k are the **partial slopes** associated with the nonrandom quantities $x_1, x_2, \ldots, x_k$. The slope, β_i, represents the expected increase in the

independent

response y corresponding to a one-unit increase in x_i when the values of all other x's are held constant. Estimates of the unknown parameters in the model are found by using the method of least squares. Using this method, we choose the estimates $\hat{\beta}_0, \hat{\beta}_1, \ldots, \hat{\beta}_k$ so as to _____ the quantity

minimize

$$\text{SSE} = \sum (y_i - \hat{y}_i)^2$$

This minimization technique leads to a set of $(k + 1)$ simultaneous equations in the unknowns $\hat{\beta}_0, \hat{\beta}_1, \ldots, \hat{\beta}_k$, which are easily solved using any multiple regression analysis computer program. Such programs are usually available at any computing facility and provide not only the estimates of the regression parameters, but also additional information required for prediction, estimation, and hypothesis testing. These programs require only that the user provide the proper commands to activate the program and then submit the data in the proper format.

In this section, we will analyze two data sets using the Minitab multiple regression program and interpret the results of the analyses.

Example 13.4
An agricultural economist is interested in predicting the cotton harvest using the number of cotton bolls per quadrat counted during the middle of the growing season and the number of damaging insects per quadrat present during a critical time in the development of the plant. He collected data on the response y, the yield in bales of cotton, x_1, hundreds of cotton bolls per quadrate counted during mid-season, and x_2, the insect count per quadrate. If we can expect a straight line relationship between cotton yield, y, and each of the two predictor variables, x_1 and x_2, write a linear model that relates y and the predictor variables x_1 and x_2.

Solution
We might expect the yield y to (increase, decrease) linearly as x_1, the number of cotton bolls, increases and x_2, the number of damaging insects, decreases. The simplest model that relates y and the predictors x_1 and x_2 is

increase

$$E(y) = \beta_0 + \beta_1 x_1 + \beta_2 x_2$$

When x_2 is held constant, we can write

$$E(y) = (\beta_0 + \beta_2 x_2) + \beta_1 x_1$$

with β_1 the rate of increase in $E(y)$ for a one-unit increase in the boll count, while the intercept, which depends on x_2, is given as $\beta_0(x_2) = \beta_0 + \beta_2 x_2$. For different values of x_2, $E(y)$ would plot as a series of parallel lines, each with slope β_1, as shown in Figure 13.1. Since yield is expected to decrease as the number of damaging insects increases, the lines in the figure are drawn assuming β_2 is negative.

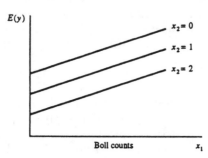

Figure 13.1

When x_1 is held constant, we can write the model as

$$E(y) = (\beta_0 + \beta_1 x_1) + \beta_2 x_2$$

which plots as a series of parallel lines with negative slope

β_2
$\beta_0 + \beta_1 x_1$

_____ and varying intercepts that depend on x_1. The intercept is given by $\beta_0(x_1) = $ _____. With the assumption that β_2
is negative, a plot of $E(y)$ for three values of x_1 is given in Figure
13.2.

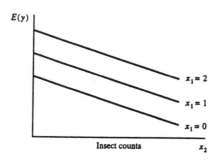

Figure 13.2

An alternative model that is a linear function of x_1 and x_2 but
allows for differing slopes as well as differing intercepts is given by

$$E(y) = \beta_0 + \beta_1 x_1 + \beta_2 x_2 + \beta_3 x_1 x_2$$

In this case, if x_2 is held constant, we can write

$$E(y) = (\beta_0 + \beta_2 x_2) + (\beta_1 + \beta_3 x_2)x_1$$

$\beta_0 + \beta_2 x_2$;
$\beta_1 + \beta_3 x_2$

where (_____) is the intercept and (_____) is the slope.
$E(y)$ would plot as a series of lines with changing slopes and
intercepts. If β_3 is negative, the plots would appear as in Figure 13.3.

If x_1 is held constant, a similar plot would result with varying
negative slopes. Hence, both models allow for a straight line relationship between y and each of the predictor variables x_1 and x_2, but the
model given by

$$E(y) = \beta_0 + \beta_1 x_1 + \beta_2 x_2 + \beta_3 x_1 x_2$$

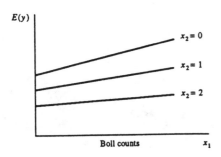

Figure 13.3

is more flexible because it allows for differing slopes and intercepts as either x_1 or x_2 is held constant.

Example 13.5
Data on cotton yield, the number of cotton bolls, and the number of damaging insects as described in Example 13.4 follows:

y	x_1	x_2
21	5.5	11
17	2.8	20
20	4.7	13
19	4.3	12
15	3.7	18
23	6.1	10
20	4.5	12

These data were analyzed using the Minitab multiple regression com mand **Stat → Regression → Regression**, with the values of the independent variable y stored in column 1 (C1) as a function of the three variables x_1, x_2, and x_1x_2 stored in columns 2, 3, and 4 (C2 C3 C4), respectively. The regression analysis is based on the model

$$E(y) = \beta_0 + \beta_1 x_1 + \beta_2 x_2 + \beta_3 x_1 x_2$$

Explain the output of the multiple regression computer printout that follows.

Regression Analysis

```
The regression equation is
y = 11.0 + 4.44 x1 + 0.649 x2 - 0.352 x1x2

Predictor      Coef       StDev         T        P
Constant     10.983       7.784      1.41    0.253
x1            4.437       1.619      2.74    0.071
x2           0.6490      0.4707      1.38    0.262
x1x2        -0.3515      0.1447     -2.43    0.093

S = 0.9364     R-Sq = 93.6%     R-Sq(adj) = 87.3%

Analysis of Variance

Source           DF       SS        MS        F        P
Regression        3   38.798    12.933    14.75    0.027
Residual Error    3    2.631     0.877
Total             6   41.429

Source        DF     Seq SS
x1             1    32.109
x2             1     1.512
x1x2           1     5.178
```

10.983
−0.3515

1.619

Solution
In explaining the output from the Minitab regression analysis, we will discuss the items in the order in which they appear on the printout.

1. *Fitted regression equation.* The regression equation that was fitted to these data is given by

$$y = 11.0 + 4.44x1 + 0.649x2 - 0.352x1x2$$

2. *Individual coefficients.* The estimates of the model parameters are found in the column labeled *Predictor*. The estimate of the intercept or constant is $b_0 = $ _____, while $b_1 = 4.437$, $b_2 = 0.6490$, and $b_3 = $ _____. Correct to *three significant digits*, the fitted regression equation was given above.

   The estimated standard error of the regression coefficients is given in the column labeled *StDev*. For example, the standard error of b_1 is

   $$SE(b_1) = \text{_____}$$

   and the standard error of $\hat{\beta}_3$ is 0.1447. The standard errors of the regression coefficients can be used to test hypotheses concerning individual coefficients and/or to construct confidence interval estimates for the coefficients.

   For example, in testing the hypothesis H_0: $\beta_1 = 0$ against H_a: $\beta_1 \neq 0$, we use the statistic

   $$t = \frac{b_1 - 0}{SE(b_1)}$$

   which has a Student's t distribution with degrees of freedom equal to $n - 2 = 3$ found in the *Error* line and the *df* column in the Analysis of Variance portion of the printout. The calculated values of the t statistics are given in the column labeled *t-ratio*, and their *two-tailed* observed significance levels or *p-values* are

given under the heading P. In the test of H_0: $\beta_1 = 0$ against
H_a: $\beta_1 \neq 0$, the calculated value of t is _____, with a significance level of .071. Hence, we could reject H_0 if we were willing
to use a value of α (greater, less) than or equal to .071.

2.74

greater

A $100(1 - \alpha)\%$ confidence interval estimate for β_1 is given by

$$b_1 \pm t_{\alpha/2}\text{SE}(b_1)$$

with 3 degrees of freedom and $\alpha = .05$, $t_{.025} =$ _____.
Therefore, the 95% confidence interval estimate for β_1 is

3.182

$$4.437 \pm 3.182(\text{_____})$$

1.619

or 4.44 ± 5.15, which includes the value of zero. This supports
our earlier finding that we (could, could not) reject the hypothesis
H_0: $\beta_1 = 0$ at the $\alpha = .05$ level of significance.

could not

It is important that tests and confidence interval estimates
concerning individual regression coefficients be put in the proper
perspective. In a multiple regression analysis, the regression coefficients are properly called *partial regression coefficients* because
they are determined in conjunction with other _____ in the
model and only partially determine the value of the response y. In
fact, the partial regression coefficients in general (would,
would not) be the same as those found using several simple linear
regression models, each with one predictor variable or just one
combination of predictor variables such as $x_1 x_2$. Estimates of the
partial regression coefficients are correlated with each other to
the extent that the underlying predictor variables share the same
predictive information. Therefore, a test of H_0: $\beta_1 = 0$ versus
H_a: $\beta_1 \neq 0$ is actually testing whether the term x_1 contributes
significant information in predicting y if, in fact, the terms x_2 and
$x_1 x_2$ are already in the model. In our example, b_1, b_2, and b_3 are
not significant at the .05 level of significance, but, as we shall see,
the model taken as a whole explains a large portion of the variability in the response y.

variables

would not

3. *The estimate of σ, the standard deviation of the points about the
   line.* The values of SSE and $s^2 = \text{MSE}$ are found in the *Error* line
   of the *Analysis of Variance* table in the columns labeled *SS* and
   MS, respectively. From the printout, we see that SSE = _____
   and $s^2 = \text{MSE} = 0.877$ with _____ degrees of freedom. The
   estimate of σ is $s = \sqrt{0.877} =$ _____, the entry labeled S
   following the column of predictors.

2.631
3
0.9364

4. *R-Sq.* The quantity R^2, called the *coefficient of determination*, measures how well the model fits the data. In a regression analysis, the
   sum of squared deviations, S_{yy}, can be decomposed as

$$\sum (y_i - \bar{y})^2 = \sum (y_i - \hat{y}_i)^2 + \sum (\hat{y}_i - \bar{y})^2$$

where $\sum (y_i - \bar{y})^2 = S_{yy}$, $\sum (y_i - \hat{y}_i)^2 = \text{SSE}$, the sum of squares
for error, and $\sum (\hat{y}_i - \bar{y})^2 = \text{SSR}$, the sum of squares for regression.
SSE measures the discrepancy between observed and predicted values

good

better
large

93.6

smaller

Total

Total

zero

of y, so that small values of SSE are indicative of a (poor, good) model fit. SSR measures the difference between the multiple regression predictor $\hat{y}$ and the simple predictor $\bar{y}$, which ignores the predictor variables entirely. Large values of SSR indicate that the predictor of $\hat{y}$ is a (poorer, better) predictor than $\bar{y}$. Since $S_{yy} = \text{SSE} + \text{SSR}$, if SSE is small, then SSR is (small, large), and vice versa. The value of R^2 is determined as

$$R^2 = \left(\frac{\text{SSR}}{S_{yy}} \right) 100\% = \left(1 - \frac{\text{SSE}}{S_{yy}} \right) 100\%$$

and always lies between 0 and 100%. Hence, R^2 represents the proportion of the variation in y that is explained by regression. From the printout, we see that $R^2 = 93.6\%$, so that

_____ % of the variation in y is explained by the regression model.

As independent variables are added one at a time to the predictor list for a fixed set of y values, the value of R^2 will either stay the same or increase; it will *never decrease* when additional variables are included in the model. Therefore, R^2 can be artificially inflated by including a large number of predictors in the model. For this reason, *R-Sq(adj)* is an alternative formulation that adjusts for the number of parameters that appear in the model by using mean squares rather than sums of squares in its calculation. MSE and $S_{yy}/(n-1)$ are used in place of SSE and S_{yy} in the second formula given above, so that

$$R^2_{\text{adj}} = \left(1 - \frac{\text{MSE}}{S_{yy}/(n-1)} \right) 100\%$$

The calculated value of the adjusted R^2 is 87.3%, R^2-adjusted will always be (smaller, larger) than unadjusted R^2.

5. *The analysis of variance.* The *Analysis of Variance* table shows how the variation in y is decomposed into its component parts. The column labeled *SOURCE* shows that the *Total* variation in y, given by S_{yy}, can be decomposed into unexplained *Error* variation and explained variation due to *Regression*. The *DF* column gives the degrees of freedom with each source of variation, and the sum of squares (*SS*) column gives the calculated sum of squares for each source of variation. In these two columns, the entries for *Regression* and *Error* add to the

_____ line. The entries in the mean square (*MS*) column are found by dividing each sum of squares entry by its degrees of freedom. No mean square is calculated for the _____ line.

If the model contributes information for the prediction of y, at least one of the model parameters β_1, β_2, or β_3 will differ from

_____. In testing the hypothesis H_0: $\beta_1 = \beta_2 = \beta_3 = 0$ against the alternative hypothesis H_a: at least one of β_1, β_2, or β_3 differs from zero, we use the statistic

$$F = \frac{\text{MSR}}{\text{MSE}} = \frac{R^2/k}{(1-R^2)/[n-(k+1)]}$$

where MSR is the mean square due to regression and MSE is the mean square due to error. MSR is found in the *Regression* line, and MSE is found in the *Error* line of the mean square (MS) column. MSR has degrees of freedom equal to the number of parameters in the model, $df_1 = k$, *excluding* the intercept, whereas MSE has degrees of freedom equal to the number of observations minus the number of parameters in the model *including* the intercept, or $df_2 = n - (k+1)$.

    Since SSR is the variation explained by the model and SSE is the variation unexplained by the model, a (small, large) value of MSR compared with MSE is cause to reject H_0. Therefore, we reject H_0 for (small, large) values of F. In this problem $F = 14.75$ with $df_1 = 3$ and $df_2 = 7 - 4 = 3$ degrees of freedom; F is found by dividing 12.933 by .877. Since this value exceeds the tabulated value of F with $df_1 = 3$, $df_2 = 3$, and $\alpha = .05$, given as $F = 9.28$ in Table 6 of your text, we reject H_0 and conclude that the model (does, does not) contribute significant information in predicting y. The observed significance level (p-value) for this test is

_____.

large

large

does

.027

6. *Sequential sum of squares.* Notice that none of the partial regression coefficients is significant at the .05 level of significance, but taken together, they account for _____ of the variation in the values of y. The table below the Analysis of Variance provides the sequential sum of squares (SEQ SS) and lists the additional contribution to the sum of squares for regression of each variable (or, combination of variables) in the order they were selected for the model through the Minitab **Regression** command. In this problem, x_1 was the first variable entered and accounted for $(32.109/41.429) =$ _____, or 77.5% of the variation in y. The variable x_2 was entered next and accounted for an additional $(1.512/41.429) =$ _____, or 3.6% of the variation in y. Entering $x_1 x_2$ accounted for an additional $(5.178/41.429) =$ _____, or 12.5% of the variation in y. Notice that within rounding errors, the sequential sums of squares add up to the sum of squares for _____, and the three percentages add up to _____.

93.6%

.7750

.036

.125

regression
R^2

7. *Checking the regression assumptions.* As you did in Chapters 11 and 12, you can use computer-generated **residual plots** to make sure that the regression assumptions have been satisfied. These plots will also signal a problem if an incorrect model has been fit to the data. The residual plots for this example are shown in Figure 13.4. These plots (do, do not) show any serious violation of assumptions, and you can proceed to use this model for its main purpose—estimation and prediction of y.

do not

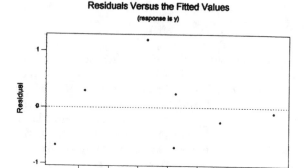

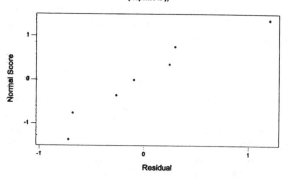

Figure 13.4

8. *Estimation and prediction.* The prediction equation

$$\hat{y} = 11.0 + 4.44x_1 + 0.649x_2 - 0.352x_1x_2$$

can be used as a point estimator for either estimating the average value of $y - E(y)$—for given values of x_1 and x_2 or for predicting a particular value of y for given values of x_1 and x_2. For example, if the agronomist counted 400 cotton bolls per quadrat ($x_1 = 4$) and 12 damaging insects per quadrat ($x_2 = 12$) during midseason, his estimate of the yield y in bales of cotton would be

$$\hat{y} = 11.0 + 4.44(4) + 0.649(12) - 0.352(4)(12)$$

19.65

$$= \underline{\hspace{2cm}}$$

To obtain confidence and/or prediction intervals, the values of x_1 and x_2 are entered using the **Options** Dialog box in Minitab, producing the printout below.

```
Predicted Values

  Fit  StDev Fit        95.0% CI          95.0% FI
19.647      0.927   (  16.697,  22.597)  ( 15.454,  23.841)
```

The first column shows the estimated value of y labeled
Fit = _____, which matches the hand calculation. Notice
that the prediction interval is (wider, narrower) than the confidence interval.

19.647
wider

13.4 A POLYNOMIAL REGRESSION MODEL

Sometimes the response variable y is a function of only one independent variable x, but the relationship between y and x may not be
linear. In this case, it may be possible to fit a **polynomial model** of the
form

$$y = a + bx + cx_2 + dx_3 + \ldots$$

to the data. The multiple regression model allows you to fit this type
of model, once the data has been properly entered into the computer.

Example 13.6
An agricultural economist interested in California cotton production
gathered the following data on the mean number of cotton bolls per
plant during the growing season in the San Joaquin Valley of California.
Here y is the mean number of bolls per plant and x is the time measured in weeks.

y	110	470	1040	1100	1000	820
x	1	4	7	9	12	15

Use a multiple regression program to fit a second-degree polynomial to
these data and discuss the computer output from the program.

Solution
1. A second-degree polynomial model is given by

$$E(y) = \beta_0 + \beta_1 x + \beta_2 x^2$$

In this case $E(y)$ is a general linear model with $x_1 = x$ and $x_2 = x^2$.
If a multiple regression program requires that the data be entered
as (y, x_1, x_2) for each of the n data points, the user would enter the
triples (y, x, x^2) for each of the n observations. For example, the
first data point would be $(110, 1, 1)$ and the last would be $(820, 15,
225)$. The Minitab program allows you to enter the values of y and
x into two columns of the Data window and to use the **Calc** →
Calculator command to generate the third column as C3 = $x*x$.
The three columns of the Minitab Data window are shown below,
along with the regression printout.

Data Display

Row	y	x	x-sq
1	110	1	1
2	470	4	16
3	1040	7	49
4	1100	9	81
5	1000	12	144
6	820	15	225

Regression Analysis

The regression equation is
y = - 176 + 244 x - 11.9 x-sq

Predictor	Coef	StDev	T	P
Constant	-175.5	125.4	-1.40	0.256
x	244.22	35.97	6.79	0.007
x-sq	-11.878	2.172	-5.47	0.012

S = 106.7   R-Sq = 95.5%   R-Sq(adj) = 92.5%

Analysis of Variance

Source	DF	SS	MS	F	P
Regression	2	727600	363800	31.97	0.009
Residual Error	3	34134	11378		
Total	5	761733			

Source	DF	Seq SS
x	1	387292
x-sq	1	340308

95.5%

95.5

2. The value of R^2 on the printout is _____, which means that _____ % of the variation in the mean number of cotton bolls is accounted for by the quadratic model.

3. In the test of the hypothesis H_0: $\beta_1 = \beta_2 = 0$ against the alternative that at least one of β_1 or β_2 differs from zero, the value of $F = MSR/MSE$ found in the Analysis of Variance section under F is _____, which is highly significant with a *p-value* of .009. Therefore, we (can, cannot) conclude that x and x^2 contain significant information for predicting y.

31.97

can

4. In the individual analysis of variables section we see that the partial regression coefficients corresponding to x and x^2 (x-sq) are both significant at the .05 level of significance. For example, in the test of whether the quadratic term x^2 contributes significant information for predicting y, the test of H_0: $\beta_2 = 0$ produced a value of $t = $ _____, which is significant at the .012 level of significance. The value of t can be verified by calculating

−5.47

−11.878

$$t = \frac{b_2}{SE(b_2)} \qquad \text{where} \qquad b_2 = \underline{\hspace{2cm}} \quad \text{and}$$

2.172

$$SE(b_2) = \underline{\hspace{2cm}}.$$

5. With the estimated coefficients b_0, b_1, and b_2 found in the column labeled *Predictor*, the prediction equation is

$$\hat{y} = -175.5 + 244.22x - 11.878x^2$$

6. Some computer facilities may have a program written specifically to perform polynomial regression analysis. For these it is sufficient to enter the pairs (y, x) because the values of x^2, x^3, and other power terms are generated within the program itself. However, any multiple regression program can be used to fit a polynomial model. The Minitab command **Stat** $\rightarrow$ **Regression** $\rightarrow$ **Fitted Line Plot** will fit either a linear, quadratic or cubic model along with a plot of the prediction curve and the observed data values (Figure 13.5). Notice that the curve appears to fit the observed data points very well.

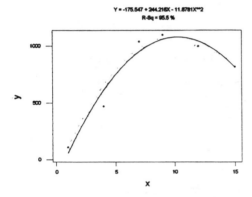

Figure 13.5

7. When the quadratic model was used to fit the response curve, the normal probability plot and the plot of the residuals versus the response y in Figure 13.6 does not reveal any violations of assumptions based on the $n = 6$ observations. It appears that four deviations are positive and two are negative with no apparent pattern, and that the absolute values of the deviations for small values of y are not systematically different from those for _____ values of y.

large

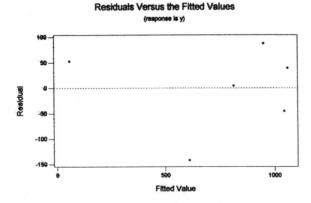

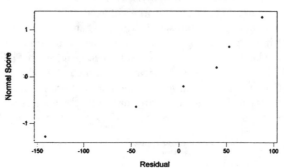

Figure 13.6

However, when a simple linear model given by

$$E(y) = \alpha + \beta x$$

was fitted to the data in this example, a plot of the residuals against $\hat{y}$ displays a nonrandom pattern that typically results when a linear model is fitted and the response follows a quadratic model. See Figure 13.7.

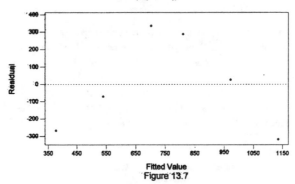

Figure 13.7

quadratic

The residuals from this model plotted against the predictor variable x = time reveal a _____ trend, reflecting the contribution to the quadratic term in the initial model. See Figure 13.8.

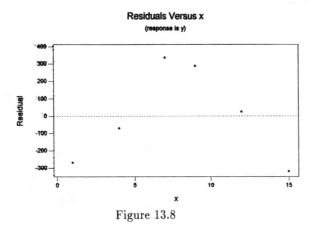

Residuals Versus x
(response is y)

Figure 13.8

Self-Correcting Exercises 13B

1. In order to study the relationship of advertising and capital investment on corporate profits, the following data, recorded in units of $100,000, was collected for ten medium-sized firms within the same year. The variable y represents profit for the year, x_1 represents capital investment, and x_2 represents advertising expenditure.

y	x_1	x_2
15	25	4
16	1	5
2	6	3
3	30	1
12	29	2
1	20	0
16	12	4
18	15	5
13	6	4
2	16	2

These data were analyzed using the Minitab **Regression** command based on the model

$$E(y) = \beta_0 + \beta_1 x_1 + \beta_2 x_2.$$

The Minitab printout and residual plots follow.

Regression Analysis

The regression equation is
y = - 8.18 + 0.292 x1 + 4.43 x2

Predictor	Coef	StDev	T	P
Constant	-8.177	4.206	-1.94	0.093
x1	0.2921	0.1357	2.15	0.068
x2	4.4343	0.8002	5.54	0.001

S = 3.303   R-Sq = 82.3%   R-Sq(adj) = 77.2%

Analysis of Variance

Source	DF	SS	MS	F	P
Regression	2	355.22	177.61	16.28	0.002
Residual Error	7	76.38	10.91		
Total	9	431.60			

Source	DF	Seq SS
x1	1	20.16
x2	1	335.05

Normal Probability Plot of the Residuals
(response is y)

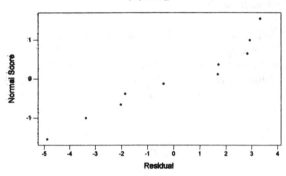

Residuals Versus the Fitted Values
(response is y)

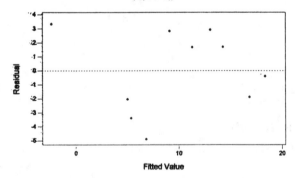

a. Find the values of S_{yy}, SSR, SSE, s^2 and s on the printout.

b. Find R^2 and interpret its value.

c. Verify the value of R^2 using the entries S_{yy}, SSR, and SSE.

d. Do the data provide sufficient evidence to indicate that the model contributes significant information in predicting y?

Use $\alpha = .05$.

e. Test the hypothesis $H_0: \beta_2 = 0$ against $H_a: \beta_2 \neq 0$ at the .05 level of significance. Verify the value of t on the printout by using the appropriate entries in the Coef and StDev columns.

f. Write the prediction equation relating $\hat{y}$ and the predictor variables x_1 and x_2.

g. Estimate the yearly corporate profits for a medium-sized firm whose capital investment was $2,200,000 and whose advertising expenditure was $400,000.

h. What do the residual plots tell you about the validity of the regression assumptions?

2. A chemical company interested in maximizing the output of a chemical process by selection of the reaction temperature recorded the following data where y is the yield in kilograms and x is the coded temperature:

y	7.5	8.1	8.8	10.9	12.5	11.8	11.1	10.4	9.5
x	-4	-3	-2	-1	0	1	2	3	4

In chemical reactions, the amount of the substance produced may increase until a critical temperature is reached, at which point the amount of substance produced begins to decrease due to its decomposition by the increasing temperature. Anticipating that this would be the case, the model

$$E(y) = \beta_0 + \beta_1 x + \beta_2 x^2$$

was fitted using the Minitab **Regression** command. The computer printout follows.

Regression Analysis

```
The regression equation is
y = 11.4 + 0.340 x - 0.205 x-sq

Predictor        Coef       StDev          T         P
Constant       11.4346     0.3734      30.62     0.000
x               0.34000    0.09539      3.56     0.012
x-sq           -0.20519    0.04210     -4.87     0.003

S = 0.7389     R-Sq = 85.9%     R-Sq(adj) = 81.2%

Analysis of Variance

Source          DF          SS         MS         F        P
Regression       2      19.9043     9.9522     18.23    0.003
Residual Error   6       3.2757     0.5459
Total            8      23.1800

Source          DF      Seq SS
x                1       6.9360
x-sq             1      12.9683
```

a. Find R^2 on the printout and interpret its value.

b. Calculate R^2 directly using the entries S_{yy}, SSR, and SSE.

c. Is there sufficient evidence to indicate that the model contributes significant information in predicting y at the .05 level of significance?

d. Test for significant curvature in the fitted response by testing H_0: $\beta_2 = 0$ against H_1: $\beta_2 \neq 0$ with $\alpha = .05$.

e. Use the fitted prediction equation to predict the yield when the coded temperature is $x = 1$.

Solutions to SCE 13B

1. a. Refer to the *Analysis of Variance* section of the printout. In the column labeled SS, find $S_{yy} = 431.60$, SSE $= 76.38$ and SSR $= 355.22$. Then $s^2 = $ MSE $= 10.91$ in the MS column, $s = \sqrt{s^2}$ is found on the printout as 3.303.

b. The entry R-Sq appears under the individual variables section and is given as 82.3%, meaning that 82.3% of the variation in the response y can be explained by regression.

c. $R^2 = \left(\dfrac{\text{SSR}}{S_{yy}}\right) 100\% = \left(\dfrac{355.22}{431.60}\right) 100\% = 82.3\%$,

which agrees with the value given on the printout.

d. H_0: $\beta_1 = \beta_2 = 0$; H_a: at least one nonzero β_i; test statistic:

$F = \dfrac{\text{MSR}}{\text{MSE}} = 16.28$ with *p-value* equal to 0.002. H_0 is rejected.

The model contributes significant information for predicting y.

e. H_0: $\beta_2 = 0$; H_a: $\beta_2 \neq 0$; test statistic: $t = \dfrac{b_2}{\text{SE}(b_2)} = \dfrac{4.4343}{.8002}$

$= 5.54$ with *p-value* $= 0.001$. Reject H_0 and conclude that $\beta_2 \neq 0$.

f. Correct to three digit accuracy, the *regression equation* is

$$\hat{y} = -8.18 + 0.292x_1 + 4.43x_2.$$

g. When $x_1 = 22$ and $x_2 = 4$, $\hat{y} = -8.177 + 6.427 + 17.737$
$= 15.987$.

h. No obvious violation of regression assumptions.

2. a. The entry R-Sq $= 85.9\%$ implies that 85.9% of the total variation can be explained by regression.

b. $R^2 = \dfrac{\text{SSR}}{S_{yy}} = \dfrac{19.90431}{23.1800} = .8587$ which agrees with the printout.

c. H_0: $\beta_1 = \beta_2 = 0$; test statistic: $F = \dfrac{\text{MSR}}{\text{MSE}} = 18.23$ with

p-value $= .003$. Reject H_0. The model contributes significant information.

d. H_0: $\beta_2 = 0$; H_a: $\beta_2 \neq 0$; test statistic: $t = \dfrac{b_2}{\text{SE}(b_2)} = -4.87$

with *p-value* $= .003$. Reject H_0. There is significant curvature.

e. When $x = 1$, $\hat{y} = 11.43463 + .34 - .20519 = 11.56944$.

13.5 USING QUANTITATIVE AND QUALITATIVE VARIABLES IN A REGRESSION MODEL

Variables are classified as being either quantitative or qualitative. A quantitative variable takes values that correspond to the points on the real line. If a variable is not quantitative, then it is said to be qualitative. Variables such as advertising expenditure, number or age of employees, per unit production cost, and number of delivery trucks are quantitative variables, whereas geographic region, plant site, and kind of stock are examples of _____ variables. Although predictor variables can be quantitative or qualitative, a dependent variable must be _____ in order to satisfy the assumptions given in Section 13.2.

    The intensity setting of an independent variable is called a _____. The levels of a quantitative independent variable correspond to the number of distinct values that the variable assumes in an investigation. For example, if an experimenter interested in maximizing the output of a chemical process observed the process when the temperature was set at 100°F, 200°F, and 300°F, the independent variable "temperature" was observed at _____ levels. The levels of a qualitative independent variable are defined by describing them. For example, the independent variable "occupational groups" might be described as white-collar workers, blue-collar workers, service workers, and farm workers. If all four groups were included in an investigation, the qualitative variable "occupational groups" would be taken to have _____ levels. Similarly, if an investigation were conducted in three regions, the qualitative variable "regions" would have _____ levels.

    It is necessary to differentiate between quantitative and qualitative variables to be included in a regression analysis because these variables are entered into a regression model in different ways. Quantitative variables, in general, are entered directly into a regression equation, whereas qualitative variables are entered through the use of dummy variables, which in effect produce different response curves at each setting of the qualitative independent variable.

    When two or more quantitative independent variables appear in a regression model, the resulting response function produces a graph called a response surface in three or more dimensions. These graphs become difficult to produce when three or more independent variables are included in the model. A model involving quantitative variables is said to be a first-order model if each independent variable appears in the model with power _____. The model

$$E(y) = \beta_0 + \beta_1 x_1 + \beta_2 x_2 + \ldots + \beta_k x_k$$

is a _first-order model_ involving _____ independent variables because the model is linear in each x. The graph of a first-order model is a response plane, which means that the surface is "flat" but has some directional tilt with respect to its axes. Second-order linear models in k quantitative predictor variables include all the terms in a first-order

qualitative

quantitative

level

three

four
three

one

_k_

model, all cross-product terms such as $x_1x_2, x_1x_3, x_2x_3, \ldots x_{k-1}x_k$, and all pure quadratic terms $x_1^2, x_2^2, \ldots, x_k^2$. A *second-order model* with two predictor variables is given as

$$E(y) = \beta_0 + \beta_1x_1 + \beta_2x_2 + \beta_3x_1^2 + \beta_4x_1x_2 + \beta_5x_2^2$$

The quadratic terms x_1^2 and x_2^2 allow for curvature, whereas the cross-product or interaction term x_1x_2 allows for warping or twisting of the response surface.

Two predictor variables are said to **interact** if the change in $E(y)$ corresponding to a change in one predictor variable depends on the value of the other variable.

In Examples 13.4 and 13.5, the model included the interaction term x_1x_2. The resulting graphs in Example 13.4 show how the change in $E(y)$ as x_1 changes depends on the value of x_2, and vice versa.

Qualitative variables are entered into a regression model using dummy variables. For each independent qualitative variable in the model, the number of dummy variables required is one less than the

levels

number of _____ associated with that qualitative variable. The following example will demonstrate how this technique is implemented.

Example 13.7
An investigator is interested in predicting the strength of particle board (y) as a function of the size of the particles (x_1) and two types of bonding compounds. If the basic response is expected to be a quadratic function of particle size, write a linear model that incorporates the qualitative variable "bonding compound" into the predictor equation.

Solution
The basic response equation for a specific type of bonding compound would be

$$E(y) = \beta_0 + \beta_1x_1 + \beta_2x_1^2$$

Since the qualitative variable "bonding compound" is at two levels, one dummy variable is needed to incorporate this variable into the model. Define the dummy variable x_2 as follows:

$x_2 = 1$ if bonding compound 2

0

$x_2 = $ _____ if not

The expanded model would now be written as

$$E(y) = \beta_0 + \beta_1x_1 + \beta_2x_1^2 + \beta_3x_2 + \beta_4x_1x_2 + \beta_5x_1^2x_2$$

1. When $x_2 = 0$, the response has been measured using bonding compound 1, and the resulting equation is

$$E(y) = \beta_0 + \beta_1 \underline{\hspace{2cm}} + \beta_2 \underline{\hspace{2cm}}$$

$x_1; \; x_1^2$

2. When $x_2 = 1$, the response has been measured using bonding compound 2, and the resulting equation is

$$E(y) = (\beta_0 + \beta_3) + (\underline{\hspace{1.5cm}})x_1 + (\beta_2 + \beta_5)x_1^2$$

$\beta_1 + \beta_4$

3. The use of the dummy variable x_2 has allowed us to simultaneously describe two quadratic response curves for each of the two bonding compounds. Notice that β_3, β_4, and β_5 measure the differences between the intercepts, the linear components, and the quadratic components, respectively, for the two bonding compounds.

4. Had another bonding compound been included in the investigation, x_3, a second dummy variable, would be defined as

$x_3 = \underline{\hspace{2cm}}$ if bonding compound 3

1

$x_3 = \underline{\hspace{2cm}}$ if not

0

and the model would be expanded to include the terms x_3, $x_1 x_3$, and $\underline{\hspace{2cm}}$ to produce in effect a third quadratic response curve for compound 3.

$x_1^2 x_3$

The formulation of the model is perhaps the most important aspect of a regression analysis because the fit of the model will depend not only on the independent variables included in the model, but also on the way in which the variables are introduced into the model. If, for example, the response increases with some variable x, achieves a maximum, and then begins to decrease, both linear and quadratic terms in x should be included in the model. Failure to include a term in x^2 may cause the model to fit poorly and/or fail to predict the response y for all values of x. Accurate formulation of a model requires experience and a knowledge of the mechanism underlying the response of interest. The latter is sometimes achieved by running several exploratory investigations and combining this information within a more elaborate model.

Self-Correcting Exercises 13C

1. Suppose the response y is related to two predictor variables x_1 and x_2.
   a. Write a first-order model relating $E(y)$ and the variables x_1 and x_2.
   b. Describe the graph relating $E(y)$ and x_1 when x_2 is held constant.
   c. Describe the graph relating $E(y)$ and x_2 when x_1 is held constant.
   d. How can the first-order model be extended in order to allow both the slope and the intercept relating $E(y)$ and x_1 to vary with the value of x_2?

2. Suppose y is related to three predictor variables x_1, x_2, and x_3.
   a. Write a first-order model relating $E(y)$ and x_1, x_2, and x_3.
   b. Write a second-order model relating $E(y)$ and x_1, x_2, and x_3.
   c. What is the effect of including x_1^2, x_2^2, and x_3^2 in the model in part b?
   d. Which terms in the model in part b are interaction terms? What effect do they have on the response surface?

3. Consider a situation in which the output (y) of an industrial plant is linearly related to the number of individuals employed (x_1) and the area in which the plant is located. In describing two areas we define the dummy variable x_2 as

$$x_2 = 1 \quad \text{if area 2}$$

$$x_2 = 0 \quad \text{if not.}$$

Write a linear model relating output to x_1 and x_2 if we assume that the relationship between y and x_1 is linear for both areas.

4. Refer to Exercise 3. Suppose that three areas were involved in the experiment. Define a second dummy variable x_3 as

$$x_3 = 1 \quad \text{if area 3}$$

$$x_3 = 0 \quad \text{if not.}$$

Write a linear model relating output to x_1, x_2, and x_3 if we assume that the relationship between y and x_1 is linear for all areas.

5. Suppose that an experiment as described in Exercise 4 is conducted and the following prediction equation is obtained:

$$\hat{y} = 2 + x_1 + x_2 + 3x_1x_2 + 2x_3 + x_1x_3.$$

Graph the three prediction lines for each of areas 1, 2, and 3.

Solutions to SCE 13C

1. a. $E(y) = \beta_0 + \beta_1 x_1 + \beta_2 x_2$.
   b. When x_2 is constant, the equation is that of a straight line with slope β_1 and intercept $\beta_0 + \beta_2 x_2$.
   c. When x_1 is constant, the equation is that of a straight line with slope β_2 and intercept $\beta_0 + \beta_1 x_1$.
   d. If the term $\beta_3 x_1 x_2$ is added to the model, the slope and intercept will vary as x_2, since the model will be $E(y) = (\beta_0 + \beta_2 x_2) + (\beta_1 + \beta_3 x_2)x_1$. The line has slope $\beta_1 + \beta_3 x_2$ and intercept $\beta_0 + \beta_2 x_2$.

2. a. $E(y) = \beta_0 + \beta_1 x_1 + \beta_2 x_2 + \beta_3 x_3$.
   b. $E(y) = \beta_0 + \beta_1 x_1 + \beta_2 x_2 + \beta_3 x_3 + \beta_4 x_1 x_2 + \beta_5 x_1 x_3 + \beta_6 x_2 x_3$
   $$+ \beta_7 x_1^2 + \beta_8 x_2^2 + \beta_9 x_3^2.$$

   c. The quadratic terms allow for curvature.

d. The interaction terms are $\beta_4 x_1 x_2$, $\beta_5 x_1 x_3$, $\beta_6 x_2 x_3$. They allow for warping or twisting of the response surface.

3. $E(y) = \beta_0 + \beta_1 x_1 + \beta_2 x_2 + \beta_3 x_1 x_2$. Note that if $x_2 = 0$, $E(y) = \beta_0 + \beta_1 x_1$, while if $x_2 = 1$, $E(y) = (\beta_0 + \beta_2) + (\beta_1 + \beta_3) x_1$. The relationship between x_1 and $E(y)$ is linear for both areas, though not the same in slope or intercept.

4. Refer to Exercise 3. The third area involves the addition of two terms. That is,
   $E(y) = \beta_0 + \beta_1 x_1 + \beta_2 x_2 + \beta_3 x_1 x_2 + \beta_4 x_3 + \beta_5 x_1 x_3$.
   Then, for area 1, $x_2 = 0$, $x_3 = 0$ and

   $$E(y) = \beta_0 + \beta_1 x_1$$

   For area 2, $x_2 = 1$, $x_3 = 0$ and

   $$E(y) = (\beta_0 + \beta_2) + (\beta_1 + \beta_3) x_1$$

   For area 3, $x_2 = 0$, $x_3 = 1$ and

   $$E(y) = (\beta_0 + \beta_4) + (\beta_1 + \beta_5) x_1.$$

5. Refer to Exercise 4, with $\beta_0 = 2$, $\beta_1 = 1$, $\beta_2 = 1$, $\beta_3 = 3$, $\beta_4 = 2$, $\beta_5 = 1$. The three lines are

   $$E(y) = 2 + x_1 \quad \text{for area 1,}$$

   $$E(y) = 3 + 4x_1 \quad \text{for area 2,}$$

   $$E(y) = 4 + 2x_1 \quad \text{for area 3.}$$

13.6 TESTING SETS OF REGRESSION COEFFICIENTS

In previous sections we presented procedures for testing the contribution of individual independent variables in predicting a response y, and

all

the joint contribution of _____ the independent variables in the model in predicting y. Interpretation of the results of tests concerning individual parameters in the model was difficult because of the possible presence in the model of other independent variables that contribute similar or perhaps identical information in the prediction of the response y. In this section a more general version of the procedure is given for testing the joint contribution of a set of independent variables in predicting y.

The rationale in implementing this procedure is quite simple. A regression model utilizing all the independent variables of interest is fitted. To test the contribution of any group of these independent variables, a second model with these variables deleted is fitted and

difference

the _____ in the two sums of squares for error is found. This difference is used to assess the additional contribution of the deleted variables above and beyond the information contained in the varia-

were not

bles that (were, were not) deleted from the model.

This procedure is formalized in the following way. Suppose we have k predictor variables, $x_1, x_2, \ldots, x_r, x_{r+1}, \ldots, x_k$ available for predicting the response y. For the *complete* or full model,

$$E(y) = \beta_0 + \beta_1 x_1 + \cdots + \beta_r x_r + \beta_{r+1} x_{r+1} + \cdots + \beta_k x_k$$

Testing whether the variables $x_{r+1}, x_{r+2}, \cdots, x_k$ contribute additional significant information in predicting y is equivalent to testing the hypothesis

0

$$H_0: \ \beta_{r+1} = \beta_{r+2} = \cdots = \beta_k = \underline{\hspace{1cm}}$$

When H_0 is true, the reduced model is

$$E(y) = \beta_0 + \beta_1 x_1 + \cdots + \beta_r x_r$$

reduced

Whenever terms are added to the model, SSE is _____.
Hence, if SSE_1 is the sum of squares for error with the *reduced* model involving r predictor variables and SSE_2 is the sum of squares for

larger

error with the *complete* model, then SSE_1 will be (smaller, larger) than SSE_2. If the difference $(SSE_1 - SSE_2)$ is significantly large, we conclude that at least one of the variables $x_{r+1} x_{r+2}, \cdots, x_k$ contributes significant information beyond that contained in the variables $x_1, x_2, \ldots, x_r$.

In testing the hypothesis $H_0: \ \beta_{r+1} = \beta_{r+2} = \cdots = \beta_k = 0$, we use the test statistic given as

$$F = \frac{MS(\text{drop})}{MSE_2}$$

$n - k - 1$

where $MS(\text{drop}) = (SSE_1 - SSE_2)/(k - r)$ and $MSE_2 = SSE_2/$ (_____). When the random errors are normally and indepen-

$k - r; \ n - k - 1$

dently distributed with mean 0 and variance σ^2, this statistic has an F distribution with $df_1 = ($_____$)$ and $df_2 = ($_____$)$ degrees of freedom. If H_0 is false and one or more of the variables tested contribute significant additional information in predicting y, then $MS(\text{drop})$ would tend to be significantly larger than MSE_2.

Therefore, the test is one-tailed, and H_0 is rejected if the observed value
of F exceeds a (left, right)-tailed critical value of F.

right

Example 13.8

In order to study the relationship of advertising and capital investment
with corporate profits, the following data, recorded in units of
$100,000, were collected for ten medium-sized firms in the same year.
The variable y represents profit for the year, x_1 represents capital
investment, and x_2 represents advertising expenditures.

y	x_1	x_2
15	25	4
16	1	5
2	6	3
3	30	1
12	29	2
1	20	0
16	12	4
18	15	5
13	6	4
2	16	2

Using the model

$$y = \beta_0 + \beta_1 x_1 + \beta_2 x_2 + \epsilon$$

find the least squares prediction equation for these data.

Solution

1. A regression analysis implemented using the Minitab **Regression**
   command produced the following printout for these data:

Regression Analysis

```
The regression equation is
y = - 8.18 + 0.292 x1 + 4.43 x2

Predictor       Coef       StDev          T        P
Constant      -8.177       4.206      -1.94    0.093
x1            0.2921      0.1357       2.15    0.068
x2            4.4343      0.8002       5.54    0.001

S = 3.303      R-Sq = 82.3%      R-Sq(adj) = 77.2%

Analysis of Variance

Source          DF          SS         MS        F        P
Regression       2      355.22     177.61    16.28    0.002
Residual Error   7       76.38      10.91
Total            9      431.60

Source      DF      Seq SS
x1           1       20.16
x2           1      335.05
```

2. The least squares prediction equation is

$$\hat{y} = \underline{\hspace{2cm}} + .2921x_1 + 4.4343x_2$$

3. Since $F = 16.28$ with $p\text{-}value = \underline{\hspace{2cm}}$, the model contributes significant information for the prediction of y.

Example 13.9
Refer to Example 13.9 in which the complete model was given as

$$y = \beta_0 + \beta_1 x_1 + \beta_2 x_2 + \epsilon$$

Test the hypothesis H_0: $\beta_2 = 0$ versus H_a: $\beta_2 \neq 0$ using the testing procedure presented in this section.

Solution
1. If H_0 is true, the reduced model is

$$y = \beta_0 + \beta_1 x_1 + \epsilon$$

the simple linear regression model discussed in Chapter 12. The computer printout of the regression analysis using this model appears below:

Regression Analysis

```
The regression equation is
y = 12.2 - 0.149 x1

Predictor       Coef        StDev           T         P
Constant       12.189       4.439         2.75     0.025
x1             -0.1493      0.2385       -0.63     0.549

S = 7.171      R-Sq = 4.7%      R-Sq(adj) = 0.0%

Analysis of Variance

Source           DF          SS           MS         F         P
Regression        1        20.16        20.16      0.39     0.549
Residual Error    8       411.44        51.43
Total             9       431.60
```

In the analysis of variance portion of the printout, we find that the sum of squares for error in the reduced model is

$$SSE_1 = \underline{\hspace{2cm}} \text{ with } \underline{\hspace{2cm}} \text{ degrees of freedom.}$$

2. From the regression analysis printout using the complete model, we find $SSE_2 = \underline{\hspace{2cm}}$ and $MSE_2 = \underline{\hspace{2cm}}$ with 7 degrees of freedom. Then

$$SS(drop) = SSE_1 - SSE_2$$

$$= 411.44 - 76.38$$

$$= \underline{\hspace{2cm}}$$

with $8 - 7 = 1$ degree of freedom. In this case MS(drop) is the same as SS(drop).

--- margin annotations ---

−8.177

.002

411.44; 8

76.38; 10.91

335.06

3. To test H_0: $\beta_2 = 0$ versus H_a: $\beta_2 \neq 0$, calculate

$$F = \frac{MS(drop)}{MSE_2}$$

$$= \frac{335.06}{10.91}$$

$$= \underline{\hspace{2cm}}$$ 30.74

With $\alpha = .05$, the critical value of F based on $df_1 = 1$ and $df_2 = 7$
degrees of freedom is $F_{.05} = \underline{\hspace{2cm}}$. Hence, we reject H_0 and   5.59
conclude that the variable advertising expenditure (does, does not)   does
contribute significant information in predicting y.

4. In comparing this result with the *t-ratio* for this same test in the individual analysis of variables portion of the computer printout for the
complete model for x_2, we find $t^2 = (5.54)^2 = \underline{\hspace{2cm}}$, which,   30.69
within rounding errors, is the same as the value we have just
computed.

The general testing procedure just described is more appropriately
applied when we are interested in assessing the joint contribution of
several variables in predicting a response y. It is worth pointing out,
however, that in applying this procedure to one predictor variable as we
have done, we produced results identical to those obtained directly from
the computer printout for the complete model. This should help clarify
and unify the test procedure that produces the *t-ratio* in the individual
analysis of variables portion of the computer printout.

13.7 INTERPRETING RESIDUAL PLOTS

The deviations between the observed values of y and their predicted
values are called $\underline{\hspace{2cm}}$. Plots of the residuals against y, $\hat{y}$, or   residuals
the individual independent variables often reveal departures from distributional assumptions and may show relationships between the response y and the independent variables that are not included in the
model selected for analysis.

When there is no indication that the underlying model is incorrect,
a plot of the residuals against the values of $\hat{y}$ should appear as a
$\underline{\hspace{2cm}}$ scatter of points above and below the horizontal axis, as   random
illustrated in Figure 13.9.

Some kinds of data violate the assumption of a common variance
for all values of y. For data that follow a Poisson distribution, the variance is equal to the $\underline{\hspace{2cm}}$, and we would expect to see larger   mean
deviations associated with larger values of y. In the same manner,
binomial proportions have a variance given by $p(1-p)/n$, a quantity
that takes its maximum at $p = \underline{\hspace{2cm}}$ and tends to 0 when p .5
tends to 0 or 1. Figure 13.16 in your text illustrates what residual
plots for these kinds of data look like.

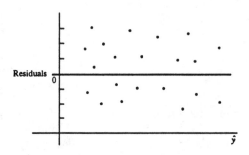

Figure 13.9

Example 13.9
Examine the residual plots which resulted when the complete and reduced models were fit to the data in Example 13.7. Do these plots confirm your conclusions about the importance of x_2, advertising expenditure, in the model?

Solution
In Example 13.7, the model

$$E(y) = \beta_0 + \beta_1 x_1 + \beta_2 x_2$$

was fitted to the data for which y was yearly profit, x_1 was capital investment, and x_2 was advertising expenditures. The residuals from this model plotted against $\hat{y}$ shown in Figure 13.10 reveal that about half are positive and half are negative, and that large deviations are not necessarily associated with large values of y.

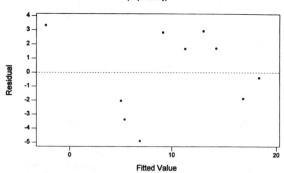

Figure 13.10 (Complete Model)

However, when the model

$$E(y) = \beta_0 + \beta_1 x_1$$

was fitted to the data, plots of the residuals against y and against x_2 shown in Figure 13.11 indicate that a linear term in x_2 is needed in the model.

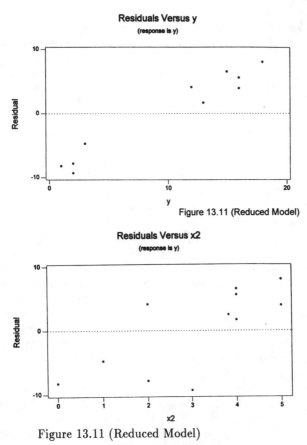

Figure 13.11 (Reduced Model)

Figure 13.11 (Reduced Model)

Most computer regression packages have options that give the residuals and the predicted values when requested and will provide plots or the data for input into a plotting program. It is good policy to request or construct residual plots in a regression analysis in order to verify whether the data satisfy the assumptions for a regression analysis, and to determine whether there are any trends or systematic patterns in the residuals that might indicate the need for other predictors in the model.

13.8 MISINTERPRETATIONS IN A REGRESSION ANALYSIS

In a multiple regression problem, the regression coefficients are called *partial* regression coefficients because they are determined in conjunction with other variables in the model and only partially determine the value of y. Further, the values of these partial regression coefficients (would, would not) in general be the same as those found by using several simple linear regression models, each with one independent variable. Estimates of the partial regression coefficients are correlated

would not

with each other to the extent that the underlying independent variables share the same predictive information.

When the independent variables included in a regression analysis are correlated among themselves, the values of the estimated β's in the model take into account the amount of shared and independent information available in the x's for estimating the response y. In this situation, individual tests of the regression coefficients are of little value. More information concerning the utility of the independent variables $x_1, x_2, \ldots, x_k$ in predicting y can be obtained by testing the hypothesis

$$H_0: \ \beta_1 = \beta_2 = \cdots = \beta_k = 0$$

A test of this hypothesis was given in Section 13.4.

When two or more of the independent variables are highly correlated with each other, we are confronted with the problem of *multicollinearity*. Multicollinearity is the technical way of saying, for example, that if one is given pairs of values for two independent variables that are highly correlated with each other, the pairs of values will exhibit a strong linear relationship when plotted on graph paper. When the correlation is very high, the points will plot as

straight; line

almost a _____ _____. Hence, we say that these variables are collinear, and, for all practical purposes, one is working with one independent variable. When this situation is repeated for several pairs of independent variables, we refer to the problem as multicollinearity.

Many investigators prefer to use a *stepwise regression program*, which at each step adds an independent variable to the regression model only if its inclusion significantly reduces SSE below the value achieved without the variable included. In this way, the investigator can look at the stepwise decrease in SSE and assess the additional contribution of the independent variable just added, above and beyond the contribution of those variables already in the model.

13.9 SUMMARY

Multiple regression analysis is an extension of simple linear regression analysis to accommodate situations in which the response y is a function of a number of independent predictor variables $x_1, x_2, \ldots, x_k$. The procedures associated with the simple linear model have analogues in the case of a multiple regression model. Hence, any

can

simple linear regression problem (can, cannot) be analyzed using multiple regression techniques.

Although identical in concept, simple and multiple regression analyses differ in two important aspects. First, simple linear

without

regression analysis can be done (with, without) the use of a computer; for multiple regression analysis, this is generally not the case. However, multiple regression analysis programs are available at most computing facilities. Second, very few real-life situations (can, cannot) be adequately described by a simple linear regression model. Multiple regression analysis provides greater utility and latitude in data analysis by allowing the user to include k independent predictor variables in the regression model.

can

KEY CONCEPTS AND FORMULAS

I.   *The General Linear Model*
 1. $y = \beta_0 + \beta_1 x_1 + \beta_2 x_2 + \ldots + \beta_k x_k + \epsilon$
 2. The random error ϵ has a normal distribution with mean 0 and variance σ^2.

II. *Method of Least Squares*
 1. Estimates $b_0, b_1, \ldots, b_k$, for $\beta_0, \beta_1, \ldots, \beta_k$, are chosen to minimize SSE, the sum of squared deviations about the regression line, $\widehat{y} = b_0 + b_1 x_1 + b_2 x_2 + \ldots + b_k x_k$.
 2. Least-squares estimates are produced by computer.

III. *Analysis of Variance*
 1. Total SS = SSR + SSE, where Total SS = S_{yy}. ANOVA table is produced by computer.
 2. Best estimate of σ^2 is

 $$MSE = \frac{SSE}{n - k - 1}$$

IV. *Testing, Estimation, and Prediction*
 1. A test for the significance of the regression, H_0: $\beta_1 = \beta_2 = \ldots = \beta_k = 0$, can be implemented using the analysis of variance F test:

 $$F = \frac{MSR}{MSE}$$

 2. The strength of the relationship between x and y can be measured using

 $$R^2 = \frac{MSR}{Total\ SS}$$

 which gets closer to 1 as the relationship gets stronger.
 3. Use residual plots to check for nonnormality, inequality of variances, and an incorrectly fit model.
 4. Significance tests for the partial regression coefficients can be performed using the Student's t test:

 $$t = \frac{b_i - \beta_i}{SE(b_i)} \qquad \text{with error } df = (n - k - 1)$$

 5. Confidence intervals can be generated by computer to estimate the average value of y, $E(y)$, for a given value of x. Computer-generated prediction intervals can be used to predict a particular observation y for a given value of x. For a given x, prediction intervals are always wider than confidence intervals.

V. *Model Building*

1. The number of terms in a regression model cannot exceed the number of observations in the data set and should be considerably less!

2. To account for a curvilinear effect in a *quantitative* variable, use a second-order polynomial model. For a cubic effect, use a third-order polynomial model.

3. To add a *qualitative* variable with k categories, use $(k-1)$ dummy or indicator variables.

4. There may be interactions between two quantitative variables or between a quantitative and qualitative variable. Interaction terms are entered as $\beta x_i x_j$.

5. Compare models using $R^2(\text{adj})$.

Exercises

1. A manufacturer, concerned about the number of defective items being produced within his plant, recorded the number of defective items produced on a given day (y) by each of 10 machine operators, recording also the average output per hour (x_1) for each operator and the time from the last machine servicing (x_2) in weeks. The data were

y	x_1	x_2
13	20	3
1	15	2
11	23	1.5
2	10	4
20	30	1
15	21	3.5
27	38	0
5	18	2
26	24	5
1	16	1.5

The following computer output resulted when these data were analyzed using the Minitab **Regression** command based on the model

$$E(y) = \beta_0 + \beta_1 x_1 + \beta_2 x_2.$$

Regression Analysis

The regression equation is
y = - 28.4 + 1.46 x1 + 3.84 x2

Predictor	Coef	StDev	T	P
Constant	-28.3906	0.8273	-34.32	0.000
x1	1.46306	0.02699	54.20	0.000
x2	3.8446	0.1426	26.97	0.000

S = 0.5484   R-Sq = 99.8%   R-Sq(adj) = 99.7%

Analysis of Variance

Source	DF	SS	MS	F	P
Regression	2	884.79	442.40	1470.84	0.000
Residual Error	7	2.11	0.30		
Total	9	886.90			

Source	DF	Seq SS
x1	1	666.04
x2	1	218.76

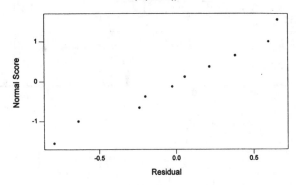

Normal Probability Plot of the Residuals
(response is y)

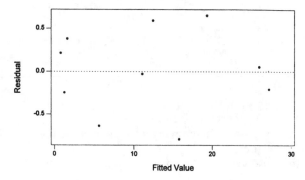

Residuals Versus the Fitted Values
(response is y)

a. Interpret R^2 and comment on the fit of the model.

b. Is there sufficient evidence to indicate that the model contributes significant information in predicting y at the .01 level of significance?

c. What is the prediction equation relating $\hat{y}$ and x_1 when $x_2 = 4$?

d. Use the fitted prediction equation to predict the number of defective items produced for an operator whose average output per hour is 25 and whose machine was serviced three weeks ago.

e. What do the residual plots tell you about the validity of the regression assumptions?

2. An experiment was conducted to investigate the relationship between the degree of metal corrosion and the length of time the metal is exposed to the action of soil acids. In the data which follow, y is the percentage corrosion and x is the exposure time measured in weeks.

y	0.1	0.3	0.5	0.8	1.2	1.8	2.5	3.4
x	1	2	3	4	5	6	7	8

The following computer output resulted in fitting the model:

$$E(y) = \beta_0 + \beta_1 x + \beta_2 x^2$$

Regression Analysis

```
The regression equation is
y = 0.196 - 0.100 x + 0.0619 x-sq

Predictor       Coef        StDev          T         P
Constant      0.19643      0.07395       2.66      0.045
x            -0.10000      0.03770      -2.65      0.045
x-sq          0.061905     0.004089     15.14      0.000

S = 0.05300      R-Sq = 99.9%      R-Sq(adj) = 99.8%

Analysis of Variance

Source          DF         SS          MS         F         P
Regression       2       9.4210      4.7105    1676.61    0.000
Residual Error   5       0.0140      0.0028
Total            7       9.4350

Source          DF      Seq SS
x                1      8.7771
x-sq             1      0.6438
```
---

a. What percent of the total variation is explained by the quadratic regression of y on x?

b. Is the regression of y on x and x^2 significant at the $\alpha = .05$ level of significance?

c. Is the linear regression coefficient significant at the .05 level of significance?

d. Is the quadratic regression coefficient significant at the .05 level of significance?

e. Fitting the model with the linear term omitted given by $E(y) = \beta_0 + \beta_2 x^2$ resulted in the output that follows. What could you say about the contribution of the linear term in x in explaining the total variation in y?

Regression Analysis

The regression equation is
y = 0.0164 + 0.0513 x-sq

Predictor	Coef	StDev	T	P
Constant	0.01643	0.04160	0.39	0.707
x-sq	0.051317	0.001256	40.84	0.000

S = 0.07507   R-Sq = 99.6%   R-Sq(adj) = 99.6%

Analysis of Variance

Source	DF	SS	MS	F	P
Regression	1	9.4012	9.4012	1668.24	0.000
Residual Error	6	0.0338	0.0056		
Total	7	9.4350			

---

f. Does the residual plot below confirm your conclusions in part e? Explain.

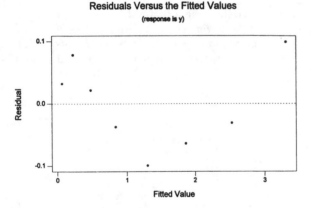

Residuals Versus the Fitted Values
(response is y)

3. In a study to examine the relationship between the time required to complete a construction project and several pertinent independent variables, an analyst compiled a list of four variables that might be useful in predicting the time to completion. These four variables were size of the contract (in $1000 unit)$(x_1)$, number of workdays adversely affected by the weather (x_2), number of subcontractors involved in the project (x_4), and a variable (x_3) that measured the presence or absence of a workers' strike during the construction. In particular,

$x_3 = 0$ if no strike

$x_3 = 1$ if strike

Fifteen construction projects were randomly chosen, and each of the four variables as well as the time to completion were measured. The data are given in the following table:

y	x_1	x_2	x_3	x_4
29	60	7	0	7
15	80	10	0	8
60	100	8	1	10
10	50	14	0	5
70	200	12	1	11
15	50	4	0	3
75	500	15	1	12
30	75	5	0	6
45	750	10	0	10
90	1200	20	1	12
7	70	5	0	3
21	80	3	0	6
28	300	8	0	8
50	2600	14	1	13
30	110	7	0	4

An analysis of these data using a first-order model in x_1, x_2, x_3, and x_4 produced the following computer printout. Give a complete analysis of the printout and interpret your results. What can you say about the apparent contribution of x_1 and x_2 in predicting y?

Regression Analysis

```
The regression equation is
y = - 1.6 - 0.00784 x1 + 0.68 x2 + 28.0 x3 + 3.49 x4

Predictor      Coef       StDev        T        P
Constant      -1.59       11.66      -0.14    0.894
x1        -0.007843     0.006230     -1.26    0.237
x2           0.6753      0.9998       0.68    0.515
x3           28.01       11.37        2.46    0.033
x4           3.489        1.935       1.80    0.102

S = 11.84     R-Sq = 84.7%     R-Sq(adj) = 78.6%

Analysis of Variance

Source           DF        SS        MS       F        P
Regression        4      7770.3    1942.6    13.85    0.000
Residual Error   10      1403.0     140.3
Total            14      9173.3

Source      DF      Seq SS
x1           1      1860.9
x2           1      2615.3
x3           1      2838.0
x4           1       456.0
```

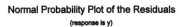

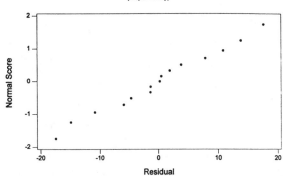

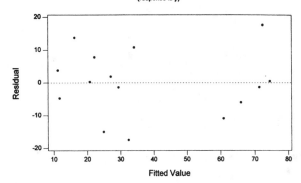

4. A particular savings and loan corporation is interested in determining how well the amount of money in family savings accounts
   can be predicted using the three independent variables, annual
   income, number in the family unit, and area in which the family
   lives. Suppose that there are two specific areas of interest to
   the corporation. The following data were collected, where

y = amount in all savings accounts

x_1 = annual income

x_2 = number in family unit

x_3 = 0 if area 1; 1 if not

Both y and x_1 were recorded in units of $1000.

y	x_1	x_2	x_3
0.5	19.2	3	0
0.3	23.8	6	0
1.3	28.6	5	0
0.2	15.4	4	0
5.4	30.5	3	1
1.3	20.3	2	1
12.8	34.7	2	1
1.5	25.2	4	1
0.5	18.6	3	1
15.2	45.8	2	1

The following computer printout resulted when the data was analyzed using Minitab.

Regression Analysis

```
The regression equation is
y = - 3.11 + 0.503 x1 - 1.61 x2 - 1.15 x3

Predictor        Coef        StDev          T        P
Constant       -3.112        3.600      -0.86    0.421
x1            0.50314      0.07670       6.56    0.001
x2            -1.6126       0.6579      -2.45    0.050
x3             -1.155        1.791      -0.64    0.543

S = 1.896     R-Sq = 92.2%     R-Sq(adj) = 88.4%

Analysis of Variance

Source           DF          SS          MS         F        P
Regression        3     256.621      85.540     23.78    0.001
Residual Error    6      21.579       3.597
Total             9     278.200

Source           DF      Seq SS
x1                1     229.113
x2                1      26.012
x3                1       1.496
```

---

a. Interpret R^2 and comment on the fit of the model.
b. Test for a significant regression of y on x_1, x_2, and x_3 at the .05 level of significance.
3. Test the hypothesis: $H_0: \beta_3 = 0$ against $H_a: \beta_3 \neq 0$ using a significance level of $\alpha = .05$. Comment on the results of your test.
4. What can be said about the utility of x_3 as a predictor variable in this problem?

Chapter 14
Analysis of Categorical Data

14.1 THE MULTINOMIAL EXPERIMENT

Examine the following experimental situations for any general similarities:

1. Two hundred people are classified according to their blood type, and the number of people in each blood type group is recorded.
2. A sample of 100 items is randomly selected from a production line. Each item is classified as belonging to one of three groups: acceptables, seconds, or rejects. The number in each group is recorded.
3. A random sample of 50 books is taken from the local library. Each book is assigned to one of four categories: science, art, fiction, or other. The number of books in each category is recorded.

These situations are similar in that classes or categories are defined and the number of items that fall into each category is recorded. Hence, these experiments result in categorical or _____ data and have the following general characteristics, which define the _____ experiment:

count

multinomial

The Multinomial Experiment

1. The experiment consists of n identical trials.
2. The outcome of each trial falls into one of k classes or cells.
3. The probability that the outcome of a single trial falls into cell i is p_i, $i = 1, 2, \ldots, k$, where p_i is _____ from trial to trial and

$$\sum p_i = \underline{\hspace{2cm}}$$

constant

1

4. The trials are _____.
5. We are interested in $O_1, O_2, O_3, \ldots, O_k$, where O_i is the number of trials in which the outcome falls into cell i and

$$\sum O_i = \underline{\hspace{2cm}}$$

independent

n

2

p_2

O_2

$np_2 = E_2$

The binomial experiment is a special case of the multinomial experiment. This can be seen by letting $k = $ _____ and noting the following correspondences:

Binomial	Multinomial $(k = 2)$
n	n
p	p_1
q	_____
x	O_1
$n - x$	
$E(x) = np$	$E(n_1) = np_1 = E_1$
$E(n - x) = nq$	$E(n_2) = $ _____

For the multinomial experiment, we wish to make inferences about the associated population parameters $p_1, p_2, \ldots, p_k$. A statistic that allows us to make inferences of this sort was developed by the British statistician Karl Pearson in around 1900.

14.2 PEARSON'S CHI-SQUARE STATISTIC

For a multinomial experiment consisting of n trials with known (or hypothesized) cell probabilities p_i, $i = 1, 2, \ldots, k$, we can find the expected number of items that fall into the ith cell by using

$$E_i = np_i, \quad i = 1, 2, \ldots, k$$

The cell probabilities are rarely known in practical situations. Consequently, we wish to estimate or test hypotheses concerning their values. If the hypothesized cell probabilities given are the correct values, then the *observed* number of items that fall into each of the cells, O_i, should differ only slightly from the expected number $E_i = np_i$. Pearson's statistic (given below) utilizes the squares of the deviations of the observed from the expected number in each cell.

Pearson's Chi-Square Test Statistic

$$X^2 = \sum \frac{[O_i - E_i]^2}{E_i}$$

summed over k cells with $E_i = np_i$.

1.25

Note that the deviations are divided by the expected number so that the deviations are weighted according to whether the expected number is large or small. A deviation of 5 from an expected number of 20 contributes $(5)^2/20 = $ _____ to X^2, whereas a deviation of 5 from an expected number of 10 contributes $(5)^2/10 =$ _____, or *twice* as much, to X^2.

When n, the number of trials, is large, this statistic has an approximate χ^2 distribution, provided that the expected numbers in each cell are not too small. We will require as a rule of thumb that $E_i \geq$ _____. This requirement can be satisfied by combining cells that have small expected numbers until every cell has an expected number of at least _____. For small deviations from the expected cell counts, the value of the statistic would be (large, small), supporting the hypothesized cell probabilities. However, for large deviations from the expected counts, the value of the statistic would be (large, small), and the hypothesized values of the cell probabilities would be _____. Hence, a one-tailed test is used, rejecting H_0 when X^2 is _____.

To find the critical value of χ^2 used for testing, the degrees of freedom must be known. Since the degrees of freedom change as Pearson's chi-square statistic is applied to different situations, the degrees of freedom will be specified for each application that follows. In general, the degrees of freedom are equal to the number of cells less one degree of freedom for each independent linear restriction placed upon the cell probabilities. One linear restriction that will always be present is that

$$p_1 + p_2 + p_3 + \cdots + p_k = \underline{\hspace{2cm}}$$

Other restrictions may be imposed by the necessity to estimate certain unknown cell parameters or by the method of sampling used in the collection of the data.

(margin answers: 2.50; 5; 5; small; large; rejected; large; 1)

14.3 TESTING SPECIFIED CELL PROBABILITIES: THE GOODNESS-OF-FIT TEST

Let us consider the following problems concerning cell probabilities in a multinomial experiment.

Example 14.1

Previous enrollment records at a large university indicate that of the total number of persons who apply for admission, 60% are admitted unconditionally, 5% are admitted on a trial basis, and the remainder are refused admission. Of 500 applications to date for the coming year, 329 applicants have been admitted unconditionally, 43 have been admitted on a trial basis, and the remainder have been refused admission. Do these data indicate a departure from previous admission rates?

Solution

This experiment consists of classifying 500 applicants into one of three cells: cell 1 (unconditional admission), cell 2 (conditional admission), and cell 3 (admission refused), where from previous observation, $p_1 = .60$, $p_2 = .05$, and $p_3 = .35$. The expected cell counts are calculated as follows:

300

.05; 25

500; .35; 175

$$E_1 = np_1 = 500(.60) = \underline{\hspace{3cm}}$$

$$E_2 = np_2 = 500(\underline{\hspace{2cm}}) = \underline{\hspace{2cm}}$$

$$E_3 = np_3 = \underline{\hspace{2cm}} \ (\underline{\hspace{2cm}}) = \underline{\hspace{2cm}}$$

Tabulating the results, we have the following:

| | Admissions | | | |
	Unconditional	Conditional	Refused	Total
Observed (O_i)	329	43	128	500
Expected (E_i)	300	25	175	500

Using Pearson's chi-square statistic, we can test the null hypothesis that the cell probabilities remain as before against the alternative hypothesis that at least one cell probability is different from those specified. The value of X^2 will be compared with a critical value of $\chi^2_{.05}$ based on the degrees of freedom found as follows. The degrees of freedom are equal to the number of cells ($k = 3$) less one degree of freedom for the linear restriction $p_1 + p_2 + p_3 = n = 1$. Therefore, the degrees of freedom are $k - 1 = 3 - 1 = 2$ and

5.991

$$\chi^2_{.05} = \underline{\hspace{3cm}}.$$

Formalizing this discussion, we have the following statistical test of the enrollment data:

1. H_0: $p_1 = .60$, $p_2 = .05$, $p_3 = .35$
2. H_a: at least one value of p_i is different from that specified by H_0
3. Test statistic:

$$X^2 = \sum \frac{(O_i - E_i)^2}{E_i}$$

5.991

4. Rejection region: Reject H_0 if $X^2 \geq \chi^2_{.05} = \underline{\hspace{3cm}}$. Now we calculate the test statistic.

$$X^2 = \frac{(329 - 300)^2}{300} + \frac{(43 - 25)^2}{25} + \frac{(128 - 175)^2}{175}$$

$$= 2.803 + 12.96 + 12.62$$

$$= 28.383$$

reject

We have found $X^2 = 28.383 > 5.991$; hence, we (reject; do not reject) H_0. It appears that the admissions to date are not following the previously stated rates.

.256

Some light can be shed on the situation by looking at the sample estimates of the cell probabilities: $\hat{p}_1 = 329/500 = .658$, $\hat{p}_2 = 43/500 = .086$, and $\hat{p}_3 = \underline{\hspace{2cm}}$. Notice that the percentage of unconditional admissions has risen slightly, the number of conditional admissions has increased, and the percentage refused admission has decreased at the expense of the first two categories. A final judgment would have to be made when admissions are closed and final figures are in.

Example 14.2
A botanist performs a secondary cross of petunias involving independent factors that control leaf shape and flower color, where the factor A represents red color, a represents white color, B represents round leaves, and b represents long leaves. According to the Mendelian model, the plants should exhibit the characteristics AB, Ab, aB, and ab in the ratio 9:3:3:1. Of 160 experimental plants, the following numbers were observed: AB, 95; Ab, 30; aB, 28; ab, 7. Is there sufficient evidence to refute the Mendelian model at the $\alpha = .01$ level?

Solution
Translating the ratios into proportions, we have the following:

$$P(AB) = p_1 = \frac{9}{16}$$

$$P(Ab) = p_2 = \frac{3}{16}$$

$$P(aB) = p_3 = \frac{3}{16}$$

$$P(ab) = p_4 = \frac{1}{16}$$

The data are tabulated as follows:

Cell	AB	Ab	aB	ab
Expected (E_i)	90	30	30	10
Observed (O_i)	95	30	28	7

Perform a statistical test of the Mendelian model by using the chi-square statistic.

1. H_0: $p_1 = 9/16$, $p_2 = 3/16$, $p_3 = 3/16$, $p_4 = 1/16$
2. H_a: $p_i \neq p_{i0}$ for at least one value of $i = 1, 2, 3, 4$
3. Test statistic:

$$X^2 = \sum \frac{[O_i - E_i]^2}{E_i}$$

4. Rejection region: With 3 degrees of freedom, we will reject H_0 if

$$X^2 > \chi^2_{.01} = \underline{\hspace{2cm}}$$

11.3449

Now we calculate

$$X^2 = \frac{(95 - 90)^2}{90} + \frac{(30 - 30)^2}{30} + \frac{(28 - 30)^2}{30} + \frac{(7 - 10)^2}{10}$$

$$= .2778 + .0000 + .1333 + .9000$$

$$= \underline{\hspace{2cm}}$$

1.3111

do not reject
is not

Since $X^2 = 1.3111 < 11.3449$, (reject, do not reject) H_0. There (is, is not) sufficient evidence to refute the Mendelian model.

Self-Correcting Exercises 14A

1. A company specializing in kitchen products offers a refrigerator in five different models. A random sample of $n = 250$ sales has produced the following data:

Model	1	2	3	4	5
Number sold	62	48	56	39	45

Test the hypothesis that there is no model preference at the $\alpha = .05$ level of significance. (*Hint*: If there is no model preference, then $p_1 = p_2 = p_3 = p_4 = p_5 = 1/5$.)

2. The number of Caucasians possessing the four blood types **A**, **B**, **AB**, and **O** are said to be in the proportions .41, .12, .03, and .44, respectively. Would the observed frequencies of 90, 16, 10, and 84, respectively, furnish sufficient evidence to refute the given proportions at the $\alpha = .05$ level of significance?

Solutions to SCE 14A

1. H_0: $p_1 = p_2 = p_3 = p_4 = p_5 = 1/5$; H_a: at least one of these equalities is incorrect.
   $E_i = np_i = 250(1/5) = 50$ for $i = 1, 2, \ldots, 5$. With $k - 1 = 5 - 1 = 4$ degrees of freedom, reject H_0 if $X^2 > \chi^2_{.05} = 9.49$. Test statistic:

$$X^2 = \sum \frac{[O_i - E_i]^2}{E_i}$$

$$= [(62 - 50)^2 + (48 - 50)^2 + (56 - 50)^2 + (39 - 50)^2 + (45 - 50)^2]/50$$

$$= \frac{144 + 4 + 36 + 121 + 25}{50} = 6.6.$$

Do not reject H_0. We cannot say there is a model preference.

2. H_0: $p_1 = .41$, $p_2 = .12$, $p_3 = .03$, $p_4 = .44$; H_a: at least one equality is incorrect.

	Blood Type			
	A	*B*	*AB*	*O*
Observed O_i	90	16	10	84
Expected E_i	82	24	6	88

With $k - 1 = 3$ degrees of freedom, reject H_0 if $X^2 > \chi^2_{.05} = 7.81$. Test statistic:

$$X^2 = \frac{(90-82)^2}{82} + \frac{(16-24)^2}{24} + \frac{(10-6)^2}{6} + \frac{(84-88)^2}{88}$$

$$= .7805 + 2.6667 + 2.6667 + .1818 = 6.30.$$

Do not reject H_0. There is insufficient evidence to refute the given proportions.

14.4 CONTINGENCY TABLES

We now examine the problem of determining whether independence exists between two methods for classifying observed data. If we were to classify people first according to their hair color and second according to their complexion, would these methods of classification be independent of each other? We might classify students first according to the college in which they are enrolled and second according to their grade point average. Would these two methods of classification be independent? In each problem we are asking whether one method of classification is *contingent* on another. We investigate this problem by displaying our data according to the two methods of classification in an array called a _____ table.   contingency

Example 14.3
A criminologist studying criminal offenders who have a record of one or more arrests is interested in knowing whether the educational achievement level of the offender influences the frequency of arrests. He has classified his data using four educational achievement level classifications:

   A: completed 6th grade or less
   B: completed 7th, 8th, or 9th grade
   C: completed 10th, 11th, or 12th grade
   D: education beyond 12th grade

Number of Arrests	Educational Achievement				Total
	A	*B*	*C*	*D*	
1	55 (45.39)	40 (43.03)	43 (43.03)	30 (36.55)	168
2	15 (21.61)	25 (20.49)	18 (20.49)	22 (17.40)	80
3 or more	7 (10.00)	8 (9.48)	12 (9.48)	10 (8.05)	37
Total	77	73	73	62	285

The contingency table shows the number of offenders in each cell together with the expected cell frequency (in parentheses). The expected frequencies are obtained as follows:

1. Define p_A as the unconditional probability that a criminal offender will have completed grade 6 or less. Define p_B, p_C, and p_D in a similar manner.
2. Define p_1, p_2, and p_3 to be the unconditional probabilities that the offender has 1, 2, or 3 or more arrests, respectively.

$P(A) \cdot P(B)$

If two events A and B are independent, then $P(A \cap B) =$ _____. Hence, if the two classifications are independent, a cell probability will equal the product of the two respective unconditional row and column probabilities. For example, the probability that an offender who has completed grade 6 is arrested 3 or more times is

$$p_{A3} = p_A \cdot p_3$$

whereas the probability that a person with a tenth-grade education is arrested twice is

$p_C \cdot p_2$

$$p_{C2} = \underline{\qquad}$$

Since the row and column probabilities are unknown, they must be estimated from the sample data. The estimators for these probabilities are defined in terms of r_i, the row totals, c_j, the column totals, and n:

$$\hat{p}_A = \frac{c_1}{n} = \frac{77}{285}$$

$$\hat{p}_B = \frac{c_2}{n} = \frac{73}{285}$$

$$\hat{p}_C = \frac{c_3}{n} = \frac{73}{285}$$

$$\hat{p}_D = \frac{c_4}{n} = \frac{62}{285}$$

$$\hat{p}_1 = \frac{r_1}{n} = \frac{168}{285}$$

$$\hat{p}_2 = \frac{r_2}{n} = \frac{80}{285}$$

$$\hat{p}_3 = \frac{r_3}{n} = \frac{37}{285}$$

If the observed cell frequency for the cell in row i and column j is denoted by O_{ij}, then an estimate for the expected cell number in the ijth cell under the hypothesis of independence can be calculated by using the estimated cell probabilities:

$$E_{ij} = n(p_{ij}) = n(p_i)(p_j)$$

and

$$\hat{E}_{ij} = n\left(\frac{r_i}{n}\right)\left(\frac{c_j}{n}\right) = \frac{r_i c_j}{n}$$

The expected cell numbers enclosed in parentheses for the contingency table are found in this way. For example,

$$\hat{E}_{11} = \frac{(168)(77)}{285} = 45.39$$

$$\hat{E}_{12} = \frac{(168)(73)}{285} = 43.03$$

$$\hat{E}_{34} = \frac{(37)(62)}{285} = \underline{\hspace{2cm}}$$

8.05

$$\hat{E}_{24} = \frac{(\underline{\hspace{1.5cm}})(\underline{\hspace{1.5cm}})}{285} = \underline{\hspace{2cm}}$$

80; 62; 17.40

The chi-square statistic can now be calculated as

$$X^2 = \sum \frac{(O_{ij} - \hat{E}_{ij})^2}{\hat{E}_{ij}}$$

$$= \frac{(55 - 45.39)^2}{45.39} + \frac{(40 - 43.03)^2}{43.03} + \cdots + \frac{(12 - 9.48)^2}{9.48}$$

$$+ \frac{(10 - 8.05)^2}{8.05}$$

$$= 10.23$$

Recall that the number of degrees of freedom associated with the X^2 statistic equals the number of cells less one degree of freedom for each independent linear restriction on the cell probabilities. The first restriction is that $\sum p_i = \underline{\hspace{2cm}}$; hence, $\underline{\hspace{2cm}}$ degree of freedom is lost here. Then $(r - 1)$ independent linear restrictions have been placed on the cell probabilities because of the estimation of $(r - 1)$ row probabilities. Note that we need only estimate $(r - 1)$ independent row probabilities because their sum must equal $\underline{\hspace{2cm}}$. In like manner, $(c - 1)$ degrees of freedom are lost because of the estimation of the column probabilities.

n; one

one (1)

Since there are rc cells, the number of degrees of freedom for testing X^2 in an $r \times c$ contingency table is

$$rc - (1) - (r - 1) - (c - 1)$$

which can be factored algebraically as

$$(r - 1)(c - 1)$$

In short, the number of degrees of freedom for an $r \times c$ contingency table, where all expected cell frequencies must be estimated from sample data (that is, from estimated row and column probabilities), is the number of rows minus one times the number of columns minus one.

For the problem concerning criminal offenders, the degrees of freedom are

$$(r - 1)(c - 1) = (\underline{\hspace{2cm}})(\underline{\hspace{2cm}}) = \underline{\hspace{2cm}}$$

2; 3; 6

We can now formalize the test of the hypothesis of independence of the two methods of classification at the $\alpha = .05$ level.

1. H_0: the two classifications are independent
2. H_a: the classifications are not independent
3. Test statistic:

$$X^2 = \sum \frac{(O_{ij} - \hat{E}_{ij})^2}{\hat{E}_{ij}}$$

4. Rejection region: With 6 degrees of freedom, we will reject H_0 if $X^2 > \chi^2_{.05} = \underline{\hspace{2cm}}$. The calculation of X^2 produces the value $X^2 = 10.23$.

12.5916

do not reject
do not

Since $X^2 < 12.5916$, (reject, do not reject) H_0. The data (do, do not) present sufficient evidence to indicate that educational achievement and the number of arrests are dependent.

Example 14.4
As an alternative to flextime, many companies allow employees to do some of their work at home. Individuals in a random sample of 300 workers were classified according to salary and number of workdays per week spent at home. The results of are given in the next table.

Family Income (in $1000)	Workdays at Home per Week			Total
	Less than one	At least one, but not all	All at home	
Under 25	38 (36.27)	16 (21.08)	14 (10.65)	68
25 to 50.999	54 (49.07)	26 (28.52)	12 (14.41)	92
50 to 74.999	35 (35.20)	22 (20.46)	9 (10.34)	66
75	33 (39.47)	29 (22.94)	12 (11.59)	74
Total	160	93	47	300

Source: Adapted from: 'Home-based workers in the US: 1997," by J.J. Kuenzi and C. A. Roschovsky, *Current Population Reports*, US Department of Commerce, Census Bureau.

The number in parenthesis is the estimated expected cell number. Is there sufficient evidence to indicate that family income and work-at-home status are independent?

Solution
The estimated cell counts have been found using

$$\hat{E}_{ij} = \frac{r_i c_j}{n}$$

and are given in the parentheses within each cell. The degrees of freedom are $(r-1)(c-1) = $ _____ .

6

1. H_0: the two classifications are independent
2. H_a: the classifications are not independent
3. Test statistic:

$$X^2 = \sum \frac{(O_{ij} - \hat{E}_{ij})^2}{\hat{E}_{ij}}$$

4. Rejection region: With 6 degrees of freedom, we will reject H_0 if

$$X^2 > \chi^2_{.05} = \underline{\hspace{2cm}} .$$

12.5916

Calculate X^2:

$$X^2 = \frac{(38 - 36.27)}{36.27} + \frac{(16 - 21.08)^2}{21.08} + \cdots + \frac{(12 - 11.59)^2}{11.59}$$

$$= \underline{\hspace{2cm}} .$$

6.447

The decision is (<u>reject, do not reject</u>) H_0. Therefore, we can conclude that income and work–at–home status (<u>are, are not</u>) independent classifications. From Table 5, with six degrees of freedom, the observed value of $X^2 = 6.447$ is less than $\chi^2_{.10} = 10.6446$. Hence the observed level of significance is

do not reject
are

$$p\text{-}value > \underline{\hspace{2cm}} .$$

.10

The null hypothesis could not be rejected for any value of α less than .10.

Packaged computer programs for performing a contingency table analysis are available at most computer facilities, or the analysis can be done on a desktop computer with appropriate statistical software. These programs produce the same basic information. Familiarity with a printout from one program will generally allow you to interpret results from other programs.

Example 14.5
Refer to Example 14.3. The following computer printout resulted when the data were analyzed using Minitab.

Chi-Square Test

Expected counts are printed below observed counts

	A	B	C	D	Total
1	55	40	43	30	168
	45.39	43.03	43.03	36.55	
2	15	25	18	22	80
	21.61	20.49	20.49	17.40	
3	7	8	12	10	37
	10.00	9.48	9.48	8.05	
Total	77	73	73	62	285

```
Chi-Sq =  2.035 +  0.214 +  0.000 +  1.173 +
          2.024 +  0.992 +  0.303 +  1.214 +
          0.898 +  0.230 +  0.672 +  0.473 = 10.227
DF = 6,  P-Value = 0.115
```

The Minitab printout gives the observed and estimated expected cell counts in the body of the table. The chi-square statistic is calculated below the table, with the individual elements in the sum listed as $(O_{ij} - \hat{E}_j)^2/\hat{E}_{ij}$. For example, the contribution to X^2 made by row 1, column 1 is

55 − 45.39; 2.035

$$\frac{(O_{11} - \hat{E}_{11})^2}{\hat{E}_{11}} = \frac{(\underline{})^2}{45.39} = \underline{}$$

Chi-Sq
is; .115

Then $X^2 = 10.227$ is given by the label _____. Notice that the *p-value* (is, is not) given on the Minitab printout as _____ and the results are not significant.

Self-Correcting Exercises 14B

1. On the basis of the following data, is there a significant relationship between levels of income and political party affiliation at the $\alpha = .05$ level of significance?

Party Affiliation	Income		
	Low	Average	High
Republican	33	85	27
Democrat	19	71	56
Other	22	25	13

2. Three hundred people were interviewed to determine their opinions regarding a uniform driving code for all states:

Sex	Opinion	
	For	*Against*
Male	114	60
Female	87	39

Is there sufficient evidence to indicate that the opinion expressed is dependent on the sex of the person interviewed?

Solutions to SCE 14B

1. H_0: independence of classifications; H_a: classifications are not independent. Expected and observed cell counts are given in the table.

Party Affiliation	Income			Total
	Low	*Average*	*High*	*Total*
Republican	33 (30.57)	85 (74.77)	27 (39.66)	145
Democrat	19 (30.78)	71 (75.29)	56 (39.93)	146
Other	22 (12.65)	25 (30.94)	13 (16.41)	60
Total	74	181	96	351

With $(r - 1)(c - 1) = 2(2) = 4$ degrees of freedom, reject H_0 if $X^2 > \chi_{.05}^2 = 9.49$. Test statistic:

$$X^2 = \frac{(2.43)^2}{30.57} + \frac{(10.23)^2}{74.77} + \ldots + \frac{(-3.41)^2}{16.41} = 25.61.$$

Reject H_0. There is a significant relationship between income levels and political party affiliation.

2. H_0: opinion independent of sex; H_a: opinion dependent on sex. Expected and observed cell counts are given in the table.

Sex	Opinion		Total
	For	*Against*	*Total*
Male	114 (116.58)	60 (57.42)	174
Female	87 (84.42)	39 (41.58)	126
Total	201	99	300

With $(r - 1)(c - 1) = 1$ degree of freedom and $\alpha = .05$, reject H_0 if $X^2 > \chi_{.05}^2 = 3.84$. The test statistic is

$$X^2 = \frac{(-2.58)^2}{116.58} + \frac{(2.58)^2}{57.42} + \frac{(2.58)^2}{84.42} + \frac{(-2.58)^2}{41.58} = .4119.$$

Do not reject H_0. There is insufficient evidence to show that opinion is dependent on sex.

14.5 $r \times c$ TABLES WITH FIXED ROW OR COLUMN TOTALS: TESTS OF HOMOGENEITY

To avoid having rows or columns that are absolutely empty, it is sometimes desirable to fix the row or column totals of a contingency table in the design of the experiment. In Example 14.4, the plan could have been to randomly sample 100 families in each of the four income brackets, thereby ensuring that each of the income brackets would be represented in the sample. On the other hand, a random sample of 80 families in each of the family size categories could have been taken so that all family size categories would appear in the overall sample.

When using fixed row or column totals, the number of independent linear restrictions on the cell probabilities is the same as for an $r \times c$ contingency table. Therefore, the data are analyzed in the same way that an $r \times c$ contingency table is analyzed, using the X^2 statistic based on $(r - 1)(c - 1)$ degrees of freedom. In the following example, we examine a case in which the column totals are fixed in advance.

Example 14.6
Fifty fifth-grade students from each of four city schools were given a standardized fifth-grade reading test. After grading, each student was rated as satisfactory or not satisfactory in reading ability, with the following results:

	School			
	1	2	3	4
Not satisfactory	7	10	13	6

Is there sufficient evidence to indicate that the percentage of fifth-grade students with an unsatisfactory reading ability varies from school to school?

Solution
The preceding table displays only half the pertinent information. Extend the table to include the satisfactory category, allowing space to write in the expected cell frequencies.

9; 9; 9
43
41; 41; 41

	School				
	1	2	3	4	Total
Satisfactory	7 (9)	10 (___)	13 (___)	6 (___)	36
Not satisfactory	(___)	40 (41)	37 (___)	44 (___)	164
Total	50	50	50	50	200

By fixing the column total at 50, we have made certain that the unconditional probability of observing a student from each of the schools is constant and equal to _____ for each school.

        1/4

    If the percentage of unsatisfactory tests does not vary from school to school, then the probability of observing an unsatisfactory reading grade is the same for each school and equal to a common value p. Therefore, the unconditional probability of observing an unsatisfactory grade is p and, in like manner, the probability of observing a satisfactory grade is $1 - p =$ _____. If the percentage of unsatisfactory grades is the same for the four schools, then the probability of observing an unsatisfactory grade for a student in the jth school will be

        q

$$p_{1j} = \left(\frac{1}{4}\right)p \text{ for } j = 1, 2, 3, 4$$

The probability of observing a satisfactory grade in the jth school will be

$$p_{2j} = \left(\frac{1}{4}\right)q \text{ for } j = 1, 2, 3, 4$$

However, if the probability of an unsatisfactory grade varies from school to school, then

$$p_{1j} \neq \left(\frac{1}{4}\right)p \quad \text{and} \quad p_{2j} \neq \left(\frac{1}{4}\right)q$$

for at least one value of j, where $j = 1, 2, 3, 4$. But this is the same as asking whether the row and column classifications are independent. Hence, the test is equivalent to a test of the independence of two classifications based on $(r - 1)(c - 1)$ degrees of freedom.

    Proceeding with the required test, we have the following:

1. H_0: $p_1 = p_2 = p_3 = p_4 = p$
2. H_a: at least one proportion differs from at least one other
3. Test statistic:

$$X^2 = \sum \frac{(O_{ij} - \hat{E}_{ij})^2}{\hat{E}_{ij}}$$

4. Rejection region: For $(2 - 1)(4 - 1) =$ _____ degrees of freedom, we will reject H_0 if $X^2 > \chi^2_{.05} =$ _____.

        3
        7.81473

    To calculate the value of the test statistic, we must first find the estimated expected cell counts:

$$\hat{E}_{11} = \hat{E}_{12} = \hat{E}_{13} = \frac{36(50)}{200} = \underline{\hspace{2cm}}$$

        9

$$\hat{E}_{21} = \hat{E}_{22} = \hat{E}_{23} = \frac{164(50)}{200} = \underline{\hspace{2cm}}$$

        41

Then

13 − 9

37 − 41

30; 30

4.0650

cannot

$$X^2 = \frac{(7-9)^2}{9} + \frac{(10-9)^2}{9} + \frac{(\underline{\hspace{2cm}})^2}{9} + \frac{(6-9)^2}{9}$$

$$+ \frac{(43-41)^2}{41} + \frac{(40-41)^2}{41} + \frac{(\underline{\hspace{2cm}})^2}{41} + \frac{(44-41)^2}{41}$$

$$= \frac{\overline{\hspace{2cm}}}{9} + \frac{\overline{\hspace{2cm}}}{41} = 3.3333 + .7317$$

$$= \underline{\hspace{2cm}}$$

Since $X^2 = 4.0650 < \chi^2_{.05} = 7.8147$, we (can, cannot) reject the hypothesis that reading ability for the fifth-graders as measured by this test does not vary from school to school. The Minitab χ^2-test for these data is given in the following printout.

Chi-Square Test

Expected counts are printed below observed counts

	1	2	3	4	Total
1	7	10	13	6	36
	9.00	9.00	9.00	9.00	
2	43	40	37	44	164
	41.00	41.00	41.00	41.00	
Total	50	50	50	50	200

Chi-Sq = 0.444 + 0.111 + 1.778 + 1.000 +
         0.098 + 0.024 + 0.390 + 0.220 = 4.065
DF = 3, P-Value = 0.255

When we fixed the column totals in this example, we found that testing for independence of the row and column categories was, in fact, equivalent to a test of the equality of four binomial parameters, p_1, p_2, p_3, and p_4. Such tests are called **tests of homogeneity** of several binomial populations. With three or more row categories with fixed column totals, the test becomes a test of homogeneity of several multinomial populations.

Self-Correcting Exercises 14C

1. A survey of voter sentiment was conducted in four mid-city political wards to compare the fraction of voters favoring a "city manager" form of government. Random samples of 200 voters were polled in each of the four wards with results as follows.

	Ward			
	1	2	3	4
Favor	75	63	69	58
Against	125	137	131	142

Can you conclude that the fractions favoring the city manager form of government differ in the four wards?

2. A personnel manager of a large company investigating employee satisfaction with their assigned jobs collected the following data for 200 employees in each of four job categories.

Satisfaction	Categories				
	I	II	III	IV	Totals
High	40	60	52	48	200
Medium	103	87	82	88	360
Low	57	53	66	64	240
Totals	200	200	200	200	800

Use the Minitab printout given below. Do these data indicate that the satisfaction scores are dependent on the job categories? (Use $\alpha = .05$.)

Chi-Square Test

```
Expected counts are printed below observed counts

          I       II      III      IV      Total
  1      40       60       52       48       200
       50.00    50.00    50.00    50.00

  2     103       87       82       88       360
       90.00    90.00    90.00    90.00

  3      57       53       66       64       240
       60.00    60.00    60.00    60.00

Total   200      200      200      200       800

Chi-Sq =  2.000 +  2.000 +  0.080 +  0.080 +
          1.878 +  0.100 +  0.711 +  0.044 +
          0.150 +  0.817 +  0.600 +  0.267 = 8.727
DF = 6, P-Value = 0.190
```

Solutions to SCE 14C

1. $H_0: p_1 = p_2 = p_3 = p_4 = p$; $H_a: p_i \neq p$ for at least one i, $i = 1, 2, 3, 4$.

	1	2	3	4	Total
	\multicolumn{4}{c}{**Ward**}				
Favor	75	63	69	58	265
	(66.25)	(66.25)	(66.25)	(66.25)	
Against	125	137	131	142	535
	(133.75)	(133.75)	(133.75)	(133.75)	
Total	200	200	200	200	800

With $(r-1)(c-1) = 3$ degrees of freedom and $\alpha = .05$, reject H_0 if $X^2 > \chi_{.05}^2 = 7.81$. Test statistic:

$$X^2 = \frac{8.75^2}{66.25} + \frac{(-3.25)^2}{66.25} + \ldots + \frac{(8.25)^2}{133.75} = \frac{162.75}{66.25} + \frac{166.75}{133.75} = 3.673.$$

Do not reject H_0. There is insufficient evidence to suggest a difference from ward to ward.

2. H_0: independence of classifications; H_a: dependence of classifications. With $(r-1)(c-1) = 6$ degrees of freedom, reject H_0 if $X^2 > \chi_{.05}^2 = 12.59$. Test statistic:

$$X^2 = \frac{(-10)^2}{50} + \frac{10^2}{50} + \ldots + \frac{4^2}{60} = 8.73. \quad \text{Do not reject } H_0.$$

14.6 OTHER APPLICATIONS

The specific uses of the chi-square test that we have dealt with in this chapter can be divided into two categories:

1. The first category is called "goodness-of-fit tests," whereby observed frequencies are compared with hypothesized frequencies that depend on the hypothesized cell probabilities for a multi-nomial probability distribution. A decision is made as to whether the data fit the hypothesized model.

2. The second category is called "tests of independence," whereby a decision is made as to whether two methods of classifying the observations are statistically independent. If it is decided that the classifications are independent, then the probability that an observation would be classified as belonging to a specific row classification would be constant across the columns, and vice versa. If it is decided that the classifications are not independent, then the implication is that the probability that an observation would be classified as belonging to a specific row classification varies from column to column.

To illustrate the general nature of the goodness-of-fit test, we could test whether a set of data comes from any specified distribution such as the normal distribution with mean μ and variance σ^2, or from a binomial distribution based on n trials with probability of success p, or perhaps from a Poisson distribution with mean μ. Binomial data produce their own natural grouping that corresponds to the cells of a multinomial experiment if one counts the number of zeros, ones, twos, and so on that occur in the data. If the expected cell frequencies are less than the required number, cells can be combined before using the chi-square statistic. Data from a normal distribution, on the other hand, do not produce an inherent natural grouping and must be grouped as in a frequency histogram. In conjunction with a table of normal curve areas and the hypothesized normal distribution, the boundary points for the histogram should be chosen so that each "cell" has approximately the same probability and an expected frequency greater than _____. Grouping the same data accordingly, one can compare the "observed" group frequencies with the theoretical ones using Pearson's chi-square statistic. If population parameters need to be estimated, the point estimates given in earlier chapters are used and _____ degree of freedom subtracted for each independent estimate.

5

one

Tests of independence of two methods of classification are easily extended to three or more classifications by first estimating the expected cell frequencies and applying the chi-square statistic with the proper degrees of freedom. For example, in testing the independence of three classifications with c_1, c_2, and c_3 categories in the respective classifications, the test statistic would be

$$X^2 = \sum \frac{(O_{ijk} - \hat{E}_{ijk})^2}{\hat{E}_{ijk}}$$

which has an approximate χ^2 distribution with $(c_1 - 1)(c_2 - 1) \cdot (c_3 - 1)$ degrees of freedom.

Other applications involving the χ^2 test are usually specifically tailored solutions to special problems. An example would be the test of a linear trend in a binomial proportion observed over time, as discussed in the text. Modifications such as this usually require the use of calculus and are beyond the scope of this text.

An alternative approach to the analysis of contingency tables is based on *log-linear models*, in which the logarithm of the cell probabilities $ln\ p_{ij}$ is assumed to be a linear function of row and column parameters. The analysis is somewhat more complicated than that presented here, and almost always requires the use of a computer and appropriate computer programs. However, the analysis of log-linear models is very flexible and hence can accommodate a wide variety of restrictions and sampling situations. An excellent presentation of log-linear modeling can be found in *Discrete Multivariate Analysis: Theory and Practice*, by Y. M. M. Bishop, S. E. Fienberg, and P. W. Holland (Massachusetts Institute of Technology Press, 1975).

Self-Correcting Exercises 14D

1. A company producing wire rope has recorded the number of "breaks" occurring for a given type of wire rope within a 4-hour period. These records were kept for fifty 4-hour periods. If x is the number of "breaks" recorded for each 4-hour period and μ is the mean number of "breaks" for a 4-hour period, does the following Poisson model adequately describe these data when $\mu = 2$?

$$P(X = x) = \frac{\mu^x e^{-\mu}}{x!}, \quad x = 0, 1, 2, \ldots.$$

x	0	1	2	3 or more
Number observed	4	15	16	15

Hint: Find $P(X = 0)$, $P(X = 1)$ and $P(X = 2)$. Use the fact that

$$P(x \geq 3) = 1 - P(X = 0) - P(X = 1) - P(X = 2)$$

After finding the expected cell numbers, you can test the model by applying Pearson's chi-square test.

2. In standardizing a score, the mean is subtracted and the result divided by the standard deviation. If 100 scores are so standardized and then grouped, test, at the $\alpha = .05$ level of significance, whether these scores were drawn from the standard normal distribution.

Interval	Frequency
less than -1.5	8
-1.5 to $-.5$	20
$-.5$ to $.5$	40
$.5$ to 1.5	29
greater than 1.5	3

Solutions to SCE 14D

1. With $e^{-2} = .135335$, $P(X = 0) = .135335$, $P(X = 1)$ $= 2e^{-2} = .270670$, $P(X = 2) = 2e^{-2} = .270670$, and $P(x \geq 3) = 1 - P(X \leq 2) = .323325$. The observed and expected cell counts are shown below.

O_i	4	15	16	15
E_i	6.77	13.53	13.53	16.17

With $k - 1 = 3$ degrees of freedom, reject H_0 if $X^2 > \chi^2_{.05} = 7.81$. Test statistic:

$$X^2 = \frac{(-2.77)^2}{6.77} + \frac{(1.47)^2 + (2.47)^2}{13.53} + \frac{(-1.17)^2}{16.17}$$

$$= 1.133 + .611 + .085 = 1.829.$$

Do not reject the model.

2. Expected numbers are $E_i = np_i = 100p_i$, where
$p_i = P$(observation falls in cell i | score drawn from the standard normal distribution). Hence, using Table 3,

$$p_1 = P(z < -1.5) = .0668$$

$$p_2 = P(-1.5 < z < -.5) = .2417$$

$$p_3 = P(-.5 < z < .5) = 2(.1915) = .3830$$

$$p_4 = .2417$$

$$p_5 = .0668$$

Observed and expected cell counts are shown below.

O_i	8	20	40	29	3
E_i	6.68	24.17	38.30	24.17	6.68

With $k - 1 = 4$ degrees of freedom, reject H_0 if $X^2 > \chi^2_{.05}$ = 9.49. Test statistic:

$$X^2 = \frac{1.32^2}{6.68} + \frac{(-4.17)^2}{24.17} + \ldots + \frac{(-3.68)^2}{6.68}$$

$$= .2608 + .7194 + .0755 + .9635 + 2.0273 = 4.047.$$
Do not reject the model.

14.7 ASSUMPTIONS

In order that the statistic

$$X^2 = \sum \frac{(O_i - E_i)^2}{E_i}$$

has an approximate χ^2 distribution, the following assumptions are made:

1. The cell counts, $O_1, O_2, \ldots, O_k$, must satisfy the conditions of a _____ experiment (or several multinomial experiments).

2. All expected cell counts should be at least _____. Although valid multinomial data arise under various sampling plans, in order to be confident in the use of the χ^2 statistic, the sample size should be large enough to ensure that all the expected cell counts are 5 or more. This is a conservative figure; some authors have stated that some expected cell counts can be as small as 1. By asking for expected cell counts of 5 or more, we automatically satisfy experimental situations in which these counts can in fact be allowed to be less than 5. In so doing, we should realize that sensitivity may be sacrificed for the sake of safety and simplicity.

multinomial

5

KEY CONCEPTS AND FORMULAS

I. *The Multinomial Experiment*
   1. There are n identical trials and each outcome falls into one of k categories.
   2. The probability of falling into category i is p_i and remains constant from trial to trial.
   3. The trials are independent, $\sum p_i = 1$, and we measure O_i, the number of observations in each of the k categories.

II. *Pearson's Chi-Square Statistic*

$$X^2 = \sum \frac{(O_i - E_i)^2}{E_i} \qquad \text{where } E_i = np_i$$

which has an approximate chi-square distribution with *degrees of freedom* determined by the application.

III. *The Goodness-of-Fit Test*
   1. This is one-way classification with cell probabilities specified in H_0.
   2. Use the chi-square statistic with $E_i = np_i$ calculated with the hypothesized probabilities.
   3. $df = k - 1 - (\text{Number of parameters estimated in order to find } E_i)$
   4. If H_0 is rejected, investigate the nature of the differences using the sample proportions.

IV. *Contingency Tables*
   1. A two-way classification with n observations categorized into $r \times c$ cells of a two-way table using two different methods of classification is called a contingency table.
   2. The test for independence of classification methods uses the chi-square statistic

$$X^2 = \sum \frac{(O_{ij} - \widehat{E}_{ij})^2}{\widehat{E}_{ij}} \qquad \text{with } \widehat{E}_{ij} = \frac{r_i c_j}{n} \qquad \text{and}$$
$$df = (r-1)(c-1)$$

   3. If the null hypothesis of independence of classifications is rejected, investigate the nature of the dependence using conditional proportions within either the rows or columns of the contingency table.

V. *Fixing Row or Column Totals*
   1. When either the row or column totals are fixed, the test of independence of classifications becomes a test of the homogeneity of cell probabilities for several multinomial experiments.
   2. Use the same chi-square statistic as for contingency tables.
   3. The large-sample z tests for one and two binomial proportions are special cases of the chi-square statistic.

VI. *Assumptions*
   The cell counts satisfy the conditions of a multinomial experiment, or a set of multinomial experiments with fixed sample

sizes. All expected cell counts must equal or exceed five in order that the chi-square approximation is valid.

Exercises

1. What are the characteristics of a multinomial experiment?

2. Do the following situations possess the properties of a multinomial experiment?
   a. A large number of red, white, and blue flower seeds are thoroughly mixed and a sample of $n = 30$ seeds is taken. The numbers of red, white, and blue flower seeds are recorded.
   b. A game of chance consists of picking three balls at random from an urn containing one white, three red, and six black balls. The game "pays" according to the number of white, red, and black balls chosen.
   c. Four production lines are checked for defectives during an 8-hour period and the number of defectives for each production line recorded.

3. The probability of receiving grades of A, B, C, D, and F are .07, .15, .63, .10, and .05, respectively, in a certain humanities course. In a class of 120 students:
   a. What is the expected number of A's?
   b. What is the expected number of B's?
   c. What is the expected number of C's?

4. A department store manager claims that her store has twice as many customers on Fridays and Saturdays than on any other day of the week (the store is closed on Sundays). That is, the probability that a customer visits the store Friday is 2/8, the probability that a customer visits the store Saturday is 2/8, while the probability that a customer visits the store on each of the remaining weekdays is 1/8. During an average week, the following numbers of customers visited the store:

   Monday: 95 Thursday: 75
   Tuesday: 110 Friday: 181
   Wednesday: 125 Saturday: 214

   Can the manager's claim be refuted at the $\alpha = .05$ level of significance?

5. If the probability of a female birth is 1/2, according to the binomial model, in a family containing four children, the probability of 0, 1, 2, 3, or 4 female births is 1/16, 4/16, 6/16, 4/16, and 1/16, respectively. A sample of 80 families each containing four children resulted in the following data:

Female births	0	1	2	3	4
Number of families	7	18	33	16	6

Do the data contradict the binomial model with $p = 1/2$ at the $\alpha = .05$ level of significance?

6. A serum thought to be effective in preventing colds was administered to 500 individuals. Their records for 1 year were compared to those of 500 untreated individuals, with the following results.

	No Colds	One Cold	More Than One Cold
Treated	252	146	102
Untreated	224	136	140

Test the hypothesis that the two classifications are independent, at the $\alpha = .05$ level of significance.

7. A manufacturer wished to know whether the number of defectives produced varied for four different production lines. A random sample of 100 items was selected from each line and the number of defectives recorded:

Production lines	1	2	3	4
Defectives	8	12	7	9

Do these data produce sufficient evidence to indicate that the percentage of defectives varies from line to line?

8. In a random sample of 50 male and 50 female undergraduates, each member was asked if he or she was for, against, or indifferent to the practice of having unannounced in-class quizzes. Do the following data indicate that attitude toward this practice is dependent on the sex of the student interviewed?

	Male	Female
For	20	10
Against	15	30
Indifferent	15	10

9. In an experiment performed in a laboratory, a ball is bounced with a container whose bottom (or floor) has holes just large enough for the ball to pass through. The ball is allowed to bounce until it passes through one of the holes. For each of 100 trials, the number of bounces until the ball falls through one of the holes is recorded. If x is the number of bounces until the ball does fall through a hole, does the model
$$P(X = x) = (.6)(.4)^x, \quad x = 0, 1, 2, 3, \ldots$$
adequately describe the following data?

x	0	1	2	3 or more
Number observed	65	28	4	3

Hint: First find $P(X = 0)$, $P(X = 1)$, $P(X = 2)$, and $P(x \geq 3)$ from which the expected numbers for the cells can be calculated using np_0, np_1, np_2, and so on. Then a goodness-of-fit test will adequately answer the question posed.

Chapter 15
Nonparametric Statistics

15.1 INTRODUCTION

In earlier chapters we tested various hypotheses concerning populations in terms of their parameters. These tests represent a group of tests that are called _____ tests because they involve specific parameters such as means, variances, or proportions. To apply the techniques of Chapters 8 and 9, a large number of observations were required to assure the approximate _____ of the statistics used in testing. In Chapters 10, 11, 12, and 13, it was assumed that the sampled populations had _____ distributions. Furthermore, if two or more populations were studied in the same experiment, it was necessary to assume that these populations had a common _____. In this chapter we will be concerned with hypotheses that do not involve population parameters directly but deal rather with the form of the distribution. The hypothesis that two distributions are identical and the hypothesis that one distribution has typically larger values than the other are nonparametric statements of H_0 and H_a.

    Nonparametric tests are appropriate in many situations where one or more of the following conditions exist:

1. Nonparametric methods can be used when the form of the distribution is unknown, so that descriptive parameters may be of little use.
2. Nonparametric techniques are particularly appropriate if the measurement scale is a rank ordering.
3. If a response can be measured on a continuous scale, a nonparametric method may nevertheless be desirable because of its relative simplicity when compared with its parametric analogue.
4. Most parametric tests require that the sampled population satisfy certain assumptions. When an experimenter cannot reasonably expect that these assumptions are met, a nonparametric test is a valid alternative.

    The following hypotheses are appropriate for nonparametric tests:

H_0: a given population is normally distributed
H_0: populations I and II have the same distribution

Since these hypotheses are less specific than those required for parametric tests, we might expect a nonparametric test to be (more, less)

(margin answers) parametric

normality

normal

variance

less

efficient than a corresponding parametric test when all the conditions required for the use of the parametric test are met.

In this chapter, we will present nonparametric tests for

1. comparing two population means using both the unpaired (independent samples) and paired (dependent samples) designs;
2. comparing more than two population means using the completely randomized and the randomized block designs;
3. measuring the association between two variables.

15.2 THE WILCOXON RANK SUM TEST: INDEPENDENT RANDOM SAMPLES

If an experimenter has two independent (not related) random samples in which the measurement scale is at least rank ordering, he can test the hypothesis that the two underlying distributions are identical versus the alternative that they are not identical using one of two equivalent nonparametric tests:

1. The Wilcoxon Rank Sum Test
2. The Man-Whitney U Test

In both of these tests, the actual observations are replaced by the *ranks* of the observations, using the following logic.

If two independent random samples are drawn from the same population (this is the case if H_0 is true), then we really have one large sample of size $n = n_1 + n_2$. If all measurements were ranked from small (1) to large (n), and each observation from sample 1 was replaced with a 1 and each observation from sample 2 was replaced with a 2, we would expect to find the 1's and 2's randomly mixed in the ranking positions. If H_0 is false and the second sample comes from a population whose values tend to be larger than those of the first population, the 2's would tend to occupy the _____ ranks. However, if the second sample comes from a population whose values tend to be smaller than those of the first, then the 2's would appear in the _____ rank positions.

higher

lower

A statistic that reflects the positions of the n_1 and n_2 sample values in the total ranking is the sum of the ranks occupied by the first sample (T_1) or the sum of the ranks occupied by the second sample (T_2). The stronger the discrepancy between T_1 and T_2, the greater is the evidence to indicate that the samples have been drawn from two different populations. Notice that the sum of the $N = n_1 + n_2$ ranks is $T_1 + T_2 = \dfrac{N(N+1)}{2}$, so that if you know one of the rank sums, you can easily find the other. This means that only _____ of these two rank sums provides all the necessary information for the test. The Wilcoxon Rank Sum test uses this information to test for a difference in the population frequency distributions that give rise to the sample observations.

one

To implement the **Wilcoxon Rank Sum test:**

1. Rank the $N = n_1 + n_2$ observations from smallest to _____. largest

2. Designate n_1 as the _smaller_ of the two sample sizes.

3. Calculate T_1, the sum of the ranks of the first sample.

4. Calculate $T_1^* = n_1(n_1 + n_2 + 1) - T_1$, which is the sum of the ranks in the first sample _if the observations had been ranked from largest to smallest._

5. The test statistic T is the smaller of T_1 and T_1^* and will depend on the nature of the alternative hypothesis.

Example 15.1

Five sample observations for each of two samples are given below. In the spaces provided, fill in the rank of each of the ten observations and calculate T_1 and T_2, and T_1^*.

Sample 1: 19(_____) 20(_____) 16(_____) 12(_____) 23(_____)   5; 6; 2; 1; 9

Sample 2: 17(_____) 21(_____) 22(_____) 25(_____) 18(_____)   3; 7; 8; 10; 4

$T_1 = 5 + 6 + 2 + 1 + 9 =$ _____   23

$T_2 = 3 + 7 + 8 +$ _____ $+$ _____ $=$ _____   10; 4; 32

As a check on our calculations, notice that

$$T_1 + T_2 = \text{_____} = \frac{10(11)}{2} = 55$$ $\dfrac{N(N+1)}{2}$

Finally, calculate

$$T_1^* = n_1(n_1 + n_2 + 1) - T_1$$

$$= 5(\text{_____}) - \text{_____} = \text{_____}.$$ 11; 23; 32

For this particular example, since $n_1 = n_2 = 5$, the value of T_1^* (is, is not) the same as the value of T_2. However, this (will, will not)   is; will not
be the case when the sample sizes are unequal!

In order to use either T_1 or T_1^* as the test statistic to test the null hypothesis that the two populations are identical against an alternative that might be one- or two-tailed, it is necessary to determine an appropriate rejection region based on the probability distribution for T_1 or T_1^*. Table 7 in the text gives the critical values, of the test statistic, say $T = a$, such that $P(T \le a) \le \alpha$ when n_1 and n_2 are ten or less. Since only lower-tailed critical values are tabulated, it is desirable to design the test statistic to take advantage of this lower-tailed rejection region.

1. When the alternative hypothesis is that the two populations differ in distribution, a two-tailed rejection region is appropriate. In that case, we calculate T_1 and T_1^*. Since T_1 is the sum of the ranks of the

large
large

first sample (ranked from smallest to largest) and T_1^* is the rank sum of the first sample (ranked from largest to smallest), then T_1 will be small if T_1^* is (small, large) and T_1^* will be small if T_1 is (small, large). Hence, the smaller of the two is used as the test statistic T, and you should reject H_0 if the smaller of T_1 and T_1^* is *too small*. In particular, H_0 will be rejected if the smaller of T_1 and T_1^* if less than T_0. Since only the lower portion of the rejection region is used (there is, in fact, another portion of the rejection region in the upper tail of the probability distribution), the one-tailed probabilities—.05, .025, .01, and .005—are doubled to obtain α.

small; large
T_1

2. When the alternative hypothesis is that the distribution of 1's lies to the left of the distribution of 2's, the rank sum T_1 for sample 1 will be (small, large) and T_1^* will be (small, large). You will use $T = (T_1, \overline{T_1^*})$ as the test statistic and reject H_0 if T is less than the critical value a specified in Table 7 for a one-tailed test.

large; small
T_1^*

3. When the alternative hypothesis is that the distribution of 1's lies to the right of the distribution of 1's, the rank sum T_1 for sample 1 will be (small, large) and T_1^* will be (small, large). You will use $T = (T_1, \overline{T_1^*})$ as the test statistic and will reject H_0 if T is less than the critical value a specified in Table 7 for a one-tailed test.

Example 15.2
Use the data in Example 15.1 to test H_0: the frequency distributions for populations 1 and 2 are identical against H_a: the population frequency distributions are not identical.

Solution

23; smaller

1. From Example 15.1, the Wilcoxon Rank Sum statistic is equal to _____, the _____ of T_1 and T_1^*.

5
17
do not reject

2. For a two-tailed test with $\alpha = .05$, the critical value of T is selected so that $P(T \leq a) \leq \alpha/2 = .025$, found in Table 7(b) for $n_1 = n_2 = $ _____. Hence, you will reject H_0 if T is less than or equal to _____.

3. Since $T = 23$ is greater than 17, we (reject, do not reject) the null hypothesis of identical population frequency distributions.

Example 15.3
Before filling several new teaching positions at the high school, the principal formed a review board consisting of five teachers who were asked to interview the 12 applicants and rank them in order of merit. Seven of the 12 applicants held college degrees but had limited teaching experience. Of the remaining 5 applicants, all had college degrees and substantial experience. The review board's rankings are given next:

Limited Experience	Substantial Experience
4	1
6	2
7	3
9	5
10	8
11	
12	

Do these rankings indicate that the review board considers experience a prime factor in the selection of the best candidates?

Solution

1. In testing the null hypothesis that the underlying populations are identical versus the alternative hypothesis that the population consisting of applicants who have substantial experience is better qualified (will receive low ranks), we require a _____-tailed test.

   one

2. Identify the smaller sample (substantial experience) as sample 1 with $n_1 = 5$ and $n_2 = 7$. If the alternative hypothesis is true, you would expect T_1 to be (small, large) and T_1^* to be (small, large).

   small; large

3. Calculate

   $$T_1 = 1 + 2 + 3 + 5 + 8 = \underline{\qquad}$$

   19

   $$T_1^* = n_1(n_1 + n_2 + 1) - T_1$$

   $$= 5(12 + 1) - \underline{\qquad} = \underline{\qquad}$$

   19; 46

   and

   $$T = \min(T_1, T_1^*) = \underline{\qquad}.$$

   19

4. For a one-tailed test with $\alpha = .05$, use Table 7(a) to find the critical value $a = \underline{\qquad}$ and the rejection region is $T \leq 21$.

   21

5. Since the observed value $T = \underline{\qquad}$ falls in the rejection region, we (reject, do not reject) H_0 and conclude that the review board (does, does not) consider the applicants with teaching experience to be more highly qualified than those without.

   19
   reject
   does

The Minitab command **Stat → Nonparametrics → Mann-Whitney** can be used to implement the Wilcoxon Rank Sum test, and will produce the printout shown below.

Mann-Whitney Confidence Interval and Test

```
Substantial N =   5    Median =     3.000
Limited    N =   7    Median =     9.000
Point estimate for ETA1-ETA2 is     -5.000
96.5 Percent CI for ETA1-ETA2 is (-9.000,-1.003)
W = 19.0
Test of ETA1 = ETA2  vs  ETA1 < ETA2 is significant at 0.0174
```

19; 0.0174
rejection

normal

Minitab reports the Wilcoxon Rank Sum T Statistic as $W =$
_____ and gives its exact *p-value* as _____. This
confirms the (rejection, non-rejection) of H_0 at the 5% significance
level.

Table 7 gives critical values for T for sample sizes of $n_1 \le n_2 =$
3, 4, ..., 15. However, when n_1 and n_2 are large (some researchers
indicate that samples of size $n_1 = n_2 = 4$ is large enough), the distribution of T can be approximated by a _____ distribution

with mean $E(T) = \dfrac{n_1 n_2 (n_1 + n_2 + 1)}{2}$ and variance

$\sigma_T^2 = \dfrac{n_1 n_2 (n_1 + n_2 + 1)}{12}$. Therefore, as a test statistic we can use

$$z = \frac{T - E(T)}{\sigma_T}$$

with the appropriate one- or two-tailed rejection region expressed in
terms of z, the standard normal random variable.

Example 15.4
A manufacturer uses a large amount of a certain chemical. Since
there are just two suppliers of this chemical, the manufacturer wishes
to test whether the percentage of contaminants is the same for the
two sources against the alternative that there is a difference in the
percentages of contaminants for the two suppliers. Data from independent random samples are given below:

Supplier	Contaminant Percentages				
1	.86	.69	.72	.65	1.13
	.65	1.18	.45	1.41	.50
	1.04	.41			
2	.55	.40	.22	.58	.16
	.07	.09	.16	.26	.36
	.20	.15			

Solution
1. We combine the obtained contaminant percentages in a single
    ordered arrangement and identify each percentage by supplier
    number:

Percentage	.07	.09	.15	.16	.16	.20	.22	.26
Rank	1	2	3	4.5	4.5	6	7	8
Supplier	2	2	2	2	2	2	2	2
Percentage	.36	.40	.41	.45	.50	.55	.58	.65
Rank	9	10	11	12	13	14	15	16.5
Supplier	2	2	1	1	1	2	2	1
Percentage	.65	.69	.72	.86	1.04	1.13	1.18	1.41
Rank	16.5	18	19	20	21	22	23	24
Supplier	1	1	1	1	1	1	1	1

2. Since the sample sizes of $n_1 = 12$ and $n_2 = 12$ are large enough, we can use the normal approximation to the distribution of T. The manufacturer, in asking whether there is a difference between the two suppliers, has specified a _____ tailed test. Therefore, we would reject H_0 if T were either too large or too small. (For a two-tailed test using the normal approximation, we are at liberty to use either T_1 or T_1^* as the value of T to be tested.)

two-

3. Calculate

$$T_1 = 11 + 12 + \ldots + 23 + 24 = \underline{\hspace{2cm}}$$

216

$$T_1^* = n_1(n_1 + n_2 + 1) - T_1$$

$$= 12(\underline{\hspace{1.5cm}}) - \underline{\hspace{1.5cm}} = \underline{\hspace{1.5cm}}$$

25; 216; 84

4. Then

$$E(T) = \frac{n_1(n_1 + n_2 + 1)}{2} = \frac{12(\underline{\hspace{1cm}})}{2} = \underline{\hspace{2cm}}$$

25; 150

$$\sigma_T^2 = \frac{n_1 n_2(n_1 + n_2 + 1)}{12} = \frac{12(12)(\underline{\hspace{1cm}})}{12} = \underline{\hspace{2cm}}$$

25; 300

and the test statistic is

$$z = \frac{T_1^* - E(T)}{\sigma_T} = \frac{84 - 150}{\sqrt{300}} = \underline{\hspace{2cm}}$$

−3.81

Hence, we would conclude that there (is, is not) a significant difference in the contaminant percentages for the two suppliers.

is

Had we used T_1 as the value of T, our result would have been

$$z = \frac{216 - 150}{\sqrt{300}} = \frac{66}{17.32} = \underline{\hspace{2cm}}$$

3.81

and we would have arrived at the same conclusion.

Use of the Wilcoxon Rank Sum test eliminates the need for the restrictive assumptions of the Student's t test, which requires that the samples be randomly drawn from _____ populations that have _____ variances.

normal
equal

15.3 A COMPARISON OF STATISTICAL TESTS

The conditions of an experiment are often such that two or more different tests would be valid for testing the hypotheses of interest. How could we compare the efficiencies of two such tests? Statisticians examine the power of a test and use power as a measure of efficiency. The power of a test is defined to be _____. If β is the probability that H_0 is accepted when H_a is true, then the complement of this event, $1 - \beta$, is the probability that H_0 is _____ when H_a is true. Since the object of a statistical test is to reject H_0 when it is

$1 - \beta$

rejected

false

$\underline{\hspace{2cm}}$, $1 - \beta$ represents the probability that the test will perform its designated task.

One method of comparing two tests that use the same sample size and the same significance level (α) is to compare their powers for alternatives of concern to the experimenter. The most common method of comparing two tests is to find the relative efficiency of one test with respect to the other. Since the sample sizes represent a measure of the costs of the tests in question, we would choose the

fewer

test that requires (fewer, more) sample observations to achieve the same level of significance (α) and the same power ($1 - \beta$) as the

more

(more, less) efficient test. If n_A and n_B denote the sample sizes required for tests A and B to achieve the same specified values of α and $1 - \beta$ for a specific alternative hypothesis, then the relative

n_B/n_A

efficiency of test A with respect to test B is $\underline{\hspace{2cm}}$. If this

more

ratio is greater than 1, test A is said to be (more, less) efficient than B.

15.4 THE SIGN TEST FOR A PAIRED EXPERIMENT

signs

The sign test is based on the $\underline{\hspace{2cm}}$ of the observed differences. Thus, in a paired-difference experiment, we may observe in each pair only whether the first element is larger (or smaller) than the second. If the first element is larger (smaller), we assign a plus (minus) sign to the difference. We define the test statistic x to be the

plus

number of $\underline{\hspace{2cm}}$ signs observed.

It is worth emphasizing that the sign test *does not* require a numerical measure of a response but merely a statement of which of two responses in a matched pair is larger. Thus, the sign test is a convenient and even necessary tool in many psychological investigations. If within a given pair it is impossible to tell which response is larger (a tie occurs), the pair is omitted. Thus, if 20 differences are analyzed and 2 of them are impossible to classify as plus or minus,

18

we base our inference on $\underline{\hspace{2cm}}$ (give number) differences.

Let p denote the probability that a difference selected at random from the population of differences would be given a plus sign. If the two population distributions are identical, the probability of a plus

1/2

sign for a given pair would equal $\underline{\hspace{2cm}}$. Then the null hypothesis, "the two populations are identical," could be stated in the form H_0: $p = 1/2$. The test statistic x will have a $\underline{\hspace{2cm}}$

binomial

distribution whether H_0 is true or not. If H_0 is true, then the number of trials n will be the number of pairs in which a difference can be detected, and the probability of success (a plus sign) on a

$p = 1/2$

given trial will be $\underline{\hspace{2cm}}$. If the alternative hypothesis is H_a:

large

$p > 1/2$, then (large, small) values of x would be placed in the rejection region. If the alternative hypothesis is H_a: $p < 1/2$, then

small

(large, small) values of x would be used in the rejection region. With

large

H_a: $p \neq 1/2$, the rejection region would include both $\underline{\hspace{2cm}}$

and _____ values of x. Alternatively, you can calculate the exact *p-value* for the test.

| small |

Example 15.5
Thirty matched pairs of schizophrenic patients were used in an experiment to determine the effect of a certain drug on sociability. In 18 of these pairs, the patient who received the drug was judged to be more sociable, whereas in 5 pairs it was not possible to detect a difference in sociability. Test to determine whether or not this drug tends to increase sociability.

Solution
Let p denote the probability that, in a matched pair selected at random, the patient who received the drug will be more sociable. Further, let x denote the number of pairs in which the drugged patient is more sociable.

1. The null hypothesis that the two populations are identical is stated as H_0: $p =$ _____.

    $1/2$

2. The alternative hypothesis that the drugged patients are more sociable can be written as H_a: _____.

    $p > 1/2$

3. The test statistic is x, the number of responses in which the drugged patient is more sociable out of the $n = 25$ pairs in which a difference is detected. For this problem, $x =$ _____.

    18

4. **Critical value approach:** Using $x = 18, 19, \ldots, 25$ as a rejection region yields the value $\alpha =$ _____. (Use the binomial tables in the text.) If $x = 17, 18, \ldots, 25$ is taken as the rejection region, $\alpha =$ _____. Assuming that $\alpha = .05$ is a satisfactory significance level, we will use the (first, second) rejection region. Since the observed value of x is $x = 18$, we agree to reject H_0 and conclude that the drug tends to increase sociability among schizophrenics.

    .022

    .054
    second

5. **p-value approach:** The *p-value* for this one-tailed test is

    $$p\text{-value} = P(x \geq 18) = 1 - P(x \leq 17)$$

    $$= 1 - .978 = \underline{\hspace{2cm}}$$

    .022

    from Table 1 in Appendix I. The results are statistically significant; H_0 is rejected at the 5% level of significance.

    We observed in Chapter 6 of the text that the normal approximation to binomial probabilities is reasonably accurate when $p = 1/2$ even when n is as small as _____. Thus, the normal distribution can ordinarily be used to approximate α and β for a given rejection region. Furthermore, when n is at least 25, the test can be based on the statistic

    10

    $$\frac{x - (n/2)}{.5\sqrt{n}}$$

which will have approximately the _____ _____ distribution when H_0 is true.

| standard normal |

Example 15.6

The productivity of 35 students was observed and measured both before and after the installation of new lighting in their classroom. The productivity of 21 of the 35 students was observed to have improved, whereas the productivity of the others appeared to show no perceptible gain as a result of the new lighting. It is necessary to determine whether or not the new lighting was effective in increasing student productivity.

Solution

Let p denote the probability that one of the 35 students selected at random exhibits increased productivity after the installation of new lighting. This constitutes a paired-difference test, where the productivity measures are paired on the students. Such pairing tends to block out student variations.

$= 1/2$

$> 1/2$

17.5
8.75

1.645

1.18

would not
has not

1. The null hypothesis is H_0: $p = $ _____.
2. The appropriate one-sided alternative hypothesis is H_a:

   p _____.
3. If x denotes the number of students who showed improved productivity after the installation of the new lighting, then x has a binomial distribution with mean $np = 35(1/2) = $ _____ and variance $npq = 35(1/2)(1/2) = $ _____. Therefore, the test statistic can be taken to be

   $$z = \frac{x - 17.5}{\sqrt{8.75}}$$

4. Reject H_0 at the $\alpha = .05$ level of significance if the calculated value of z is greater than $z_{.05} = $ _____.

Since $x = 21$,

$$z = \frac{21 - 17.5}{\sqrt{8.75}} = \underline{\hspace{2cm}}$$

Hence, we (would, would not) reject H_0; the new lighting (has, has not) improved student productivity.

Self-Correcting Exercises 15A

1. An experiment was designed to compare the durabilities of two highway paints, paint A and paint B, under actual highway conditions. An A strip and a B strip were painted across a highway at each of 30 locations. At the end of the test period, the experimenter observed the following results: At 8 locations paint A showed the least wear; at 17 locations paint B showed the least wear; and at the other 5 locations the paint samples showed the same amount of wear. Can we conclude that paint B is more durable? (Use $\alpha = .05$.)

2. In a deprivation study to test the strength of two physiological drives, ten rats who were fed the same diet according to a feeding schedule were randomly divided into two groups of five rats. Group I was deprived of water for 18 hours and group II was deprived of food for 18 hours. At the end of this time, each rat was put into a maze having the appropriate reward at the exit, and the time required to run the maze was recorded for each rat. The results follow, with time measured in seconds.

Water	Food
16.8	20.8
22.5	24.7
18.2	19.4
13.1	28.9
20.2	25.3

Is there a difference in strength of these two drives as measured by the time required to find the incentive reward. Use the Wilcoxon Rank Sum test with $a = .05$.

3. Rootstock of varieties A and B was tested for resistance to nematode intrusion. An A and a B were planted side by side in each of ten widely separated locations. At the conclusion of the experiment, all roots were brought into the laboratory for a nematode count. The results are recorded in the following table.

	Location									
	1	2	3	4	5	6	7	8	9	10
Variety A	463	268	871	730	474	432	538	305	173	592
Variety B	277	130	522	610	482	340	319	266	205	540

Can it be said that varieties A and B differ in their resistance to nematode intrusion? Use a two-tailed sign test.

4. The score on a certain psychological test, P, is used as an index of status frustration. The scale ranges from $P = 0$ (low frustration) to $P = 10$ (high frustration). This test was administered to independent random samples of seven women and eight men with the following results:

Status Frustration Score							
Women	6	10	3	8	8	7	9
Men	3	5	2	0	3	1	0	4

Use the Wilcoxon Rank Sum statistic with $\alpha \le .05$ to test whether the distribution of status frustration scores is the same for the two groups against the alternative that the status frustration scores are higher among women.

Solutions to SCE 15A

1. Let $p = P(\text{paint } A \text{ shows less wear})$ and $x = $ number of locations where paint A shows less wear. Since no numerical measure of a

response is given, the sign test is appropriate. H_0: $p = 1/2$; H_a: $p < 1/2$. Rejection region: With $n = 25$, $p = 1/2$, reject H_0 if $x \leq 8$ with $\alpha = .054$. (See Table 1e.) Observe $x = 8$; therefore, reject H_0. Paint B is more durable.

2. Rank the times from low to high. Note $n_1 = n_2 = 5$.

Water	Food
2	6
7	8
3	4
1	10
5	9

H_0: no difference in the distributions; H_a: the distributions are different. Rejection region: Use the convention of choosing the smaller of T_1 and T_1^*, so that we are only concerned with the lower portion of the rejection region. Using Table 7, reject H_0 if the smaller of T_1 and T_1^* is less than or equal to 17 with $\alpha/2 = .025$, so that $\alpha = .05$. Calculate $T_1 = 18$, $T_1^* = 5(11) - 18 = 37$. Do not reject H_0.

3. H_0: $p = \frac{1}{2}$, where $p = P(\text{variety } A \text{ exceeds variety } B)$; H_a: $p \neq \frac{1}{2}$.

   Since $x = $ number of plus signs $= 8$, $p\text{-value} = 2P(x \geq 8)$ $= 2(1 - .945) = .110$. Do not reject H_0. We cannot detect a difference between varieties A and B.

4. Rank the scores from low to high. Let $n_1 = 7$, $n_2 = 8$.

Women (1)	Men (2)
6(10)	3(6)
10(15)	5(9)
3(6)	2(4)
8(12.5)	0(1.5)
8(12.5)	3(6)
7(11)	1(3)
9(14)	0(1.5)
	4(8)

H_0: no difference in the distributions; H_a: scores are higher for women. Rejection region: Women have higher ranks if H_a is true, making

$$T_1^* = n_1(n_1 + n_2 + 1) - T_1$$
$$= 7(16) - 81 = 31 \quad \text{small.}$$

Use Table 7(a) with $\alpha \leq .05$ and reject H_0 if $T \leq 41$. Reject H_0.

15.5 THE WILCOXON SIGNED-RANK TEST FOR A PAIRED EXPERIMENT

One previously discussed nonparametric test that may be used for a paired-difference experiment is the _____ test. Although the sign test requires only the direction of the difference within each matched pair, a more efficient test is available if in addition the _____ _____ of the differences can be ranked in order of magnitude. The Wilcoxon signed-rank test uses as a test statistic, T, the (smaller, larger) sum of ranks for differences of the same sign where the differences are ranked in order of their _____ _____. In the calculation of T, zero differences are _____ and ties in the absolute values of nonzero differences are treated in the same manner as prescribed for the _____ test. Critical values of T are given in Table 8.

sign

absolute values

smaller
absolute values
omitted

Wilcoxon Rank
Sum

Example 15.7
Twelve matched pairs of brain-damaged children were formed for an experiment to determine which of two forms of physical therapy is more effective. One child was chosen at random from each pair and treated over a period of several months using therapy A; the other child was treated during this period using therapy B. There was judged to be no difference in two of the matched pairs at the end of the treatment period. The results are summarized in the following table:

Pair	Difference Favorable to Treatment	Rank for the Absolute Value of the Difference
1	A	9
2	A	5
3	B	1.5
4	A	4
5	A	1.5
6	*	*
7	A	7
8	A	8
9	*	*
10	A	10
11	A	6
12	A	3

*Zero difference

For a two-sided test with $\alpha = .05$, we should reject H_0, "treatments equally effective," when $T \leq$ _____. The sample value of T is _____. Hence, we _____ H_0.

8
1.5; reject

It can be shown that the expected value and variance of T are

$$E(T) = \frac{n(n+1)}{4}$$

pairs

$$\sigma_T^2 = \frac{n(n+1)(2n+1)}{24}$$

where n is the number of _____ in the experiment. When n is at least 25, we may use the test statistic

$$z = \frac{T - E(T)}{\sigma_T}$$

standard normal

which will have approximately the _____ _____ distribution when H_0 is true.

Example 15.8
A drug was developed for reducing cholesterol levels in heart patients. The cholesterol levels before and after drug treatment were obtained for a random sample of 25 heart patients, with the following results:

Patient	Cholesterol Level Before	Cholesterol Level After	Patient	Cholesterol Level Before	Cholesterol Level After
1	257	243	13	364	343
2	222	217	14	210	217
3	177	174	15	263	243
4	258	260	16	214	198
5	294	295	17	392	388
6	244	236	18	370	357
7	390	383	19	310	299
8	247	233	20	255	258
9	409	410	21	281	276
10	214	216	22	294	295
11	217	210	23	257	227
12	340	335	24	227	231
			25	385	374

It is necessary to determine whether or not this drug has an effect on the cholesterol levels of heart patients.

Solution
Differences, Before − After, arranged in order of their absolute values are shown below together with the corresponding ranks. Fill in the missing ranks.

Difference	Rank	Difference	Rank	
−1	2	7	14	
−1	2	−7	14	
−1	2	7	14	
−2	4.5	8	16	
−2	4.5	11	_____	17.5
3	6.5	11	_____	17.5
−3	6.5	13	19	
−4	8.5	14	_____	20.5
4	8.5	14	_____	20.5
5	11	16	_____	22
5	11	20	23	
5	11	21	24	
		30	25	

Suppose that the alternative hypothesis of interest to the experimenter is "the drug has the effect of reducing cholesterol levels in heart patients." Thus, the appropriate rejection region for $\alpha = .05$ is $z < -1.645$, where, in calculating z, we take T to be the smaller sum of ranks (the sum of ranks of the _____ differences).   **negative**

When H_0 is true,

$$E(T) = \frac{n(n+1)}{4} = \underline{\hspace{2cm}}$$ **325**

$$\sigma_T^2 = \frac{n(n+1)(2n+1)}{24} = \underline{\hspace{2cm}}$$ **1381.25**

Thus, we will reject H_0 at the $\alpha = .05$ significance level if

$$z = \frac{T - 325}{\sqrt{1381.25}} < -1.645$$

Summing the ranks of the negative differences, we obtain $T =$ _____ and hence $z =$ _____. Comparing z with its critical value, we _____ H_0 in favor of the alternative hypothesis that the drug has the effect of reducing cholesterol levels in heart patients.   **44; −7.56**   **reject**

It is interesting to see what conclusion is obtained by using the sign test. Recall that x is equal to the number of positive differences and that the test statistic

$$z = \frac{x - n/2}{\sqrt{n}/2}$$

has approximately the _____ _____ distribution when n is greater than ten and H_0: $p = 1/2$ is true. With $\alpha = .05$, the rejection region for z is z _____. But $x = 17$, so that $z =$ _____. Thus, we obtain the same conclusion as before, although the sample value of the test statistic does not penetrate as deeply into the rejection region as when the Wilcoxon signed-rank test was used. Since the Wilcoxon signed-rank test makes fuller use of the information   **standard normal**   **> 1.645**   **1.8**

efficient

available in the experiment, we say that the Wilcoxon signed-rank test is more _____ than the sign test.

The MINITAB command **Stat → Nonparametrics → 1 − Sample Wilcoxon**, followed by the column containing the differences to be analyzed, will implement Wilcoxon's signed-rank procedure. The Wilcoxon statistic reported is the rank sum for the positive differences. One-tailed tests are indicated by selecting the alternative "Greater than" for the right-tailed test. The following printout used the data from Example 15.8. Since the sum of the ranks of the positive differences and the sum of the ranks of the negative differences equal $n(n + 1)/2$, the rank sum of the positive differences is

$$T^+ = \frac{25(26)}{2} - 44$$

$$= 325 - 44$$

$$= 281$$

as reported on the printout.

Wilcoxon Signed Rank Test

```
Test of median = 0.000000 versus median  >  0.000000

                 N for   Wilcoxon          Estimated
             N   Test    Statistic    P     Median
Differences 25    25      281.0     0.001    6.000
```

Self-Correcting Exercises 15B

1. The sign test is not as efficient as the Wilcoxon signed rank test for data of the type presented in Exercise 3, Self-Correcting Exercises 15A. Analyze the data of Exercise 3, Self-Correcting Exercises, by using the two-tailed Wilcoxon signed-rank test with $\alpha = .02$. Can it be said that varieties A and B differ in their resistance to nematode intrusion?

2. The sign test is sometimes used as a "quick and easy" substitute for more powerful tests that require lengthy computations. The following differences were obtained in a paired-difference experiment: $-.93$, $.95$, $.52$, $-.26$, $-.75$, $.25$, 1.08, 1.47, $.60$, 1.20, $-.65$, $-.15$, 2.50, 1.22, $.80$, 1.27, 1.46, 3.05, $-.43$, 1.82, $-.56$, 1.08, $-.16$, 2.64. Use the sign test with $\alpha = .05$ to test H_0: $\mu_d = 0$ against the one-sided alternative H_a: $\mu_d > 0$.

3. Refer to Exercise 2. Use the large-sample Wilcoxon signed rank test with $\alpha = .05$ to test H_0: $\mu_d = 0$ against the alternative hypothesis H_a: $\mu_d > 0$. Compare (in efficiency and in computational requirements) the sign test and the Wilcoxon signed rank test as substitute tests in a paired-difference experiment.

Solutions to SCE 15B

1. H_0: no difference in distribution of number of nematodes for varieties A and B; H_a: distribution of a number of nematodes differs for varieties A and B. Rank the absolute differences from smallest to largest and calculate T, the smaller of the two (positive and negative) rank sums.

	Location									
	1	2	3	4	5	6	7	8	9	10
d_i	186	138	349	120	−8	92	219	39	−32	52
Rank $\lvert d_i \rvert$	8	7	10	6	1	5	9	3	2	4

Rejection region: With $\alpha = .02$ and a two-sided test, reject H_0 if $T \leq 5$. Since $T = 1 + 2 = 3$, reject H_0. There is a difference between A and B.

2. H_0: $p = 1/2$, where $p = P(\text{positive difference})$ and $n = 24$; H_a: $p > 1/2$. Using $\alpha = .05$ and the normal approximation, H_0 will be rejected if

$$\frac{x - .5n}{.5\sqrt{n}} > 1.645,$$

where $x =$ number of positive differences. Calculate

$$z = \frac{x - 12}{.5\sqrt{24}} = \frac{16 - 12}{2.45} = 1.63.$$

Do not reject H_0.

3. The ranks of the absolute differences are given along with their corresponding signs: -12, 13, 6, -4, -10, 3, 14.5, 20, 8, 16, -9, -1, 22, 17, 11, 18, 19, 24, -5, 21, -7, 14.5, -2, 23. With $\alpha = .05$ and a one-sided test, reject H_0 if $T \leq 92$. Since $T = 50$, reject H_0. Note that the sign test is computationally simple, but the Wilcoxon signed rank test is more efficient since it allows us to reject H_0 while the sign test did not.

15.6 THE KRUSKAL-WALLIS H TEST FOR COMPLETELY RANDOMIZED DESIGNS

To compare several populations based on independent samples from these populations, the Kruskal-Wallis H test, which uses the rank sums for each sample, is an extension of the Wilcoxon Rank Sum test. The test statistic for comparing k populations is

$$H = \frac{12}{n(n-1)} \sum \frac{T_i^2}{n_i} - 3(n+1)$$

for T_i, the rank sum of the n_i observations in the ith sample, based on the total ranking of all $n = n_1 + n_2 + \cdots + n_k$ observations.

The hypotheses to be tested using the Kruskal-Wallis H test are:

H_0: all k population distributions are identical

H_a: at least one of the k population distributions is different

If H_0 is true and all the samples are being drawn from the same population, then there should be (little, large) variation in the rank sums, $T_1, T_2, \ldots, T_k$. However, if one or more samples are from different populations, the rank sums will exhibit (little, large) variation and the value of H will increase. This statistic then always uses an upper-tailed rejection region. Furthermore, when the null hypothesis is true, the test statistic H has an approximate chi-square distribution with $k-1$ degrees of freedom. Hence, Table 5 in the text can be used to determine the appropriate rejection region for the test.

Example 15.9

Three job training programs were tested on 15 new employees by randomly assigning 5 employees to participate in each program. After completing the programs and having performed on the job for one week, the 15 were ranked according to their ability to perform the task for which they had been trained, with a high rank indicating a low ability.

	Program	
A	B	C
2	6	10
13	7	15
1	9	8
5	3	12
4	11	14

Sums ____ ____ ____

Do these rankings indicate that one program is better than another at the $\alpha = .05$ level of significance?

Solution

1. We are interested in testing whether these three samples of five measurements come from the same population, against the alternative that at least one sample comes from a population different from the others.

2. Since the data are given directly as ranks, we need only calculate the statistic H using $T_1 = 25$, $T_2 = 36$, and $T_3 = 59$ with $n_1 = n_2 = n_3 = 5$ and $n =$ _____ :

$$H = \frac{12}{n(n+1)} \sum \frac{T_i^2}{n_i} - 3(n+1)$$

little

large

25; 36; 59

15

$$= \frac{12}{15(16)}\left(\frac{25^2}{5} + \frac{36^2}{5} + \frac{59^2}{5}\right) - 3(16)$$

$$= \frac{12}{15(16)}\left(\frac{5402}{5}\right) - \underline{\hspace{2cm}} \qquad\qquad 48$$

$$= \underline{\hspace{2cm}} - 48 = \underline{\hspace{2cm}} \qquad\qquad 54.02;\ 6.02$$

3. Using Table 5 with $k - 1 = 2$ degrees of freedom and $\alpha = .05$, we find the rejection region consists of values of $H \geq \chi^2_{.05} = 5.99$. Since the calculated value of H exceeds 5.99, we (reject, do not reject) H_0 reject
and conclude that a difference (exists, does not exist) among the   exists
three job training programs. Although this is all that can be said statistically, by a look at the rank sums for the three programs, program A seems to produce employees with higher job abilities.

Example 15.10
Considered a comparison of the lengths of time required for kindergarten children to assemble a device when the children had been instructed for four different lengths of time. Four children were randomly assigned to each instructional group, but two were eliminated during the experiment because of sickness. The length of time to assemble the device was recorded for each child in the experiment.

Training Periods (hours)			
.5	1.0	1.5	2.0
8 (9.5)	9 (11.5)	4 (1.5)	4 (1.5)
14 (14)	7 (7)	6 (5)	7 (___)
9 (___)	5 (___)	7 (7)	5 (3.5)
12 (13)		8 (___)	

7

11.5; 3.5

9.5

Use the Kruskal-Wallis H test to determine whether the data present sufficient evidence to indicate a difference in the distribution of times for the four different lengths of instructional time. Use $\alpha = .01$.

Solution
1. The data are first ranked according to their magnitude. The ranks of the combined sample of $n = 20$ are shown in parentheses in the table. Fill in the missing entries. The rank sums are

$T_1 = 48$ $\qquad\qquad$ $T_3 = \underline{\hspace{2cm}}$ $\qquad\qquad$ 23

$T_2 = 22$ $\qquad\qquad$ $T_4 = 12$

with $n_1 = n_3 = 4$ and $n_2 = n_4 = 3$.

2. The test statistic is calculated as

$$H = \frac{12}{n(n+1)}\ \sum \frac{T_i^2}{n_i} - 3(n+1)$$

$$= \frac{12}{14(15)}\left(\frac{48^2}{4} + \frac{22^2}{3} + \frac{23^2}{4} + \frac{12^2}{3}\right) - 3(15)$$

52.4333; 7.4333

$$= \underline{\hspace{2cm}} - 45 = \underline{\hspace{2cm}}$$

3. Using Table 5 with $k - 1 = 3$ degrees of freedom and $\alpha = .01$, we find the rejection region is $H \geq \chi^2_{.01} = 11.3449$. Hence, we

do not reject; is not

(reject, do not reject) H_0. There (is, is not) sufficient evidence to indicate that there is a difference in the distributions of times for the four groups.

The Kruskal-Wallis test can also be implemented using the MINITAB command **Stat → Nonparametrics → Kruskal-Wallis**, followed by the column designations for the data and the group subscripts. The following printout used the data from Example 15.10. Notice that the Kruskal-Wallis statistic can be adjusted for any ties that may occur in the rankings. Although we have not discussed this problem, the adjustment has the effect of increasing the value of the statistic and hence decreasing its *p-value*. Differences in the adjusted and unadjusted statistics are small if the number of ties is not large.

Kruskal-Wallis Test

```
Kruskal-Wallis Test on Times

Trts      N     Median    Ave Rank        Z
0.5       4     10.500       12.0       2.55
1.0       3      7.000        7.3      -0.08
1.5       4      6.500        5.7      -0.99
2.0       3      5.000        4.0      -1.63
Overall  14                   7.5

H = 7.43   DF = 3   P = 0.059
H = 7.57   DF = 3   P = 0.056 (adjusted for ties)

* NOTE * One or more small samples
```

Self-Correcting Exercises 15C

1. In Exercise 1, Self-Correcting Exercises 11A, the dressed weights of five chickens fed from birth on one of three rations were recorded. The data are reproduced below.

	Rations	
1	*2*	*3*
7.1	4.9	6.7
6.2	6.6	6.0
7.0	6.8	7.3
5.6	4.6	6.2
6.4	5.3	7.1

Use the Kruskal-Wallis H test to determine whether the data present sufficient evidence to indicate a difference in the distribution of weights for the three rations. Use $\alpha = .05$.

2. In Exercise 2, Self-Correcting Exercises 11A, we considered an investigation of a citizens committee's complaint about the availability of fire protection within the county. The distance in miles to the nearest fire station was measured for each of 5 randomly selected residences in each of four areas. The data are reproduced below.

	Areas		
1	*2*	*3*	*4*
7	1	7	4
5	4	9	6
5	3	8	3
6	4	7	7
8	5	8	5

Suppose that the experimenter was not willing to assume that the distribution of distances for each of the four areas was normal. Use the Kruskal-Wallis test to determine if there is sufficient evidence to indicate a difference in the distributions of distances for the four areas. Use $\alpha = .01$.

Solutions to SCE 15C

1. The data are ranked according to magnitude:

1	*2*	*3*
7.1 (13.5)	4.9 (2)	6.7 (10)
6.2 (6.5)	6.6 (9)	6.0 (5)
7.0 (12)	6.8 (11)	7.3 (15)
5.6 (4)	4.6 (1)	6.2 (6.5)
6.4 (8)	5.3 (3)	7.1 (13.5)
$T_1 = 44$	$T_2 = 26$	$T_3 = 50$

Test statistic:

$$H = \frac{12}{15(16)}\left[\frac{44^2 + 26^2 + 50^2}{5}\right] - 3(16) = 51.12 - 48 = 3.12$$

Rejection region: With $\alpha = .05$ and $df = k - 1 = 2$, reject H_0 if

$$H \geq \chi^2_{.05} = 5.99.$$

Do not reject H_0. There is insufficient evidence to indicate a difference between rations.

2. The data are ranked according to magnitude:

1	2	3	4
7 (14.5)	1 (1)	7 (14.5)	4 (5)
5 (8.5)	4 (5)	9 (20)	6 (11.5)
5 (8.5)	3 (2.5)	8 (18)	3 (2.5)
6 (11.5)	4 (5)	7 (14.5)	7 (14.5)
8 (18)	5 (8.5)	8 (18)	5 (8.5)
$T_1 = 61$	$T_2 = 22$	$T_3 = 85$	$T_4 = 42$

Test statistic:

$$H = \frac{12}{20(21)}\left[\frac{61^2 + 22^2 + 85^2 + 42^2}{5}\right] - 3(21) = 12.39$$

Rejection region: With $\alpha = .01$ and $k - 1 = 3$ degrees of freedom, reject H_0 if $H > \chi^2_{.01} = 11.3449$. Reject H_0. There is evidence to indicate a difference among areas.

15.7 THE FRIEDMAN F_r TEST FOR RANDOMIZED BLOCK DESIGNS

The randomized block design is used in situations where the individual experimental units may be quite different, but where blocks of relatively homogeneous experimental units may be formed. In this situation, the Friedman test provides an extension of the Wilcoxon signed-rank test and makes use of the ranks of the individual measurements. Instead of determining the ranks of the observations with respect to the entire data set, as was done in the case of the Kruskal-Wallis test, the ranks of the observations are determined within each block. Tied measurements in the same block are each given the average of the ranks the measurements would have received if there had been slight differences in the observed values.

The hypotheses to be tested are:

H_0: the distributions for the k treatment populations are identical

H_a: at least one of the distributions differs from the other $k - 1$

The Friedman F_r statistic is given by

$$F_r = \frac{12}{bk(k+1)}\left(\sum T_i^2\right) - 3b(k+1)$$

treatments

blocks; rank sums

where k is the number of _____, b is the number of _____, and $T_1, T_2, \ldots, T_k$ are the _____ _____ for the k treatments. Wide discrepancies in the values of $T_1, T_2, \ldots, T_k$ occur if high or low ranks tend to be concentrated in certain treatments, which indicates that the population distributions are not identical. Discrepancies in the values of the rank sums cause

F_r to be large. For this reason, the rejection region consists of
_____ values of F_r. As with the Kruskal-Wallis test, the
the sampling distribution of the statistic F_r is approximately
_____ with _____ degrees of freedom. Thus, for signi-
ficance level α, the null hypothesis that the k treatments have the
same population frequency distributions should be rejected if
$F_r >$ _____ with $k - 1$ degrees of freedom.

large

$\chi^2;\ k - 1$

χ^2_α

Example 15.11
In an applicant screening interview, four applicants were ranked by six
panel members, with the following results:

Panel Member	Applicant			
	A_1	A_2	A_3	A_4
1	3	4	1	2
2	4	2	3	1
3	4	3	2	1
4	3	4	1	2
5	4	3	1	2
6	4	2	1	3

Do these rankings indicate agreement among the panel members with
respect to their ranking of the four applicants at the $\alpha = .05$ level of
significance?

Solution
1. Since the observations are in fact rank orderings, to test the hypothesis that the panel members' rankings represent a random rank assignment versus the alternative that there is "agreement," we need to
   find the rank sum for each applicant:

   $T_1 =$ _____; $T_2 =$ _____;

   $T_3 =$ _____; $T_4 =$ _____

   22; 18

   9; 11

2. The value of F_r is found as

   $$F_r = \frac{12}{bk(k+1)} \sum T_i^2 - 3b(k+1)$$

   $$= \frac{12}{6(4)(5)}(22^2 + 18^2 + 9^2 + 11^2) - 3(6)(5)$$

   $$= \frac{12}{120}(1010) - \underline{\hspace{2cm}} = \underline{\hspace{2cm}}$$

   90; 11

3. From Table 5 with $k - 1 = 3$ degrees of freedom and $\alpha = .05$,
   the rejection region consists of values of $F_r \geq \chi^2_{.05} = 7.81$. Since the
   calculated value of F_r exceeds 7.81, we (reject, do not reject) H_0
   and conclude that a difference (does, does not) exist among the four
   applicants.

   reject

   does

Example 15.12

In a study where the objective was to investigate methods of reducing fatigue among employees whose jobs involved a monotonous assembly procedure, 12 randomly selected employees were asked to perform their usual job under each of three trial conditions. As a measure of fatigue, the experimenter used the number of assembly line stoppages during a four-hour period for each trial condition. Do the following data indicate that employee fatigue as measured by stoppages differs for these three conditions? Use $\alpha = .05$.

Employees	Conditions		
	1	2	3
1	31 (___)	22 (___)	26 (___)
2	20 (2)	15 (1)	23 (3)
3	26 (3)	21 (2)	18 (1)
4	31 (2)	22 (1)	32 (3)
5	12 (___)	16 (___)	18 (___)
6	22 (1)	29 (2)	34 (3)
7	28 (___)	17 (___)	26 (___)
8	15 (3)	9 (1)	12 (2)
9	41 (2)	31 (1)	46 (3)
10	19 (___)	19 (___)	25 (___)
11	31 (1)	34 (2)	41 (3)
12	18 (2)	11 (1)	21 (3)
Rank sums	$T_1 = $ _____	$T_2 = $ _____	$T_3 = $ _____

Marginal answers: 3; 1; 2 — 1; 2; 3 — 3; 1; 2 — 1.5; 1.5; 3 — 24.5; 16.5; 31

Solution

1. In order to use the Friedman test as an alternative to the parametric analysis of variance, we need to rank the responses within employees and then find the rank totals for each of the three conditions. Complete any missing rank entries and find the rank sums in the table.

2. In testing for significant differences among the three conditions, we shall use Friedman's statistic, given as

$$F_r = \frac{12}{bk(k+1)}\left(\sum T_i^2 \right) - 3b(k+1)$$

For $b = $ _____, $k = $ _____, $T_1 = 24.5$, $T_2 = 16.5$, and $T_3 = 31$,

Marginal answer: 12; 3

$$F_r = \frac{12}{(12)(3)(4)}\left((24.5^2) + (16.5)^2 + (31)^2\right) - 3(12)(4)$$

$$= 152.79 - \text{_____} = \text{_____}$$

Marginal answer: 144; 8.79

3. We shall use the chi-square approximation in finding a rejection region. With $k - 1 = $ _____ degrees of freedom, $\chi^2_{.05} = $ _____. Since the observed value of $F_r = 8.79$ is _____ than 5.99, we (reject, do not reject) H_0 and

Marginal answers: 2 — 5.99 — greater; reject

conclude that there (are, are not) significant differences among   are
the three conditions at the $\alpha = .05$ level of significance.

The Minitab command **Stat → Nonparametrics → Friedman** implements the Friedman test procedure when the data are entered as for
the randomized block design. Analyzing the data of Example 15.12 produced the following printout:

Friedman Test

```
Friedman test for Stoppage by Condition blocked by Employee

S = 8.79   DF = 2   P = 0.012
S = 8.98   DF = 2   P = 0.011 (adjusted for ties)

                     Est      Sum of
Condition     N    Median     Ranks
1            12    24.750     24.5
2            12    19.917     16.5
3            12    27.083     31.0

Grand median  =    23.917
```

Self-Correcting Exercises 15D

1. Refer to Exercise 1, Self-Correcting Exercises 11B, in which the
   growth in pounds for pigs fed on four feed additives were compared, using pig litters as blocks. The data are reproduced below.

	Additive			
Litter	*1*	*2*	*3*	*4*
1	78	69	78	85
2	66	64	70	70
3	81	78	72	83
4	76	66	77	74
5	61	66	69	70

   Use Friedman's test to determine whether there is a significant
   difference among the distributions of growth for pigs fed on the
   four additives. Use $\alpha = .01$.

2. In a brand identification experiment involving four brands, 10 subjects in each of 5 geographic areas were asked to listen to an adver-
   vising jingle associated with each of the four brands and identify
   the brand through its jingle. The length of time in seconds until
   correct identification was averaged for the 10 people with the
   following results.

	Brands			
Areas	*1*	*2*	*3*	*4*
1	3.7	3.9	4.2	4.0
2	4.2	4.8	4.6	4.7
3	2.9	3.5	3.0	3.4
4	5.0	5.4	5.0	5.5
5	3.3	4.3	4.1	3.9

Use Friedman's test to determine whether there is a significant difference in the distribution of times among the four brands. Use $\alpha = .05$.

Solutions to SCE 15D

1. The responses within litters are ranked from 1 to 4.

Litter	Additive			
	1	2	3	4
1	2.5	1	2.5	4
2	2	1	3.5	3.5
3	3	2	1	4
4	3	1	4	2
5	1	2	3	4
	$T_1 = 11.5$	$T_2 = 7$	$T_3 = 14$	$T_4 = 17.5$

Test statistic:

$$F_r = \frac{12}{5(4)(5)}(11.5^2 + 7^2 + 14^2 + 17.5^2) - 3(5)(5)$$

$$= 82.02 - 75 = 7.02$$

Rejection region: With $\alpha = .01$ and $df = k - 1 = 3$, reject H_0 if

$$F_r \geq \chi^2_{.01} = = 11.3449$$

Do not reject H_0. There is insufficient evidence to indicate a difference due to additives.

2. The data are ranked within areas from 1 to 4.

Areas	Brands			
	1	2	3	4
1	1	2	4	3
2	1	4	2	3
3	1	4	2	3
4	1.5	3	1.5	4
5	1	4	3	2
	$T_1 = 5.5$	$T_2 = 17$	$T_3 = 12.5$	$T_4 = 15$

Test statistic:

$$F_r = \frac{12}{5(4)(5)}(5.5^2 + 17^2 + 12.5^2 + 15^2) - 3(5)(5)$$

$$= 84.06 - 75 = 9.06$$

Rejection region: With $\alpha = .05$ and $df = k - 1 = 3$, reject H_0 if

$$F_r \geq 7.81$$

Reject H_0. There is a difference among the four brands.

15.8 RANK CORRELATION COEFFICIENT, r_s

The Spearman rank correlation coefficient, r_s, is a numerical measure of
the association between two variables y and x. As implied in the name
of the test statistic, r_s makes use of _____, and hence the exact   ranks
value of numerical measurements on y and x need not be known. Con-
veniently, r_s is computed in exactly the same manner as _____ r
in Chapter 12.

    To determine whether two variables y and x are related, we select
a _____ sample of n experimental units (or items) from the popu-   random
lation of interest. Each of the n items is ranked first according to the
variable x and then according to the variable _____. Thus, for y
each item in the experiment, we obtain two _____. (Tied ranks   ranks
are treated as in other parts of this chapter.) Let x_i and y_i denote the
respective ranks assigned to item i. Then, as in Chapter 12,

$$r_s = \frac{S_{xy}}{\sqrt{S_{xx}S_{yy}}}$$

Example 15.13
An investigator wished to determine whether leadership ability is related to the amount of a certain hormone present in the blood. Six individuals were selected at random from the Junior Chamber of Commerce
in a large city and ranked on the characteristic leadership ability. A
determination of the hormone content for each individual was made
from blood samples. The leadership ranks and hormone measurements
are recorded in the following table. Fill in the missing hormone ranks.
Note that no difference in leadership ability could be detected for
individuals 2 and 5.

Individual	Leadership Ability Rank (y_i)	Hormone Content	Hormone Rank (x_i)	
1	6	131	1	
2	3.5	174	_____	3
3	1	189	_____	5
4	2	200	6	
5	3.5	186	_____	4
6	5	156	_____	2

To calculate r_s we form an auxiliary table. Fill in the missing
quantities.

36
10.5
25
6
3.5
10
21

Individual	y_i	y_i^2	x_i	x_i^2	$x_i y_i$
1	6	___	1	1	6
2	3.5	12.25	3	9	
3	1	1	5		5
4	2	4		36	12
5	___	12.25	4	16	14
6	5	25	2	4	
Total	___	90.50	21	91	57.5

Thus,

$$S_{xy} = 57.5 - \frac{21(21)}{6}$$

−16

$$= \underline{\hspace{2cm}}$$

$$S_{xx} = 91 - \frac{(21)^2}{6}$$

17.5

$$= \underline{\hspace{2cm}}$$

$$S_{yy} = 90.5 - \frac{(21)^2}{6}$$

17

$$= \underline{\hspace{2cm}}$$

Finally,

17

$$r_s = \frac{-16}{\sqrt{(17.5)(\underline{\hspace{1.5cm}})}} = -.93$$

Thus, high leadership ability (reflected in low rank) seems to be associated with greater amounts of hormone in the blood.

no association

The Spearman rank correlation coefficient may be used as a test statistic to test an hypothesis of _____ _____ between two characteristics. Critical values of r_s are given in Table 9 in the text. The tabulated quantities are values of r_0 such that $P(r_s > r_0)$ $= .05, .025, .01,$ or $.005$, as indicated. For a lower-tail test, reject H_0: "no association between the two characteristics" when

−r_0

$r_s < \underline{\hspace{2cm}}$.

Two-tailed tests require doubling the stated values of α, and hence, critical values for two-tailed tests may be read from Table 9 if

.10; .05; .02; .01

$\alpha = \underline{\hspace{2cm}}, \underline{\hspace{2cm}}, \underline{\hspace{2cm}},$ or $\underline{\hspace{2cm}}$.

Example 15.14
Continuing Example 15.13, we may wish to test whether leadership ability is associated with hormone level. If the experimenter had designed the experiment with the objective of demonstrating that low leadership ranks (high leadership abilities) are associated with high

hormone levels, the appropriate test would be (a lower, an upper)-
tail test.

     For $\alpha = .05$, the critical value of r_s is $r_0 = $ _____ . Hence,
we reject H_0 if $r_s < $ _____ . Since the sample value of r_s found
in Example 15.13 (does, does not) fall in the rejection region, we (do,
do not) reject H_0.

a lower

.829
$-.829$
does; do

Self-Correcting Exercises 15E

1. An interviewer was asked to rank seven applicants as to their
   suitability for a given position. The same seven applicants took a
   written examination that was designed to rate an applicant's ability
   to function in the given position. The interviewer's ranking and the
   examination score for each applicant are given below.

Applicant	Interview Rank	Examination Score
1	4	49
2	7	42
3	5	58
4	3	50
5	6	33
6	2	65
7	1	67

   Calculate the value of Spearman's rank correlation coefficient for
   these data. Test for a significant negative rank correlation at the
   $\alpha = .05$ level of significance.

2. A manufacturing plant is considering building a subsidiary plant
   in another location. Nine plant sites are currently under consideration. After considering land and building costs, zoning regulations,
   available local work force, and transportation facilities associated
   with each possible plant site, two corporate executives have independently ranked the nine possible plant sites as follows.

	Site								
	1	2	3	4	5	6	7	8	9
Executive 1	2	7	1	5	3	9	4	8	6
Executive 2	1	4	3	6	2	9	7	8	5

   a. Calculate Spearman's rank correlation coefficient between the
      two sets of rankings.
   b. Is there reason to believe the two executives are in basic agree-
      ment regarding their evaluation of the nine plant sites? (Use
      $\alpha = .05$.)

Solutions to SCE 15E

1. Rank the examination scores, and note that the interview scores are already in rank order.

Interview Rank (x_i)	Exam Rank (y_i)
4	3
7	2
5	5
3	4
6	1
2	6
1	7

$$n = 7 \qquad \sum x_i^2 = 140$$

$$\sum x_i = 28 \qquad \sum y_i^2 = 140$$

$$\sum y_i = 28 \qquad \sum x_i y_i = 88$$

$$r_s = \frac{88 - (28)^2/7}{140 - (28)^2/7} = \frac{-24}{28} = -.857.$$

To test $H_0: \rho_s = 0$; $H_a: \rho_s < 0$, the rejection region is $r_s < -.714$. Hence H_0 is rejected with $\alpha = .05$.

2. a. $\sum x_i = \sum y_i = 45 \qquad\qquad \sum x_i^2 = \sum y_i^2 = 285$

$$\sum x_i y_i = 272$$

$$r_s = \frac{272 - (45)^2/9}{285 - (45)^2/9} = \frac{47}{60} = .783.$$

b. $H_0: \rho_s = 0$; $H_a: \rho_s > 0$. With $\alpha = .05$, reject H_0 if $r_s > .600$. Reject H_0. They are in basic agreement.

KEY CONCEPTS AND FORMULAS

I. *Nonparametric Methods*
    1. These methods can be used when the data cannot be measured on a quantitative scale, or when
    2. The numerical scale of measurement is arbitrarily set by the researcher, or when
    3. The parametric assumptions such as normality or constant variance are seriously violated.
II. *Wilcoxon Rank Sum Test: Independent Random Samples*
    1. Jointly rank the two samples. Designate the smaller sample as sample 1. Then

$$T_1 = \text{Rank sum of sample 1} \qquad T_1^* = n_1(n_1 + n_2 + 1) - T_1$$

2. Use T_1 to test for population 1 to the left of population 2. Use T_1^* to test for population 1 to the right of population 2. Use the smaller of T_1 and T_1^* to test for a difference in the locations of the two populations.
3. Table 7 of Appendix I has critical values for the rejection of H_0.
4. When the sample sizes are large, use the normal approximation:

$$\mu_T = \frac{n_1(n_1 + n_2 + 1)}{2} \qquad\qquad \sigma_T^2 = \frac{n_1 n_2(n_1 + n_2 + 1)}{12}$$

$$z = \frac{T - \mu_T}{\sigma_T}$$

III. *Sign Test for a Paired Experiment*
1. Find x, the number of times that observation A exceeds observation B for a given pair.
2. To test for a difference in two populations, test H_0: $p = .5$ versus a one- or two-tailed alternative.
3. Use Table 1 of Appendix I to calculate the *p-value* for the test.
4. When the sample sizes are large, use the normal approximation:

$$z = \frac{x - .5n}{.5\sqrt{n}}$$

IV. *Wilcoxon Signed-Rank Test: Paired Experiment*
1. Calculate the differences in the paired observations. Rank the absolute values of the differences. Calculate the rank sums T^+ and T^- for the positive and negative differences, respectively. The test statistic T is the smaller of the two rank sums.
2. Table 8 in Appendix I has critical values for the rejection of H_0 for both one- and two-tailed tests.
3. When the sample sizes are large, use the normal approximation:

$$z = \frac{T^+ - [n(n+1)/4]}{\sqrt{[n(n+1)(2n+1)]/24}}$$

V. *Kruskal-Wallis H Test: Completely Randomized Design*
1. Jointly rank the n observations in the k samples. Calculate the rank sums, $T_i = $ rank sum of sample i, and the test statistic

$$H = \frac{12}{n(n+1)} \sum \frac{T_i^2}{n_i} - 3(n + 1)$$

2. If the null hypothesis of equality of distributions is false, H will be unusually large, resulting in a one-tailed test.
3. For sample sizes of five or greater, the rejection region H is based on the chi-square distribution with $(k - 1)$ degrees of freedom.

VI. *The Friedman F_r Test: Randomized Block Design*
1. Rank the responses within each block from 1 to k. Calculate the rank sums, $T_1, T_2, \ldots, T_k$, and the test statistic

$$F_r = \frac{12}{bk(k+1)} \sum T_i^2 - 3b(k+1)$$

2. If the null hypothesis of equality of treatment distributions is false, F_r will be unusually large, resulting in a one-tailed test.
3. For block sizes of five or greater, the rejection region for F_r is based on the chi-square distribution with $(k-1)$ degrees of freedom.

VII. *Spearman's Rank Correlation Coefficient*
1. Rank the responses for the two variables from smallest to largest.
2. Calculate the correlation coefficient for the ranked observations:

$$r_s = \frac{S_{xy}}{\sqrt{S_{xx}S_{yy}}} \quad \text{or} \quad r_s = 1 - \frac{6\sum d_i^2}{n(n^2-1)}$$

if there are no ties.
3. Table 9 in Appendix I gives critical values for rank correlations significantly different from 0.
4. The rank correlation coefficient detects not only significant linear correlation but also any other monotonic relationship between the two variables.

Exercises

1. For each of the following tests, state whether the test would be used for related samples or for independent samples: sign test, Wilcoxon Rank Sum T Test, Wilcoxon signed rank test.

2. About 1.2% of our combat forces in a certain area develop combat fatigue. To find identifying characteristics of men who are predisposed to this breakdown, the level of a certain adrenal chemical was measured in samples from two groups: men who had developed battle fatigue and men who had adjusted readily to combat situations. The following determinations were recorded:

Battle fatigue group	23.35	21.08	22.36	20.24
	21.69	21.54	21.26	20.71
	20.00	23.40	21.43	21.54
	22.21			
Well-adjusted group	21.66	21.85	21.01	20.54
	20.19	19.26	21.16	19.97
	20.40	19.92	20.52	19.78
	21.15			

Use a large-sample, one-tailed, Wilcoxon Rank Sum test with α approximately equal to .05 to test whether the distributions of levels of this chemical are the same in the two groups against the alternative that the mean level is higher in the combat fatigue group.

3. An experiment was designed to determine whether exposure to cigarette smoke has an effect on the length of life of beagle dogs. Twenty beagles of the same age were used in the experiment. The animals were assigned at random to one of two groups. Ten of the dogs were subjected to conditions equivalent to smoking up to 12 cigarettes each day. The other 10 acted as a control group. Recorded in the table is the number of days until death for the dogs in both groups. Since the experiment was concluded when the last of the experimental dogs died, an L is recorded for each of the dogs still living.

Experimental Group	Control Group
45	315
112	474
251	727
340	894
412	L
533	L
712	L
790	L
845	L
974	L

Use the Wilcoxon Rank Sum test and $\alpha \approx .05$ to obtain a one-tailed test of whether the experimental group has a shorter mean life than the control group.

4. The value of r (defined in Chapter 12) for the following data is .636:

x	y
.05	1.08
.14	1.15
.24	1.27
.30	1.33
.47	1.41
.52	1.46
.57	1.54
.61	2.72
.67	4.01
.72	9.63

Calculate r_s for these data. What advantage of r_s was brought out in this example?

5. A ranking of the quarterbacks in the top eight teams of the National Football League was made by polling a number of professional football coaches and sportswriters. This "true ranking" is shown below together with my ranking.

a. Calculate r_s.

b. Do the data provide evidence at the $\alpha = .05$ level of significance to indicate a positive correlation between my ranking and that of the experts?

	Quarterback							
	A	B	C	D	E	F	G	H
True ranking	1	2	3	4	5	6	7	8
My ranking	3	1	4	5	2	8	6	7

6. Eight recent college graduates have interviewed for positions within the marketing department of a large industrial organization. The organization's vice-president for marketing and the personnel director rated each candidate independently on a 0-10 assumed interval scale. Their ratings are shown below. Use a two-tailed sign test to determine if the vice-president and the personnel manager differ in their evaluations of the eight candidates. (Use $\alpha \leq .10$.)

	Candidate							
	1	2	3	4	5	6	7	8
Vice-president	3	7	6	9	7	4	3	8
Personnel manager	2	4	5	5	9	1	6	6

7. Refer to Exercise 6 and analyze the data using the two-tailed Wilcoxon test with α as close as possible to the significance level used in Exercise 6. Explain any differences in conclusions arrived at using the Wilcoxon versus the sign test. Which conclusion should we believe?

8. It is of interest to determine whether the efficiency of a certain machine operator is superior to the efficiency of another. To examine this question, the percentages of defective items produced daily by machine operators A and B are recorded over a period of time. More recorded data are available from operator A due to a recent illness experienced by operator B. The data are shown below.

	Percentage of Output Defective								
Operator A	3	2	7	6	5	5	3	7	4	6
Operator B	6	5	8	4	8	7	9	10		

The plant analyst realizes that distributions of percentages do not follow a normal distribution and is hesitant to use a t test in the analysis. Use the most powerful nonparametric test at your disposal to determine whether operator A produces a lower percentage defective than operator B. (Use $\alpha = .05$.)

9. Daily lost production from three production lines in a manufacturing operation were recorded for a ten-day period.

Day	Line		
	1	2	3
1	15	11	8
2	9	9	6
3	6	8	4
4	7	6	5
5	16	13	9
6	23	25	14
7	12	9	7
8	10	12	9
9	12	10	11
10	16	10	9

a. What type of experimental design has been used?

b. If the assumptions required for the parametric analysis of variance have been violated, use an alternative nonparametric test to determine whether there is a difference in the distributions of lost daily production for the three production lines. Use $\alpha = .01$.

10. Refer to Exercise 3, Chapter 11. Four different treatments were used for maladjusted children, and psychological scores were recorded at the end of a three-month period. The data are reproduced below.

Treatments			
1	2	3	4
112	111	140	101
92	129	121	116
124	102	130	105
89	136	106	126
97	99		119

a. Give the type of design which has been used in this experiment.

b. Use an appropriate nonparametric test to determine whether there is a significant difference in the distribution of psychological scores for the four treatments. Use $\alpha = .05$.

Answers to Exercises

Chapter 1

4. a. bar chart; pie chart
5. a. Relatively mound-shaped.

Character Stem-and-Leaf Display
```
Stem-and-leaf of Ages        N  = 42
Leaf Unit = 1.0

       2     1 89
       7     2 11333
      14     2 5677889
     (9)     3 001222344
      19     3 556789
      13     4 112344
       7     4 69
       5     5 012
       2     5 59
```
   b.

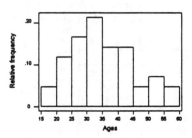

Character Stem-and-Leaf Display
6. a.
```
Stem-and-leaf of Ages        N  = 50
Leaf Unit = 1.0

       5     3 00011
      10     3 22233
      15     3 44555
      22     3 6666677
     (5)     3 88999
      23     4 00001111
      15     4 2233
      11     4 55
       9     4 6667
       5     4 88
       3     5 00
       1     5
       1     5 5
```
   d. no

e. 0.54

f. 0.40   **Character Stem-and-Leaf Display**

7. a. Stem-and-leaf of Rates   N = 32
        Leaf Unit = 0.10

```
         1      5  9
         4      6  005
         6      7  59
         8      8  77
        10      9  01
       (11)    10  01166677888
        11     11  09
         9     12  058
         6     13  3
         5     14  68
         3     15  4
         2     16  2
         1     17
         1     18  4
```

Skewed right; one possible outlier.

c.

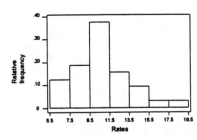

d. 26/32

e. 5/32

Chapter 2

1. a. 27; 20.2; 6.8

   b.

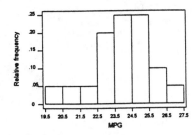

c. 0.8

d. 0.05

e. $\bar{x} = 23.96$; $s = 1.641$

f. $x = 1.85$ and $z = -2.29$. $x = 20.2$ lies more than two standard deviations form the mean and may be an outlier.

g. $m = 24.3$; $Q_1 = 22.95$; $Q_3 = 24.85$

2. no suspect outliers

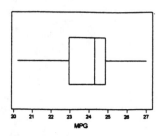

3. a. Range $= 7$

b. $m = 2$; $Q_1 = .5$; $Q_2 = 3$

c. no

d. $\bar{x} = 1.96$; $s^2 = 3.1233$; $s = 1.77$

f. .44, .32

g. No, not bell-shaped

4. a. 16%

b. 81.5%

5. a. 8

b. 2.45 to 2.95; 5.95 to 6.45.

6. $\mu = 6.6$ ounces

7. a. s is approximated as 16.

b. $\bar{x} = 136.07$; $s^2 = 292.4952$; $s = 17.1$.

c. $a = 101.82$; $b = 170.27$.

d. Yes, for approximate calculations.

e. No.

9. a. $x = 172$ is a suspected outlier; $z = 2.10$.

b. Inner fences: 85.5 and 185.5; Outer fences: 48 and 223. No outliers detected.

10. a. s is approximated as 5.

b. $\bar{x} = 11.67$; $s^2 = 13.9523$; $s = 3.74$.

c. 14/15.

12. a. Approximately .974.

b. Approximately .16.

13. Approximately .025.

14. a. Range $= 16$

c. $\bar{x} = 82.8$; $s^2 = 20.0$; $s = 4.472$

d. .68, .96; yes.

Chapter 3

4. a. slightly skewed right
   b. slightly skewed right.
   c. no discernible pattern
   d. $r = -.008$; weak
   e. $y = 124 - 0.013x$
   f. weak; slope is almost zero
5. a. slightly skewed; one unusually large observation
   b. relatively mound-shaped
   c. positive linear relationship
   d. $r = 0.475$; relatively weak
   e. $y = 69.1 + 0.552x$
   f. weak relationship in females; no relationship in males; no, since systolic blood pressure increases with age
6. a. positive linear relationship
   b. $r = .888$; strong positive relationship
   c. $y = -276 + 6.51x$; height increases by 6.51 inches; no; no
7. a. positive linear relationship
   b. $r = .945$; strong positive relationship
   c. only if these 11 states are a representative sample of all states

Chapter 4

1. a. (ABC), (ACB).
   b. (CAB), (ACB).
   c. (ABC), (ACB), (CAB).
   d. (ACB).
   e. $P(A) = 1/3$; $P(A \mid B) = 1/2$; A and B are dependent.
2. a. 1/2.
   b. 1/2.
   c. 5/6.
3. .1792.
4. .459.
5. 56.
6. 6720.
7. a. 5/32.
   b. 31/32.
8.

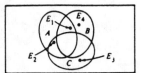

9. a. 1/6.
   b. 2/3; 1/3.
   c. 2/3; 2/3.
   d. 5/6; 5/6.
   e. no; no.

10. a. 56.
    b. 30.
    c. 15/28.
11. .41.
12. a. .328.
    b. .263.
13. .045; yes.
14. a. 14/22; 4/22; 4/22.
    b. 35/44; 8/44; 1/44.
15. a. 99/991.
    b. 892/991.
    c. 3% defective.
16.

x	$p(x)$
0	.1
1	.6
2	.3

17. $p(0) = 3/10$; $p(1) = 6/10$; $p(2) = 1/10$.

18. a.

x	$p(x)$
0	8/27
1	12/27
2	6/27
3	1/27

19. $E(x) = 11{,}250$; $\sigma = 5673.4$.
20. $E(R) = 225{,}000$; $\sigma_R = 113{,}468$.
21. .65.
22. a. $E(x) = 2$; $\sigma^2 = 1$.
23. a. 1/4.
    b. 3/16.
    c. 9/64.
    d. $p(x) = (3/4)^{x-1}(1/4)$.
    e. Yes.
24. a. $p(0) = 6/15$; $p(1) = 8/15$; $p(2) = 1/15$
    b. $E(x) = 10/15$; $\sigma^2 = .3556$
    c. 1/15.
25. $E(x) = 2.125$; $\sigma = .5995$.
26. $E(x) = 2$; $\sigma^2 = 5$.
27. $6.50.
28. $2.70.
29. No.

Chapter 5

1. .655360.
2. a. $p(x) = [20!/x!(20 - x)!](.3)^x(.7)^{20-x}$,
      $x = 0, 1, 2, \ldots, 20$.
   b. .772.
   c. .780.

3. a. 82.
   b. 76.
4. a. .122.
   b. .957.
5. a. n trials are not fixed in advance; not binomial.
   b. Binomial; $p(x) = C_x^5(1/15)^x(14/15)^{5-x}$
   c. Not binomial; p varies from trial to trial.
   d. Binomial; $p(x) = C_x^5(.6)^x(.4)^{5-x}$.
6. 1202; 1246
7. a. 1.96 to 12.24.
   b. $p > .071$.
8. a. .794.
   b. .056.
   c. 0.34 to 2.66.
9. .083
10. a. .214; .214
    b. .316; .211
11. a. .629
    b. .6
12. a. .110803
    b. .012
    c. .119
13. a. $44
    b. $2200
14. .0758

Chapter 6

1. a. .9713.
   b. .1009.
   c. .7257.
   d. .9706.
   e. .8925.
   f. .5917.
2. a. $z_0 = .70$.
   b. $z_0 = 2.13$.
   c. $z_0 = 1.645$.
   d. $z_0 = 1.55$.
3. a. .8413.
   b. .8944.
   c. .9876.
   d. .0401.
4. 87.48
5. $\mu = 10.071$.
6. a. .0139.
   b. .0668.
   c. .5764.
   d. .00000269.

7. a. .048.
    b. .0436.
8. ≈ 0; survival rate is greater than 30%.
9. a. .9838.
    b. .0000.
    c. .8686.
10. .3520.
11. a. .1635.
    b. .0192.
    c. yes, since $P[x \geq 60$ when $p = .2] = .0192$.

Chapter 7

1. .9623
2. .3413
3. .0409.
4. a. $165,000
    b. $2000
5. .9876
6. .0032
7. .0336
8. .0062
9. a. $110 \pm 2(.99)$ or 108 to 112
    b. Approximately 0
10. LCL $= 31.423$; UCL $= 31.977$
11. mean too large at hours 2, 3, and 4.
12. a. $\overline{p} = .052$; LCL $= 0$, UCL $= 188$
    b. Fourth sample indicates p too large.

Chapter 8

1. The inference; measure of goodness.
2. Unbiasedness; minimum variance.
3. 61.23 ± 1.50.
4. $23.705 \pm .012$.
5. $-.06 \pm .047$.
6. $.6 \pm .048$.
7. $2.705 \pm .009$.
8. $21.6 \pm .49$.
9. $.2 \pm .0392$.
10. Approximately 246.
11. Approximately 97.
12. Approximately 97.
13. -8 ± 4.49.
14. 19.3 ± 1.86.
15. $.56 \pm .15$; $-.2475 \pm .22$.

Chapter 9

2. $z = 2.5$; yes.
3. $p\text{-}value = .0062$
4. $z = -5.14$; reject the claim.
5. $p\text{-}value < .002$
6. $z = 2.8$; yes; $p\text{-}value = .0026$
7. $z = -3.40$; yes.
8. a. $z = 5.8$; yes.
   b. $3.0 \pm .85$.
9. $z = 7.30$; no.
10. $p\text{-}value < .001$
11. $z = -4.59$; yes; $p\text{-}value < .002$
12. $z = -1.06$; yes
13. a. $z = .80$; do not reject H_0.
    b. $p\text{-}value = .4238$
14. a. 7.2 ± 1.65
    b. $z = 2.19$; Reject H_0: $\mu = 5$
    c. For $\mu = 6$, $\beta = .6906$, $1 - \beta = .3094$; for $\mu = 7$, $\beta = .2578$, $1 - \beta = .7422$.
    d.

μ	8	9	10	12
$1 - \beta$	.9495	.9957	1.0000	1.0000

    e. $\mu > 8$

Chapter 10

1. According to the Central Limit Theorem, these statistics will be approximately normally distributed for large n.
2. i. The parent population has a normal distribution.
   ii. The sample is a random sample.
3. The number of degrees of freedom associated with a . statistic is the denominator of the estimator of σ^2.
4. Do not reject H_0: $t = -6$.
5. $2.48 < \mu < 4.92$.
6. Do not reject H_0: $t = 1.16$.
7. $-.76 < \mu_1 - \mu_2 < 3.96$.
8. Do not reject H_0: $t = 2.29$.
9. Do not reject H_0: $t = 1.48$.
10. Do not reject H_0: $\chi^2 = 8.19$.
11. $.214 < \sigma^2 < 4.387$.
12. Do not reject H_0: $F = 1.796$.
13. $.565 < \sigma_1^2/\sigma_2^2 < 5.711$.
14. Do not reject H_0: $F = 2.06$.
15. Do not reject H_0: $t = .95$; $p\text{-}value > .10$.
16. a. Reject H_0: $F = 3.88$
    b. Reject H_0: $t^* = 2.13$; $df = 13$.

Chapter 11

1. a. Completely randomized design.

   b.

Source	df	SS	MS	F
Chemicals	2	25.1667	12.5834	2.59
Error	9	43.75	4.8611	
Total	11	68.9167		

   c. No.

   d. -2.25 ± 3.53

   e. 11 ± 2.02.

   f. 39.

2. a. Randomized block design.

   b.

Source	df	SS	MS	F
Applications	3	18.9167	6.3056	9.87
Chemicals	2	62.1667	31.0833	48.65
Error	6	3.8333	0.6288	
Total	11	84.9167		

   c. $F = 48.65$; reject H_0; yes.

   d. 5.25 ± 1.38.

   e. 21.

   f. Yes.

3. a. Completely randomized design.

   b.

Source	df	SS	MS	F
Treatments	3	1052.68	350.89	1.76
Error	15	2997.95	199.86	
Total	18	4050.63		

   c. $F = 1.76$; no.

   d. -12.6 ± 19.06.

4. a.

Source	df	SS	MS	F
Programs	2	25,817.49	12,908.74	4.43
Error	13	37,851.51	2,911.65	
Total	15			

   $F = 4.43$; yes.

   b. -63.1 ± 59.9.

5. a.

Source	df	SS	MS	F
Employees	11	477.8889	43.4444	2.9444
Conditions	2	230.7222	115.3611	7.8184
Error	22	324.6111	14.7551	
Total	35			

   $F = 7.8184 > 3.44$. Reject H_0. Three is a significant difference between conditions.

   b. $F = 2.944$; yes; yes.

   c. $-9.34 < \mu_2 - \mu_3 < 2.84$.

Chapter 12

1.
	y-intercept	slope
a.	-2	3
b.	0	2
c.	-0.5	-1
d.	2.5	-1.5
e.	2	0

2. a. $\hat{y} = .86 + .71x$.

   Analysis of Variance

Source	DF	SS	MS	F
Regression	1	14.286	14.286	125.00
Residual Error	5	0.571	0.114	
Total	6	14.857		

   SSE will be zero only if all of the observed points were to lie on the fitted line.

   d. Reject H_0: $\beta_1 = 0$, since $t = 11.11$.

   e. $.56 < \beta_1 < .86$.

   f. $r = .98$.

   g. Since $r^2 = .96$, the use of the linear model rather than $\bar{y}$ as a predictor for y reduced the sum of squares for error by 96%.

   h. $.82 < y_p < 2.32$.

3. a. $\hat{y} = 8.86 - 1.27x$.

   c. Analysis of Variance

Source	DF	SS	MS	F
Regression	1	113.16	113.16	193.20
Residual Error	4	2.34	0.59	
Total	5	115.50		

   d. Reject H_0: $\beta_1 = 0$, since $t = -13.9$.

   e. $r^2 = .98$; see problem 2g.

   f. $.205 < y < 4.795$

   g. $1.633 < E(y \mid x = 5) < 3.367$

4. If $r = 1$, the observed points all lie on the fitted line having a positive slope and if $r = -1$, the observed points all lie on the fitted line having a negative slope.

5. a. $\hat{y} = 2 - .875x$.

   c. Analysis of Variance

Source	DF	SS	MS	F
Regression	1	12.250	12.250	147.00
Residual Error	3	0.250	0.083	
Total	4	12.500		

   d. Reject H_0: $\beta_1 = 0$, since $t = 12.12$.

   e. $2.01 < E(y \mid x = -1) < 3.74$

   f. $r^2 = .98$; see problem 2g.

   g. $-.77 < y < 3.02$

   h. $\bar{x}$.

6. The fitted line may not adequately describe the relationship between x and y outside the experimental region.

7. The error will be a maximum for the values of x at the extremes of the experimental region.

8. a. $\hat{y} = 7.0 + 15.4x$.

   b. Analysis of Variance

Source	DF	SS	MS	F
Regression	1	2371.6	2371.6	282.33
Residual Error	6	50.4	8.4	
Total	6	2422.0		

   c. Reject H_0: $\beta_1 = 0$, since $t = 16.7$.

   d. $43.0 < E(y \mid x = 2.5) < 48.0$

   e. $r^2 = .979$; see problem 2g.

9. a. $\hat{y} = 6.96 + 2.31x$.

   c. Analysis of Variance

Source	DF	SS	MS	F
Regression	1	45.909	45.909	376.67
Residual Error	8	0.975	0.122	
Total	9	46.884		

   d. Reject H_0: $\beta_1 = 0$, since $t = 19.25$.

   e. $r = .99$.

   f. $r^2 = .979$.

   g. $9.27 \pm .69$; $8.58 < y < 9.96$.

10. a. $\hat{y} = 20.47 - .76x$.

    b. Analysis of Variance

Source	DF	SS	MS	F
Regression	1	287.28	287.28	493.40
Residual Error	8	4.66	0.58	
Total	9	291.94		

    c. Reject H_0: $\beta_1 = 0$, since $t = -22.3$.

    d. $-.86 < \beta_1 < -.66$.

    e. $9.83 \pm .55$; $9.28 < E(y \mid x = 14) < 10.38$.

    f. $r^2 = .984$; see problem 2g.

Chapter 13

1. a. .998

   b. Yes; $F = 1470.84$.

   c. $\hat{y} = -13.01227 + 1.463306x_1$.

   d. $\hat{y} = 23.56$.

   e. no violation of assumptions

2. a. 99.9%.

   b. Yes; $F = 1676.61$.

   c. Yes, $t = -2.65$.

   d. Yes, $t = 15.14$.

   e. 0.3% of the total variation is explained by the linear term.

   f. yes

3. The contributions of x_1 and x_2 are minimal in the presence of x_3 and x_4.

4. a. .922.

   b. $F = 23.78$; reject H_0.

c. $t = -.64$; do not reject H_0.

d. See part c.

Chapter 14

2. a. Yes.

   b. No, since p_i, $i = 1, 2, 3$, changes from trial to trial.

   c. Yes.

3. a. 8.4.

   b. 18.0.

   c. 75.6.

4. Reject H_0: $\chi^2 = 16.535$.

5. Do not reject H_0: $\chi^2 = 2.300$.

6. Reject H_0: $\chi^2 = 7.97$.

7. Do not reject H_0: $p_1 = p_2 = p_3 = p_4$: $\chi^2 = 1.709$.

8. Reject H_0: $\chi^2 = 9.333$.

9. Do not reject the model; $\chi^2 = 6.156$.

Chapter 15

1. Sign test: both; Wilcoxon Rank Sum test; independent; Wilcoxon signed-rank test; related.

2. $z = -2.54 < -1.645$. Reject H_0.

3. Reject H_0 when $T \le 17$. $T = 26$. Do not reject H_0.

4. $r_s = 1$. While $r = 1$ only when the data points all lie on the same straight line, r_s will be 1 whenever y increases steadily with x.

5. a. .738.

   b. $.738 \ge .643$; reject H_0.

6. With $\alpha = .0703$, reject H_0 if $x = 0, 1, 7, 8$. $x = 6$. Do not reject H_0.

7. Reject H_0 if $T \le 6$, $\alpha = .10$. $T = 9.5$. Do not reject H_0.

8. Reject H_0 if $T \le 56$ with $\alpha = .05$. $T = 51.5$. Reject H_0. Operator A produces a lower percentage defective.

9. a. randomized block design

   b. $F_r = 12.5$, reject H_0; there is a difference in the three production lines.

10. a. completely randomized design

    b. $H = 4.86$; do not reject H_0.

Solutions to Selected Text Exercises

1: Describing Data with Graphs

1.1   **a**   The experimental unit, the individual or object on which a variable is measured, is the student.
   b   The experimental unit on which the number of errors is measured is the exam.
   c   The experimental unit is the patient.
   d   The experimental unit is the azalea plant.
   e   The experimental unit is the car.

1.5   The population of interest consists of voter opinions (for or against the candidate) <u>at the time of the election</u> for all persons voting in the election. Note that when a sample is taken (at some time prior to the election), we are not actually sampling from the population of interest. As time passes, voter opinions change. Hence, the population of voter opinions changes with time, and the sample may not be representative of the population of interest.

1.9   **a-b** The experimental unit is the pair of jeans, on which the qualitative variable "state" is measured.
   c-d Construct a statistical table to summarize the data. The pie and bar charts are shown in Figures 1.3 and 1.4.

State	Frequency	Fraction of Total	Sector Angle
CA	9	.36	129.6
AZ	8	.32	115.2
TX	8	.32	115.2

Figure 1.3

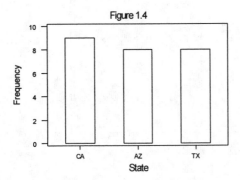

Figure 1.4

e   From the table or the chart, Texas produced $8/25 = .32$ of the jeans.

f   The highest bar represents California, which produced the most pairs of jeans.

g   Since the bars and the sectors are almost equal in size, the three states produced roughly the same number of pairs of jeans.

1.13   **a**   The total percentage of responses given in the table is only $(40 + 34 + 19)\% = 93\%$. Hence there are 7% of the opinions not recorded, which should go into a category called "Other" or "More than a few days".

b   Similar to previous exercises. The pie chart is shown below. The bar chart is probably more interesting to look at.

Figure 1.7

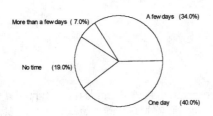

1.17 **a**   Since the variable of interest can only take the values 0, 1, or 2, the classes can be chosen as the integer values 0, 1 and 2. The table below shows the classes, their corresponding frequencies and their relative frequencies. The relative frequency histogram is shown in Figure 1.9.

Value	Frequency	Relative frequency
0	5	.25
1	9	.45
2	6	.30

Figure 1.9

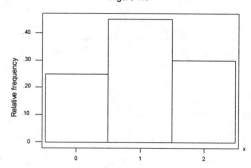

b   Using the table in part **a**, the proportion of measurements greater than 1 is the same as the proportion of "2"s, or .30.

c   The proportion of measurements less than 2 is the same as the proportion of "0"s and "1"s, or .25 + .45 = .70.

d   The probability of selecting a "2" in a random selection from these twenty measurements is 6/20 = .30.

e   There are no outliers in this relatively symmetric, mound-shaped distribution.

1.21 **a**   The test scores are graphed using a stem and leaf plot generated by *Minitab*.
Stem-and-Leaf Display: Scores

```
Stem-and-leaf of Scores    N  = 20
Leaf Unit = 1.0

  2      5 57
  5      6 123
  8      6 578
  9      7 2
 (2)     7 56
  9      8 24
  7      8 6679
  3      9 134
```

b-c   The distribution is not mound-shaped, but is rather bimodal with two peaks centered around the scores 65 and 85. This might indicate that the students are divided into two groups—those who understand the material and do well on exams, and those who do not have a thorough command of the material.

1.25 **a**   The data ranges from .2 to 5.2, or 5.0 units. Since the number of class intervals
should be between five and twenty, we choose to use eleven class intervals, with each
class interval having length .50 ($5.0/11 = .45$, which, rounded to the nearest convenient
fraction, is .50). We must now select interval boundaries such that no measurement can
fall on a boundary point. The subintervals .1 to < .6, .6 to < 1.1, and so on, are
convenient and a tally is constructed. The relative frequency histogram is shown in Figure
1.14.

Class i	Class Boundaries	Tally	f_i	Relative frequency, f_i/n
1	0.1 to < 0.6	11111 11111	10	.167
2	0.6 to < 1.1	11111 11111 11111	15	.250
3	1.1 to <1.6	11111 11111 11111	15	.250
4	1.6 to < 2.1	11111 11111	10	.167
5	2.1 to < 2.6	1111	4	.067
6	2.6 to < 3.1	1	1	.017
7	3.1 to < 3.6	11	2	.033
8	3.6 to < 4.1	1	1	.017
9	4.1 to < 4.6	1	1	.017
10	4.6 to < 5.1		0	.000
11	5.1 to < 5.6	1	1	.017

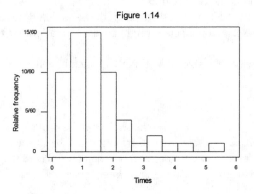

Figure 1.14

a   The distribution is skewed to the right, with several unusually large observations.
b   For some reason, one person had to wait 5.2 minutes. Perhaps the supermarket was
understaffed that day, or there may have been an unusually large number of customers in
the store.
c   The two graphs convey the same information. The stem and leaf plot allows us to
actually recreate the actual data set, while the histogram does not.

1.29   **a**   Histograms will vary from student to student. A typical histogram, generated by *Minitab* is shown below.

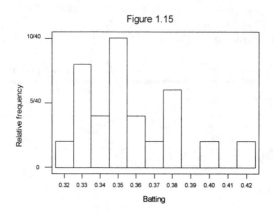

Figure 1.15

b   Since 2 of the 20 players have averages above .400, the chance is 2 out of 20 or $2/20 = .1$.

1.33   To determine whether a distribution is likely to be skewed, look for the likelihood of observing extremely large or extremely small values of the variable of interest.

a   The distribution of non-secured loan sizes might be skewed (a few extremely large loans are possible).

b   The distribution of secured loan sizes is not likely to contain unusually large or small values.

c   Not likely to be skewed.

d   Not likely to be skewed.

e   If a package is dropped, it is likely that all the shells will be broken. Hence, a few large number of broken shells is possible. The distribution will be skewed.

f   If an animal has one tick, he is likely to have more than one. There will be some "0"s with uninfected rabbits, and then a larger number of large values. The distribution will not be symmetric.

1.37   **a**   Stem and leaf displays may vary from student to student. The most obvious choice is to use the
tens digit as the stem and the ones digit as the leaf.

```
 7 | 8  9
 8 | 0  1  7
 9 | 0  1  2  4  4  5  6  6  6  8  8
10| 1  7  9
11| 2
```

The display is fairly mound-shaped, with a large peak in the middle.

1.41   Answers will vary from student to student. The students should notice that the distribution is skewed to the right with a few presidents (Truman, Cleveland, and F.D. Roosevelt) casting an unusually large number of vetoes.

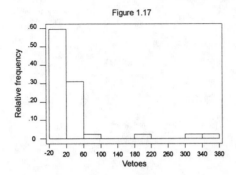

Figure 1.17

1.45   **a**   The popular vote within each state should vary depending on the size of the state. Since there are several very large states (in population) in the United States, the distribution should be skewed to the right.

b-c Histograms will vary from student to student, but should resemble the histogram generated by **Minitab** in Figure 1.24. The distribution is indeed skewed to the right, with two outliers—California and New York.

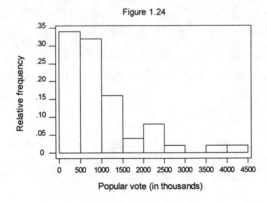

Figure 1.24

1.49   **a-b** Answers will vary from student to student. The line chart should look similar to the one shown in Figure 1.26.

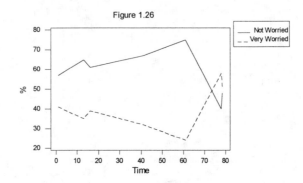

Figure 1.26

c   The percentage of people who were not worried was rising at a slow rate until September 11, 2001, at which time the percentages reversed themselves dramatically.
d   The horizontal axis on the www.gallup.com chart is not an actual time line, so that the time frame in which these changes occur may be distorted.

1.53

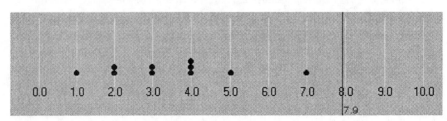

a   The distribution is somewhat mound-shaped (as much as a small set can be); there are no outliers.
b $2/10 = .2$

1.57   **a**   There are a few extremely large numbers, indicating that the distribution is probably skewed to the right.
b-c The distribution is indeed skewed right with three possible outliers—Yahoo!, Microsoft and EBay.

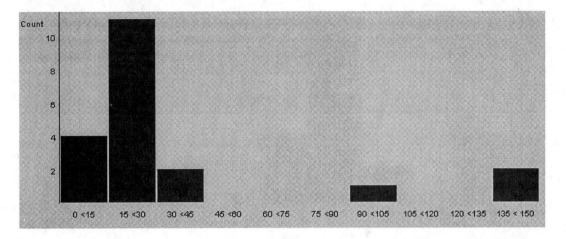

2: Describing Data with Numerical Measures

2.1   **a**   The dotplot shown below plots the five measurements along the horizontal axis. Since there are two "1"s, the corresponding dots are placed one above the other. The approximate center of the data appears to be around 2.

Character Dotplot

```
        .           :                 .                            .
   +------+------+------+------+------+----x
  0.0    1.0    2.0    3.0    4.0    5.0
        median      x̄
        mode
```

b   The mean is the sum of the measurements divided by the number of measurements, or

$$\bar{x} = \frac{\sum x_i}{n} = \frac{0 + 5 + 1 + 1 + 3}{5} = \frac{10}{5} = 2$$

To calculate the median, the observations are first ranked from smallest to largest: 0, 1, 1, 3, 5. Then since n $= 5$, the position of the median is $.5(n+1) = 3$, and the median is the 3rd ranked measurement, or $m = 1$. The mode is the measurement occurring most frequently, or mode $= 1$.

c   The three measures in part **b** are located on the dotplot. Since the median and mode are to the left of the mean, we conclude that the measurements are skewed to the right.

2.5   **a**   Although there may be a few households who own more than one VCR, the majority should own either 0 or 1. The distribution should be slightly skewed to the right.

b   Since most households will have only one VCR, we guess that the mode is 1.

c   The mean is

$$\bar{x} = \frac{\sum x_i}{n} = \frac{1 + 0 + \cdots + 1}{25} = \frac{27}{25} = 1.08$$

To calculate the median, the observations are first ranked from smallest to largest: There are six 0s, thirteen 1s, four 2s, and two 3s. Then since $n = 25$, the position of the median is $.5(n+1) = 13$, which is the 13th ranked measurement, or $m = 1$. The mode is the measurement occurring most frequently, or mode $= 1$.

d   The relative frequency histogram is shown below, with the three measures superimposed. Notice that the mean falls slightly to the right of the median and mode, indicating that the measurements are slightly skewed to the right.

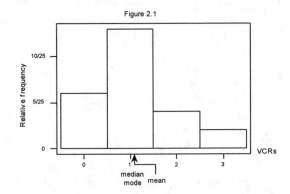

Figure 2.1

2.9   The distribution of sports salaries will be skewed to the right, because of the very high salaries of some sports figures. Hence, the median salary would be a better measure of center than the mean.

2.11   **a** $\bar{x} = \dfrac{\sum x_i}{n} = \dfrac{12}{5} = 2.4$

b   Create a table of differences, $(x_i - \bar{x})$ and their squares, $(x_i - \bar{x})^2$.

x_i	$x_i - \bar{x}$	$(x_i - \bar{x})^2$
2	-0.4	0.16
1	-1.4	1.96
1	-1.4	1.96
3	0.6	0.36
5	2.6	6.76
Total	0	11.20

Then

$$s^2 = \frac{\sum(x_i - \bar{x})^2}{n - 1} = \frac{(2 - 2.4)^2 + \cdots + (5 - 2.4)^2}{4} = \frac{11.20}{4} = 2.8$$

c   The sample standard deviation is the positive square root of the variance or

$$s = \sqrt{s^2} = \sqrt{2.8} = 1.673$$

d   Calculate $\sum x_i^2 = 2^2 + 1^2 + \cdots + 5^2 = 40$. Then

$$s^2 = \frac{\sum x_i^2 - \dfrac{(\sum x_i)^2}{n}}{n - 1} = \frac{40 - \dfrac{(12)^2}{5}}{4} = \frac{11.2}{4} = 2.8 \text{ and } s = \sqrt{s^2} = \sqrt{2.8} = 1.673.$$

The results of parts **a** and **b** are identical.

2.15 **a**   The range of the data is $R = 6 - 1 = 5$ and the range approximation with $n = 10$ is

$$s \approx \frac{R}{3} = 1.67$$

b   The standard deviation of the sample is

$$s = \sqrt{s^2} = \sqrt{\frac{\sum x_i^2 - \frac{(\sum x_i)^2}{n}}{n-1}} = \sqrt{\frac{130 - \frac{(32)^2}{10}}{9}} = \sqrt{3.0667} = 1.751$$

which is very close to the estimate from part **a**.

c-e From the dotplot below, you can see that the data set is not mound-shaped. Hence you can use Tchebysheff's Theorem, but not the Empirical Rule to describe the data.

Character Dotplot

```
     :          :          :          .          :          .
 ---+---------+---------+---------+---------+---------+---x
   1.0        2.0        3.0        4.0        5.0        6.0
```

2.17 **a**   The interval from 40 to 60 represents $\mu \pm \sigma = 50 \pm 10$. Since the distribution is relatively mound-shaped, the proportion of measurements between 40 and 60 is 68% according to the Empirical Rule and is shown in Figure 2.2.

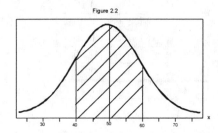

Figure 2.2

b   Again, using the Empirical Rule, the interval $\mu \pm 2\sigma = 50 \pm 2(10)$ or between 30 and 70 contains approximately 95% of the measurements.

c   Refer to Figure 2.3.

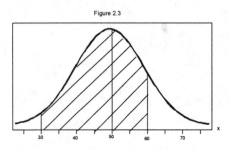

Figure 2.3

Since approximately 68% of the measurements are between 40 and 60, the symmetry of the distribution implies that 34% of the measurements are between 50 and 60. Similarly, since 95% of the measurements are between 30 and 70, approximately 47.5% are between 30 and 50. Thus, the proportion of measurements between 30 and 60 is

$$.34 + .475 = .815$$

d   From Figure 2.2, the proportion of the measurements between 50 and 60 is .34 and the proportion of the measurements which are greater than 50 is .50. Therefore, the proportion which are greater than 60 must be

$$.5 - .34 = .16$$

2.21   According to the Empirical Rule, if a distribution of measurements is approximately mound-shaped,
a   approximately 68% or .68 of the measurements fall in the interval $\mu \pm \sigma = 12 \pm 2.3$ or 9.7 to 14.3.
b   approximately 95% or .95 of the measurements fall in the interval $\mu \pm 2\sigma$ $= 12 \pm 4.6$ or 7.4 to 16.6.
c   approximately 99.7% or .997 of the measurements fall in the interval $\mu \pm 3\sigma = 12 \pm 6.9$ or 5.1 to 18.9. Therefore, approximately 0.3% or .003 will fall outside of this interval.

2.27   **a**   We choose to use 12 classes of length 1.0. The tally and the relative frequency histogram follow.

Class i	Class Boundaries	Tally	f_i	Relative frequency, f_i/n
1	2 to < 3	1	1	1/70
2	3 to < 4	1	1	1/70
3	4 to < 5	111	3	3/70
4	5 to < 6	11111	5	5/70
5	6 to < 7	11111	5	5/70
6	7 to < 8	11111 11111 11	12	12/70
7	8 to < 9	11111 11111 11111 111	18	18/70
8	9 to < 10	11111 11111 11111	15	15/70
9	10 to < 11	11111 1	6	6/70
10	11 to < 12	111	3	3/70
11	12 to < 13		0	0
12	13 to < 14	1	1	1/70

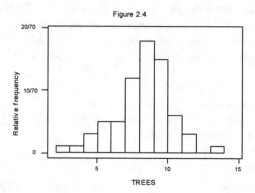

Figure 2.4

b   Calculate $n = 70$, $\sum x_i = 541$ and $\sum x_i^2 = 4453$. Then $\bar{x} = \dfrac{\sum x_i}{n} = \dfrac{541}{70} = 7.729$ is an estimate of μ.

c   The sample standard deviation is

$$s = \sqrt{\frac{\sum x_i^2 - \dfrac{(\sum x_i)^2}{n}}{n-1}} = \sqrt{\frac{4453 - \dfrac{(541)^2}{70}}{69}} = \sqrt{3.9398} = 1.985$$

The three intervals, $\bar{x} \pm ks$ for $k = 1, 2, 3$ are calculated below. The table shows the actual percentage of measurements falling in a particular interval as well as the percentage predicted by Tchebysheff's Theorem and the Empirical Rule. Note that the Empirical Rule should be fairly accurate, as indicated by the mound-shape of the histogram in Figure 2.3.

k	$\bar{x} \pm ks$	Interval	Fraction in Interval	Tchebysheff	Empirical Rule
1	7.729 ± 1.985	5.744 to 9.714	$50/70 = .71$	at least 0	$\approx .68$
2	7.729 ± 3.970	3.759 to 11.699	$67/70 = .96$	at least .75	$\approx .95$
3	7.729 ± 5.955	1.774 to 13.684	$70/70 = 1.00$	at least .89	$\approx .997$

2.31   **a**   The data in this exercise have been arranged in a frequency table.

x_i	0	1	2	3	4	5	6	7	8	9	10
f_i	10	5	3	2	1	1	1	0	0	1	1

Using the frequency table and the grouped formulas, calculate

$$\sum x_i f_i = 0(10) + 1(5) + \cdots + 10(1) = 51$$

$$\sum x_i^2 f_i = 0^2(10) + 1^2(5) + \cdots + 10^2(1) = 293$$

Then

$$\bar{x} = \frac{\sum x_i f_i}{n} = \frac{51}{25} = 2.04$$

$$s^2 = \frac{\sum x_i^2 f_i - \dfrac{\left(\sum x_i f_i\right)^2}{n}}{n-1} = \frac{293 - \dfrac{(51)^2}{25}}{24} = 7.873 \text{ and } s = \sqrt{7.873} = 2.806.$$

b-c The three intervals, $\bar{x} \pm ks$ for $k = 1, 2, 3$ are calculated in the table along with the actual proportion of measurements falling in the intervals. Tchebysheff's Theorem is satisfied and the approximation given by the Empirical Rule are fairly close for $k = 2$ and $k = 3$..

k	$\bar{x} \pm ks$	Interval	Fraction in Interval	Tchebysheff	Empirical Rule
1	2.04 ± 2.806	$-.766$ to 4.846	$21/25 = .84$	at least 0	$\approx .68$
2	2.04 ± 5.612	-3.572 to 7.652	$23/25 = .92$	at least .75	$\approx .95$
3	2.04 ± 8.418	-6.378 to 10.458	$25/25 = 1.00$	at least .89	$\approx .997$

2.35   The ordered data are:

$$2,\ 3,\ 4,\ 5,\ 6,\ 6,\ 6,\ 7,\ 8,\ 9,\ 9,\ 10,\ 22$$

For $n = 13$, the position of the median is $.5(n+1) = .5(13+1) = 7$ and $m = 6$. The positions of the quartiles are $.25(n+1) = 3.5$ and $.75(n+1) = 10.5$, so that $Q_1 = 4.5$, $Q_3 = 9$, and $IQR = 9 - 4.5 = 4.5$.
The *lower and upper fences* are:

$$Q_1 - 1.5IQR = 4.5 - 6.75 = -2.25$$
$$Q_3 + 1.5IQR = 9 + 6.75 = 15.75$$

The value $x = 22$ lies outside the upper fence and is an outlier. The box plot is shown in Figure 2.5. The lower whisker connects the box to the smallest value that is not an outlier, which happens to be the minimum value, $x = 2$. The upper whisker connects the box to the largest value that is not an outlier or $x = 10$.

Figure 2.5

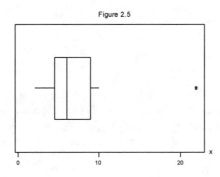

2.39   Since 26% of all adults own five or more pairs of wearable sneakers, 74% of all adults own less than five. The value $x = 5$ represents the 74th percentile. Since 61% of U.S. households have two or more television sets, 39% have less than 2. The value $x = 2$ represents the 39th percentile.

2.43   **a**   The ordered sets are shown below:

Generic					Sunmaid		
24	25	25	25	26	22	24	24
24 24							
26	26	26	26	27	25	25	27
28 28							
27	28	28	28		28	28	29
30							

For $n = 14$, the position of the median is $.5(n + 1) = .5(14 + 1) = 7.5$ and the positions of the quartiles are $.25(n + 1) = 3.75$ and $.75(n + 1) = 11.25$, so that

Generic: $m = 26$, $Q_1 = 25$, $Q_3 = 27.25$, and $IQR = 27.25 - 25 = 2.25$.
Sunmaid: $m = 26$, $Q_1 = 24$, $Q_3 = 28$, and $IQR = 28 - 24 = 4$.

b   **Generic:** *Lower and upper fences* are:
$$Q_1 - 1.5IQR = 25 - 3.375 = 21.625$$
$$Q_3 + 1.5IQR = 27.25 + 3.375 = 30.625$$
Sunmaid: *Lower and upper fences* are:
$$Q_1 - 1.5IQR = 24 - 6 = 18$$
$$Q_3 + 1.5IQR = 28 + 6 = 34$$
The box plots are shown in Figure 2.9. There are no outliers.

Figure 2.9

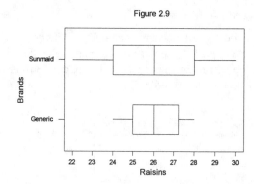

d   If the boxes are not being underfilled, the average size of the raisins is roughly the same for the two brands. However, since the number of raisins is more variable for the Sunmaid brand, it would appear that some of the Sunmaid raisins are large while others are small. The individual sizes of the generic raisins are not as variable.

2.47   The ordered data are shown below. Since $n = 50$, the position of the median is $.5(n + 1) = 25.5$, and the position of the lower and upper quartiles are $.25(n + 1) = 12.75$ and $.75(n + 1) = 38.25$.

0.2	2.0	4.3	8.2	14.7
0.2	2.1	4.4	8.3	16.7
0.3	2.4	5.6	8.7	18.0
0.4	2.4	5.8	9.0	18.0
1.0	2.7	6.1	9.6	18.4
1.2	3.3	6.6	9.9	19.2
1.3	3.5	6.9	11.4	23.1
1.4	3.7	7.4	12.6	24.0
1.6	3.9	7.4	13.5	26.7
1.6	4.1	8.2	14.1	32.3

Then $m = (6.1 + 6.6)/2 = 6.35$, $Q_1 = 2.1 + .75(2.4 - 2.1) = 2.325$, and $Q_3 = 12.6 + .25(13.5 - 12.6) = 12.825$. Then $IQR = 12.825 - 2.325 = 10.5$.

The *lower and upper fences* are:
$$Q_1 - 1.5 IQR = 2.325 - 15.75 = -13.425$$
$$Q_3 + 1.5 IQR = 12.825 + 15.75 = 28.575$$

and the box plot is shown in Figure 2.10. There is one outlier, $x = 32.3$. The distribution is skewed to the right.

Figure 2.10

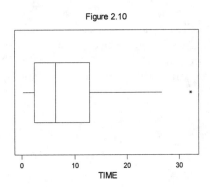

TIME

2.51   The following information is available:
$$n = 400 \quad \bar{x} = 600 \quad s^2 = 4900$$

The standard deviation of these scores is then 70, and the results of Tchebysheff's Theorem follow:

k	$\bar{x} \pm ks$	Interval	Tchebysheff
1	600 ± 70	530 to 670	at least 0
2	600 ± 140	460 to 740	at least .75
3	600 ± 210	390 to 810	at least .89

If the distribution of scores is mound-shaped, we use the Empirical Rule, and conclude that approximately 68% of the scores would lie in the interval 530 to 670 (which is $\bar{x} \pm s$). Approximately 95% of the scores would lie in the interval 460 to 740.

2.55   If the distribution is mound-shaped, then almost all of the measurements will fall in the interval $\mu \pm 3\sigma$, which is an interval 6σ in length. That is, the range of the measurements should be approximately 6σ. In this case, the range is $800 - 200 = 600$, so that $\sigma \approx 600/6 = 100$.

2.61   Notice that three (Sosa, McGwire, Bonds) of the four players have relatively symmetric distributions. The whiskers are the same length and the median line is in the middle of the box. Of the three, Bonds has the least variable distribution, with one unusual year (2001). Other than the year 2001, Bonds fairly consistently has hit between 15 and 50 homeruns per season. The other players have much more variable distributions. The distribution for Babe Ruth is slightly different from the others. The median line to the right of middle indicates a distribution skewed to the left; that there were a few seasons in which his homerun total was unusually low. In fact, the median number of homeruns for the other three players are all about 34-35, while Babe Ruth's median number of homeruns is closer to 40.

2.65   **a-b** As the value of x gets smaller, so does the mean.
c   The median does not change until the green dot is smaller than $x = 10$, at which point the green dot becomes the median.
d   The largest and smallest possible values for the median are $5 \le m \le 10$.

3:Describing Bivariate Data

3.1   **a**   The side-by-side pie charts are constructed as in Chapter 1 for each of the two groups (men and women) and are displayed in Figure 3.1 using the percentages shown in the table below.

	Group 1	Group 2	Group 3	Total
Men	23%	31%	46%	100%
Women	8%	57%	35%	100%

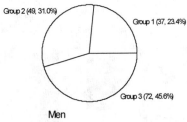

Men

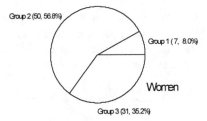

b-c The side-by-side and stacked bar charts in Figures 3.2 and 3.3 measure the frequency of occurrence for each of the three groups. A separate bar (or portion of a bar) is used for men and women.

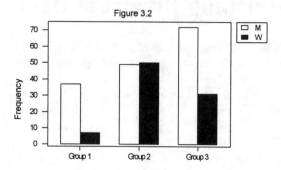

Figure 3.2

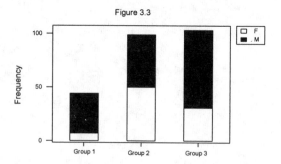

Figure 3.3

d The differences in the proportions of men and women in the three groups is most graphically portrayed by the pie charts, since the unequal number of men and women tend to confuse the interpretation of the bar charts. However, the bar charts are useful in retaining the actual frequencies of occurrence in each group, which is lost in the pie chart.

3.5 **a** The population of interest is the population of responses to the question about free time for all parents and children in the United States. The sample is the set of responses generated for the 198 parents and 200 children in the survey.

b The data can be considered bivariate if, for each person interviewed, we record the person's relationship (Parent or Child) and their response to the question (just the right amount, not enough, too much, don't know). Since the measurements are not numerical in nature, the variables are qualitative.

c The entry in a cell represents the number of people who fell into that relationship-opinion category.

d A pie chart is created for both the "parent" and the "children" categories. The size of each sector angle is proportional to the fraction of measurements falling into that category.

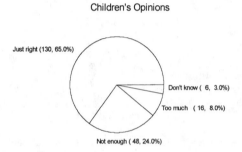

Parents' Opinions   Figure 3.9

Just right (138, 69.7%)

Don't know (6, 3.0%)

Too much (40, 20.2%)

Not enough (14, 7.1%)

Children's Opinions

Just right (130, 65.0%)

Don't know (6, 3.0%)

Too much (16, 8.0%)

Not enough (48, 24.0%)

e   Either stacked or comparative bar charts could be used, but since the height of the bar represents the frequency of occurrence (and hence is tied to the sample size), this type of chart would be misleading. The comparative pie charts are the best choice.

3.9   **a**   Similar to Exercise 3.7. The scatterplot is shown in Figure 3.14.

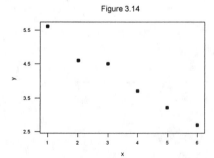

Figure 3.14

b   There appears to be a negative relationship between x and y; that is, as x increases, y decreases.

c   Use your scientific calculator to calculate the sums, sums of squares and sum of cross products for the pairs (x_i, y_i).

$$\sum x_i = 21; \ \sum y_i = 24.3; \ \sum x_i^2 = 91; \ \sum y_i^2 = 103.99; \ \sum x_i y_i = 75.3$$

Then the covariance is

$$s_{xy} = \frac{\sum x_i y_i - \dfrac{(\sum x_i)(\sum y_i)}{n}}{n-1} = \frac{75.3 - \dfrac{21(24.3)}{6}}{5} = -1.95$$

and the sample standard deviations are

$$s_x = \sqrt{\frac{\sum x_i^2 - \dfrac{(\sum x_i)^2}{n}}{n-1}} = \sqrt{\frac{91 - \dfrac{(21)^2}{6}}{5}} = 1.8708 \text{ and}$$

$$s_y = \sqrt{\frac{\sum y_i^2 - \dfrac{(\sum y_i)^2}{n}}{n-1}} = \sqrt{\frac{103.99 - \dfrac{(24.3)^2}{6}}{5}} = 1.0559$$

The correlation coefficient is

$$r = \frac{s_{xy}}{s_x s_y} = \frac{-1.95}{(1.8708)(1.0559)} = -.987$$

This value of r indicates a strong negative relationship between x and y.

3.13   **a-b** The scatterplot is shown in Figure 3.19. There is a slight positive trend between pre- and post-test scores, but the trend is not too pronounced.

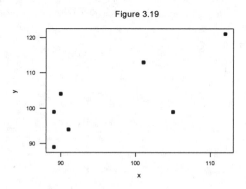

Figure 3.19

c   Calculate $n = 7$; $\sum x_i = 677$; $\sum y_i = 719$; $\sum x_i^2 = 65993$; $\sum y_i^2 = 74585$; $\sum x_i y_i = 70006$. Then the covariance is

$$s_{xy} = \frac{\sum x_i y_i - \dfrac{(\sum x_i)(\sum y_i)}{n}}{n-1} = 78.071429$$

The sample standard deviations are $s_x = 9.286447$ and $s_y = 11.056134$ so that $r = .760$. This is a relatively strong positive correlation, confirming the interpretation of the scatterplot.

3.17   **a**   For each person interviewed in the survey, the following variables are recorded: the number of working parents (quantitative), the monthly expense for a particular type of expense (quantitative) and the category of expense being recorded (qualitative).
b   The population of interest is the population of responses for families with two children in Riverside, San Bernardino, Orange and Ventura, California. Although the source of this data is not given, it is probably based on census information, in which case the data represents the entire population.
c   A comparative (side-by-side) bar chart has been used. An alternative presentation can be obtained by using comparative pie charts, with the monthly expenses divided into seven types of expenses, and compared for one and two working parent families.
d   In order to make the increases look dramatic, the vertical scale should be stretched.

3.23   **a-c** No. There seems to be a large cluster of points in the lower left hand corner showing no apparent relationship between the variables, while 7-10 data points from top left to bottom right show a negative linear relationship.
b   The pattern described in parts **a** and **c** would indicate a weak correlation:

$$r = \frac{s_{xy}}{s_x s_y} = \frac{-92.633}{\sqrt{(708.290)(9346.603)}} = -.036$$

d   Number of waste sites is only slightly affected by the size of the state. Some other possible explanatory variables might be local environmental regulations, population per square mile, or geographical region in the United States.

3.27   **a**   Calculate $n = 8$; $\sum x_i = 451$; $\sum y_i = 555$; $\sum x_i^2 = 29619$; $\sum y_i^2 = 43205$; $\sum x_i y_i = 35082$. Then the covariance is

$$s_{xy} = \frac{\sum x_i y_i - \frac{(\sum x_i)(\sum y_i)}{n}}{n-1} = 541.9821$$

The sample standard deviations are $s_x = 24.4770$ and $s_y = 25.9171$ so that $r = .8544$.
b-c The scatterplot should look like the one shown below. The correlation coefficient should be close to $r = .85$. There is a strong positive trend.

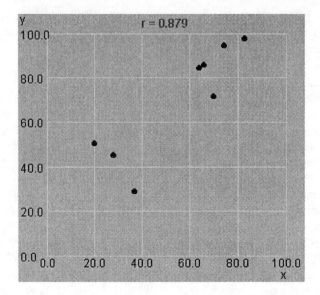

5: Several Useful Discrete Distributions

5.3 **a** $C_2^8 (.3)^2(.7)^6 = \dfrac{8(7)}{2(1)}(.09)(.117649) = .2965$

b $C_0^4(.05)^0(.95)^4 = (.95)^4 = .8145$

c $C_3^{10}(.5)^3(.5)^7 = \dfrac{10(9)(8)}{3(2)(1)} (.5)^{10} = .1172$

d $C_1^7(.2)^1(.8)^6 = 7(.2)(.8)^6 = .3670$

5.7 **a** For $n = 10$ and $p = .4$, $P(x = 4) = C_4^{10}(.4)^4(.6)^6 = .251$.

b To calculate $P(x \geq 4) = p(4) + p(5) + \cdots + p(10)$ it is easiest to write

$$P(x \geq 4) = 1 - P(x < 4) = 1 - P(x \leq 3).$$

These probabilities can be found individually using the binomial formula, or alternatively using the cumulative binomial tables in Appendix I.

$$P(x = 0) = C_0^{10}(.4)^0(.6)^{10} = .006 \quad P(x = 1) = C_1^{10}(.4)^1(.6)^9 = .040$$
$$P(x = 2) = C_2^{10}(.4)^2(.6)^8 = .121 \quad P(x = 3) = C_3^{10}(.4)^3(.6)^7 = .215$$

The sum of these probabilities gives $P(x \leq 3) = .382$ and $P(x \geq 4) = 1 - .382 = .618$.

c Use the results of parts **a** and **b**.

$$P(x > 4) = 1 - P(x \leq 4) = 1 - (.382 + .251) = .367$$

d From part **c**, $P(x \leq 4) = P(x \leq 3) + P(x = 4) = .382 + .251 = .633$.

e $\mu = np = 10(.4) = 4$

f $\sigma = \sqrt{npq} = \sqrt{10(.4)(.6)} = \sqrt{2.4} = 1.549$

5.11 **a** $P[x < 12] = P[x \leq 11] = .748$

b $P[x \leq 6] = .610$

c $P[x > 4] = 1 - P[x \leq 4] = 1 - .633 = .367$

d $P[x \geq 6] = 1 - P[x \leq 5] = 1 - .034 = .966$

e $P[3 < x < 7] = P[x \leq 6] - P[x \leq 3] = .828 - .172 = .656$

5.15 **a** $p(0) = C_0^{20}(.1)^0(.9)^{20} = .1215767 \qquad p(3) = C_3^{20}(.1)^3(.9)^{17} = .1901199$

$p(1) = C_1^{20}(.1)^1(.9)^{19} = .2701703 \qquad p(4) = C_4^{20}(.1)^4(.9)^{16} = .0897788$

$p(2) = C_2^{20}(.1)^2(.9)^{18} = .2851798$

so that

$$P[x \leq 4] = p(0) + p(1) + p(2) + p(3) + p(4) = .9568255$$

b Using Table 1, Appendix I, $P[x \leq 4]$ is read directly as .957.

c Adding the entries for $x = 0, 1, 2, 3, 4$, we have $P[x \leq 4] = .9569$

d $\mu = np = 20(.1) = 2$ and $\sigma = \sqrt{npq} = \sqrt{1.8} = 1.3416$

e For $k = 1$, $\mu \pm \sigma = 2 \pm 1.342$ or .658 to 3.342 so that

$$P[.658 \leq x \leq 3.342] = P[1 \leq x \leq 3] = .2702 + .2852 + .1901 = .7455$$

For $k = 2$, $\mu \pm 2\sigma = 2 \pm 2.683$ or $-.683$ to 4.683 so that

$$P[-.683 \le x \le 4.683] = P[0 \le x \le 4] = .9569$$

For $k = 3$, $\mu \pm 3\sigma = 2 \pm 4.025$ or -2.025 to 6.025 so that

$$P[-2.025 \le x \le 6.025] = P[0 \le x \le 6] = .9977$$

f The results are consistent with Tchebysheff's Theorem and the Empirical Rule.

5.19 Define x to be the number of alarm systems that are triggered. Then $p = P[\text{alarm is triggered}] = .99$ and $n = 9$. Since there is a table available in Appendix I for $n = 9$ and $p = .99$, you should use it rather than the binomial formula to calculate the necessary probabilities.
  a P(at least one alarm is triggered) $= P(x \ge 1) = 1 - P(x = 0) = 1 - .000 = 1.000$.
  b P(more than seven) $= P(x > 7) = 1 - P(x \le 7) = 1 - .003 = .997$
  c P(eight or fewer) $= P(x \le 8) = .086$

5.21 Define x to be the number of cars that are black. Then $p = P[\text{black}] = .1$ and $n = 25$. Use Table 1 in Appendix I.
  a $P(x \ge 5) = 1 - P(x \le 4) = 1 - .902 = .098$
  b $P(x \le 6) = .991$
  c $P(x > 4) = 1 - P(x \le 4) = 1 - .902 = .098$
  d $P(x = 4) = P(x \le 4) - P(x \le 3) = .902 - .764 = .138$
  e $P(3 \le x \le 5) = P(x \le 5) - P(x \le 2) = .967 - .537 = .430$
  f P(more than 20 *not* black) $= $ P(less than 5 black) $= P(x \le 4) = .902$

5.25 Define x to be the number of fields infested with whitefly. Then $p = P[\text{infected field}] = .1$ and $n = 100$.
  a $\mu = np = 100(.1) = 10$
  b Since n is large, this binomial distribution should be fairly mound-shaped, even though $p = .1$. Hence you would expect approximately 95% of the measurements to lie within two standard deviations of the mean with $\sigma = \sqrt{npq} = \sqrt{100(.1)(.9)} = 3$. The limits are calculated as

$$\mu \pm 2\sigma \Rightarrow 10 \pm 6 \text{ or from } 4 \text{ to } 16$$

  c From part **b**, a value of $x = 25$ would be very unlikely, assuming that the characteristics of the binomial experiment are met and that $p = .1$. If this value were actually observed, it might be possible that the trials (fields) are not independent. This could easily be the case, since an infestation in one field might quickly spread to a neighboring field. This is evidence of *contagion*.

5.29 Using $p(x) = \dfrac{\mu^x e^{-\mu}}{x!} = \dfrac{2^x e^{-2}}{x!}$,
  a $P[x = 0] = \dfrac{2^0 e^{-2}}{0!} = .135335$
  b $P[x = 1] = \dfrac{2^1 e^{-2}}{1!} = .27067$
  c $P[x > 1] = 1 - P[x \le 1] = 1 - .135335 - .27067 = .593994$

d $P[x = 5] = \dfrac{2^5 e^{-2}}{5!} = .036089$

5.33   Let x be the number of near misses during a given month. Then x has a Poisson distribution with $\mu = 5$.

a $p(0) = e^{-5} = .0067$ **b** $p(5) = \dfrac{5^5 e^{-5}}{5!} = .1755$

c $P[x \geq 5] = 1 - P[x \leq 4] = 1 - .440 = .560$ from Table 2.

5.37   The random variable x, number of bacteria, has a Poisson distribution with $\mu = 2$. The probability of interest is

$$P[x \text{ exceeds maximum count}] = P[x > 5]$$

Using the fact that $\mu = 2$ and $\sigma = 1.414$ from Exercise 5.37, most of the observations should fall within $\mu \pm 2\sigma$ or 0 to 4. Hence, it is unlikely that x will exceed 5. In fact, the exact Poisson probability is $P[x > 5] = .017$.

5.41   The formula for $p(x)$ is $p(x) = \dfrac{C_x^4 C_{3-x}^{11}}{C_3^{15}}$ for $x = 0, 1, 2, 3$

a $p(0) = \dfrac{C_0^4 C_3^{11}}{C_3^{15}} = \dfrac{165}{455} = .36$ $p(1) = \dfrac{C_1^4 C_2^{11}}{C_3^{15}} = \dfrac{220}{455} = .48$

$p(2) = \dfrac{C_2^4 C_1^{11}}{C_3^{15}} = \dfrac{66}{455} = .15$ $p(3) = \dfrac{C_3^4 C_0^{11}}{C_3^{15}} = \dfrac{4}{455} = .01$

b   The probability histogram is shown in Figure 5.2.

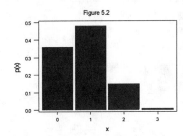

Figure 5.2

c   Using the formulas given in Section 5.4,

$$\mu = E(x) = n\left(\dfrac{M}{N}\right) = 3\left(\dfrac{4}{15}\right) = .8$$

$$\sigma^2 = n\left(\dfrac{M}{N}\right)\left(\dfrac{N-M}{N}\right)\left(\dfrac{N-n}{N-1}\right) = 3\left(\dfrac{4}{15}\right)\left(\dfrac{15-4}{15}\right)\left(\dfrac{15-3}{15-1}\right)$$

$$= .50286$$

d   Calculate the intervals
$$\mu \pm 2\sigma = .8 \pm 2\sqrt{.50286} = .8 \pm 1.418 \text{ or } -.618 \text{ to } 2.218$$
$$\mu \pm 3\sigma = .8 \pm 3\sqrt{.50286} = .8 \pm 2.127 \text{ or } -1.327 \text{ to } 2.927$$

Then,

$$P[-.618 \le x \le 2.218] = p(0) + p(1) + p(2) = .99$$
$$P[-1.327 \le x \le 2.927] = p(0) + p(1) + p(2) = .99$$

These results agree with Tchebysheff's Theorem.

5.45 **a** The random variable x has a hypergeometric distribution with $N = 8$, $M = 5$ and $n = 3$. Then

$$p(x) = \frac{C_x^5 \, C_{3-x}^3}{C_3^8} \quad \text{for } x = 0, 1, 2, 3$$

b $P(x = 3) = \dfrac{C_3^5 \, C_0^3}{C_3^8} = \dfrac{10}{56} = .1786$ **c**

$$P(x = 0) = \frac{C_0^5 \, C_3^3}{C_3^8} = \frac{1}{56} = .01786$$

d $P(x \le 1) = \dfrac{C_0^5 \, C_3^3}{C_3^8} + \dfrac{C_1^5 \, C_2^3}{C_3^8} = \dfrac{1 + 15}{56} = .2857$

5.51 Refer to Exercise 5.50 and assume that $p = .1$ instead of $p = .5$.

a $P[x = 0] = p(0) = C_0^3(.1)^0(.9)^3 = .729$ $P[x = 1] = p(1) = C_1^3(.1)^1(.9)^2 = .243$
$P[x = 2] = p(2) = C_2^3(.1)^2(.9)^1 = .027$ $P[x = 3] = p(3) = C_3^3(.1)^3(.9)^0 = .001$

b Note that the probability distribution is no longer symmetric; that is, since the probability of observing a head is so small, the probability of observing a small number of heads on three flips is increased (see Figure 5.5).

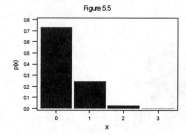

Figure 5.5

c $\mu = np = 3(.1) = .3$ and $\sigma = \sqrt{npq} = \sqrt{3(.1)(.9)} = .520$

d The desired intervals are

$$\mu \pm \sigma = .3 \pm .520 \qquad \text{or} \qquad -.220 \text{ to } .820$$
$$\mu \pm 2\sigma = .3 \pm 1.04 \qquad \text{or} \qquad -.740 \text{ to } 1.34$$

The only value of x which falls in the first interval is $x = 0$, and the fraction of measurements in this interval will be .729. The values $x = 0$ and $x = 1$ are enclosed by the second interval, so that $.729 + .243 = .972$ of the measurements fall within two standard deviations of the mean, consistent with both Tchebysheff's Theorem and the Empirical Rule.

5.55 Refer to Exercise 5.54. Redefine x to be the number of people who choose an interior number in the sample of $n = 20$. Then x has a binomial distribution with $p = .3$.

a $P[x \ge 8] = 1 - P[x \le 7] = 1 - .772 = .228$

b   Observing eight or more people choosing an interior number is not an unlikely event, assuming that the integers are all equally likely. Therefore, there is no evidence to indicate that people are more likely to choose the interior numbers than any others.

5.59   It is given that x = number of patients with a psychosomatic problem, $n = 25$, and $p = P$[patient has psychosomatic problem]. A psychiatrist wishes to determine whether or not $p = .8$.

a   Assuming that the psychiatrist is correct (that is, $p = .8$), the expected value of x is $E(x) = np = 25(.8) = 20$.

b $\sigma^2 = npq = 25(.8)(.2) = 4$

c   Given that $p = .8$, $P[x \le 14] = .006$ from Table 1 in Appendix I.

d   Assuming that the psychiatrist is correct, the probability of observing $x = 14$ or the more unlikely values, $x = 0, 1, 2, \ldots, 13$ is very unlikely. Hence, one of two conclusions can be drawn. Either we have observed a very unlikely event, or the psychiatrist is incorrect and p is actually less than .8. We would probably conclude that the psychiatrist is incorrect. The probability that we have made an incorrect decision is

$$P[x \le 14 \text{ given } p = .8] = .006$$

which is quite small.

5.61   Define x to be the number of students 30 years or older, with $n = 200$ and $p = P$[student is 30+ years] $= .25$.

a   Since x has a binomial distribution, $\mu = np = 200(.25) = 50$ and $\sigma = \sqrt{npq} = \sqrt{200(.25)(.75)} = 6.124$.

b   The observed value, $x = 35$, lies

$$\frac{35 - 50}{6.124} = -2.45$$

standard deviations below the mean. It is unlikely that $p = .25$.

5.65   **a**   The random variable x, the number of plants with red petals, has a binomial distribution with $n = 10$ and $p = P$[red petals] $= .75$.

b   Since the value $p = .75$ is not given in Table 1, you must use the binomial formula to calculate

$$P(x \ge 9) = C_9^{10}(.75)^9(.25)^1 + C_{10}^{10}(.75)^{10}(.25)^0 = .1877 + .0563 = .2440$$

c $P(x \le 1) = C_0^{10}(.75)^0(.25)^{10} + C_1^{10}(.75)^1(.25)^9 = .0000296.$

d   Refer to part **c**. The probability of observing $x = 1$ or something even more unlikely $(x = 0)$ is very small—.0000296. This is a highly unlikely event if in fact $p = .75$. Perhaps there has been a nonrandom choice of seeds, or the 75% figure is not correct for this particular genetic cross.

5.69   **a**   The distribution of x is actually hypergeometric, with $N = 1200$, $n = 20$ and M = number of defectives in the lot. However, since N is so large in comparison to n, the distribution of x can be closely approximated by the binomial distribution with $n = 20$ and $p = P$[defective].

b   If p is small, with $np < 7$, the Poisson approximation can be used.

c   If there are 10 defectives in the lot, then $p = 10/1200 = .008333$ and $\mu = .1667$. The probability that the lot is shipped is

$$P(x = 0) \approx \frac{(.1667)^0 e^{-.1667}}{0!} = .85$$

If there are 20 defectives, $p = 20/1200$ and $\mu = .3333$. Then

$$P(x = 0) \approx \frac{(.3333)^0 e^{-.3333}}{0!} = .72$$

If there are 30 defectives, $p = 30/1200$ and $\mu = .5$. Then

$$P(x = 0) \approx \frac{(.5)^0 e^{-.5}}{0!} = .61$$

5.73   Use the **Calculating Binomial Probabilities** applet. The correct answers are given below.
a $P(x < 6) = 6.0 \, (10)^{-5} = 0.00006$ **d** $P(2 < x < 6) = .5948$
b $P(x = 8) = .042$ **e** $P(x \geq 6) = 1$
c $P(x > 14) = .0207$

5.77   Define x to be the number of young adults who prefer McDonald's. Then x has a binomial distribution with $n = 100$ and $p = .5$. Use the **Calculating Binomial Probabilities** applet.
a $P(61 \leq x \leq 100) = .0176$
b $P(40 \leq x \leq 60) = .9648$
c   If 40 prefer Burger King, then 60 prefer McDonalds, and vice versa. The probability is the same as that calculated in part **b**, since $p = .5$.

6: The Normal Probability Distribution

6.1   The first few exercises are designed to provide practice for the student in evaluating areas under the normal curve. The following notes may be of some assistance.

1   Table 3, Appendix I tabulates the cumulative area under a standard normal curve to the left of a specified value of z.

2   Since the total area under the curve is one, the total area lying to the right of a specified value of z and the total area to its left must add to 1. Thus, in order to calculate a "tail area", such as the one shown in Figure 6.1, the value $z = z_0$ will be indexed in Table 3, and the area that is obtained will be subtracted from 1. Denote the area obtained by indexing $z = z_0$ in Table 3 by $A(z_0)$ and the desired area by A. Then, in the above example, $A = 1 - A(z_0)$.

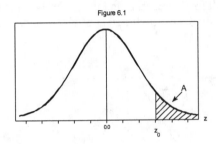

Figure 6.1

3   Note that z, similar to x, is actually a random variable which may take on an infinite number of values, both positive and negative. Negative values of z lie to the left of the mean, $z = 0$, and positive values lie to the right.

a   It is necessary to find the area to the left of $z = 1.6$. That is, $A = A(1.6) = .9452$.
b   The area to the left of $z = 1.83$ is $A = A(1.83) = .9664$.
c $A = A(.90) = .8159$
d $A = A(4.58) \approx 1$. Notice that the values in Table 3 approach 1 as the value of z increases. When the value of z is larger than $z = 3.49$ (the largest value in the table), we can assume that the area to its left is approximately 1.

6.5   Now we are asked to find the z-value corresponding to a particular area.
a   We need to find a z_0 such that $P(z > z_0) = .025$. This is equivalent to finding an indexed area of $1 - .025 = .475$. Search the interior of Table 3 until you find the four-digit number **.9750**. The corresponding z-value is **1.96**; that is, $A(1.96) = .9750$. Therefore, $z_0 = 1.96$ is the desired z-value (see Figure 6.2).

Figure 6.2

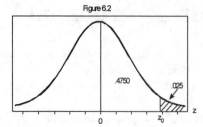

.4750

.025

z_0

0

Figure 6.3

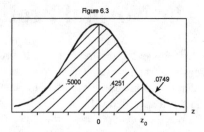

.5000

.4251

.0749

0   z_0

b   We need to find a z_0 such that $P(z < z_0) = .9251$ (see Figure 6.3). Using Table 3, we find a value such that the indexed area is .9251. The corresponding z-value is $z_0 = 1.44$.

6.9   The pth percentile of the standard normal distribution is a value of z which has area $p/100$ to its left. Since all four percentiles in this exercise are greater than the 50th percentile, the values of z will all lie to the right of $z = 0$, as shown for the 90th percentile in Figure 6.7.

Figure 6.7

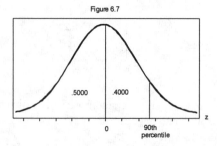

.5000   .4000

0   90th percentile

a   From Figure 6.7, the area to the left of the 90th percentile is .9000. From Table 3, the appropriate value of z is closest to $z = 1.28$ with area .8997. Hence the 90th percentile is approximately $z = 1.28$.
b   As in part **a**, the area to the left of the 95th percentile is .9500. From Table 3, the appropriate value of z is found using linear interpolation (see Exercise 6.7b) as $z = 1.645$. Hence the 95th percentile is $z = 1.645$.
c   The area to the left of the 98th percentile is .9800. From Table 3, the appropriate value of z is closest to $z = 2.05$ with area .9798. Hence the 98th percentile is approximately $z = 2.05$.
d   The area to the left of the 99th percentile is .9900. From Table 3, the appropriate value of z is closest to $z = 2.33$ with area .9901. Hence the 99th percentile is approximately $z = 2.33$.   .

6.13   The 99th percentile of the standard normal distribution was found in Exercise 6.9d to be $z = 2.33$. Since the relationship between the general normal random variable x and the standard normal z is $z = \dfrac{x - \mu}{\sigma}$, the corresponding percentile for this general normal random variable is found by solving for $x = \mu + z\sigma$:

$$2.33 = \frac{x - 35}{10}$$
$$x - 35 = 23.3 \quad \text{or} \quad x = 58.3$$

6.17   The random variable x, the height of a male human, has a normal distribution with $\mu = 69$ and $\sigma = 3.5$.

a   A height of 6' 0" represents $6(12) = 72$ inches, so that

$$P(x > 72) = P\left(z > \frac{72 - 69}{3.5}\right) = P(z > .86) = 1 - .8051 = .1949$$

b   Heights of 5' 8" and 6' 1" represent $5(12) + 8 = 68$ and $6(12) + 1 = 73$ inches, respectively. Then

$$P(68 < x < 73) = P\left(\frac{68 - 69}{3.5} < z < \frac{73 - 69}{3.5}\right) = P(-.29 < z < 1.14) = .8729 - .3859 = .4870$$

c   A height of 6' 0" represents $6(12) = 72$ inches, which has a z-value of

$$z = \frac{72 - 69}{3.5} = .86$$

This would not be considered an unusually large value, since it is less than two standard deviations from the mean.

d   The probability that a man is 6' 0" or taller was found in part **a** to be .1949, which is not an unusual occurrrence. However, if you define y to be the number of men in a random sample of size $n = 36$ who are 6' 0" or taller, then y has a binomial distribution with mean $\mu = np = 36(.1949) = 7.02$ and standard deviation $\sigma = \sqrt{npq} = \sqrt{36(.1949)(.8051)} = 2.38$. The value $y = 17$ lies

$$\frac{y - \mu}{\sigma} = \frac{17 - 7.02}{2.38} = 4.19$$

standard deviations from the mean, and would be considered an unusual occurrence for the general population of male humans. Perhaps our presidents do not represent a *random* sample from this population.

6.21   The random variable x, total weight of 8 people, has a mean of $\mu = 1200$ and a variance $\sigma^2 = 9800$. It is necessary to find $P(x > 1300)$ and $P(x > 1500)$ if the distribution of x is approximately normal. Refer to Figure 6.12.

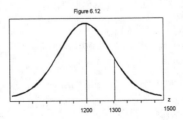

Figure 6.12

The z-value corresponding to $x_1 = 1300$ is

$$z_1 = \frac{x_1 - \mu}{\sigma} = \frac{1300 - 1200}{\sqrt{9800}} = \frac{100}{98.995} = 1.01. \text{ Hence,}$$

$$P(x > 1300) = P(z > 1.01) = 1 - A(1.01) = 1 - .8438 = .1562.$$

Similarly, the z-value corresponding to $x_2 = 1500$ is

$$z_2 = \frac{x_2 - \mu}{\sigma} = \frac{1500 - 1200}{\sqrt{9800}} = 3.03$$

and $\qquad P(x > 1500) = P(z > 3.03) = 1 - A(3.03) = 1 - .9988 = .0012.$

6.25 **a** It is given that the prime interest rate forecasts, x, are approximately normal with mean $\mu = 7.5$ and standard deviation $\sigma = 1.3$. It is necessary to determine the probability that x exceeds 10. Calculate $z = \dfrac{x - \mu}{\sigma} = \dfrac{10 - 7.5}{1.3} = 1.92$. Then

$$P(x > 10) = P(z > 1.92) = 1 - .9726 = .0274.$$

b Calculate $z = \dfrac{x - \mu}{\sigma} = \dfrac{7 - 7.5}{1.3} = -0.38$. Then

$$P(x < 9) = P(z < -0.38) = .3520.$$

6.29 Let w be the number of words specified in the contract. Then x, the number of words in the manuscript, is normally distributed with $\mu = w + 20,000$ and $\sigma = 10,000$. The publisher would like to specify w so that

$$P(x < 100,000) = .95.$$

As in Exercise 6.28, calculate

$$z = \frac{100,000 - (w + 20,000)}{10,000} = \frac{80,000 - w}{10,000}.$$

Then $\qquad P(x < 100,000) = P(z < \dfrac{80,000 - w}{10,000}) = .95.$ It is necessary that $z_0 = (80,000 - w)/10,000$ be such that

$$P(z < z_0) = .95 \quad \Rightarrow \quad A(z_0) = .9500 \quad \text{or} \quad z_0 = 1.645.$$

Hence,

$$\frac{80,000 - w}{10,000} = 1.645 \quad \text{or } w = 63,550.$$

6.31 **a** The normal approximation will be appropriate if both np and nq are greater than 5. For this binomial experiment,

$$np = 25(.3) = 7.5 \quad \text{and} \quad nq = 25(.7) = 17.5$$

and the normal approximation is appropriate.

b   For the binomial random variable, $\mu = np = 7.5$ and
$\sigma = \sqrt{npq} = \sqrt{25(.3)(.7)} = 2.291$.

c   The probability of interest is the area under the binomial probability histogram corresponding to the rectangles $x = 6, 7, 8$ and 9 in Figure 6.14.

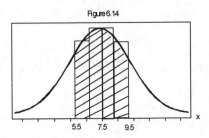

Figure 6.14

To approximate this area, use the "correction for continuity" and find the area under a normal curve with mean $\mu = 7.5$ and $\sigma = 2.291$ between $x_1 = 5.5$ and $x_2 = 9.5$. The z-values corresponding to the two values of x are

$$z_1 = \frac{5.5 - 7.5}{2.291} = -.87 \quad \text{and} \quad z_2 = \frac{9.5 - 7.5}{2.291} = .87$$

The approximating probability is
$P(5.5 < x < 9.5) = P(-.87 < z < .87) = .8078 - .1922 = .6156$.

d   From Table 1, Appendix I,

$$P(6 \le x \le 9) = P(x \le 9) - P(x \le 5) = .811 - .193 = .618$$

which is not too far from the approximate probability calculated in part **c**.

6.35   Using the binomial tables for $n = 20$ and $p = .3$, you can verify that
a $P(x = 5) = P(x \le 5) - P(x \le 4) = .416 - .238 = .178$
b $P(x \ge 7) = 1 - P(x \le 6) = 1 - .608 = .392$

6.39   **a**   The approximating probability will be $P(x > 20.5)$ where x has a normal distribution with $\mu = 50(.32) = 16$ and $\sigma = \sqrt{50(.32)(.68)} = 3.298$. Then

$$P(x > 20.5) = P(z > \frac{20.5 - 16}{3.298}) = P(z > 1.36) = 1 - .9131 = .0869$$

b   The approximating probability is

$$P(x < 14.5) = P(z < \frac{14.5 - 16}{3.298}) = P(z < -.45) = .3264$$

c   If fewer than 28 students *do not* prefer cherry, then more than $50 - 28 = 22$ do prefer cherry. The approximating probability is

$$P(x > 22.5) = P(z > \frac{22.5 - 16}{3.298}) = P(z > 1.97) = 1 - .9756 = .0244$$

d As long as your class can be assumed to be a representative sample of all Americans, the probabilities in parts **a-c** will be accurate.

6.43 Define x to be the number of elections in which the taller candidate won. If Americans are not biased by height, then the random variable x has a binomial distribution with $n = 30$ and $p = .5$. Calculate

$$\mu = np = 31(.5) = 15.5 \text{ and } \sigma = \sqrt{31(.5)(.5)} = \sqrt{7.75} = 2.784$$

a Using the normal approximation with correction for continuity, we find the area to the right of $x = 16.5$:

$$P(x > 16.5) = P\left(z > \frac{16.5 - 15.5}{2.784}\right) = P(z > .36) = 1 - .6406 = .3594$$

b Since the occurence of 17 out of 31 taller choices is not unusual, based on the results of part **a**, it appears that Americans do not consider height when casting a vote for a candidate.

6.47 **a** The desired area A_1, as shown in Figure 6.17, is found by subtracting the cumulative areas corresponding to $z = 1.56$ and $z = 0.3$, respectively.

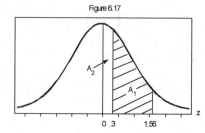

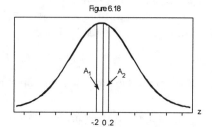

$$A_1 = A(1.56) - A(.3) = .9406 - .6179 = .3227.$$

b The desired area is shown in Figure 6.18:

$$A_1 + A_2 = A(.2) - A(-.2) = .5793 - .4207 = .1586$$

6.51 $P(-z_0 \le z \le z_0) = 2A(z_0) = .5000$. Hence, $A(-z_0) = \frac{1}{2}(1 - .5000) = .2500$. The desired value, z_0, will be between $z_1 = .67$ and $z_2 = .68$ with associated probabilities $P_1 = .2514$ and $P_2 = .2483$. Since the desired tail area, .2500, is closer to $P_1 = .2514$, we approximate z_0 as $z_0 = .67$.   The values $z = -.67$ and $z = .67$ represent the 25th and 75th percentiles of the standard normal distribution.

6.55 It is given that x is normally distributed with $\mu = 10$ and $\sigma = 3$. Let t be the guarantee time for the car. It is necessary that only 5% of the cars fail before time t (see Figure

6.20). That is,

$$P(x < t) = .05 \text{ or } P\left(z < \frac{t - 10}{3}\right) = .05$$

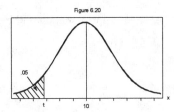

Figure 6.20

From Table 3, we know that the value of z that satisfies the above probability statement is $z = -1.645$. Hence,

$$\frac{t - 10}{3} = -1.645 \text{ or } t = 5.065 \text{ months.}$$

6.59 For this exercise, $\mu = 70$ and $\sigma = 12$. The object is to determine a particular value, x_0, for the random variable x so that $P(x < x_0) = .90$ (that is, 90% of the students will finish the examination before the set time limit). Referring to Figure 6.22 above,

$$P(x < x_0) = P\left(z \le \frac{x_0 - 70}{12}\right) = .90$$

$$A\left(\frac{x_0 - 70}{12}\right) = .90$$

Consider $z_0 = \dfrac{x_0 - 70}{12}$. Without interpolating, the approximate value for z_0 is

$$z_0 = \frac{x_0 - 70}{12} = 1.28 \quad \text{or} \quad x_0 = 85.36$$

6.63 It is given that the random variable x (ounces of fill) is normally distributed with mean μ and standard deviation $\sigma = .3$. It is necessary to find a value of μ so that $P(x > 8) = .01$. That is, an 8-ounce cup will overflow when $x > 8$, and this should happen only 1% of the time. Then,

$$P(x > 8) = P\left(z > \frac{8 - \mu}{.3}\right) = .01.$$

From Table 3, the value of z corresponding to an area (in the upper tail of the distribution) of .01 is $z_0 = 2.33$. Hence, the value of μ can be obtained by solving for μ in the following equation:

$$2.33 = \frac{8 - \mu}{.3} \quad \text{or} \quad \mu = 7.301$$

6.67 Define $x = $ number of incoming calls that are long distance
$p = P[\text{incoming call is long distance}] = .3$

$n = 200$

The desired probability is $P(x \geq 50)$, where x is a binomial random variable with

$$\mu = np = 200(.3) = 60 \text{ and } \sigma = \sqrt{npq} = \sqrt{200(.3)(.7)} = \sqrt{42} = 6.481.$$

A correction for continuity is made to include the entire area under the rectangle corresponding to $x = 50$ and hence the approximation will be

$$P(x \geq 49.5) = P(z \geq \frac{49.5 - 60}{6.481}) = P(z \geq -1.62) = 1 - .0526 = .9474$$

6.71 The random variable x, the gestation time for a human baby is normally distributed with $\mu = 278$ and $\sigma = 12$.

a   From Exercise 6.51, the values (rounded to two decimal places) $z = -.67$ and $z = .67$ represent the 25th and 75th percentiles of the standard normal distribution. Converting these values to their equivalents for the general random variable x using the relationship $x = \mu + z\sigma$, you have:

The lower quartile: $x = -.67(12) + 278 = 269.96$ and
The upper quartile: $x = .67(12) + 278 = 286.04$

b   If you consider a month to be approximately 30 days, the value $x = 6(30) = 180$ is unusual, since it lies

$$z = \frac{x - \mu}{\sigma} = \frac{180 - 278}{12} = -8.167$$

standard deviations below the mean gestation time.

6.75 Use either the **Normal Distribution Probabilities** or the **Normal Probabilities and z-scores** applets.

a $P(-2.0 < z < 2.0) = .9772 - .0228 = .9544$
b $P(-2.3 < z < -1.5) = .0668 - .0107 = .0561$

6.79 **a**   Use the **Normal Distribution Probabilities** applet. Enter 5 as the mean and 2 as the standard deviation, and the appropriate lower and upper boundaries for the probabilities you need to calculate. The probability is read from the applet as **Prob = 0.9651.**
b   Use the **Normal Probabilities and z-scores** applet. Enter 5 as the mean, 2 as the standard deviation and $x = 7.5$. Choose **One-tail** from the dropdown list and read the probability as **Prob = 0.1056.**
c   Use the **Normal Probabilities and z-scores** applet. Enter 5 as the mean, 2 as the standard deviation and $x = 0$. Choose **Cumulative** from the dropdown list and read the probability as **Prob = 0.0062.**

6.83 **a**   It is given that the scores on a national achievement test were approximately normally distributed with a mean of 540 and standard deviation of 110. It is necessary to determine how far, in standard deviations, a score of 680 departs from the mean of 540. Calculate

$$z = \frac{x - \mu}{\sigma} = \frac{680 - 540}{110} = 1.27.$$

b   To find the percentage of people who scored higher than 680, we find the area under the standardized normal curve greater than 1.27. Using Table 3, this area is equal to

$$P(x > 680) = P(z > 1.27) = 1 - .8980 = .1020$$

Thus, approximately 10.2% of the people who took the test scored higher than 680. (The applet uses three decimal place accuracy and shows $z = 1.273$ with **Prob** $= 0.1016$.)

7: Sampling Distributions

7.1   You can select a simple random sample of size $n = 20$ using Table 7 in Appendix I. First choose a starting point and consider the first three digits in each number. Since the experimental units have already been numbered from 000 to 999, the first 20 can be used. The three digits OR the (three digits $- 500$) will identify the proper experimental unit. For example, if the three digits are 742, you should select the experimental unit numbered $742 - 500 = 242$. The probability that any three digit number is selected is $2/1000 = 1/500$. One possible selection for the sample size $n = 20$ is

242	134	173	128	399
056	412	188	255	388
469	244	332	439	101
399	156	028	238	231

7.5   If all of the town citizenry is likely to pass this corner, a sample obtained by selecting every tenth person is probably a fairly random sample.

7.9   Use a randomization scheme similar to that used in Exercise 7.1. Number each of the 50 rats from 01 to 50. To choose the 25 rats who will receive the dose of MX, select 25 two-digit random numbers from Table 7. Each two-digit number OR the (two digits $- 50$) will identify the proper experimental unit.

7.13   Regardless of the shape of the population from which we are sampling, the sampling distribution of the sample mean will have a mean μ equal to the mean of the population from which we are sampling, and a standard deviation equal to $\sigma/\sqrt{n}$.

   a $\mu = 10$; $\sigma/\sqrt{n} = 3/\sqrt{36} = .5$

   b $\mu = 5$; $\sigma/\sqrt{n} = 2/\sqrt{100} = .2$

   c $\mu = 120$; $\sigma/\sqrt{n} = 1/\sqrt{8} = .3536$

7.16-17   For a population with $\sigma = 1$, the standard error of the mean is

$$\sigma/\sqrt{n} = 1/\sqrt{n}$$

The values of $\sigma/\sqrt{n}$ for various values of n are tabulated below and plotted in Figure 7.2. Notice that the standard error *decreases* as the sample size *increases*.

n	1	2	4	9	16	25	100
$SE(\bar{x}) = \sigma/\sqrt{n}$	1.00	.707	.500	.333	.250	.200	.100

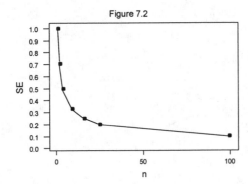

Figure 7.2

7.19 **a** Age of equipment, technician error, technician fatigue, equipment failure, difference in chemical purity, contamination from outside sources, and so on.
b The variability in the average measurement is measured by the standard error, $\sigma/\sqrt{n}$. In order to decrease this variability you should increase the sample size n.

7.23 **a** The population from which we are randomly sampling $n = 35$ measurements is not necessarily normally distributed. However, the sampling distribution of $\bar{x}$ does have an approximate normal distribution, with mean μ and standard deviation $\sigma/\sqrt{n}$. The probability of interest is

$$P(\,|\,\bar{x} - \mu\,|\, < 1) = P(-1 < (\bar{x} - \mu) < 1).$$

Since $z = \dfrac{\bar{x} - \mu}{\sigma/\sqrt{n}}$ has a standard normal distribution, we need only find $\sigma/\sqrt{n}$ to approximate the above probability. Though σ is unknown, it can be approximated by $s = 12$ and $\sigma/\sqrt{n} \approx 12/\sqrt{35} = 2.028$. Then

$$P(\,|\,\bar{x} - \mu\,|\, < 1) = P(-1/2.028 < z < 1/2.028)$$
$$= P(-.49 < z < .49) = .6879 - .3121 = .3758$$

b No. There are many possible values for x, the actual percent tax savings, as given by the probability distribution for x.

7.27 **a** $p = .3;$ $SE(\hat{p}) = \sqrt{\dfrac{pq}{n}} = \sqrt{\dfrac{.3(.7)}{100}} = .0458$

b $p = .1;$ $SE(\hat{p}) = \sqrt{\dfrac{pq}{n}} = \sqrt{\dfrac{.1(.9)}{400}} = .015$

c $p = .6;$ $SE(\hat{p}) = \sqrt{\dfrac{pq}{n}} = \sqrt{\dfrac{.6(.4)}{250}} = .0310$

7.31 The values $SE = \sqrt{pq/n}$ for $n = 100$ and various values of p are tabulated and graphed below. Notice that SE is maximum for $p = .5$ and becomes very small for p near zero and one.

p	.01	.10	.30	.50	.70	.90	.99
$SE(\hat{p})$	.0099	.03	.0458	.05	.0458	.03	.0099

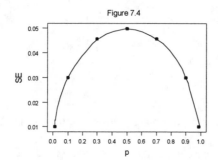

Figure 7.4

7.35   **a**   The random variable $\hat{p}$, the sample proportion of brown M&Ms in a package of $n = 55$, has a binomial distribution with $n = 55$ and $p = .30$. Since $np = 16.5$ and $nq = 38.5$ are both greater than 5, this binomial distribution can be approximated by a normal distribution with mean $p = .30$ and $SE = \sqrt{\dfrac{.30(.70)}{55}} = .06179$.

b $P(\hat{p} < .2) = P(z < \dfrac{.2 - .3}{.06179}) = P(z < -1.62) = .0526$

c $P(\hat{p} > .35) = P(z > \dfrac{.35 - .3}{.06179}) = P(z > .81) = 1 - .7910 = .2090$

d   From the Empirical Rule (and the general properties of the normal distribution), approximately 95% of the measurements will lie within 2 (or 1.96) standard deviations of the mean:

$$p \pm 2SE \ \Rightarrow \ .3 \pm 2(.06179)$$
$$.3 \pm .12 \quad \text{or} \quad .18 \text{ to } .42$$

7.37   Similar to Exercise 7.36.

a   The upper and lower control limits are

$$\text{UCL} = \overline{\overline{x}} + 3\,\frac{s}{\sqrt{n}} = 155.9 + 3\frac{4.3}{\sqrt{5}} = 155.9 + 5.77 = 161.67$$

$$\text{LCL} = \overline{\overline{x}} - 3\,\frac{s}{\sqrt{n}} = 155.9 - 3\frac{4.3}{\sqrt{5}} = 155.9 - 5.77 = 150.13$$

b   The control chart is constructed by plotting two horizontal lines, one the upper control limit and one the lower control limit (see Figure 7.15 in the text). Values of $\overline{x}$ are plotted, and should remain within the control limits. If not, the process should be checked.

7.43   Calculate $\overline{p} = \dfrac{\Sigma \, \widehat{p}_i}{k} = \dfrac{.14 + .21 + \cdots + .26}{30} = .197$. The upper and lower control limits for the p chart are then

$$UCL = \overline{p} + 3\sqrt{\frac{\overline{p}(1-\overline{p})}{n}} = .197 + 3\sqrt{\frac{.197(.803)}{100}} = .197 + .119 = .316$$

$$LCL = \overline{p} - 3\sqrt{\frac{\overline{p}(1-\overline{p})}{n}} = .197 - 3\sqrt{\frac{.197(.803)}{100}} = .197 - .119 = .078$$

7.46   **a** $C_2^4 = \dfrac{4!}{2!2!} = 6$ samples are possible.

b-c The 6 samples along with the sample means for each are shown below.

Sample	Observations	$\overline{x}$
1	6, 1	3.5
2	6, 3	4.5
3	6, 2	4.0
4	1, 3	2.0
5	1, 2	1.5
6	3, 2	2.5

d   Since each of the 6 distinct values of $\overline{x}$ are equally likely (due to random sampling), the sampling distribution of $\overline{x}$ is given as

$$p(\overline{x}) = \frac{1}{6} \quad \text{for} \quad \overline{x} = 1.5, 2, 2.5, 3.5, 4, 4.5$$

The sampling distribution is shown in Figure 7.5.

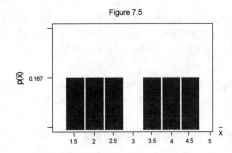

Figure 7.5

e   The population mean is $\mu = (6 + 1 + 3 + 2)/4 = 3$. Notice that none of the samples of size $n = 2$ produce a value of $\overline{x}$ exactly equal to the population mean.

7.51   **a**   To divide a group of 20 people into two groups of 10, use Table 7 in Appendix I. Assign an identification number from 01 to 20 to each person. Then select ten two digit numbers from the random number table to identify the ten people in the first group. (If the number is greater than 20, subtract multiples of 20 from the random number until you obtain a number between 01 and 20.)

b   Although it is not possible to select an actual random sample from this hypothetical population, the researcher must obtain a sample that *behaves like* a random sample. A large database of some sort should be used to ensure a fairly representative sample.

c   The researcher has actually selected a *convenience sample*; however, it will probably behave like a simple random sample, since a person's enthusiasm for a paid job should not affect his response to this psychological experiment.

7.57   **a**   Since each cluster (a city block) is censused, this is an example of cluster sampling.

b   This is a 1-in-10 systematic sample.

c   The wards are the strata, and the sample is a stratified sample.

d   This is a 1-in-10 systematic sample.

e   This is a simple random sample from the population of all tax returns filed in the city of San Bernardino, California.

7.61   **a**   The average proportion of defectives is

$$\bar{p} = \frac{.04 + .02 + \cdots + .03}{25} = .032$$

and the control limits are

$$\text{UCL} = \bar{p} + 3\sqrt{\frac{\bar{p}(1 - \bar{p})}{n}} = .032 + 3\sqrt{\frac{.032(.968)}{100}} = .0848$$

and

$$\text{LCL} = \bar{p} - 3\sqrt{\frac{\bar{p}(1 - \bar{p})}{n}} = .032 - 3\sqrt{\frac{.032(.968)}{100}} = -.0208$$

If subsequent samples do not stay within the limits, UCL $= .0848$ and LCL $= 0$, the process should be checked.

b   From part **a**, we must have $\hat{p} > .0848$.

c   An erroneous conclusion will have occurred if in fact $p < .0848$ and the sample has produced $\hat{p} = .15$ by chance. One can obtain an upper bound on the probability of this particular type of error by calculating $P(\hat{p} \geq .15$ when $p = .0848)$.

7.65   Answers will vary from student to student. Paying cash for opinions will not necessarily produce a random sample of opinions of all Pepsi and Coke drinkers.

7.69   The theoretical mean and standard deviation of the sampling distribution of $\bar{x}$ when $n = 4$ are

$$\mu = 3.5 \quad \text{and} \qquad \sigma/\sqrt{n} = 1.708/\sqrt{4} = .854$$

b-c Answers will vary from student to student. The distribution should be relatively uniform with mean and standard deviation close to those given in part **a**.

8: Large-Sample Estimation

8.1   The margin of error in estimation provides a practical upper bound to the difference between a particular estimate and the parameter which it estimates. In this chapter, the margin of error is $1.96 \times$ (standard error of the estimator).

8.5   The margin of error is $1.96 \, \text{SE} = 1.96 \dfrac{\sigma}{\sqrt{n}}$, where σ can be estimated by the sample standard deviation s for large values of n.

    a $1.96\sqrt{\dfrac{4}{50}} = .554$ **b** $1.96\sqrt{\dfrac{4}{500}} = .175$ **c**

$1.96\sqrt{\dfrac{4}{5000}} = .055$

8.9   For the estimate of p given as $\hat{p} = x/n$, the margin of error is $1.96 \, \text{SE} = 1.96\sqrt{\dfrac{pq}{n}}$. Use the estimated value given in the exercise for p.

    a $1.96\sqrt{\dfrac{(.1)(.9)}{100}} = .0588$ **b** $1.96\sqrt{\dfrac{(.3)(.7)}{100}} = .0898$

    c $1.96\sqrt{\dfrac{(.5)(.5)}{100}} = .098$ **d** $1.96\sqrt{\dfrac{(.7)(.3)}{100}} = .0898$

    e $1.96\sqrt{\dfrac{(.9)(.1)}{100}} = .0588$

    f   The largest margin of error occurs when $p = .5$.

8.13   The point estimate of μ is $\bar{x} = 39.8°$ and the margin of error with $s = 17.2$ and $n = 50$ is

$$1.96 \, \text{SE} = 1.96\frac{\sigma}{\sqrt{n}} \approx 1.96\frac{s}{\sqrt{n}} = 1.96\frac{17.2}{\sqrt{50}} = 4.768$$

8.17   **a**   The point estimate for p is $\hat{p} = \dfrac{x}{n} = .78$ and the margin of error is approximately

$$1.96\sqrt{\frac{\widehat{pq}}{n}} = 1.96\sqrt{\frac{.78(.22)}{1000}} = .026$$

    b   The poll's margin of error does not agree with the results of part **a**, because the sampling error was reported using the maximum margin of error using $p = .5$:

$$1.96\sqrt{\frac{\widehat{pq}}{n}} = 1.96\sqrt{\frac{.5(.5)}{1000}} = .031 \text{ or } \pm 3.1\%$$

8.21 Similar to Exercise 8.20, with a 90% confidence interval for μ given as

$$\bar{x} \pm 1.645 \frac{\sigma}{\sqrt{n}}$$

where σ can be estimated by the sample standard deviation s for large values of n.

a $.84 \pm 1.645\sqrt{\dfrac{.086}{125}} = .84 \pm .043$ or $.797 < \mu < .883$

b $21.9 \pm 1.645\sqrt{\dfrac{3.44}{50}} = 21.9 \pm .431$ or $21.469 < \mu < 22.331$

c   Intervals constructed in this manner will enclose the true value of μ 90% of the time in repeated sampling. Hence, we are fairly confident that these particular intervals will enclose μ.

8.25 The width of a 95% confidence interval for μ is given as $1.96\dfrac{\sigma}{\sqrt{n}}$. Hence,

a   When $n = 100$, the width is $2(1.96\dfrac{10}{\sqrt{100}}) = 2(1.96) = 3.92$.

b   When $n = 200$, the width is $2(1.96\dfrac{10}{\sqrt{200}}) = 2(1.386) = 2.772$.

c   When $n = 400$, the width is $2(1.96\dfrac{10}{\sqrt{400}}) = 2(.98) = 1.96$.

8.29 With $n = 40, \bar{x} = 3.7$ and $s = .5$ and $\alpha = .01$, a 99% confidence interval for μ is approximated by

$$\bar{x} \pm 2.58\frac{s}{\sqrt{n}} = 3.7 \pm 2.58\frac{.5}{\sqrt{40}} = 3.7 \pm .204 \text{ or } 3.496 < \mu < 3.904$$

In repeated sampling, 99% of all intervals constructed in this manner will enclose μ. Hence, we are fairly certain that this particular interval contains μ. (In order for this to be true, the sample must be randomly selected.)

8.33   **a**   The point estimate for p is $\hat{p} = \dfrac{x}{n} = \dfrac{68}{500} = .136$, and the approximate 95% confidence interval for p is

$$\hat{p} \pm 1.96\sqrt{\frac{\hat{p}\hat{q}}{n}} = .136 \pm 1.96\sqrt{\frac{.136(.864)}{500}} = .136 \pm .030$$

or $.106 < p < .166$.

b   In order to increase the accuracy of the confidence interval, you must decrease its width. You can accomplish this by (1) increasing the sample size n, or (2) decreasing $z_{\alpha/2}$ by decreasing the confidence coefficient.

8.35   **a**   When estimating the difference $\mu_1 - \mu_2$, the $(1 - \alpha)100\%$ confidence interval is

$(\bar{x}_1 - \bar{x}_2) \pm z_{\alpha/2}\sqrt{\dfrac{\sigma_1^2}{n_1} + \dfrac{\sigma_2^2}{n_2}}$. Estimating σ_1^2 and σ_2^2 with s_1^2 and s_2^2, the approximate 95%

confidence interval is

$$(12.7 - 7.4) \pm 1.96\sqrt{\frac{1.38}{35} + \frac{4.14}{49}} = 5.3 \pm .690 \quad \text{or} \quad 4.61 < \mu_1 - \mu_2 < 5.99.$$

b   Since the value $\mu_1 - \mu_2 = 0$ is not in the confidence interval, it is not likely that $\mu_1 = \mu_2$. You should conclude that there is a difference in the two population means.

8.39   **a**   The parameter to be estimated is μ, the mean score for the posttest for all BACC classes. The 95% confidence interval is approximately

$$\bar{x} \pm 1.96\frac{s}{\sqrt{n}} = 18.5 \pm 1.96\frac{8.03}{\sqrt{365}} = 18.5 \pm .824 \quad \text{or} \quad 17.676 < \mu < 19.324$$

b   The parameter to be estimated is μ, the mean score for the posttest for all traditional classes. The 95% confidence interval is approximately

$$\bar{x} \pm 1.96\frac{s}{\sqrt{n}} = 16.5 \pm 1.96\frac{6.96}{\sqrt{298}} = 16.5 \pm .790 \quad \text{or} \quad 15.710 < \mu < 17.290$$

c   Now we are interested in the difference between posttest means, $\mu_1 - \mu_2$, for BACC versus traditional classes. The 95% confidence interval for $\mu_1 - \mu_2$ is approximately

$$(\bar{x}_1 - \bar{x}_2) \pm 1.96\sqrt{\frac{s_1^2}{n_1} + \frac{s_2^2}{n_2}}$$

$$(18.5 - 16.5) \pm 1.96\sqrt{\frac{8.03^2}{365} + \frac{6.96^2}{298}}$$

$$2.0 \pm 1.142 \quad \text{or} \quad .858 < (\mu_1 - \mu_2) < 3.142$$

d   Since the confidence interval is part **c** has two positive endpoints, it does not contain the value $\mu_1 - \mu_2 = 0$. Hence, it is not likely that the means are equal. It appears that there is a real difference in the mean scores.

8.43   Refer to Exercise 8.18.

a   The 95% confidence interval for $\mu_1 - \mu_2$ is approximately

$$(\bar{x}_1 - \bar{x}_2) \pm 1.96\sqrt{\frac{s_1^2}{n_1} + \frac{s_2^2}{n_2}}$$

$$(120 - 110) \pm 1.96\sqrt{\frac{17.5^2}{50} + \frac{16.5^2}{50}}$$

$$10 \pm 6.667 \quad \text{or} \quad 3.333 < (\mu_1 - \mu_2) < 16.667$$

b   The 99% confidence interval for $\mu_1 - \mu_2$ is approximately

$$(\bar{x}_1 - \bar{x}_2) \pm 2.58\sqrt{\frac{s_1^2}{n_1} + \frac{s_2^2}{n_2}}$$

$$(95 - 110) \pm 2.58\sqrt{\frac{10^2}{50} + \frac{16.5^2}{50}}$$

$$-15 \pm 7.040 \quad \text{or} \quad -22.040 < (\mu_1 - \mu_2) < -7.960$$

c   Neither of the intervals contain the value $(\mu_1 - \mu_2) = 0$. If $(\mu_1 - \mu_2) = 0$ is contained in the confidence interval, then it is not unlikely that μ_1 could equal μ_2, implying no difference in the average room rates for the two hotels. This would be of interest to the experimenter.

d   Since neither confidence interval contains the value $\mu_1 - \mu_2 = 0$, it is not likely that the means are equal. You should conclude that there is a difference in the average room rates for the Marriott and the Wyndham and also for the Radisson and the Wyndham chains.

8.47   **a**   Calculate $\hat{p}_1 = \dfrac{x_1}{n_1} = \dfrac{12}{56} = .214$ and $\hat{p}_2 = \dfrac{x_2}{n_2} = \dfrac{8}{32} = .25$. The approximate 95% confidence interval is

$$(\hat{p}_1 - \hat{p}_2) \pm 1.96 \sqrt{\frac{\hat{p}_1\hat{q}_1}{n_1} + \frac{\hat{p}_2\hat{q}_2}{n_2}}$$

$$(.214 - .25) \pm 1.96 \sqrt{\frac{.214(.786)}{56} + \frac{.25(.75)}{32}}$$

$$-.036 \pm .185 \quad \text{or} \quad -.221 < (p_1 - p_2) < .149$$

b   Since the value $p_1 - p_2 = 0$ is in the confidence interval, it is possible that $p_1 = p_2$. You should not conclude that there is a difference in the proportion of red candies in plain and peanut M&Ms.

8.51   Calculate $\hat{p}_1 = \dfrac{x_1}{n_1} = \dfrac{126}{180} = .7$ and $\hat{p}_2 = \dfrac{x_2}{n_2} = \dfrac{54}{100} = .54$. The approximate 90% confidence interval is

$$(\hat{p}_1 - \hat{p}_2) \pm 1.645 \sqrt{\frac{\hat{p}_1\hat{q}_1}{n_1} + \frac{\hat{p}_2\hat{q}_2}{n_2}}$$

$$(.7 - .54) \pm 1.645 \sqrt{\frac{.7(.3)}{180} + \frac{.54(.46)}{100}}$$

$$.16 \pm .099 \quad \text{or} \quad .061 < (p_1 - p_2) < .259$$

Intervals constructed in this manner enclose the true value of $p_1 - p_2$ 90% of the time in repeated sampling. Hence, we are fairly certain that this particular interval encloses $p_1 - p_2$.

8.55   For the difference $\mu_1 - \mu_2$ in the population means for two quantitative populations, the 95% upper confidence bound uses $z_{.05} = 1.645$ and is calculated as

$$(\bar{x}_1 - \bar{x}_2) + 1.645 \sqrt{\frac{s_1^2}{n_1} + \frac{s_2^2}{n_2}} = (12 - 10) + 1.645 \sqrt{\frac{5^2}{50} + \frac{7^2}{50}}$$

$$2 + 2.00 \quad \text{or} \quad (\mu_1 - \mu_2) < 4$$

8.59   In this exercise, the parameter of interest is $p_1 - p_2$, $n_1 = n_2 = n$, and $B = .05$. Since no prior knowledge is available about p_1 and p_2, we assume the largest possible variation, which occurs if $p_1 = p_2 = .5$. Then

$$z_{\alpha/2} \times (\text{std error of } \hat{p}_1 - \hat{p}_2) \leq B$$

$$z_{.01}\sqrt{\frac{p_1 q_1}{n_1} + \frac{p_2 q_2}{n_2}} \le .05 \Rightarrow 2.33\sqrt{\frac{(.5)(.5)}{n} + \frac{(.5)(.5)}{n}} \le .05$$

$$\sqrt{n} \ge \frac{2.33\sqrt{.5}}{.05} \Rightarrow n \ge 1085.78 \quad \text{or} \quad n_1 = n_2 = 1086$$

8.63   Similar to Exercise 8.58 using $s \approx R/4 = 104/4 = 26$.

$$2.58\sqrt{\frac{\sigma_1^2}{n_1} + \frac{\sigma_2^2}{n_2}} \le 5 \Rightarrow 2.58\sqrt{\frac{26^2}{n} + \frac{26^2}{n}} \le 5$$

$$\sqrt{n} \ge \frac{2.58\sqrt{1352}}{5} \Rightarrow n \ge 359.98 \quad \text{or} \quad n_1 = n_2 = 360$$

8.64   **a**   For the difference $\mu_1 - \mu_2$ in the population means this year and ten years ago, the 99% lower confidence bound uses $z_{.01} = 2.33$ and is calculated as

$$(\bar{x}_1 - \bar{x}_2) - 2.33\sqrt{\frac{s_1^2}{n_1} + \frac{s_2^2}{n_2}} = (73 - 63) - 2.33\sqrt{\frac{25^2}{400} + \frac{28^2}{400}}$$

$$10 - 4.37 \quad \text{or} \quad (\mu_1 - \mu_2) > 5.63$$

b   Since the difference in the means is positive, you can conclude that there has been a decrease in the average per-capita beef consumption over the last ten years.

8.71   **a**   The point estimate of μ is $\bar{x} = 29.1$ and the margin of error in estimation with $s = 3.9$ and $n = 64$ is

$$1.96\sigma_{\bar{x}} = 1.96\frac{\sigma}{\sqrt{n}} \approx 1.96\frac{s}{\sqrt{n}} = 1.96\left(\frac{3.9}{\sqrt{64}}\right) = .9555.$$

b   The approximate 90% confidence interval is

$$\bar{x} \pm 1.645\frac{s}{\sqrt{n}} = 29.1 \pm 1.645\frac{3.9}{\sqrt{64}} = 29.1 \pm .802 \quad \text{or} \quad 28.298 < \mu < 29.902$$

Intervals constructed in this manner enclose the true value of μ 90% of the time in repeated sampling. Therefore, we are fairly certain that this particular interval encloses μ.

c   The approximate 90% lower confidence bound is

$$\bar{x} - 1.28\frac{s}{\sqrt{n}} = 29.1 - 1.28\frac{3.9}{\sqrt{64}} = 28.48 \quad \text{or} \quad \mu > 28.48$$

d   With $B = .5$, $\sigma \approx 3.9$, and $1 - \alpha = .95$, we must solve for n in the following inequality:

$$1.96 \frac{\sigma}{\sqrt{n}} \leq B \qquad \Rightarrow \qquad 1.96 \frac{3.9}{\sqrt{n}} \leq .5$$

$$\sqrt{n} \geq 15.288 \; \Rightarrow \; n \geq 233.723 \; \text{ or } \; n \geq 234$$

8.75 Assuming maximum variation with $p = .5$, solve

$$1.645 \sqrt{\frac{pq}{n}} \leq .025$$

$$\sqrt{n} \geq \frac{1.645 \sqrt{.5(.5)}}{.025} = 32.9 \Rightarrow \quad n \geq 1082.41 \quad \text{or } n \geq 1083$$

8.79 **a** Define sample #1 as the sample of 482 women and sample #2 as the sample of 356 men. Then $\widehat{p}_1 = .5$ and $\widehat{p}_2 = .75$.

b The approximate 95% confidence interval is

$$(\widehat{p}_1 - \widehat{p}_2) \pm 1.96 \sqrt{\frac{\widehat{p}_1 \widehat{q}_1}{n_1} + \frac{\widehat{p}_2 \widehat{q}_2}{n_2}}$$

$$(.5 - .75) \pm 1.96 \sqrt{\frac{.5(.5)}{482} + \frac{.75(.25)}{356}}$$

$$-.25 \pm .063 \quad \text{or} \quad -.313 < (p_1 - p_2) < -.187$$

c Since the value $p_1 - p_2 = 0$ is not in the confidence interval, it is unlikely that $p_1 = p_2$. You should conclude that there is a difference in the proportion of women and men on Wall Street who have children. In fact, since all the probable values of $p_1 - p_2$ are negative, the proportion of men on Wall Street who have children appears to be larger than the proportion of women.

8.83 Assume that $\sigma = 2.5$ and the desired bound is .5. Then

$$1.96 \frac{\sigma}{\sqrt{n}} \leq B \quad \Rightarrow \quad 1.96 \frac{2.5}{\sqrt{n}} \leq .5 \Rightarrow \quad n \geq 96.04 \quad \text{or} \quad n \geq 97$$

8.87 **a** If you use $p = .8$ as a conservative estimate for p, the margin of error is approximately

$$\pm 1.96 \sqrt{\frac{.8(.2)}{750}} = \pm .029$$

b To reduce the margin of error in part **a** to $\pm .01$, solve for n in the equation

$$1.96 \sqrt{\frac{.8(.2)}{n}} = .01 \; \Rightarrow \; \sqrt{n} = \frac{1.96(.4)}{.01} = 78.4 \; \Rightarrow \; n = 6146.56 \; \text{ or } \; n = 6147$$

8.91   It is assumed that $p = .2$ and that the desired bound is .01. Hence,

$$1.96\sqrt{\frac{pq}{n}} \le .01 \quad \Rightarrow \quad \sqrt{n} \ge \frac{1.96\sqrt{.05(.95)}}{.01} = 42.72$$

$$n \ge 1824.76 \quad \text{or} \quad n \ge 1825$$

8.95   **a**   The approximate 95% confidence interval for μ is

$$\bar{x} \pm 1.96\frac{s}{\sqrt{n}} = 2.962 \pm 1.96\frac{.529}{\sqrt{69}} = 2.962 \pm .125$$

or $2.837 < \mu < 3.087$.

b   In order to cut the interval in half, the sample size must increase by 4. If this is done, the new half-width of the confidence interval is

$$1.96\frac{\sigma}{\sqrt{4n}} = \frac{1}{2}(1.96\frac{\sigma}{\sqrt{n}}).$$

Hence, in this case, the new sample size is $4(69) = 276$.

8.99   Use the **Interpreting Confidence Intervals** applet. Answers will vary, but the widths of all the intervals should be the same. Most of the simulations will show between 8 and 10 intervals that work correctly.

8.103   Use the **Exploring Confidence Intervals** applet.
a-b Move the slider on the right side of the applet to change the sample size. Increasing the sample size results a smaller standard error and in a narrower interval.
c   By increasing the sample size n, you obtain more information and can obtain this more precise estimate of μ without sacrificing confidence.

9: Large-Sample Tests of Hypotheses

9.1   **a**   The critical value that separates the rejection and nonrejection regions for a right-tailed test based on a z-statistic will be a value of z (called z_α) such that $P(z > z_\alpha) = \alpha = .01$. That is, $z_{.01} = 2.33$ (see Figure 9.1). The null hypothesis H_0 will be rejected if $z > 2.33$.

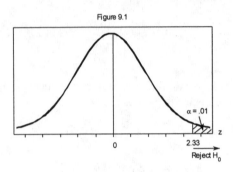

Figure 9.1

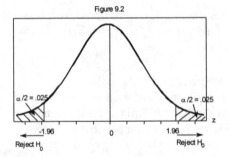

Figure 9.2

b   For a two-tailed test with $\alpha = .05$, the critical value for the rejection region cuts off $\alpha/2 = .025$ in the two tails of the z distribution in Figure 9.2, so that $z_{.025} = 1.96$. The null hypothesis H_0 will be rejected if $z > 1.96$ or $z < -1.96$ (which you can also write as $|z| > 1.96$).
c   Similar to part **a**, with the rejection region in the lower tail of the z distribution. The null hypothesis H_0 will be rejected if $z < -2.33$.
d   Similar to part **b**, with $\alpha/2 = .005$. The null hypothesis H_0 will be rejected if $z > 2.58$ or $z < -2.58$ (which you can also write as $|z| > 2.58$).

9.4   In this exercise, the parameter of interest is μ, the population mean. The objective of the experiment is to show that the mean exceeds 2.3.
a   We want to prove the alternative hypothesis that μ is, in fact, greater than 2.3. Hence, the alternative hypothesis is

$$H_a: \mu > 2.3$$

and the null hypothesis is

$$H_0: \mu = 2.3$$

b   The best estimator for μ is the sample average $\bar{x}$, and the test statistic is

$$z = \frac{\bar{x} - \mu_0}{\sigma/\sqrt{n}}$$

which represents the distance (measured in units of standard deviations) from $\bar{x}$ to the hypothesized mean μ. Hence, if this value is large in absolute value, one of two conclusions may be drawn. Either a very unlikely event has occurred, or the hypothesized mean is incorrect. Refer to part **a**. If $\alpha = .05$, the critical value of z that separates the rejection and non-rejection regions will be a value (denoted by z_0) such that

$$P(z > z_0) = \alpha = .05$$

That is, $z_0 = 1.645$ (see Figure 9.5). Hence, H_0 will be rejected if $z > 1.645$.

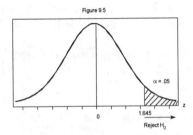

Figure 9.5

c   The standard error of the mean is found using the sample standard deviation s to approximate the population standard deviation σ:

$$SE = \frac{\sigma}{\sqrt{n}} \approx \frac{s}{\sqrt{n}} = \frac{.29}{\sqrt{35}} = .049$$

d   To conduct the test, calculate the value of the test statistic using the information contained in the sample. Note that the value of the true standard deviation, σ, is approximated using the sample standard deviation s.

$$z = \frac{\bar{x} - \mu_0}{\sigma/\sqrt{n}} \approx \frac{\bar{x} - \mu_0}{s/\sqrt{n}} = \frac{2.4 - 2.3}{.049} = 2.04$$

The observed value of the test statistic, $z = 2.04$, falls in the rejection region and the null hypothesis is rejected. There is sufficient evidence to indicate that $\mu > 2.3$.

9.9   **a**   In order to make sure that the average weight was one pound, you would test

$$H_0 : \mu = 1 \quad \text{versus} \quad H_a : \mu \neq 1$$

b-c   The test statistic is

$$z = \frac{\bar{x} - \mu_0}{\sigma/\sqrt{n}} \approx \frac{\bar{x} - \mu_0}{s/\sqrt{n}} = \frac{1.01 - 1}{.18/\sqrt{35}} = .33$$

with p-value $= P(|z| > .33) = 2(.3707) = .7414$. Since the p-value is greater than .05, the null hypothesis should not be rejected. The manager should report that there is insufficient evidence to indicate that the mean is different from 1.

9.13　**a**　The hypothesis to be tested is

$$H_0 : \mu = 110 \quad \text{versus} \quad H_a : \mu < 110$$

and the test statistic is

$$z = \frac{\bar{x} - \mu_0}{\sigma/\sqrt{n}} \approx \frac{\bar{x} - \mu_0}{s/\sqrt{n}} = \frac{107 - 110}{13/\sqrt{100}} = -2.31$$

with p-value $= P(z < -2.31) = .0104$. To draw a conclusion from the p-value, use the guidelines for statistical significance in Section 9.3. Since the p-value is between .01 and .05, the test results are significant at the 5% level, but not at the 1% level.

b　If $\alpha = .05$, H_0 can be rejected and you can conclude that the average score improvement is less than claimed. This would be the most beneficial way for the competitor to state these conclusions.

c　If you worked for the *Princeton Review*, it would be more beneficial to conclude that there was *insufficient evidence at the 1% level* to conclude that the average score improvement is less than claimed.

9.17　**a**　The hypothesis of interest is one-tailed:

$$H_0: \mu_1 - \mu_2 = 0 \quad \text{versus} \quad H_a: \mu_1 - \mu_2 > 0$$

b　The test statistic, calculated under the assumption that $\mu_1 - \mu_2 = 0$, is

$$z \approx \frac{(\bar{x}_1 - \bar{x}_2) - 0}{\sqrt{\dfrac{s_1^2}{n_1} + \dfrac{s_2^2}{n_2}}} = \frac{6.9 - 5.8}{\sqrt{\dfrac{(2.9)^2}{35} + \dfrac{(1.2)^2}{35}}} = 2.074$$

The rejection region, with $\alpha = .05$, is $z > 1.645$ and H_0 is rejected. There is evidence to indicate that $\mu_1 - \mu_2 > 0$, or $\mu_1 > \mu_2$. That is, there is reason to believe that Vitamin C reduces the mean time to recover.

9.19　**a**　The hypothesis of interest is two-tailed:

$$H_0: \mu_1 - \mu_2 = 0 \quad \text{versus} \quad H_a: \mu_1 - \mu_2 \neq 0$$

The test statistic, calculated under the assumption that $\mu_1 - \mu_2 = 0$, is

$$z \approx \frac{(\bar{x}_1 - \bar{x}_2) - 0}{\sqrt{\dfrac{s_1^2}{n_1} + \dfrac{s_2^2}{n_2}}} = \frac{34.1 - 36}{\sqrt{\dfrac{(5.9)^2}{100} + \dfrac{(6.0)^2}{100}}} = -2.26$$

with p-value $= P(|z| > 2.26) = 2(.0119) = .0238$. Since the p-value is less than .05, the null hypothesis is rejected. There is evidence to indicate a difference in the mean lead levels for the two sections of the city.

b   From Section 8.6, the 95% confidence interval for $\mu_1 - \mu_2$ is approximately

$$(\bar{x}_1 - \bar{x}_2) \pm 1.96 \sqrt{\frac{s_1^2}{n_1} + \frac{s_2^2}{n_2}}$$

$$(34.1 - 36) \pm 1.96 \sqrt{\frac{(5.9)^2}{100} + \frac{(6.0)^2}{100}}$$

$$-1.9 \pm 1.65 \quad \text{or} \quad -3.55 < (\mu_1 - \mu_2) < -.25$$

c   Since the value $\mu_1 - \mu_2 = 5$ or $\mu_1 - \mu_2 = -5$ is not in the confidence interval in part **b**, it is not likely that the difference will be more than 5 ppm, and hence the statistical significance of the difference is not of practical importance to the engineers.

9.23   **a**   The hypothesis of interest is two-tailed:

$$H_0: \mu_1 - \mu_2 = 0 \quad \text{versus} \quad H_a: \mu_1 - \mu_2 \neq 0$$

and the test statistic is

$$z \approx \frac{(\bar{x}_1 - \bar{x}_2) - 0}{\sqrt{\frac{s_1^2}{n_1} + \frac{s_2^2}{n_2}}} = \frac{.94 - 2.8}{\sqrt{\frac{1.2^2}{36} + \frac{2.8^2}{26}}} = -3.18$$

with p-value $= P(|z| > 3.18) = 2(.0007) = .0014$. Since the p-value is less than .05, the null hypothesis is rejected. There is evidence to indicate a difference in the mean concentrations for these two types of sites.

b   The 95% confidence interval for $\mu_1 - \mu_2$ is approximately

$$(\bar{x}_1 - \bar{x}_2) \pm 1.96 \sqrt{\frac{s_1^2}{n_1} + \frac{s_2^2}{n_2}}$$

$$(.94 - 2.8) \pm 1.96 \sqrt{\frac{1.2^2}{36} + \frac{2.8^2}{26}}$$

$$-1.86 \pm 1.15 \quad \text{or} \quad -3.01 < (\mu_1 - \mu_2) < -.71$$

Since the value $\mu_1 - \mu_2 = 0$ does not fall in the interval in part **b**, it is not likely that $\mu_1 = \mu_2$. There is evidence to indicate that the means are different, confirming the conclusion in part **a**.

9.27   **a**   The two sets of hypotheses both involve a different binomial parameter p:

$$H_0: p = .6 \quad \text{versus} \quad H_a: p \neq .6 \quad \text{(part c)}$$

$$H_0: p = .5 \quad \text{versus} \quad H_a: p < .5 \quad \text{(part b)}$$

b   For the second test in part **a**, $x = 35$ and $n = 75$, so that $\hat{p} = \dfrac{x}{n} = \dfrac{35}{75} = .4667$, the test statistic is

$$z = \frac{\hat{p} - p_0}{\sqrt{\frac{p_0 q_0}{n}}} = \frac{.4667 - .5}{\sqrt{\frac{.5(.5)}{75}}} = -.58$$

Since no value of α is specified in advance, we calculate p-value $= P(z < -.58) = .2810$. Since this p-value is greater than .10, the null hypothesis is not rejected. There is insufficient evidence to contradict the claim.

c For the first test in part **a**, $x = 49$ and $n = 75$, so that $\hat{p} = \dfrac{x}{n} = \dfrac{49}{75} = .6533$, the test statistic is

$$z = \frac{\hat{p} - p_0}{\sqrt{\dfrac{p_0 q_0}{n}}} = \frac{.6533 - .6}{\sqrt{\dfrac{.6(.4)}{75}}} = .94$$

with p-value $= P(|z| > .94) = 2(.1736) = .3472$. Since this p-value is greater than .10, the null hypothesis is not rejected. There is insufficient evidence to contradict the claim.

9.31 The hypothesis of interest is

$$H_0: p = .45 \quad \text{versus} \quad H_a: p \neq .45$$

With $\hat{p} = \dfrac{x}{n} = \dfrac{32}{80} = .4$, the test statistic is

$$z = \frac{\hat{p} - p_0}{\sqrt{\dfrac{p_0 q_0}{n}}} = \frac{.40 - .45}{\sqrt{\dfrac{.45(.55)}{80}}} = -.90$$

The rejection region is two-tailed with $\alpha = .01$, or $|z| > 2.58$ and H_0 is not rejected. There is insufficient evidence to dispute the newspaper's claim.

9.37 **a** The hypothesis of interest is

$$H_0: p_1 - p_2 = 0 \quad \text{versus} \quad H_a: p_1 - p_2 < 0$$

Calculate $\hat{p}_1 = .36$, $\hat{p}_2 = .60$, and $\hat{p} = \dfrac{n_1 \hat{p}_1 + n_2 \hat{p}_2}{n_1 + n_2} = \dfrac{18 + 30}{50 + 50} = .48$. The test statistic is then

$$z = \frac{\hat{p}_1 - \hat{p}_2}{\sqrt{\hat{p}\hat{q}\left(\dfrac{1}{n_1} + \dfrac{1}{n_2}\right)}} = \frac{.36 - .60}{\sqrt{.48(.52)(1/50 + 1/50)}} = -2.40$$

The rejection region with $\alpha = .05$ is $z < -1.645$ and H_0 is rejected. There is evidence of a difference in the proportion of survivors for the two groups.

b From Section 8.7, the approximate 95% confidence interval is

$$(\hat{p}_1 - \hat{p}_2) \pm 1.96\sqrt{\frac{\hat{p}_1 \hat{q}_1}{n_1} + \frac{\hat{p}_2 \hat{q}_2}{n_2}}$$

$$(.36 - .60) \pm 1.96\sqrt{\frac{.36(.64)}{50} + \frac{.60(.40)}{50}}$$

$$-.24 \pm .19 \quad \text{or} \quad -.43 < (p_1 - p_2) < -.05$$

9.39   The hypothesis of interest is

$$H_0: p_1 - p_2 = 0 \quad \text{versus} \quad H_a: p_1 - p_2 \neq 0$$

Calculate $\hat{p}_1 = \dfrac{12}{56} = .214, \quad \hat{p}_2 = \dfrac{8}{32} = .25,$ and $\hat{p} = \dfrac{x_1 + x_2}{n_1 + n_2} = \dfrac{12 + 8}{56 + 32} = .227$

The test statistic is then

$$z = \frac{\hat{p}_1 - \hat{p}_2}{\sqrt{\hat{p}\hat{q}\left(\dfrac{1}{n_1} + \dfrac{1}{n_2}\right)}} = \frac{.214 - .25}{\sqrt{.227(.773)(1/56 + 1/32)}} = -.39$$

The rejection region with $\alpha = .05$ is $|z| > 1.96$ and H_0 is not rejected. There is insufficient evidence to indicate a difference in the proportion of red M&Ms for the plain and peanut varieties. These results match the conclusions of Exercise 8.47.

9.43   The hypothesis of interest is

$$H_0: p_1 - p_2 = 0 \quad \text{versus} \quad H_a: p_1 - p_2 > 0$$

Calculate $\hat{p}_1 = \dfrac{93}{121} = .769, \quad \hat{p}_2 = \dfrac{119}{199} = .598,$ and

$\hat{p} = \dfrac{x_1 + x_2}{n_1 + n_2} = \dfrac{93 + 119}{121 + 199} = .6625.$ The test statistic is then

$$z = \frac{\hat{p}_1 - \hat{p}_2}{\sqrt{\hat{p}\hat{q}\left(\dfrac{1}{n_1} + \dfrac{1}{n_2}\right)}} = \frac{.769 - .598}{\sqrt{.6625(.3375)(1/121 + 1/199)}} = 3.14$$

with p-value $= P(z > 3.14) = 1 - .9992 = .0008.$ Since the p-value is less than .01, the results are reported as highly significant at the 1% level of significance. There is evidence to confirm the researcher's conclusion.

9.47   The power of the test is $1 - \beta = P(\text{reject } H_0 \text{ when } H_0 \text{ is false}).$ As μ gets farther from μ_0, the power of the test increases.

9.51   **a-b** Since it is necessary to prove that the average pH level is less than 7.5, the hypothesis to be tested is one-tailed:

$$H_0: \mu = 7.5 \quad \text{versus} \quad H_a: \mu < 7.5$$

d   The test statistic is

$$z = \frac{\bar{x} - \mu}{\sigma/\sqrt{n}} \approx \frac{\bar{x} - \mu}{s/\sqrt{n}} = \frac{-.2}{.2/\sqrt{30}} = -5.477$$

and the rejection region with $\alpha = .05$ is $z < -1.645.$ The observed value, $z = -$ falls in the rejection region and H_0 is rejected. We conclude that the average pH l less than 7.5.

9.55   Let p_1 be the proportion of defectives produced by machine A and p_2 be the proportion of defectives produced by machine B.   The hypothesis to be tested is

$$H_0: p_1 - p_2 = 0 \quad \text{versus} \quad H_a: p_1 - p_2 \neq 0$$

Calculate $\widehat{p}_1 = \dfrac{16}{200} = .08$, $\widehat{p}_2 = \dfrac{8}{200} = .04$, and $\widehat{p} = \dfrac{x_1 + x_2}{n_1 + n_2} = \dfrac{16 + 8}{200 + 200} = .06$.
The test statistic is then

$$z = \frac{\widehat{p}_1 - \widehat{p}_2}{\sqrt{\widehat{pq}\left(\dfrac{1}{n_1} + \dfrac{1}{n_2}\right)}} = \frac{.08 - .04}{\sqrt{.06(.94)(1/200 + 1/200)}} = 1.684$$

The rejection region with $\alpha = .05$ is $|z| > 1.96$ and H_0 is not rejected. There is insufficient evidence to indicate that the machines are performing differently in terms of the percentage of defectives being produced.

9.59   The hypothesis to be tested is

$$H_0: \mu_1 - \mu_2 = 0 \quad \text{versus} \quad H_a: \mu_1 - \mu_2 > 0$$

and the test statistic is

$$z \approx \frac{(\overline{x}_1 - \overline{x}_2) - 0}{\sqrt{\dfrac{s_1^2}{n_1} + \dfrac{s_2^2}{n_2}}} = \frac{10 - 8}{\sqrt{\dfrac{4.3}{40} + \dfrac{5.7}{40}}} = 4$$

rejection region, with $\alpha = .05$, is one-tailed or $z > 1.645$ and the null hypothesis is
ed. There is sufficient evidence to indicate a difference in the two means. Hence,
clude that diet I has a greater mean weight loss than diet II.

pothesis to be tested is

$$H_0: \mu_1 - \mu_2 = 0 \quad \text{versus} \quad H_a: \mu_1 - \mu_2 > 0$$

tic is

$$z \approx \frac{(\overline{x}_1 - \overline{x}_2) - 0}{\sqrt{\dfrac{s_1^2}{n_1} + \dfrac{s_2^2}{n_2}}} = \frac{240 - 227}{\sqrt{\dfrac{980}{200} + \dfrac{820}{200}}} = 4.33$$

$\alpha = .05$, is one-tailed or $z > 1.645$ and the null hypothesis is
nce in mean yield for the two types of spray.
nfidence interval for $\mu_1 - \mu_2$ is approximately

$$1.96\sqrt{\dfrac{s_1^2}{n_1} + \dfrac{s_2^2}{n_2}}$$

$$6\sqrt{\dfrac{980}{200} + \dfrac{820}{200}}$$

or $7.12 < (\mu_1 - \mu_2) < 18.88$

5.477,
evel is

9.67   **a**   The parameter of interest is μ, the average daily wage of workers in a given industry. A sample of $n = 40$ workers has been drawn from a particular company within this industry and $\bar{x}$, the sample average, has been calculated. The objective is to determine whether this company pays wages different from the total industry. That is, assume that this sample of forty workers has been drawn from a hypothetical population of workers. Does this population have as an average wage $\mu = 54$, or is μ different from 54? Thus, the hypothesis to be tested is

$$H_0: \mu = 54 \quad \text{versus} \quad H_a: \mu \neq 54$$

b-c The test statistic is

$$z \approx \frac{\bar{x} - \mu}{s/\sqrt{n}} = \frac{51.50 - 54}{11.88/\sqrt{40}} = -1.331$$

and the **Large-Sample Test of a Population Mean** applet gives p-value $= .1832$. (Using Table 3 will produce a p-value of .1836.)

d   Since $\alpha = .01$ is smaller than the p-value, .1832, H_0 cannot be rejected and we cannot conclude that the company is paying wages different from the industry average.

e   Since n is greater than 30, the Central Limit Theorem will guarantee the normality of $\bar{x}$ regardless of whether the original population was normal or not.

10: Inference from Small Samples

10.1   Refer to Table 4, Appendix I, indexing df along the left or right margin and t_α across the top.

    a $t_{.05} = 2.015$ with 5 df **b** $t_{.025} = 2.306$ with 8 df

    c $t_{.10} = 1.330$ with 18 df **d** $t_{.025} \approx 1.96$ with 30 df

10.5   **a**   Using the formulas given in Chapter 2, calculate $\sum x_i = 70.5$ and $\sum x_i^2 = 499.27$. Then

$$\bar{x} = \frac{\sum x_i}{n} = \frac{70.5}{10} = 7.05$$

$$s^2 = \frac{\sum x_i^2 - \dfrac{(\sum x_i)^2}{n}}{n-1} = \frac{499.27 - \dfrac{(70.5)^2}{10}}{9} = .249444 \text{ and } s = .4994$$

b   With $df = n - 1 = 9$, the appropriate value of t is $t_{.005} = 3.250$ (from Table 4) and the 99% confidence interval is

$$\bar{x} \pm t_{.005}\frac{s}{\sqrt{n}} \;\Rightarrow\; 7.05 \pm 3.250\sqrt{\frac{.249444}{10}} \;\Rightarrow\; 7.05 \pm .513$$

or $6.537 < \mu < 7.563$. Intervals constructed using this procedure will enclose μ 99% of the time in repeated sampling. Hence, we are fairly certain that this particular interval encloses μ.

c   The hypothesis to be tested is

$$H_0: \mu = 7.5 \quad \text{versus} \quad H_a: \mu < 7.5$$

and the test statistic is

$$t = \frac{\bar{x} - \mu}{s/\sqrt{n}} = \frac{7.05 - 7.5}{\sqrt{\dfrac{.249444}{10}}} = -2.849$$

The rejection region with $\alpha = .01$ and $n - 1 = 9$ degrees of freedom is located in the lower tail of the t-distribution and is found from Table 4 as $t < -t_{.01} = -2.821$. Since the observed value of the test statistic falls in the rejection region, H_0 is rejected and we conclude that μ is less than 7.5.

d   Notice that the 99% confidence interval for μ does indeed include the value $\mu = 7.5$. However, the *one-tailed test of hypothesis* allows us to reject the hypothesis that $\mu = 7.5$ with $\alpha = .01$. These seemingly contradictory results are explained by the fact that the test of hypothesis is one-tailed, while the confidence interval is two-sided. If the test in part **c** had been two-tailed, or if you had used a one-sided confidence bound in part **b**, you would have had identical conclusions.

10.9   Similar to previous exercises. The hypothesis to be tested is

$$H_0: \mu = 100 \qquad H_a: \mu < 100$$

Calculate $\bar{x} = \dfrac{\sum x_i}{n} = \dfrac{1797.095}{20} = 89.85475$

$$s^2 = \dfrac{\sum x_i^2 - \dfrac{(\sum x_i)^2}{n}}{n-1} = \dfrac{165,697.7081 - \dfrac{(1797.095)^2}{20}}{19} = 222.115067 \text{ and } s = 14.8998$$

The test statistic is

$$t = \dfrac{\bar{x} - \mu}{s/\sqrt{n}} = \dfrac{89.85475 - 100}{\dfrac{14.8998}{\sqrt{20}}} = -3.044$$

The critical value of t with $\alpha = .01$ and $n - 1 = 19$ degrees of freedom is $t_{.01} = 2.539$ and the rejection region is $t < -2.539$. The null hypothesis is rejected and we conclude that μ is less than 100 DL.

10.13 a   The hypothesis to be tested is

$$H_0: \mu = 25 \qquad H_a: \mu < 25$$

The test statistic is

$$t = \dfrac{\bar{x} - \mu_0}{s/\sqrt{n}} = \dfrac{20.3 - 25}{\dfrac{5}{\sqrt{21}}} = -4.31$$

The critical value of t with $\alpha = .05$ and $n - 1 = 20$ degrees of freedom is $t_{.05} = 1.725$ and the rejection region is $t < -1.725$. Since the observed value falls in the rejection region, H_0 is rejected, and we conclude that the pre-treatment mean is less than 25.

b   The 95% confidence interval based on $df = 20$ is

$$\bar{x} \pm t_{.025}\dfrac{s}{\sqrt{n}} \Rightarrow 26.6 \pm 2.086\dfrac{7.4}{\sqrt{21}} \Rightarrow 26.6 \pm 3.37$$

or $23.23 < \mu < 29.97$.

c   The pre-treatment mean looks considerably smaller than the other two means.

10.17 a $s^2 = \dfrac{(n_1 - 1)s_1^2 + (n_2 - 1)s_2^2}{n_1 + n_2 - 2} = \dfrac{9(3.4) + 3(4.9)}{10 + 4 - 2} = 3.775$

b $s^2 = \dfrac{(n_1 - 1)s_1^2 + (n_2 - 1)s_2^2}{n_1 + n_2 - 2} = \dfrac{11(18) + 20(23)}{12 + 21 - 2} = 21.2258$

10.23 a   The hypothesis to be tested is

$$H_0: \mu_1 - \mu_2 = 0 \quad \text{versus} \quad H_a: \mu_1 - \mu_2 \neq 0$$

From the Minitab printout, the following information is available:

$$\bar{x}_1 = .896 \qquad s_1^2 = (.400)^2 \qquad n_1 = 14$$
$$\bar{x}_2 = 1.147 \qquad s_2^2 = (.679)^2 \qquad n_2 = 11$$

and the test statistic is

$$t = \frac{(\bar{x}_1 - \bar{x}_2) - 0}{\sqrt{s^2\left(\dfrac{1}{n_1} + \dfrac{1}{n_2}\right)}} = -1.16$$

The rejection region is two-tailed, based on $n_1 + n_2 - 2 = 23$ degrees of freedom. With $\alpha = .05$, from Table 4, the rejection region is $|t| > t_{.025} = 2.069$ and H_0 is not rejected. There is not enough evidence to indicate a difference in the population means.

b It is not necessary to bound the p-value using Table 4, since the exact p-value is given on the printout as P-Value $= .26$.

c If you check the ratio of the two variances using the rule of thumb given in this section you will find:

$$\frac{\text{larger } s^2}{\text{smaller } s^2} = \frac{(.679)^2}{(.400)^2} = 2.88$$

which is less than three. Therefore, it is reasonable to assume that the two population variances are equal.

10.25 a Check the ratio of the two variances using the rule of thumb given in this section:

$$\frac{\text{larger } s^2}{\text{smaller } s^2} = \frac{2.78095}{.17143} = 16.22$$

which is greater than three. Therefore, it is not reasonable to assume that the two population variances are equal.

b You should use the unpooled t test with Satterthwaite's approximation to the degrees of freedom for testing

$$H_0: \mu_1 - \mu_2 = 0 \quad \text{versus} \quad H_a: \mu_1 - \mu_2 \neq 0$$

The test statistic is

$$t = \frac{(\bar{x}_1 - \bar{x}_2) - 0}{\sqrt{\dfrac{s_1^2}{n_1} + \dfrac{s_2^2}{n_2}}} = \frac{3.73 - 4.8}{\sqrt{\dfrac{2.78095}{15} + \dfrac{.17143}{15}}} = -2.412$$

with

$$df = \frac{\left(\dfrac{s_1^2}{n_1} + \dfrac{s_2^2}{n_2}\right)^2}{\dfrac{\left(\dfrac{s_1^2}{n_1}\right)^2}{n_1 - 1} + \dfrac{\left(\dfrac{s_2^2}{n_2}\right)^2}{n_2 - 1}} = \frac{(.185397 + .0114287)^2}{.002455137 + .00000933} = 15.7$$

With $df \approx 15$, the p-value for this test is bounded between .02 and .05 so that H_0 can be rejected at the 5% level of significance. There is evidence of a difference in the mean number of uncontaminated eggplants for the two disinfectants.

10.29 The test statistic is

$$t = \frac{\overline{d} - \mu_d}{s_d/\sqrt{n}} = \frac{.3 - 0}{\sqrt{\frac{.16}{10}}} = 2.372$$

with $n - 1 = 9$ degrees of freedom. The p-value is then

$$P(|t| > 2.372) = 2P(t > 2.372) \quad \text{so that} \quad P(t > 2.372) = \frac{1}{2}p\text{-value}$$

Since the value $t = 2.372$ falls between two tabled entries for $df = 9$ ($t_{.025} = 2.262$ and $t_{.01} = 2.821$), you can conclude that

$$.01 < \frac{1}{2}p\text{-value} < .025$$
$$.02 < p\text{-value} < .05$$

Since the p-value is less than $\alpha=.05$, the null hypothesis is rejected and we conclude that there is a difference in the two population means.

10.33 **a**   It is necessary to use a paired-difference test, since the two samples are not random and independent. The hypothesis of interest is

$$H_0:\mu_1 - \mu_2 = 0 \quad \text{or} \quad H_0:\mu_d = 0$$

$$H_a:\mu_1 - \mu_2 \neq 0 \quad \text{or} \quad H_a:\mu_d \neq 0$$

The table of differences, along with the calculation of $\overline{d}$ and s_d^2, is presented below.

d_i	d_i^2
.1	.01
.1	.01
0	.00
.2	.04
$-.1$	.01
.3	.07

$$\overline{d} = \frac{\sum d_i}{n} = \frac{.3}{5} = .06$$

$$s_d^2 = \frac{\sum d_i^2 - \frac{(\sum d_i)^2}{n}}{n - 1} = \frac{.07 - \frac{(.3)^2}{5}}{4} = .013$$

The test statistic is

$$t = \frac{\bar{d} - \mu_d}{s_d/\sqrt{n}} = \frac{.06 - 0}{\sqrt{\frac{.013}{5}}} = 1.177$$

with $n - 1 = 4$ degrees of freedom. The rejection region with $\alpha = .05$ is $|t| > t_{.025} = 2.776$, and H_0 is not rejected. We cannot conclude that the means are different.

b The p-value is

$$P(|t| > 1.177) = 2P(t > 1.177) > 2(.10) = .20.$$

c A 95% confidence interval for $\mu_1 - \mu_2 = \mu_d$ is

$$\bar{d} \pm t_{.025}\frac{s_d}{\sqrt{n}} \Rightarrow .06 \pm 2.776\sqrt{\frac{.013}{5}} \Rightarrow .06 \pm .142$$

or $-.082 < (\mu_1 - \mu_2) < .202$.

d In order to use the paired-difference test, it is necessary that the n paired observations be randomly selected from normally distributed populations.

10.37 a Each subject was presented with both signs in random order. If his reaction time in general is high, both responses will be high; if his reaction time in general is low, both responses will be low. The large variability from subject to subject will mask the variability due to the difference in sign types. The paired-difference design will eliminate the subject to subject variability.

b The hypothesis of interest is

$$H_0:\mu_1 - \mu_2 = 0 \qquad \text{or} \quad H_0:\mu_d = 0$$

$$H_a:\mu_1 - \mu_2 \neq 0 \qquad \text{or} \quad H_a:\mu_d \neq 0$$

The table of differences, along with the calculation of $\bar{d}$ and s_d^2, is presented below.

Driver	1	2	3	4	5	6	7	8	9	10	Totals
d_i	122	141	97	107	37	56	110	146	104	149	1069

$$\bar{d} = \frac{\sum d_i}{n} = \frac{1069}{10} = 106.9$$

$$s_d^2 = \frac{\sum d_i^2 - \frac{(\sum d_i)^2}{n}}{n - 1} = \frac{126,561 - \frac{(1069)^2}{10}}{9} = 1364.98889 \text{ and } s_d = 36.9458$$

and the test statistic is

$$t = \frac{\bar{d} - \mu_d}{s_d/\sqrt{n}} = \frac{106.9 - 0}{\frac{36.9458}{\sqrt{10}}} = 9.150$$

Since $t = 9.150$ with $df = n - 1 = 9$ is greater than the largest tabled value $t_{.005}$,

$$p\text{-value} < 2(.005) = .01$$

for this two tailed test and H_0 is rejected. We can conclude that the means are different.

c   The 95% confidence interval for $\mu_1 - \mu_2 = \mu_d$ is

$$\overline{d} \pm t_{.005}\frac{s_d}{\sqrt{n}} \Rightarrow 106.9 \pm 2.262\frac{36.9458}{\sqrt{10}} \Rightarrow 106.9 \pm 26.428$$

or $80.472 < (\mu_1 - \mu_2) < 133.328$.

10.41   **a**   Use the Minitab printout given in the text. The hypothesis of interest is

$$H_0\colon \mu_A - \mu_B = 0 \qquad H_a\colon \mu_A - \mu_B > 0$$

and the test statistic is

$$t = \frac{\overline{d} - \mu_d}{s_d/\sqrt{n}} = \frac{1.488 - 0}{\dfrac{1.491}{\sqrt{8}}} = 2.82$$

The p-value shown in the printout is $p\text{-value} = .013$. Since the p-value is less than .05, H_0 is rejected at the 5% level of significance. We conclude that assessor A gives higher assessments than assessor B.

b   A 95% confidence interval for $\mu_1 - \mu_2 = \mu_d$ is

$$\overline{d} \pm t_{.025}\frac{s_d}{\sqrt{n}} \Rightarrow 1.488 \pm 2.365\frac{1.491}{\sqrt{8}} \Rightarrow 1.488 \pm 1.247$$

or $.241 < (\mu_1 - \mu_2) < 2.735$.

c   In order to apply the paired-difference test, the 8 properties must be randomly and independently selected and the assessments must be normally distributed.

d   Yes. If the individual assessments are normally distributed, then the mean of four assessments will be normally distributed. Hence, the difference $x_A - \overline{x}$ will be normally distributed and the t test on the differences is valid as in **c**.

10.45   For this exercise, $s^2 = .3214$ and $n = 15$. A 90% confidence interval for σ^2 will be

$$\frac{(n-1)s^2}{\chi^2_{\alpha/2}} < \sigma^2 < \frac{(n-1)s^2}{\chi^2_{(1-\alpha/2)}}$$

where $\chi^2_{\alpha/2}$ represents the value of χ^2 such that 5% of the area under the curve (shown in Figure 10.3) lies to its right. Similarly, $\chi^2_{(1-\alpha/2)}$ will be the χ^2 value such that an area .95 lies to its right.

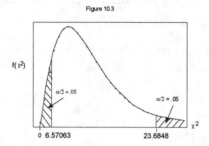

Figure 10.3

Hence, we have located one-half of α in each tail of the distribution. Indexing $\chi^2_{.05}$ and $\chi^2_{.95}$ with $n - 1 = 14$ degrees of freedom in Table 5 yields

$$\chi^2_{.05} = 23.6848 \quad \text{and} \quad \chi^2_{.95} = 6.57063$$

and the confidence interval is

$$\frac{14(.3214)}{23.6848} < \sigma^2 < \frac{14(.3214)}{6.57063} \quad \text{or} \quad .190 < \sigma^2 < .685$$

10.49 **a**   The hypothesis to be tested is

$$H_0: \mu = 5 \qquad H_a: \mu \neq 5$$

Calculate $\bar{x} = \dfrac{\sum x_i}{n} = \dfrac{19.96}{4} = 4.99$

$$s^2 = \frac{\sum x_i^2 - \dfrac{(\sum x_i)^2}{n}}{n - 1} = \frac{99.6226 - \dfrac{(19.96)^2}{4}}{3} = .0074$$

and the test statistic is

$$t = \frac{\bar{x} - \mu_0}{s/\sqrt{n}} = \frac{4.99 - 5}{\sqrt{\dfrac{.0074}{4}}} = -.232$$

The rejection region with $\alpha = .05$ and $n - 1 = 3$ degrees of freedom is found from Table 4 as $|t| > t_{.025} = 3.182$. Since the observed value of the test statistic does not fall in the rejection region, H_0 is not rejected. There is insufficient evidence to show that the mean differs from 5 mg/cc.

b   The manufacturer claims that the range of the potency measurements will equal .2. Since this range is given to equal 6σ, we know that $\sigma \approx .0333$. Then

$$H_0: \sigma^2 = (.0333)^2 = .0011 \quad H_a: \sigma^2 > .0011$$

The test statistic is $\chi^2 = \dfrac{(n - 1)s^2}{\sigma_0^2} = \dfrac{3(.0074)}{.0011} = 20.18$

and the one-tailed rejection region with $\alpha = .05$ and $n - 1 = 3$ degrees of freedom is

$$\chi^2 > \chi^2_{.05} = 7.81$$

H_0 is rejected; there is sufficient evidence to indicate that the range of the potency will exceed the manufacturer's claim.

10.53 The hypothesis of interest is $H_0: \sigma = 150$ $H_a: \sigma < 150$
Calculate

$$(n-1)s^2 = \sum x_i^2 - \frac{(\sum x_i)^2}{n} = 92,305,600 - \frac{(42,812)^2}{20} = 662,232.8$$

and the test statistic is $\chi^2 = \frac{(n-1)s^2}{\sigma_0^2} = \frac{662,232.8}{150^2} = 29.433$. The one-tailed rejection region with $\alpha = .01$ and $n - 1 = 19$ degrees of freedom is $\chi^2 < \chi^2_{.99} = 7.63273$, and H_0 is not rejected. There is insufficient evidence to indicate that he is meeting his goal.

10.57 The hypothesis of interest is $H_0: \sigma_1^2 = \sigma_2^2$ versus $H_a: \sigma_1^2 \neq \sigma_2^2$
and the test statistic is

$$F = \frac{s_1^2}{s_2^2} = \frac{71^2}{69^2} = 1.059.$$

The critical values of F for various values of α are given below using $df_1 = 15$ and $df_2 = 14$.

α	.10	.05	.025	.01	.005
F_α	2.01	2.46	2.95	3.66	4.25

Hence,

$$p\text{-value} = 2P(F > 1.059) > 2(.10) = .20$$

Since the p-value is so large, H_0 is not rejected. There is no evidence to indicate that the variances are different.

10.61 For each of the three tests, the hypothesis of interest is

$$H_0: \sigma_1^2 = \sigma_2^2 \quad \text{versus} \quad H_a: \sigma_1^2 \neq \sigma_2^2$$

and the test statistics are

$$F = \frac{s_1^2}{s_2^2} = \frac{3.98^2}{3.92^2} = 1.03 \quad F = \frac{s_1^2}{s_2^2} = \frac{4.95^2}{3.49^2} = 2.01 \quad \text{and} \quad F = \frac{s_1^2}{s_2^2} = \frac{16.9^2}{4.47^2} = 14.29$$

The critical values of F for various values of α are given below using $df_1 = 9$ and $df_2 = 9$.

α	.10	.05	.025	.01	.005
F_α	2.44	3.18	4.03	5.35	6.54

Hence, for the first two tests,

$$p\text{-value} > 2(.10) = .20$$

while for the last test,

$$p\text{-value} < 2(.005) = .01.$$

There is no evidence to indicate that the variances are different for the first two tests, but H_0 is rejected for the third variable. The two-sample t-test with a pooled estimate of σ^2 cannot be used for the third variable.

10.65 Paired observations are used to estimate the difference between two population means in preference to an estimation based on independent random samples selected from the two populations because of the increased information caused by blocking the observations. We expect blocking to create a large reduction in the standard deviation, if differences do exist among the blocks.

Paired observations are not always preferable. The degrees of freedom that are available for estimating σ^2 are less for paired than for unpaired observations. If there were no difference between the blocks, the paired experiment would then be less beneficial.

10.69 Since it is necessary to determine whether the injected rats drink more water than noninjected rats, the hypothesis to be tested is

$$\text{H}_0\text{: } \mu = 22.0 \qquad \text{H}_a\text{: } \mu > 22.0$$

and the test statistic is

$$t = \frac{\bar{x} - \mu_0}{s/\sqrt{n}} = \frac{31.0 - 22.0}{\dfrac{6.2}{\sqrt{17}}} = 5.985.$$

Using the *critical value approach*, the rejection region with $\alpha = .05$ and $n - 1 = 16$ degrees of freedom is located in the upper tail of the t-distribution and is found from Table 4 as $t > t_{.05} = 1.746$. Since the observed value of the test statistic falls in the rejection region, H_0 is rejected and we conclude that the injected rats do drink more water than the noninjected rats. The 90% confidence interval is

$$\bar{x} \pm t_{.05} \frac{s}{\sqrt{n}} \quad \Rightarrow \quad 31.0 \pm 1.746 \frac{6.2}{\sqrt{17}} \quad \Rightarrow \quad 31.0 \pm 2.625$$

or $28.375 < \mu < 33.625$.

10.72 The student may use the rounded values for $\bar{x}$ and s given in the display, or he may wish to calculate $\bar{x}$ and s and use the more exact calculations for the confidence intervals. The calculations follow.

a $\bar{x} = \dfrac{\sum x_i}{n} = \dfrac{1845}{10} = 184.5$

$s^2 = \dfrac{\sum x_i^2 - \dfrac{(\sum x_i)^2}{n}}{n - 1} = \dfrac{344,567 - \dfrac{(1845)^2}{10}}{9} = 462.7222$

$s = 21.511$ and the 95% confidence interval is

$$\bar{x} \pm t_{.025} \frac{s}{\sqrt{n}} \quad \Rightarrow \quad 184.5 \pm 2.262 \frac{21.511}{\sqrt{10}} \quad \Rightarrow \quad 184.5 \pm 15.4$$

or $169.1 < \mu < 199.9$.

b $\bar{x} = \dfrac{\sum x_i}{n} = \dfrac{730}{10} = 73.0$

$$s^2 = \frac{\sum x_i^2 - \dfrac{(\sum x_i)^2}{n}}{n-1} = \frac{53514 - \dfrac{(730)^2}{10}}{9} = 24.8889$$

$s = 4.989$ and the 95% confidence interval is

$$\bar{x} \pm t_{.025} \frac{s}{\sqrt{n}} \quad \Rightarrow \quad 73.0 \pm 2.262 \frac{4.989}{\sqrt{10}} \quad \Rightarrow \quad 73.0 \pm 3.57$$

or $69.43 < \mu < 76.57$.

c $\bar{x} = \dfrac{\sum x_i}{n} = \dfrac{25.42}{10} = 2.542$

$$s^2 = \frac{\sum x_i^2 - \dfrac{(\sum x_i)^2}{n}}{n-1} = \frac{65.8398 - \dfrac{(25.42)^2}{10}}{9} = .13579556$$

$s = .3685$ and the 95% confidence interval is

$$\bar{x} \pm t_{.025} \frac{s}{\sqrt{n}} \quad \Rightarrow \quad 2.54 \pm 2.262 \frac{.3685}{\sqrt{10}} \quad \Rightarrow \quad 2.54 \pm .26$$

or $2.28 < \mu < 2.80$.

d   No. The relationship between the confidence intervals is not the same as the relationship between the original measurements.

10.75 Use the computing formulas or your scientific calculator to calculate

$$\bar{x} = \frac{\sum x_i}{n} = \frac{322.1}{13} = 24.777$$

$$s^2 = \frac{\sum x_i^2 - \dfrac{(\sum x_i)^2}{n}}{n-1} = \frac{8114.59 - \dfrac{(322.1)^2}{13}}{12} = 11.1619$$

$s = 3.3409$ and the 95% confidence interval is

$$\bar{x} \pm t_{.025} \frac{s}{\sqrt{n}} \quad \Rightarrow \quad 24.777 \pm 2.179 \frac{3.3409}{\sqrt{13}} \quad \Rightarrow \quad 24.777 \pm 2.019$$

or $22.578 < \mu < 26.796$.

10.79 **a**   The range of the first sample is 47 while the range of the second sample is only 16. There is probably a difference in the variances.

b   The hypothesis of interest is

$$H_0: \sigma_1^2 = \sigma_2^2 \quad \text{versus} \quad H_a: \sigma_1^2 \neq \sigma_2^2$$

Calculate $s_1^2 = \dfrac{177,294 - \dfrac{(838)^2}{4}}{3} = 577.6667$ $\qquad s_2^2 = \dfrac{192,394 - \dfrac{(1074)^2}{6}}{5} = 29.6$

and the test statistic is

$$F = \frac{s_1^2}{s_2^2} = \frac{577.6667}{29.6} = 19.516$$

The critical values with $df_1 = 3$ and $df_2 = 5$ are shown below from Table 6.

α	.10	.05	.025	.01	.005
F_α	3.62	5.41	7.76	12.06	16.53

Hence,

$$p\text{-value} = 2P(F > 19.516) < 2(.005) = .01$$

Since the p-value is smaller than .01, H_0 is rejected at the 1% level of signficance. There is a difference in variability.

c   Since the Student's t test requires the assumption of equal variances, it would be inappropriate in this instance. You should use the unpooled t test with Satterthwaite's approximation to the degrees of freedom.

10.83   A paired-difference test is used, since the two samples are not random and independent. The hypothesis of interest is

$$H_0: \mu_1 - \mu_2 = 0 \qquad H_a: \mu_1 - \mu_2 > 0$$

and the table of differences, along with the calculation of $\bar{d}$ and s_d^2, is presented below.

Pair	1	2	3	4	Totals
d_i	-1	5	11	7	22

$\bar{d} = \dfrac{\sum d_i}{n} = \dfrac{22}{4} = 5.5$ $\qquad s_d^2 = \dfrac{\sum d_i^2 - \dfrac{(\sum d_i)^2}{n}}{n-1} = \dfrac{196 - \dfrac{(22)^2}{4}}{3} = 25$ and $s_d = 5$

and the test statistic is

$$t = \frac{\bar{d} - \mu_d}{s_d/\sqrt{n}} = \frac{5.5 - 0}{\dfrac{5}{\sqrt{4}}} = 2.2$$

The one-tailed p-value with $df = 3$ can be bounded between .05 and .10. Since this value is greater than .10, H_0 is not rejected. The results are not significant; there is insufficient evidence to indicate that lack of school experience has a depressing effect on IQ scores.

10.87   The object is to determine whether or not there is a difference between the mean responses for the two different stimuli to which the people have been subjected. The samples are independently and randomly selected, and the assumptions necessary for the t test of Section 10.4 are met. The hypothesis to be tested is

$$H_0: \mu_1 - \mu_2 = 0 \qquad H_a: \mu_1 - \mu_2 \neq 0$$

and the preliminary calculations are as follows:

$$\bar{x}_1 = \frac{15}{8} = 1.875 \text{ and } \bar{x}_2 = \frac{21}{8} = 2.625$$

$$s_1^2 = \frac{33 - \frac{(15)^2}{8}}{7} = .69643 \text{ and } s_2^2 = \frac{61 - \frac{(21)^2}{8}}{7} = .83929$$

Since the ratio of the variances is less than 3, you can use the pooled t test. The pooled estimator of σ^2 is calculated as

$$s^2 = \frac{(n_1 - 1)s_1^2 + (n_2 - 1)s_2^2}{n_1 + n_2 - 2} = \frac{4.875 + 5.875}{14} = .7679$$

and the test statistic is

$$t = \frac{(\bar{x}_1 - \bar{x}_2) - 0}{\sqrt{s^2\left(\frac{1}{n_1} + \frac{1}{n_2}\right)}} = \frac{1.875 - 2.625}{\sqrt{.7679\left(\frac{1}{8} + \frac{1}{8}\right)}} = -1.712$$

The two-tailed rejection region with $\alpha = .05$ and $df = 14$ is $|t| > t_{.025} = 2.145$, and H_0 is not rejected. There is insufficient evidence to indicate that there is a difference in the means.

10.91 **a**   The Minitab printouts below show the 90% confidence intervals for the paired and unpaired analyses.

Paired T-Test and CI: A, B

```
Paired T for A - B
                N       Mean      StDev    SE Mean
A               9     1.2367     0.0644     0.0215
B               9     0.9778     0.0494     0.0165
Difference      9     0.2589     0.0805     0.0268
90% CI for mean difference: (0.2090, 0.3088)
T-Test of mean difference = 0 (vs not = 0): T-Value = 9.64   P-Value = 0.000
```

Two-Sample T-Test and CI: A, B

```
Two-sample T for A vs B
     N     Mean     StDev    SE Mean
A    9    1.2367    0.0644     0.021
B    9    0.9778    0.0494     0.016
Difference = mu A - mu B
Estimate for difference:  0.2589
90% CI for difference: (0.2116, 0.3061)
T-Test of difference = 0 (vs not =): T-Value = 9.56   P-Value = 0.000   DF =
16
Both use Pooled StDev = 0.0574
```

The width of the paired interval, $.2090 < \mu_1 - \mu_2 < .3088$ is .0998, while the width of the unpaired interval, $.2116 < \mu_1 - \mu_2 < .3061$ is .0945. You have actually *lost* information by pairing when in fact the data were not collected in a paired manner.

b   No. You must use the statistical method of analysis which is correct for the type of experiment (paired or unpaired) which you have designed!

10.95   It is possible to test the null hypothesis $H_0: \sigma_1^2 = \sigma_2^2$ against any one of three alternative hypotheses:

$$(1) \ H_a: \sigma_1^2 \neq \sigma_2^2 \quad (2) \ H_a: \sigma_1^2 < \sigma_2^2 \quad (3) \ H_a: \sigma_1^2 > \sigma_2^2$$

a   The first alternative would be preferred by the manager of the dairy. He does not know anything about the variability of the two machines and would wish to detect departures from equality of the type, $\sigma_1^2 < \sigma_2^2$ or $\sigma_1^2 > \sigma_2^2$. These alternatives are implied in (1).

b   The salesman for company A would prefer that the experimenter select the second alternative. Rejection of the null hypothesis would imply that his machine had smaller variability. Moreover, even if the null hypothesis were not rejected, there would be no evidence to indicate that the variability of the company A machine was greater than the variability of the company B machine.

c   The salesman for company B would prefer the third alternative for a similar reason.

10.99   A paired-difference test is used. To test $H_0: \mu_2 - \mu_1 = 0$ versus $H_a: \mu_2 - \mu_1 > 0$, where μ_2 is the mean reaction time after injection and μ_1 is the mean reaction time before injection, calculate the differences $(x_2 - x_1)$:

$$6, \ 1, \ 6, \ 1$$

Then $\bar{d} = \dfrac{\sum d_i}{n} = \dfrac{14}{4} = 3.5 \quad s_d^2 = \dfrac{74 - \dfrac{(14)^2}{4}}{3} = 8.33 \text{ and } \ s_d = 2.88675$

and the test statistic is

$$t = \frac{\bar{d} - \mu_d}{s_d / \sqrt{n}} = \frac{3.5 - 0}{\dfrac{2.88675}{\sqrt{4}}} = 2.425$$

For a one-tailed test with $df = 3$, the rejection region with $\alpha = .05$ is $t > t_{.05} = 2.353$, and H_0 is rejected. We conclude that the drug significantly increases reaction time.

10.103   The underlying populations are ratings and can only take on the finite number of values, 1, 2, ..., 9, 10. Neither population has a normal distribution, but both are discrete. Further, the samples are not independent, since the same person is asked to rank each car design. Hence, two of the assumptions required for the Student's t test have been violated.

10.107   The Minitab printout below shows the summary statistics for the two samples:

```
                N       Mean      StDev   SE Mean
Method 1   5     137.0     10.2      4.5
Method 2   5     147.20    7.36      3.3
```

Since the ratio of the two sample variances is less than 3, you can use the pooled t test to compare the two methods of measurement, using the remainder of the Minitab printout below:

Two-Sample T-Test and CI: Method 1, Method 2

```
Two-sample T for Method 1 vs Method 2
Difference = mu Method 1 - mu Method 2
Estimate for difference:  -10.20
95% CI for difference: (-23.15, 2.75)
T-Test of difference = 0 (vs not =): T-Value = -1.82   P-Value = 0.107   DF =
8
Both use Pooled StDev = 8.88
```

The test statistic is $t = -1.82$ with p-value $= .107$ and the results are not signficant. There is insufficient evidence to declare a difference in the two population means.

10.111 Use the **Interpreting Confidence Intervals** applet. Answers will vary from student to student. The widths of the ten intervals will not be the same, since the value of s changes with each new sample. The student should find that approximately 95% of the intervals in the first applet contain μ, while roughly 99% of the intervals in the second applet contain μ.

10.115 Use the **Two Sample t Test: Independent Samples** applet. The hypothesis to be tested concerns the difference between mean recovery rates for the two surgical procedures. Let μ_1 be the population mean for Procedure I and μ_2 be the population mean for Procedure II. The hypothesis to be tested is

$$H_0: \mu_1 - \mu_2 = 0 \qquad H_a: \mu_1 - \mu_2 \neq 0$$

Since the ratio of the variances is less than 3, you can use the pooled t test. Enter the appropriate statistics into the applet and you will find that test statistic is

$$t = \frac{(\bar{x}_1 - \bar{x}_2) - 0}{\sqrt{s^2\left(\dfrac{1}{n_1} + \dfrac{1}{n_2}\right)}} = -3.33$$

with a two-tailed p-value of .0030. Since the p-value is very small, H_0 can be rejected for any value of α greater than .003 and the results are judged highly significant. There is sufficient evidence to indicate a difference in the mean recovery rates for the two procedures.

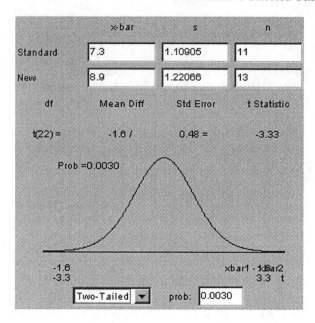

11: The Analysis of Variance

11.1   In comparing 6 populations, there are $k - 1$ degrees of freedom for treatments and $n = 6(10) = 60$. The ANOVA table is shown below.

Source	df
Treatments	5
Error	54
Total	59

11.4   Similar to Exercise 11.1. With $n = 4(6) = 24$ and $k = 4$, the sources of variation and associated *df* are shown below.

Source	df
Treatments	3
Error	20
Total	23

11.7   The following preliminary calculations are necessary:

$$T_1 = 14 \quad T_2 = 19 \quad T_3 = 5 \quad G = 38$$

a $CM = \dfrac{(\sum x_{ij})^2}{n} = \dfrac{(38)^2}{14} = 103.142857$

   Total

   $SS = \sum x_{ij}^2 - CM = 3^2 + 2^2 + \cdots + 2^2 + 1^2 - CM = 130 - 103.142857 = 26.8571$

b $SST = \sum \dfrac{T_i^2}{n_i} - CM = \dfrac{14^2}{5} + \dfrac{19^2}{5} + \dfrac{5^2}{4} - CM = 117.65 - 103.142857 = 14.5071$

   and $MST = \dfrac{SST}{k-1} = \dfrac{14.5071}{2} = 7.2536$.

c   By subtraction, $SSE = $ Total $SS - SST = 26.8571 - 14.5071 = 12.3500$ and the degrees of freedom, by subtraction, are $13 - 2 = 11$. Then

$$MSE = \dfrac{SSE}{11} = \dfrac{12.3500}{11} = 1.1227$$

d   The information obtained in parts **a-c** is consolidated in an ANOVA table.

Source	df	SS	MS
Treatments	2	14.5071	7.2536
Error	11	12.3500	1.1227
Total	13	26.8571	

e   The hypothesis to be tested is

$H_0: \mu_1 = \mu_2 = \mu_3$ versus $H_a:$ at least one pair of means are different

f   The rejection region for the test statistic $F = \dfrac{MST}{MSE} = \dfrac{7.2536}{1.1227} = 6.46$ is based on an F-distribution with 2 and 11 degrees of freedom. The critical values of F for bounding the *p*-value for this one-tailed test are shown below.

α	.10	.05	.025	.01	.005
F_α	2.86	3.98	5.26	7.21	8.91

Since the observed value $F = 6.46$ is between $F_{.01}$ and $F_{.025}$,

$$.01 < p\text{-value} < .025$$

and H_0 is rejected at the 5% level of significance. There is a difference among the means.

11.10 a The following preliminary calculations are necessary:

$$T_1 = 380 \quad T_2 = 199 \quad T_3 = 261 \quad G = 840$$

$$\text{CM} = \frac{(\sum x_{ij})^2}{n} = \frac{(840)^2}{11} = 64,145.4545$$

Total SS $= \sum x_{ij}^2 - \text{CM} = 65,286 - \text{CM} = 1140.5455$

$$\text{SST} = \sum \frac{T_i^2}{n_i} - \text{CM} = \frac{380^2}{5} + \frac{199^2}{3} + \frac{261^2}{3} - \text{CM} = 641.87883$$

Calculate MS $= $ SS/df and consolidate the information in an ANOVA table.

Source	df	SS	MS
Treatments	2	641.8788	320.939
Error	8	498.6667	62.333
Total	10	1140.5455	

b The hypothesis to be tested is

$$H_0: \mu_1 = \mu_2 = \mu_3 \qquad H_a: \text{at least one pair of means are different}$$

and the F test to detect a difference in mean student response is

$$F = \frac{\text{MST}}{\text{MSE}} = 5.15.$$

The rejection region with $\alpha = .05$ and 2 and 8 df is $F > 4.46$ and H_0 is rejected. There is a significant difference in mean response due to the three different methods.

11.13 a We would be reasonably confident that the data satisfied the normality assumption because each measurement represents the average of 10 continuous measurements. The Central Limit Theorem assures us that this mean will be approximately normally distributed.

b We have a completely randomized design with four treatments, each containing 6 measurements. The analysis of variance table is given in the Minitab printout. The F test is

$$F = \frac{\text{MST}}{\text{MSE}} = \frac{6.580}{.115} = 57.38$$

with p-value $= .000$ (in the column marked "P"). Since the p-value very small (less than .01), H_0 is rejected. There is a significant difference in the mean leaf length among the four locations with P $< .01$ or even P $< .001$.

c The hypothesis to be tested is $H_0: \mu_1 = \mu_4$ versus $H_a: \mu_1 \neq \mu_4$ and the test statistic is

$$t = \frac{\overline{x}_1 - \overline{x}_4}{\sqrt{MSE\left(\dfrac{1}{n_1} + \dfrac{1}{n_4}\right)}} = \frac{6.0167 - 3.65}{\sqrt{.115\left(\dfrac{1}{6} + \dfrac{1}{6}\right)}} = 12.09$$

The p-value with $df = 20$ is $2P(t > 12.09)$ is bounded (using Table 4) as

$$p\text{-value} < 2(.005) = .01$$

and the null hypothesis is rejected. We conclude that there is a difference between the means.

d The 99% confidence interval for $\mu_1 - \mu_4$ is

$$(\overline{x}_1 - \overline{x}_4) \pm t_{.005}\sqrt{MSE\left(\frac{1}{n_1} + \frac{1}{n_4}\right)}$$

$$(6.0167 - 3.65) \pm 2.845\sqrt{.115\left(\frac{1}{6} + \frac{1}{6}\right)}$$

$$2.367 \pm .557 \quad \text{or} \quad 1.810 < \mu_1 - \mu_4 < 2.924$$

d When conducting the t tests, remember that the stated confidence coefficients are based on random sampling. If you looked at the data and only compared the largest and smallest sample means, the randomness assumption would be disturbed.

11.19 a $w = q_{.05}(4,12)\dfrac{s}{\sqrt{5}} = 4.20\dfrac{s}{\sqrt{5}} = 1.878s$

b $w = q_{.01}(6,12)\dfrac{s}{\sqrt{8}} = 6.10\dfrac{s}{\sqrt{8}} = 2.1567s$

11.23 The design is completely randomized with 3 treatments and 5 replications per treatment. The Minitab printout below shows the analysis of variance for this experiment.

One-way ANOVA: mg/dl versus Lab

```
Analysis of Variance for mg/dl
Source      DF        SS        MS        F        P
Lab          2      42.6      21.3     0.60    0.562
Error       12     422.5      35.2
Total       14     465.0
                             Individual 95% CIs For Mean
                             Based on Pooled StDev
Level        N      Mean     StDev   --+---------+---------+---------+----
1            5    108.86      7.47                  (-----------*----------)
2            5    105.04      6.01      (----------*-----------)
3            5    105.60      3.70      (----------*-----------)
                                       --+---------+---------+---------+----
Pooled StDev =     5.93               100.0     105.0     110.0     115.0

Tukey's pairwise comparisons
Family error rate = 0.0500
Individual error rate = 0.0206
Critical value = 3.77
Intervals for (column level mean) - (row level mean)
            1                  2
2       -6.184
        13.824

3       -6.744           -10.564
        13.264             9.444
```

a   The analysis of variance F test for $H_0 : \mu_1 = \mu_2 = \mu_3$ is $F = .60$ with p-value $= .562$. The results are not significant and H_0 is not rejected. There is insufficient evidence to indicate a difference in the treatment means.

b   Since the treatment means are not significantly different, there is no need to use Tukey's test to search for the pairwise differences. Notice that all three intervals generated by Minitab contain zero, indicating that the pairs cannot be judged different.

11.25   Refer to Exercise 11.24. The given sums of squares are inserted and missing entries found by subtraction. The mean squares are found as $MS = SS/df$.

Source	df	SS	MS	F
Treatments	2	11.4	5.70	4.01
Blocks	5	17.1	3.42	2.41
Error	10	14.2	1.42	
Total	17	42.7		

11.29   Use Minitab to obtain an ANOVA printout, or use the following calculations:

$$CM = \frac{(\sum x_{ij})^2}{n} = \frac{(113)^2}{12} = 1064.08333$$

$$\text{Total SS} = \sum x_{ij}^2 - CM = 6^2 + 10^2 + \cdots + 14^2 - CM = 1213 - CM = 148.91667$$

$$SST = \frac{\sum T_j^2}{3} - CM = \frac{22^2 + 34^2 + 27^2 + 30^2}{3} - CM = 25.58333$$

$$SSB = \frac{\sum B_i^2}{4} - CM = \frac{33^2 + 25^2 + 55^2}{4} - CM = 120.66667 \text{ and}$$

$$SSE = \text{Total SS} - SST - SSB = 2.6667$$

Calculate $MS = SS/df$ and consolidate the information in an ANOVA table.

Source	df	SS	MS	F
Treatments	3	25.5833	8.5278	19.19
Blocks	2	120.6667	60.3333	135.75
Error	6	2.6667	0.4444	
Total	11	148.9167		

a   To test the difference among treatment means, the test statistic is

$$F = \frac{MST}{MSE} = \frac{8.528}{.4444} = 19.19$$

and the rejection region with $\alpha = .05$ and 3 and 6 df is $F > 4.76$. There is a significant difference among the treatment means.

b   To test the difference among block means, the test statistic is

$$F = \frac{MSB}{MSE} = \frac{60.3333}{.4444} = 135.75$$

and the rejection region with $\alpha = .05$ and 2 and 6 df is $F > 5.14$. There is a significant difference among the block means.

c   With $k = 4$, $df = 6$, $n_t = 3$,

$$\omega = q_{.01}(4,6)\frac{\sqrt{\text{MSE}}}{\sqrt{n_t}} = 7.03\sqrt{\frac{.4444}{3}} = 2.71$$

The ranked means are shown below.

$$
\begin{array}{cccc}
7.33 & 9.00 & 10.00 & 11.33 \\
\overline{x}_1 & \overline{x}_3 & \overline{x}_4 & \overline{x}_2
\end{array}
$$

d   The 95% confidence interval is

$$(\overline{x}_A - \overline{x}_B) \pm t_{.025}\sqrt{\text{MSE}\left(\frac{2}{b}\right)}$$

$$(7.333 - 11.333) \pm 2.447\sqrt{.4444\left(\frac{2}{3}\right)}$$

$$-4 \pm 1.332 \quad \text{or} \quad -5.332 < \mu_A - \mu_B < -2.668$$

e   Since there is a significant difference among the block means, blocking has been effective. The variation due to block differences can be isolated using the randomized block design.

11.33   Similar to previous exercises. The Minitab printout for this randomized block experiment is shown below.

Two-way ANOVA: y versus Blocks, Chemicals

```
Analysis of Variance for y
Source      DF        SS        MS        F         P
Blocks      2      7.1717    3.5858    40.21     0.000
Chemical    3      5.2000    1.7333    19.44     0.002
Error       6      0.5350    0.0892
Total      11     12.9067
```

```
                        Individual 95% CI
Blocks        Mean    ---------+---------+---------+---------+-
1            10.87    (----*-----)
2            12.70                          (----*-----)
3            12.22                      (-----*----)
                      ---------+---------+---------+---------+-
                        11.20     11.90     12.60     13.30
```

```
                        Individual 95% CI
Chemical      Mean    -+---------+---------+---------+---------+
1            11.40       (------*------)
2            12.33                    (------*------)
3            11.20    (------*------)
4            12.80                         (------*------)
                      -+---------+---------+---------+---------+
                        10.80     11.40     12.00     12.60     13.20
```

Both the treatment and block means are significantly different. Since the four chemicals represent the treatments in this experiment, Tukey's test can be used to determine where the differences lie:

$$w = q_{.05}(4,6)\frac{\sqrt{MSE}}{\sqrt{n_t}} = 4.90\sqrt{\frac{.0892}{3}} = .845$$

The ranked means are shown below. 11.20   11.40   12.33   12.80

$\overline{x}_3$ $\overline{x}_1$ $\overline{x}_2$ $\overline{x}_4$

The chemicals falls into two significantly different groups—A and C versus B and D.

11.37 A randomized block design has been used with "estimators" as treatments and "construction job" as the block factor. The analysis of variance table is found in the Minitab printout below.

Two-way ANOVA: Cost versus Estimator, Job

```
Analysis of Variance for Cost
Source       DF        SS         MS        F         P
Estimator     2     10.862     5.431      7.20     0.025
Job           3     37.607    12.536     16.61     0.003
Error         6      4.528     0.755
Total        11     52.997
                            Individual 95% CI
Estimator      Mean   --------+---------+---------+---------+--------+---
A             32.61   (--------*--------)
B             34.89                          (--------*--------)
C             34.19                    (--------*--------)
                       --------+---------+---------+---------+--------+---
                       32.40      33.60      34.80      36.00
```

Both treatments and blocks are significant. The treatment means can be further compared using Tukey's test with

$$w = q_{.05}(3,6)\frac{\sqrt{MSE}}{\sqrt{n_t}} = 4.34\sqrt{\frac{.755}{4}} = 1.885$$

The ranked means are shown below.

32.6125   34.1875   34.8875

$\overline{x}_A$ $\overline{x}_C$ $\overline{x}_B$

Estimators A and B show a significant difference in average costs.

11.39 **a-b** There are $4 \times 5 = 20$ treatments and $4 \times 5 \times 3 = 60$ total observations.
   c   In a factorial experiment, variation due to the interaction $A \times B$ is isolated from SSE. The sources of variation and associated degrees of freedom are given below.

Source	df
A	3
B	4
A × B	12
Error	40
Total	59

11.43 **a** Similar to Exercise 11.42. Based on the fact that the mean response for the two levels of factor B behaves very differently depending on the level of factor A under investigation, there is a strong interaction present between factors A and B.

b The test statistic for interaction is $F = MS(AB)/MSE = 37.85$ with p-value $= .000$ from the Minitab prinout. There is evidence of a significant interaction. That is, the effect of factor A depends upon the level of factor B at which A is measured.

c In light of this type of interaction, the main effect means (averaged over the levels of the other factor) differ only slightly. Hence, a test of the main-effect terms produces a non-significant result.

d No. A significant interaction indicates that the effect of one factor depends upon the level of the other. Each factor-level combination should be investigated individually.

e Answers will vary.

11.47 **a** The design is a 2×4 factorial experiment with $r = 5$ replications. There are two factors, Gender and School, one at two levels and one at four levels.

b The analysis of variance table can be found using a computer printout or the following calculations:

Gender	Schools 1	2	3	4	Total
Male	2919	3257	3330	2461	11967
Female	3082	3629	3344	2410	12465
Total	6001	6886	6674	4871	24432

$$CM = \frac{24432^2}{40} = 14923065.6 \qquad \text{Total}$$

$$SS = 15281392 - CM = 358326.4$$

$$SSG = \frac{11967^2 + 12465^2}{20} - CM = 6200.1$$

$$SS(Sc) = \frac{6001^2 + 6886^2 + 6674^2 + 4871^2}{10} - CM = 246725.8$$

$$SS(G \times Sc) = \frac{2919^2 + 3257^2 + \cdots + 2410^2}{5} - SSG - SS(Sc) - CM = 10574.9$$

Source	df	SS	MS	F
G	1	6200.1	6200.100	2.09
Sc	3	246725.8	82241.933	27.75
G × Sc	3	10574.9	3524.967	1.19
Error	32	94825.6	2963.300	
Total	39	358326.4		

c   The test statistic is $F = \text{MS(GSc)/MSE} = 1.19$ and the rejection region is $F > 2.92$ (with $\alpha = .05$). Alternately, you can bound the p-value $> .10$. Hence, H_0 is not rejected. There is insufficient evidence to indicate interaction between gender and schools.

d   You can see in the interaction plot that there is a small difference between the average scores for male and female students at schools 1 and 2, but no difference to speak of at the other two schools. The interaction is not significant.

Interaction Plot - Data Means for Score

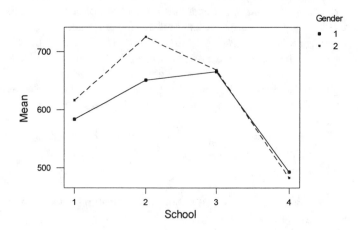

e   The test statistic for testing gender is $F = 2.09$ with $F_{.05} = 4.17$ (or p-value $> .10$). The test statistic for schools is $F = 27.75$ with $F_{.05} = 2.92$ (or p-value $< .005$). There is a significant effect due to schools. Using Tukey's method of paired comparisons with $\alpha = .01$, calculate

$$w = q_{.01}(4,32)\frac{\sqrt{\text{MSE}}}{\sqrt{n_t}} = 4.80\sqrt{\frac{2963.3}{10}} = 82.63$$

The ranked means are shown below.

$$
\begin{array}{cccc}
487.1 & 600.1 & 667.4 & 688.6 \\
\overline{x}_4 & \overline{x}_1 & \overline{x}_3 & \overline{x}_2
\end{array}
$$

11.51   The objective is to determine whether or not mean reaction time differs for the five stimuli. The four people used in the experiment act as blocks, in an attempt to isolate the variation from person to person. A randomized block design is used, and the analysis of variance table is given in the printout.

a   The F statistic to detect a difference due to stimuli is

$$F = \frac{\text{MST}}{\text{MSE}} = 27.78$$

with p-value $= .000$. There is a significant difference in the effect of the five stimuli.

b   The treatment means can be further compared using Tukey's test with

$$w = q_{.05}(5,12)\frac{\sqrt{\text{MSE}}}{\sqrt{n_t}} = 4.51\sqrt{\frac{.00708}{4}} = .190$$

The ranked means are shown below.

E	A	B	D	C
.525	.7	.8	1.025	1.05

<hr>

c   The F test for blocks produces $F = 6.59$ with p-value $= .007$. The block differences are signficant; blocking has been effective.

11.55   This is similar to previous exercises. The completed ANOVA table is shown below.

Source	df	SS	MS	F
A	1	1.14	1.14	6.51
B	2	2.58	1.29	7.37
A × B	2	0.49	0.245	1.40
Error	24	4.20	0.175	
Total	29	8.41		

a   The test statistic is $F = \text{MS(AB)/MSE} = 1.40$ and the rejection region is $F > 3.40$. There is insufficient evidence to indicate an interaction.

b   Using Table 6 with $df_1 = 2$ and $df_2 = 24$, the following critical values are obtained.

α	.10	.05	.025	.01	.005
F_α	2.54	3.40	4.32	5.61	6.66

The observed value of F is less than $F_{.10}$, so that p-value $> .10$.

c   The test statistic for testing factor A is $F = 6.51$ with $F_{.05} = 4.26$. There is evidence that factor A affects the response.

d   The test statistic for factor B is $F = 7.37$ with $F_{.05} = 3.40$. Factor B also affects the response.

11.59 a   The design is a randomized block design, with weeks representing blocks and stores as treatments.

b   The Minitab computer printout is shown below.

Two-way ANOVA: Total versus Week, Store

```
Analysis of Variance for Total
Source       DF        SS        MS         F        P
Week          3     571.7     190.6      8.27    0.003
Store         4     684.6     171.2      7.43    0.003
Error        12     276.4      23.0
Total        19    1532.7
```

c   The F test for treatments is $F = 7.43$ with p-value $= .003$. The p-value is small enough to allow rejection of H_0. There is a significant difference in the average weekly totals for the five supermarkets.

d   With $k = 5$, $df = 12$, $n_t = 4$,

$$\omega = q_{.05}(5,12)\frac{\sqrt{MSE}}{\sqrt{n_t}} = 4.51\sqrt{\frac{23.0}{4}} = 10.81$$

The ranked means are shown below.

1	5	4	3	2
240.23	249.19	252.18	254.87	256.99

12: Linear Regression and Correlation

12.1   The line corresponding to the equation $y = 2x + 1$ can be graphed by locating the y values corresponding to $x = 0, 1$, and 2.

$$\text{When } x = 0, y = 2(0) + 1 = 1$$
$$\text{When } x = 1, y = 2(1) + 1 = 3$$
$$\text{When } x = 2, y = 2(2) + 1 = 5$$

The graph is shown in Figure 12.1.

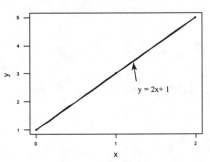

Flgure12.1

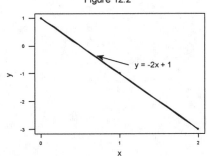

Figure 12.2

Note that the equation is in the form

$$y = \alpha + \beta x$$

Thus, the slope of the line is $\beta = 2$ and the y-intercept is $\alpha = 1$.

12.5   A deterministic mathematical model is a model in which the value of a response y is exactly predicted from values of the variables that affect the response. On the other hand, a probabilistic mathematical model is one that contains random elements with specific probability distributions. The value of the response y in this model is not exactly determined.

12.9   **a**   Calculate

$$\sum x_i = 850.8 \qquad \sum y_i = 3.755 \qquad \sum x_i y_i = 443.7727$$
$$\sum x_i^2 = 101,495.78 \qquad \sum y_i^2 = 1.941467 \qquad n = 9$$

Then

$$S_{xy} = \sum x_i y_i - \frac{(\sum x_i)(\sum y_i)}{n} = 88.80003333$$

$$S_{xx} = \sum x_i^2 - \frac{(\sum x_i)^2}{n} = 21,066.82$$

$$S_{yy} = \sum y_i^2 - \frac{(\sum y_i)^2}{n} = .3747976$$

b-c $b = \dfrac{S_{xy}}{S_{xx}} = \dfrac{88.800033}{21066.82} = .00421516$ and $a = \bar{y}$
$- b\bar{x} = .41722 - .00421516(94.5333) = .0187$
and the least squares line is

$$\hat{y} = a + bx = .0187 + .0042x.$$

The graph of the least squares line and the nine data points are shown in Figure 12.6.

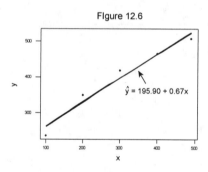

Figure 12.6

$\hat{y} = 195.90 + 0.67x$

d   When $x = 100$, the value for y can be predicted using the least squares line as

$$\hat{y} = .0187 + .0042(100) = .44$$

e   Using the additivity properties for the sums of squares and degrees of freedom for an analysis of variance, and the fact that MS $=$ SS/df, the completed ANOVA table is shown below.

Analysis of Variance

Source	DF	SS	MS	F	P
Regression	1	0.37431	0.37431	5334.84	0.000
Residual Error	7	0.00049	0.00007		
Total	8	0.37480			

12.13 **a**   The hypothesis to be tested is

$$H_0: \beta = 0 \text{ versus } H_a: \beta \neq 0$$

and the test statistic is a Student's t, calculated as

$$t = \frac{b - \beta_0}{\sqrt{MSE/S_{xx}}} = \frac{1.2 - 0}{\sqrt{0.533/10}} = 5.20$$

The critical value of t is based on $n - 2 = 3$ degrees of freedom and the rejection region for $\alpha = .05$ is $|t| > t_{.025} = 3.182$. Since the observed value of t falls in the rejection region, we reject H_0 and conclude that $\beta \neq 0$. That is, x is useful in the prediction of y.

b   From the ANOVA table in Exercise 12.6, calculate

$$F = \frac{MSR}{MSE} = \frac{14.4}{.5333} = 27.00$$

which is the square of the t statistic from part **a**: $t^2 = (5.20)^2 = 27.0$.

c   The critical value of t from part **a** was $t_{.025} = 3.182$, while the critical value of F from part **b** with $df_1 = 1$ and $df_2 = 3$ is $F_{.05} = 10.13$. Notice that the relationship between the two critical values is

$$F = 10.13 = (3.182)^2 = t^2$$

12.17 **a**   The dependent variable (to be predicted) is y = cost and the independent variable is x = distance.

b   Preliminary calculations:

$$\sum x_i = 21,530 \qquad \sum y_i = 5052 \qquad \sum x_i y_i = 7,569,999$$
$$\sum x_i^2 = 37,763,314 \qquad \sum y_i^2 = 1,695,934 \qquad n = 18$$

Then

$$S_{xy} = \sum x_i y_i - \frac{(\sum x_i)(\sum y_i)}{n} = 1,527,245.667$$

$$S_{xx} = \sum x_i^2 - \frac{(\sum x_i)^2}{n} = 12,011,041.78$$

$b = \dfrac{S_{xy}}{S_{xx}} = .127153$ and $a = \bar{y}$

$-\, b\bar{x} = 280.6667 - .127153(1196.1111) = 128.57699$

and the least squares line is

$$\hat{y} = a + bx = 128.57699 + .127153x.$$

c   The plot is shown in Figure 12.7. The line appears to fit well through the 18 data points.

Figure 12.7

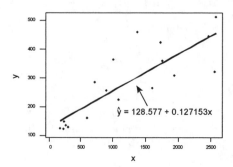

$$\hat{y} = 128.577 + 0.127153x$$

d   Calculate Total SS $= S_{yy} = \sum y_i^2 - \dfrac{(\sum y_i)^2}{n} = 1,695,934 - \dfrac{(5052)^2}{18} = 278,006.$

Then

$$\text{SSE} = S_{yy} - \frac{(S_{xy})^2}{S_{xx}} = 278,006 - \frac{(1,527,245.667)^2}{12,011,041.78} = 83811.41055$$

and MSE $= \dfrac{\text{SSE}}{n-2} = \dfrac{83811.41055}{16} = 5238.213.$ The hypothesis to be tested is

$$H_0\colon \beta = 0 \quad \text{versus} \quad H_a\colon \beta \neq 0$$

and the test statistic is

$$t = \frac{b - \beta_0}{\sqrt{\text{MSE}/S_{xx}}} = \frac{.127153 - 0}{\sqrt{5238.213/12,011,041.78}} = 6.09$$

The critical value of t is based on $n - 2 = 16$ degrees of freedom and the rejection region for $\alpha = .05$ is $|t| > t_{.025} = 2.120$, and H_0 is rejected. There is evidence at the 5% level to indicate that x and y are linearly related. That is, the regression model $y = \alpha + \beta x + \epsilon$ is useful in predicting cost y.

12.21   Use a plot of residuals versus fits. The plot should appear as a random scatter of points, free of any patterns.

12.25   **a**   If you look carefully, there appears to be a slight curve to the five points.
b   The fit of the regression line, measured as $r^2 = .959$ indicates that 95.9% of the overall variation can be explained by the straight line model.
c   When we look at the residuals there is a strong curvilinear pattern that has not been explained by the straight line model. The relationship between time in months and number of books appears to be curvilinear.

12.29   **a**   Although very slight, the student might notice a slight curvature to the data points.
b   The fit of the linear model is very good, assuming that this is *indeed* the correct model for this data set.

c   The normal probability plot follows the correct pattern for the assumption of normality. However, the residuals show the pattern of a quadratic curve, indicating that a quadratic rather than a linear model may have been the correct model for this data.

12.33 If the value of r is positive, then the least squares line slopes upward to the right. Similarly, if the value of r is negative, the line slopes downward to the right. The coefficient of correlation r will be zero only when b is zero (see Section 12.8 of the text). Moreover, the least squares equation when $b = 0$ is given by $\hat{y} = a$. The variable x has no effect on the value of y and there is no linear correlation between x and y. Finally, r will equal ± 1 only when SSE $= 0$; that is, all points fall exactly on the fitted line.

12.35 a   Refer to Figure 12.9. The sample correlation coefficient will be positive.

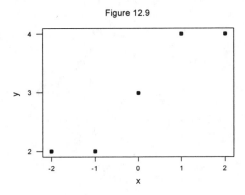

Figure 12.9

b   Calculate

$$S_{xy} = \sum x_i y_i - \frac{(\sum x_i)(\sum y_i)}{n} = 6 - \frac{0(15)}{5} = 6$$

$$S_{xx} = \sum x_i^2 - \frac{(\sum x_i)^2}{n} = 10 - \frac{0^2}{5} = 10$$

$$S_{yy} = \sum y_i^2 - \frac{(\sum y_i)^2}{n} = 49 - \frac{15^2}{5} = 4$$

Then $r = \dfrac{S_{xy}}{\sqrt{S_{xx}S_{yy}}} = \dfrac{6}{\sqrt{40}} = .9487$ and $r^2 = (.9487)^2 = .9000.$ Approximately 90% of the total sum of squares of deviations was reduced by using the least squares equation instead of $\bar{y}$ as a predictor of y.

12.39 When the pre-test score x is high, the post-test score y should also be high. There should be a positive correlation.

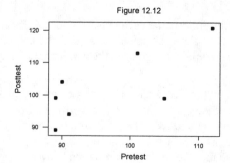

Figure 12.12

Calculate

$$S_{xy} = \sum x_i y_i - \frac{(\sum x_i)(\sum y_i)}{n} = 70,006 - \frac{677(719)}{7} = 468.42857$$

$$S_{xx} = \sum x_i^2 - \frac{(\sum x_i)^2}{n} = 65,993 - \frac{677^2}{7} = 517.42857$$

$$S_{yy} = \sum y_i^2 - \frac{(\sum y_i)^2}{n} = 74,585 - \frac{719^2}{7} = 733.42857$$

Then $\quad r = \dfrac{S_{xy}}{\sqrt{S_{xx}S_{yy}}} = \dfrac{468.42857}{\sqrt{517.42857(733.42857)}} = .760.$

The test of hypothesis is

$$H_0: \rho = 0 \qquad H_a: \rho > 0$$

and the test statistic is

$$t = \frac{r\sqrt{n-2}}{\sqrt{1-r^2}} = \frac{.760\sqrt{5}}{\sqrt{1-(.760)^2}} = 2.615$$

The rejection region for $\alpha = .05$ is $t > t_{.05} = 2.015$ and H_0 is rejected. There is sufficient evidence to indicate positive correlation.

12.43 **a**   Since neither of the two variables, amount of sodium or number of calories, is controlled, the methods of correlation rather than linear regression analysis should be used.

b   Use a computer program, your scientific calculator or the computing formulas given in the text to calculate the correlation coefficient r. The Minitab printout for this data set is shown below.

Correlations: C1, C2
```
Pearson correlation of C1 and C2 = 0.981
P-Value = 0.003
```

There is evidence of a highly significant correlation, since the *p*-value is so small. The correlation is positive.

12.47 Answers will vary. The Minitab output for this linear regression problem is shown below.

Regression Analysis: y versus x

```
The regression equation is
y = 21.9 + 15.0 x
Predictor           Coef        StDev          T          P
Constant          21.867        3.502       6.24      0.000
x                14.9667        0.9530      15.70      0.000

S = 3.691        R-Sq = 96.1%      R-Sq(adj) = 95.7%

Analysis of Variance
Source              DF           SS         MS          F          P
Regression           1       3360.0     3360.0     246.64      0.000
Residual Error      10        136.2       13.6
Total               11       3496.2
```

Correlations: x,y

```
Pearson correlation of y and x = 0.980
P-Value = 0.000
```

a   The correlation coefficient is $r = .980$.

b   The coefficient of determination is $r^2 = .961$ (or 96.1%).

c   The least squares line is $\hat{y} = 21.9 + 15.0x$.

d   We wish to estimate the mean percentage of kill for an application of 4 pounds of nematicide per acre. Since the percent kill y is actually a binomial percentage, the variance of y will change depending on the value of p, the proportion of nematodes killed for a particular application rate. The residual plot versus the fitted values shows this phenomenon as a "football-shaped" pattern. The normal probability plot also shows some deviation from normality in the tails of the plot. A transformation may be needed to assure that the regression assumptions are satisfied.

12.51 **a**   Use a computer program, your scientific calculator or the computing formulas given in the text to calculate the correlation coefficient r.

$$S_{xy} = \sum x_i y_i - \frac{(\sum x_i)(\sum y_i)}{n} = 1,233,987 - \frac{5028(2856)}{12} = 37,323$$

$$S_{xx} = \sum x_i^2 - \frac{(\sum x_i)^2}{n} = 2,212,178 - \frac{5028^2}{12} = 105,446$$

$$S_{yy} = \sum y_i^2 - \frac{(\sum y_i)^2}{n} = 723,882 - \frac{2856^2}{12} = 44,154$$

Then $r = \dfrac{S_{xy}}{\sqrt{S_{xx}S_{yy}}} = \dfrac{37,323}{\sqrt{105,446(44,154)}} = .5470.$ The test of hypothesis is

$$H_0: \rho = 0 \qquad H_a: \rho > 0$$

and the test statistic is

$$t = \frac{r\sqrt{n-2}}{\sqrt{1-r^2}} = \frac{.5470\sqrt{10}}{\sqrt{1-(.5470)^2}} = 2.066$$

with p-value $= P(t > 2.066)$ bounded as

$$.05 < p\text{-value} < .10$$

If the experimenter is willing to tolerate a p-value this large, then H_0 can be rejected. Otherwise, you would declare the results not significant; there is insufficient evidence to indicate that bending stiffness and twisting stiffness are positively correlated.

b $r^2 = (.5470)^2 = .2992$ so that 29.9% of the total variation in y can be explained by the independent variable x.

12.55 **a**   The plot is shown below. Notice that the relationship may be linear, but it is fairly weak.

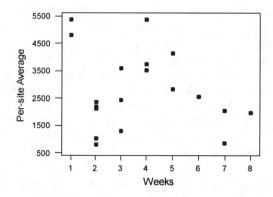

Regression Analysis: Per-site Average versus Weeks
```
The regression equation is
Per-site Average = 3372 - 155 Weeks

Predictor        Coef        StDev            T          P
Constant        3372.0       686.7         4.91      0.000
Weeks           -155.3       161.2        -0.96      0.349
S = 1439        R-Sq = 5.2%       R-Sq(adj) = 0.0%

Analysis of Variance
Source           DF           SS           MS          F          P
Regression        1       1921795      1921795       0.93      0.349
Residual Error   17      35181701      2069512
Total            18      37103496
```

b   From the printout, $r^2 = .052$. Only about 5% of the overall variation in y is explained by using the linear model.

c   From the printout, the regression equation is $\hat{y} = 3372.0 - 155.3x$ and the regression is not significant ($t = -.96$ with p-value $= .349$).

d   Since the regression is not significant, it is not appropriate to use the regression line for estimation or prediction.

12.60 **a** Use a computer, your scientific calculator or the computing formulas to find the correlation between x and y. The Minitab correlation printout below shows $r = .231$ with p-value $= .549$ which is not significant at the 5% level of significance. You cannot conclude that there is a significant positive correlation between median rate and score for "budget" hotels.

Correlations: Rate, Score
```
Pearson correlation of Rate and Score = 0.231
P-Value = 0.549
```

b-c Use the **Correlation and the Scatterplot** applet. There is a random scatter of points, with no outliers. The student's plot should look similar to the Minitab plot shown below.

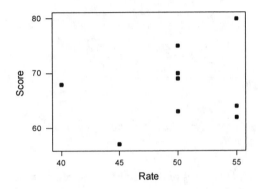

13: Multiple Regression Analysis

13.1 **a** When $x_2 = 2$, $E(y) = 3 + x_1 - 2(2) = x_1 - 1$.
When $x_2 = 1$, $E(y) = 3 + x_1 - 2(1) = x_1 + 1$.
When $x_2 = 0$, $E(y) = 3 + x_1 - 2(0) = x_1 + 3$.

These three straight lines are graphed in Figure 13.1.

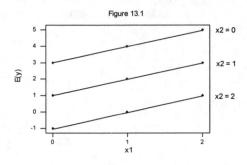

Figure 13.1

b Notice that the lines are parallel (they have the same slope).

13.5 **a** The model is quadratic.
b Since $R^2 = .815$, the sum of squares of deviations is reduced by 81.5% using the quadratic model rather than $\bar{y}$ to predict y.
c The hypothesis to be tested is

$$H_0: \beta_1 = \beta_2 = 0 \qquad H_a: \text{at least one } \beta_i \text{ differs from zero}$$

and the test statistic is

$$F = \frac{MSR}{MSE} = 37.37$$

which has an F distribution with $df_1 = k = 2$ and $df_2 = n - k - 1 = 20 - 2 - 1 = 17$. The p-value given in the printout is P $= .000$ and H_0 is rejected. There is evidence that the model contributes information for the prediction of y.

13.9 **a** Rate of increase is measured by the slope of a line tangent to the curve; this line is given by an equation obtained as dy/dx, the derivative of y with respect to x. In particular,

$$\frac{dy}{dx} = \frac{d}{dx}(\beta_0 + \beta_1 x + \beta_2 x^2) = \beta_1 + 2\beta_2 x$$

which has slope $2\beta_2$. If β_2 is negative, then the rate of increase is decreasing. Hence, the hypothesis of interest is

$$H_0: \beta_2 = 0, \quad H_a: \beta_2 < 0$$

b The individual t-test is $t = -8.11$ as in Exercise 13.8b. However, the test is one-tailed, which means that the p-value is half of the amount given in the printout. That is, p-value $= \frac{1}{2}(.000) = .000$. Hence, H_0 is again rejected. There is evidence to indicate a decreasing rate of increase.

13.12   a   Answers will vary. The student should notice that the F test for the overall utility of the model is not significant ($F = 1.74$ with p-value $= .244$), which means that the fit is not very good. Also, $R^2 = 49.9\%$ indicates that only 49.9% of the total variation in the experiment is accounted for by using the four predictor variables in a first order model.
b   If one of the variables is to be omitted, you should choose the variable with the smallest individual t-value (and the largest p-value). Hence, you should eliminate $x_1 =$ price and refit the model.

13.17   a   The variable x_2 must be the quantitative variable, since it appears as a quadratic term in the model. Qualitative variables appear only with exponent 1, although they may appear as the coefficient of another quantitative variable with exponent 2 or greater.
b   When $x_1 = 0$, $\hat{y} = 12.6 + 3.9x_2^2$ while when $x_1 = 1$,

$$\hat{y} = 12.6 + .54(1) - 1.2x_2 + 3.9x_2^2$$

$$= 13.14 - 1.2x_2 + 3.9x_2^2$$

c   The graph in Figure 13.9 shows the two parabolas.

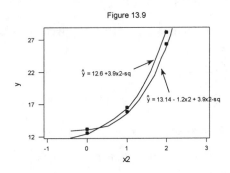

Figure 13.9

13.21   a   From the printout, the prediction equation is $\hat{y} = 8.585 + 3.8208x - 0.21663x^2$.
b R^2 is labeled "R-sq" or $R^2 = .944$. Hence, 94.4% of the total variation is accounted for by using x and x^2 in the model.
c   The hypothesis of interest is

$$H_0: \beta_1 = \beta_2 = 0 \qquad H_a: \text{at least one } \beta_i \text{ differs from zero}$$

and the test statistic is $F = 33.44$ with p-value $= .003$. Hence, H_0 is rejected, and we conclude that the model contributes significant information for the prediction of y.
d   The hypothesis of interest is

$$H_0: \beta_2 = 0 \qquad H_a: \beta_2 \neq 0$$

and the test statistic is $t = -4.93$ with p-value $= .008$. Hence, H_0 is rejected, and we conclude that the quadratic model provides a better fit to the data than a simple linear model.

e   The pattern of the diagnostic plots does not indicate any obvious violation of the regression assumptions.

13.25 a   The model is

$$y = \beta_0 + \beta_1 x_1 + \beta_2 x_2 + \beta_3 x_1^2 + \beta_4 x_1 x_2 + \beta_5 x_1^2 x_2 + \epsilon$$

and the Minitab printout is shown below.

Regression Analysis
```
The regression equation is
y = 4.5 + 6.39 x1 - 50.9 x2 + 17.1 x1x2 + 0.132 x1sq - 0.502 x1sqx2

Predictor        Coef        StDev            T         P
Constant         4.51        42.24         0.11     0.916
x1               6.394        5.777         1.11     0.275
x2             -50.85        56.21        -0.90     0.371
x1x2            17.064        7.101         2.40     0.021
x1sq             0.1318       0.1687        0.78     0.439
x1sqx2          -0.5025       0.1992       -2.52     0.016
S = 71.69        R-Sq = 76.8%       R-Sq(adj) = 73.8%

Analysis of Variance
Source            DF          SS           MS         F         P
Regression         5      664164       132833     25.85     0.000
Residual Error    39      200434         5139
Total             44      864598
```

b   The fitted prediction model uses the coefficients given in the column marked "Coef" in the printout:

$$\hat{y} = 4.51 + 6.394x_1 - 50.85x_2 + 17.064x_1x_2 + .1318x_1^2 - .5025x_1^2x_2$$

The F test for the model's utility is $F = 25.85$ with P $= .000$ and $R^2 = .768$. The model fits quite well.

c   If the dolphin is female, $x_2 = 0$ and the prediction equation becomes

$$\hat{y} = 4.51 + 6.394x_1 + .1318x_1^2$$

d   If the dolphin is male, $x_2 = 1$ and the prediction equation becomes

$$\hat{y} = -46.34 + 23.458x_1 - .3707x_1^2$$

e   The hypothesis of interest is

$$H_0:\ \beta_4 = 0 \qquad H_a:\ \beta_4 \neq 0$$

and the test statistic is $t = .78$ with p-value $= .439$. H_0 is not rejected and we conclude that the quadratic term is not important in predicting mercury concentration for female dolphins.

13.27 a-b The data is plotted below. It appears to be a curvilinear relationship, which could be described using the quadratic model $y = \beta_0 + \beta_1 x + \beta_2 x^2 + \epsilon$.

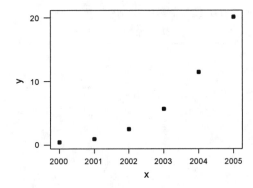

c   The Minitab printout is shown below.

Regression Analysis: y versus x, x-sq

```
The regression equation is
y = 4114749 - 4113 x + 1.03 x-sq
```

Predictor	Coef	StDev	T	P
Constant	4114749	343582	11.98	0.001
x	-4113.4	343.2	-11.99	0.001
x-sq	1.02804	0.08568	12.00	0.001

```
S = 0.5235      R-Sq = 99.7%      R-Sq(adj) = 99.5%
```

```
Analysis of Variance
```

Source	DF	SS	MS	F	P
Regression	2	297.16	148.58	542.11	0.000
Residual Error	3	0.82	0.27		
Total	5	297.98			

d   The hypothesis of interest is

$$H_0: \ \beta_1 = \beta_2 = 0$$

and the test statistic is $F = 542.11$ with p-value $= .000$. H_0 is rejected and we conclude that the model provides valuable information for the prediction of y.

e $R^2 = .997$. Hence, 99.7% of the total variation is accounted for by using x and x^2 in the model.

f   The residual plots are shown below. There is no reason to doubt the validity of the regression assumptions.

Residuals Versus the Fitted Values

(response is y)

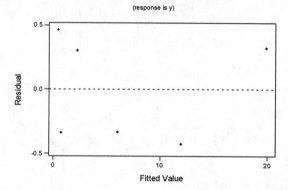

Normal Probability Plot of the Residuals

(response is y)

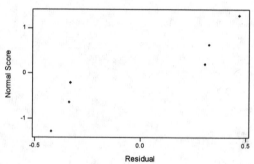

14: Analysis of Categorical Data

14.3 For a test of specified cell probabilities, the degrees of freedom are $k - 1$. Use Table 5, Appendix I:

   a $df = 6$; $\chi^2_{.05} = 12.59$; reject H_0 if $X^2 > 12.59$

   b $df = 9$; $\chi^2_{.01} = 21.666$; reject H_0 if $X^2 > 21.666$

   c $df = 13$; $\chi^2_{.005} = 29.8194$; reject H_0 if $X^2 > 29.8194$

   d $df = 2$; $\chi^2_{.05} = 5.99$; reject H_0 if $X^2 > 5.99$

14.7 One thousand cars were each classified according to the lane which they occupied (one through four). If no lane is preferred over another, the probability that a car will be driven in lane i, $i = 1, 2, 3, 4$ is $1/4$. The null hypothesis is then

$$H_0: \ p_1 = p_2 = p_3 = p_4 = \frac{1}{4}$$

and the test statistic is

$$X^2 = \sum \frac{(O_i - E_i)^2}{E_i}$$

with $E_i = np_i = 1000(1/4) = 250$ for $i = 1, 2, 3, 4$. A table of observed and expected cell counts follows:

Lane	1	2	3	4
O_i	294	276	238	192
E_i	250	250	250	250

Then

$$X^2 = \frac{(294 - 250)^2}{250} + \frac{(276 - 250)^2}{250} + \frac{(238 - 250)^2}{250} + \frac{(192 - 250)^2}{250}$$

$$= \frac{6120}{250} = 24.48$$

The rejection region with $k - 1 = 3 \ df$ is $X^2 > \chi^2_{.05} = 7.81$. Since the observed value of X^2 falls in the rejection region, we reject H_0. There is difference in preference for the four lanes.

14.11 Similar to previous exercises. The hypothesis to be tested is

$$H_0: \ p_1 = p_2 = \cdots = p_{12} = \frac{1}{12}$$

versus H_a: at least one p_i is different from the others
with

$$E_i = np_i = 400(1/12) = 33.333.$$

The test statistic is

$$X^2 = \frac{(38 - 33.33)^2}{33.33} + \cdots + \frac{(35 - 33.33)^2}{33.33} = 13.58$$

The upper tailed rejection region is with $\alpha = .05$ and $k - 1 = 11$ df is $X^2 > \chi^2_{.05} = 19.675$. The null hypothesis is not rejected and we cannot conclude that the proportion of cases varies from month to month.

14.15 Refer to Section 14.4 of the text. For a 3×5 contingency table with $r = 3$ and $c = 5$, there are $(r - 1)(c - 1) = (2)(4) = 8$ degrees of freedom.

14.19 **a** The hypothesis of independence between attachment pattern and child care time is tested using the chi-square statistic. The contingency table, including column and row totals and the estimated expected cell counts, follows.

Attachment	Child Care			Total
	Low	Moderate	High	
Secure	24 (24.09)	35 (30.97)	5 (8.95)	64
Anxious	11 (10.91)	10 (14.03)	8 (4.05)	29
Total	111	51	297	459

The test statistic is

$$X^2 = \frac{(24 - 24.09)^2}{24.09} + \frac{(35 - 30.97)^2}{30.97} + \cdots + \frac{(8 - 4.05)^2}{4.05} = 7.267$$

and the rejection region is $X^2 > \chi^2_{.05} = 5.99$ with 2 df. H_0 is rejected. There is evidence of a dependence between attachment pattern and child care time.
b The value $X^2 = 7.267$ is between $\chi^2_{.05}$ and $\chi^2_{.025}$ so that $.025 < p\text{-value} < .05$. The results are significant.

14.22 **a** The hypothesis of independence between type of sport and year is tested using the chi-square statistic. The contingency table, including column and row totals and the estimated expected cell counts, follows.

Period	Sport						Total
	Basketball	Track	Golf	Soccer	Lacrosse	Rowing	
1986-87	275 (219.93)	215 (182.41)	99 (108.11)	54 (116.65)	34 (40.49)	27 (36.41)	704
1999-00	317 (372.07)	276 (308.59)	192 (182.89)	260 (197.35)	75 (68.51)	71 (61.59)	1191
Total	592	491	291	314	109	98	1895

The test statistic is

$$X^2 = \frac{(275 - 219.93)^2}{219.93} + \frac{(215 - 182.41)^2}{182.41} + \cdots + \frac{(71 - 61.59)^2}{61.59} = 91.49$$

The test statistic is very large, compared to the largest value in Table 5 ($\chi^2_{.005} = 16.75$), so that $p\text{-value} < .005$ and H_0 is rejected. There is evidence of a difference in the distribution of women participating in these sports for the two periods of time.

b   You can see that the number of women participating in the "non-traditional" women's sports (soccer, lacrosse and rowing) was higher than expected in 1999-00.

14.25   Because a set number of Americans in each sub-population were each fixed at 200, we have a contingency table with fixed rows. The table, with estimated expected cell counts appearing in parentheses, follows.

	Yes	No	Total
White-American	40 (62)	160 (138)	200
African-American	56 (62)	144 (138)	200
Hispanic-American	68 (62)	132 (138)	200
Asian-American	84 (62)	116 (138)	200
Total	248	552	800

The test statistic is

$$X^2 = \frac{(40 - 62)^2}{62} + \frac{(56 - 62)^2}{62} + \cdots + \frac{(116 - 138)^2}{138} = 24.31$$

and the rejection region with 3 df is $X^2 > 11.3449$. H_0 is rejected and we conclude that the incidence of parental support is dependent on the sub-population of Americans.

14.29   If the housekeeper actually has no preference, he or she has an equal chance of picking any of the five floor polishes. Hence, the null hypothesis to be tested is

$$H_0: p_1 = p_2 = p_3 = p_4 = p_5 = \frac{1}{5}$$

The values of O_i are the actual counts observed in the experiment, and $E_i = np_i = 100(1/5) = 20$.

Polish	A	B	C	D	E
O_i	27	17	15	22	19
E_i	20	20	20	20	20

Then

$$X^2 = \frac{(27 - 20)^2}{20} + \frac{(17 - 20)^2}{20} + \cdots + \frac{(19 - 20)^2}{20} = 4.40$$

The p-value with $df = k - 1 = 4$ is greater than .10 and H_0 is not rejected. We cannot conclude that there is a difference in the preference for the five floor polishes. Even if this hypothesis **had** been rejected, the conclusion would be that at least one of the values of the p_i was significantly different from 1/6. However, this does not imply that p_i is necessarily greater than 1/6. Hence, we could not conclude that polish A is superior.

If the objective of the experiment is to show that polish A is superior, a better procedure would be to test an hypothesis as follows:

$$H_0: p_1 = 1/6 \qquad H_a: p_1 > 1/6$$

From a sample of $n = 100$ housewives, $x = 27$ are found to prefer polish A. A z-test can be performed on the single binomial parameter p_1.

14.33 **a** Let p be the proportion of New York voters who think that the World Trade Center should be rebuilt in some form. The hypothesis to be tested is

$$H_0 : p = .5 \text{ versus } H_a : p > .5$$

and the test statistic is

$$z = \frac{\hat{p} - p_0}{\sqrt{\frac{p_0 q_0}{n}}} = \frac{.62995 - .5}{\sqrt{\frac{.5(.5)}{1262}}} = 9.233$$

The p-value is $P(z > 9.233) \approx 0$ and H_0 is rejected. There is sufficient evidence to indicate that a majority of New York voters want to see the World Trade Center rebuilt.

b The data can also be analysed using a goodness of fit test. The hypothesis to be tested is

$$H_0 : p_1 = .5;\ p_2 = .5 \quad \text{versus } H_a : p_1 > .5$$

The expected cell counts are

$$E_1 = np_1 = 1262(.5) = 631 \text{ and } E_2 = 631$$

and the test statistic is

$$X^2 = \frac{(795 - 631)^2}{631} + \frac{(467 - 631)^2}{631} = 85.2488$$

Since the value of X^2 is always positive, we need to first check the direction of the difference, concluding that the sample number who agree is indeed more than 50%. Then, to adjust for the one-tailed test, the p-value is half of the one-tailed p-value or $\frac{1}{2}P(X^2 > 85.2488) < \frac{1}{2}(.005) = .0025$. The null hypothesis is rejected, as in part **a**.

c Notice that the results in parts **a** and **b** are identical, and that $z^2 = (9.233)^2 = 85.248$ which agrees to within rounding error with the value of X^2.

14.37 The 2×2 contingency table is analysed as in previous exercises. The Minitab printout is shown below.

Chi-Square Test: Had access, Did not

```
Expected counts are printed below observed counts

        Had access  Did not    Total
  1        224        276       500
         276.00     224.00
  2        328        172       500
         276.00     224.00
Total      552        448       1000

Chi-Sq =  9.797 + 12.071 +
          9.797 + 12.071 = 43.737
```

```
DF = 1, P-Value = 0.000
```
The test statistic is $X^2 = 43.737$ with p-value $= .000$ and the null hypothesis is rejected. There is evidence of a significant difference in the number of workers who have Internet access last year and this.

14.42   In order to perform a chi-square "goodness of fit" test on the given data, it is necessary that the values O_i and E_i are known for each of the five cells. The O_i (the number of measurements falling in the i-th cell) are given. However, $E_i = np_i$ must be calculated. Remember that p_i is the probability that a measurement falls in the i-th cell. The hypothesis to be tested is

H_0: the experiment is binomial   versus H_a: the experiment is not binomial

Let x be the number of successes and p be the probability of success on a single trial. Then, assuming the null hypothesis to be true,

$$p_0 = P(x = 0) = C_0^4 p^0 (1-p)^4 \qquad p_1 = P(x = 1) = C_1^4 p^1 (1-p)^3$$
$$p_2 = P(x = 2) = C_2^4 p^2 (1-p)^2 \qquad p_3 = P(x = 3) = C_3^4 p^3 (1-p)^1$$
$$p_4 = P(x = 4) = C_4^4 p^4 (1-p)^0$$

Hence, once an estimate for p is obtained, the expected cell frequencies can be calculated using the above probabilities. Note that each of the 100 experiments consists of four trials and hence the complete experiment involves a total of 400 trials.

The best estimator of p is $\hat{p} = x/n$ (as in Chapter 9). Then,

$$\hat{p} = \frac{x}{n} = \frac{\text{number of successes}}{\text{number of trials}} = \frac{0(11) + 1(17) + 2(42) + 3(12) + 4(9)}{400} = \frac{1}{2}.$$

The experiment (consisting of four trials) was repeated 100 times. There are a total of 400 trials in which the result "no successes in four trials" was observed 11 times, the result "one success in four trials" was observed 17 times, and so on. Then

$$p_0 = C_0^4 (1/2)^0 (1/2)^4 = 1/16 \qquad p_1 = C_1^4 (1/2)^1 (1/2)^3 = 4/16$$
$$p_2 = C_2^4 (1/2)^2 (1/2)^2 = 6/16 \qquad p_3 = C_3^4 (1/2)^3 (1/2)^1 = 4/16$$
$$p_4 = C_4^4 (1/2)^4 (1/2)^0 = 1/16$$

The observed and expected cell frequencies are shown in the following table.

x	0	1	2	3	4
O_i	11	17	42	21	9
E_i	6.25	25.00	37.50	25.00	6.25

and the test statistic is

$$X^2 = \frac{(11 - 6.25)^2}{6.25} + \frac{(17 - 25.00)^2}{25.00} + \cdots + \frac{(9 - 6.25)^2}{6.25} = 8.56$$

In order to bound the p-value or set up a rejection region, it is necessary to determine the appropriate degrees of freedom associated with the test statistic. Two degrees of freedom are lost because:

1   The cell probabilities are restricted by the fact that $\sum p_i = 1$.

2   The binomial parameter p is unknown and must be estimated before calculating the expected cell counts.

The number of degrees of freedom is equal to $k - 1 - 1 = k - 2 = 3$. With $df = 3$, the p-value for $X^2 = 8.56$ is between .025 and .05 and the null hypothesis can be rejected at the 5% level of signficance. We conclude that the experiment in question does not fulfill the requirements for a binomial experiment.

14.45   The null hypothesis to be tsted is

$$H_0: \ p_1 = p_2 = p_3 = \frac{1}{3}$$

and the test statistic is

$$X^2 = \sum \frac{(O_i - E_i)^2}{E_i}$$

with $E_i = np_i = 200(1/3) = 66.67$ for $i = 1, 2, 3$. A table of observed and expected cell counts follows:

Entrance	1	2	3
O_i	83	61	56
E_i	66.67	66.67	66.67

Then

$$X^2 = \frac{(84 - 66.67)^2}{66.67} + \frac{(61 - 66.67)^2}{66.67} + \frac{(56 - 66.67)^2}{66.67} = 6.190$$

With $df = k - 1 = 2$, the p-value is between .025 and .05 and we can reject H_0 at the 5% level of significance. There is difference in preference for the three doors. A 95% confidence interval for p_1 is given as

$$\frac{x_1}{n} \pm z_{.025}\sqrt{\frac{\hat{p}_1\hat{q}_1}{n}} \ \Rightarrow \ \frac{83}{200} \pm 1.96\sqrt{\frac{.415(.585)}{200}} \ \Rightarrow \ .415 \pm .068$$

or $.347 < p_1 < .483$.

14.49   Since the cards for each of the three holidays will be either "humorous" or "not humorous", the table actually consists of two rows and three columns, and is shown with estimated expected and observed cell counts in the Minitab printout below.

Chi-Square Test

```
Expected counts are printed below observed counts
        Father's Mother's Valentine's Total
   1       100       125       120       345
         115.00    115.00    115.00
   2       400       375       380      1155
         385.00    385.00    385.00
Total      500       500       500      1500
Chi-Sq =  1.957 +   0.870 +   0.217 +
          0.584 +   0.260 +   0.065 =   3.953
DF = 2, P-Value = 0.139
```

The test statistic for the equality of the three population proportions is $X^2 = 3.953$ with p-value $= .139$ and H_0 is not rejected. There is insufficient evidence to indicate a difference in the proportion of humorous cards for the three holidays.

14.53 a   The 2×3 contingency table is analysed as in previous exercises. The Minitab printout below shows the observed and estimated expected cell counts, the test statistic and its associated p-value.

Chi-Square Test: 3 or fewer, 4 or 5, 6 or more

```
Expected counts are printed below observed counts

         3 or fewer   4 or 5   6 or more      Total
    1        49          43        34          126
            37.89       42.63     45.47
    2        31          47        62          140
            42.11       47.37     50.53
 Total       80          90        96          266

Chi-Sq =  3.254 +  0.003 +  2.895 +
          2.929 +  0.003 +  2.605 =  11.690
DF = 2, P-Value = 0.003
```

The results are highly significant (p-value $= .003$) and we conclude that there is a difference in the susceptibility to colds depending on the number of relationships you have.

b   The proportion of people with colds is calculated conditionally for each of the three groups, and is shown in the table below.

	Three or fewer	Four or five	Six or more
Cold	$\dfrac{49}{80} = .61$	$\dfrac{43}{90} = .48$	$\dfrac{34}{96} = .35$
No cold	$\dfrac{31}{80} = .39$	$\dfrac{47}{90} = .52$	$\dfrac{62}{96} = .65$
Total	1.00	1.00	1.00

As the researcher suspects, the susceptibility to a cold seems to decrease as the number of relationships increases!

14.57   Use the first **Goodness-of-Fit** applet. Enter the observed values into the three cells in the first row, and the expected cell counts will automatically appear in the second row. The value of $X^2 = 18.5$ with p-value $= .0001$ provide sufficient evidence to reject H_0 and conclude that customers have a preference for one of the three brands (in this case, Brand II).

	Green	Red	Blue	Total
Observed	115	120	65	300
Expected	100.0	100.0	100.0	300

Display Data/Null   Reset Original Data

Observed Frequencies

ChiSq(2) = 18.5, p-value = 1.0E-4

14.61 The data is analysed as a 2×3 contingency table with estimated expected cell counts shown in parentheses. Use the **Chi-Square Test of Independence** applet. Your results should agree with the hand calculations shown below.

		Opinion		
Party	1	2	3	Total
Republican	114 (120.86)	53 (48.10)	17 (15.03)	184
Democrat	87 (80.14)	27 (31.89)	8 (9.97)	122
Total	201	80	25	306

The test statistic is

$$X^2 = \frac{(114 - 120.86)^2}{120.86} + \frac{(53 - 48.10)^2}{48.10} + \cdots + \frac{(8 - 9.97)^2}{9.97} = 2.87$$

With $df = 2$, the p-value is greater than .10 (the applet reports p-value $= .2378$) and H_0 is not rejected. There is no evidence that party affiliation has any effect on opinion.

15: Nonparametric Statistics

15.1   **a**   If distribution 1 is shifted to the right of distribution 2, the rank sum for sample 1 (T_1) will tend to be large. The test statistic will be T_1^*, the rank sum for sample 1 if the observations had been ranked from large to small. The null hypothesis will be rejected if T_1^* is unusually small.

b   From Table 7a with $n_1 = 6$, $n_2 = 8$ and $\alpha = .05$, H_0 will be rejected if $T_1^* \le 31$.

c   From Table 7c with $n_1 = 6$, $n_2 = 8$ and $\alpha = .01$, H_0 will be rejected if $T_1^* \le 27$.

15.5   If H_a is true and population 1 lies to the right of population 2, then T_1 will be large and T_1^* will be small. Hence the test statistic will be T_1^* and the large sample approximation can be used. Calculate

$$T_1^* = n_1(n_1 + n_2 + 1) - T_1 = 12(27) - 193 = 131$$

$$\mu_T = \frac{n_1(n_1 + n_2 + 1)}{2} = \frac{12(26 + 1)}{2} = 162$$

$$\sigma_T^2 = \frac{n_1 n_2(n_1 + n_2 + 1)}{12} = \frac{12(14)(27)}{12} = 378$$

The test statistic is

$$z = \frac{T_1 - \mu_T}{\sigma_T} = \frac{131 - 162}{\sqrt{378}} = -1.59$$

The rejection region with $\alpha = .05$ is $z < -1.645$ and H_0 is not rejected. There is insufficient evidence to indicate a difference in the two population distributions.

15.9   Similar to previous exercises. The data, with corresponding ranks, are shown in the following table.

Deaf (1)	Hearing (2)
2.75 (15)	0.89 (1)
2.14 (11)	1.43 (7)
3.23 (18)	1.06 (4)
2.07 (10)	1.01 (3)
2.49 (14)	0.94 (2)
2.18 (12)	1.79 (8)
3.16 (17)	1.12 (5.5)
2.93 (16)	2.01 (9)
2.20 (13)	1.12 (5.5)
$T_1 = 126$	

Calculate

$$T_1 = 126$$
$$T_1^* = n_1(n_1 + n_2 + 1) - T_1 = 9(19) - 126 = 45$$

The test statistic is

$$T = min(T_1, T_1^*) = 45$$

With $n_1 = n_2 = 9$, the two-tailed rejection region with $\alpha = .05$ is found in Table 7b to be $T_1^* \leq 62$. The observed value, $T = 45$, falls in the rejection region and H_0 is rejected. We conclude that the deaf children do differ from the hearing children in eye-movement rate.

15.13 **a** If a paired difference experiment has been used and the sign test is one-tailed (H_a: $p > .5$), then the experimenter would like to show that one population of measurements lies above the other population. An exact practical statement of the alternative hypothesis would depend on the experimental situation.

b It is necessary that α (the probability of rejecting the null hypothesis when it is true) take values less than $\alpha = .15$. Assuming the null hypothesis to be true, the two populations are identical and consequently, $p = P(A$ exceeds B for a given pair of observations) is $1/2$. The binomial probability was discussed in Chapter 5. In particular, it was noted that the distribution of the random variable x is symmetrical about the mean np when $p = 1/2$. For example, with $n = 25$, $P(x = 0) = P(x = 25)$. Similarly, $P(x = 1) = P(x = 24)$ and so on. Hence, the lower tailed probabilities tabulated in Table 1, Appendix I will be identical to their upper tailed equivalent probabilities. The values of α available for this upper tailed test and the corresponding rejection regions are shown below.

Rejection Region	α
$x \geq 20$	.002
$x \geq 19$	.007
$x \geq 18$	.022
$x \geq 17$	.054
$x \geq 16$	.115

15.17 **a** If assessors A and B are equal in their property assessments, then p, the probability that A's assessment exceeds B's assessment for a given property, should equal $1/2$. If one of the assessors tends to be more conservative than the other, then either $p > 1/2$ or $p < 1/2$. Hence, we can test the equivalence of the two assessors by testing the hypothesis

$$H_0: p = 1/2 \quad \text{versus} \quad H_a: p \neq 1/2$$

using the test statistic x, the number of times that assessor A exceeds assessor B for a particular property assessment. To find a two-tailed rejection region with α close to .05, use Table 1 with $n = 8$ and $p = .5$. For the rejection region $\{x = 0, x = 8\}$ the value of α is $.004 + .004 = .008$, while for the rejection region $\{x = 0, 1, 7, 8\}$ the value of α is $.035 + .035 = .070$ which is closer to .05. Hence, using the rejection region $\{x \leq 1$ or

$x \geq 7\}$, the null hypothesis is not rejected, since x = number of properties for which A exceeds B = 6. The p-value for this two-tailed test is

$$p\text{-value} = 2P(x \geq 6) = 2(1 - .855) = .290$$

Since the p-value is greater than .10, the results are not significant; H_0 is not rejected (as with the critical value approach).

b The t statistic used in Exercise 10.43 allows the experimenter to reject H_0, while the sign test fails to reject H_0. This is because the sign test used less information and makes fewer assumptions than does the t test. If all normality assumptions are met, the t test is the more powerful test and can reject when the sign test cannot.

15.21 a H_0: population distributions 1 and 2 are identical
H_a: the distributions differ in location
b Since Table 8, Appendix I gives critical values for rejection in the lower tail of the distribution, we use the smaller of T^+ and T^- as the test statistic.
c From Table 8 with $n = 30$, $\alpha = .05$ and a two-tailed test, the rejection region is $T \leq 137$.
d Since $T^+ = 249$, we can calculate

$$T^- = \frac{n(n+1)}{2} - T^+ = \frac{30(31)}{2} - 249 = 216.$$

The test statistic is the smaller of T^+ and T^- or $T = 216$ and H_0 is not rejected. There is no evidence of a difference between the two distributions.

15.25 a The hypothesis to be tested is

$$H_0\text{: population distributions 1 and 2 are identical}$$
$$H_a\text{: the distributions differ in location}$$

and the test statistic is T, the rank sum of the positive (or negative) differences. The ranks are obtained by ordering the differences according to their absolute value. Define d_i to be the difference between a pair in populations 1 and 2 (i.e., $x_{1i} - x_{2i}$). The differences, along with their ranks (according to absolute magnitude), are shown in the following table.

d_i	.1	.7	.3	$-.1$	.5	.2	.5
Rank $\|d_i\|$	1.5	7	4	1.5	5.5	3	5.5

The rank sum for positive differences is $T^+ = 26.5$ and the rank sum for negative differences is $T^- = 1.5$ with $n = 7$. Consider the smaller rank sum and determine the appropriate lower portion of the two-tailed rejection region. Indexing $n = 7$ and $\alpha = .05$ in Table 8, the rejection region is $T \leq 2$ and H_0 is rejected. There is a difference in the two population locations.

b The results do not agree with those obtained in Exercise 15.16. We are able to reject H_0 with the more powerful Wilcoxon test.

15.29 **a** The paired data are given in the exercise. The differences, along with their ranks (according to absolute magnitude), are shown in the following table.

d_i	1	2	-1	1	3	1	-1	3	-2	3	1	0
Rank $|d_i|$	3.5	7.5	3.5	3.5	10	3.5	3.5	10	7.5	10	2.5	--

Let $p = P(A$ exceeds B for a given intersection) and $x =$ number of intersections at which A exceeds B. The hypothesis to be tested is

$$H_0: p = 1/2 \quad \text{versus} \quad H_a: p \neq 1/2$$

using the sign test with x as the test statistic.

Critical value approach: Various two tailed rejection regions are tried in order to find a region with $\alpha \approx .05$. These are shown in the following table.

Rejection Region	α
$x \leq 1; x \geq 10$	.012
$x \leq 2; x \geq 9$	.066
$x \leq 3; x \geq 8$	.226

We choose to reject H_0 if $x \leq 2$ or $x \geq 9$ with $\alpha = .066$. Since $x = 8$, H_0 is not rejected. There is insufficient evidence to indicate a difference between the two methods.

_p_-value approach: For the observed value $x = 8$, calculate the two-tailed p-value:

$$p\text{-value} = 2P(x \geq 8) = 2(1 - .887) = .226$$

Since the p-value is greater than .10, H_0 is not rejected.

b To use the Wilcoxon signed rank test, we use the ranks of the absolute differences shown in the table above. Then $T^+ = 51.5$ and $T^- = 14.5$ with $n = 11$. Indexing $n = 11$ and $\alpha = .05$ in Table 8, the lower portion of the two-tailed rejection region is $T \leq 11$ and H_0 is not rejected, as in part **a**.

15.31 **a** Since the experiment has been designed as a paired experiment, there are three tests available for testing the differences in the distributions with and without imagery—(1) the paired difference t test; (2) the sign test or (3) the Wilcoxon signed rank test. In order to use the paired difference t test, the scores must be approximately normal; since the number of words recalled has a binomial distribution with $n = 25$ and unknown recall probability, this distribution may not be approximately normal.

b Using the **sign test**, the hypothesis to be tested is

$$H_0: p = 1/2 \quad \text{versus} \quad H_a: p > 1/2$$

For the observed value $x = 0$, calculate the two-tailed p-value:

$$p\text{-value} = 2P(x \leq 0) = 2(.000) = .000$$

The results are highly significant; H_0 is rejected and we conclude that there is a difference in the recall scores with and without imagery.

Using the **Wilcoxon signed-rank test**, the differences will all be positive ($x = 0$ for the sign test), so that

and

$$T^+ = \frac{n(n+1)}{2} = \frac{20(21)}{2} = 210 \quad \text{and} \quad T^- = 210 - 210 = 0$$

Indexing $n = 20$ and $\alpha = .01$ in Table 8, the lower portion of the two-tailed rejection region is $T \le 37$ and H_0 is rejected.

15.35 Similar to Exercise 15.32. The data with corresponding ranks in parentheses are shown below.

Age			
$10 - 19$	$20 - 39$	$40 - 59$	$60 - 69$
29 (21)	24 (8)	37 (39)	28 (18)
33 (29.5)	27 (15)	25 (10.5)	29 (21)
26 (12.5)	33 (29.5)	22 (5.5)	34 (34)
27 (15)	31 (24)	33 (29.5)	36 (37.5)
39 (40)	21 (3)	28 (18)	21 (3)
35 (36)	28 (18)	26 (12.5)	20 (1)
33 (29.5)	24 (8)	30 (23)	25 (10.5)
29 (21)	34 (34)	34 (34)	24 (8)
36 (37.5)	21 (3)	27 (15)	33 (29.5)
22 (5.5)	32 (25.5)	33 (29.5)	32 (25.5)
$T_1 = 247.5$	$T_2 = 168$	$T_3 = 216.5$	$T_4 = 188$
$n_1 = 10$	$n_2 = 10$	$n_3 = 10$	$n_4 = 10$

a The test statistic, based on the rank sums, is

$$H = \frac{12}{n(n+1)} \sum \frac{T_i^2}{n_i} - 3(n+1)$$
$$= \frac{12}{40(41)} \left[\frac{(247.5)^2}{10} + \frac{(168)^2}{10} + \frac{(216.5)^2}{10} + \frac{(188)^2}{10} \right] - 3(41) = 2.63$$

The rejection region with $\alpha = .01$ and $k - 1 = 3$ df is based on the chi-square distribution, or $H > \chi_{.01}^2 = 11.35$. The null hypothesis is not rejected. There is no evidence of a difference in location.

b Since the observed value $H = 2.63$ is less than $\chi_{.10}^2 = 6.25$, the p-value is greater than .10.

c From Exercise 11.49, $F = .87$ with 3 and 36 df. Again, the p-value is greater than .10 and the results are the same.

15.39 Similar to Exercise 15.38. The ranks of the data are shown below.

Block	Treatment			
	1	2	3	4
1	4	1	2	3
2	4	1.5	1.5	3
3	4	1	3	2
4	4	1	2	3
5	4	1	2.5	2.5
6	4	1	2	3
7	4	1	3	2
8	4	1	2	3
	$T_1 = 32$	$T_2 = 8.5$	$T_3 = 18$	$T_4 = 21.5$

a   The test statistic is

$$F_r = \frac{12}{bk(k+1)} \sum T_i^2 - 3b(k+1)$$

$$= \frac{12}{8(4)(5)} \left[(32)^2 + (8.5)^2 + 18^2 + (21.5)^2 \right] - 3(8)(5) = 21.19$$

and the rejection region is $F_r > \chi^2_{.05} = 7.81$. Hence, H$_0$ is rejected and we conclude that there is a difference among the four treatments.

b   The observed value, $F_r = 21.19$, exceeds $\chi^2_{.005}$, p-value $< .005$.

c-e The analysis of variance is performed as in Chapter 11. The ANOVA table is shown below.

Source	df	SS	MS	F
Treatments	3	198.34375	66.114583	75.43
Blocks	7	220.46875	31.495536	
Error	21	18.40625	0.876488	
Total	31	437.40625		

The analysis of variance F test for treatments is $F = 75.43$ and the approximate p-value with 3 and 21 df is p-value $< .005$. The result is identical to the parametric result.

15.43   Table 9, Appendix I gives critical values r_0 such that $P(r_s \geq r_0) = \alpha$. Hence, for an upper-tailed test, the critical value for rejection can be read directly from the table.

a $r_s \geq .425$ **b** $r_s \geq .601$

15.47 a   The two variables (rating and distance) are ranked from low to high, and the results are shown in the following table.

Voter	x	y	Voter	x	y
1	7.5	3	7	6	4
2	4	7	8	11	2
3	3	12	9	1	10
4	12	1	10	5	9
5	10	8	11	9	5.5
6	7.5	11	12	2	5.5

Calculate

$$\sum x_i y_i = 442.5 \qquad \sum x_i^2 = 649.5 \qquad \sum y_i^2 = 649.5$$
$$n = 12 \qquad \sum x_i = 78 \qquad \sum y_i = 78$$

Then

$$S_{xy} = 422.5 - \frac{78^2}{12} = -84.5 \qquad S_{xx} = 649.5 - \frac{78^2}{12} = 142.5 \qquad S_{yy} = 649.5 - \frac{78^2}{12} = 142.5$$

and

$$r_s = \frac{S_{xy}}{\sqrt{S_{xx}S_{yy}}} = \frac{-84.5}{142.5} = -.593.$$

b   The hypothesis of interest is H_0: no correlation versus H_a: negative correlation. Consulting Table 9 for $\alpha = .05$, the critical value of r_s, denoted by r_0, is $-.497$. Since the value of the test statistic is less than the critical value, the null hypothesis is rejected. There is evidence of a significant negative correlation between rating and distance.

15.51   Refer to Exercise 15.50. To test for positive correlation with $\alpha = .05$, index .05 in Table 9, and the rejection region is $r_s \geq .600$. We reject the null hypothesis of no association and conclude that a positive correlation exists between the teacher's ranks and the ranks of the IQs.

15.55   **a**   Define $p = P$(response for stimulus 1 exceeds that for stimulus 2) and $x =$ number of times the response for stimulus 1 exceeds that for stimulus 2. The hypothesis to be tested is

$$H_0: p = 1/2 \quad \text{versus} \quad H_a: p \neq 1/2$$

using the sign test with x as the test statistic. Notice that for this exercise $n = 9$, and the observed value of the test statistic is $x = 2$. Various two tailed rejection regions are tried in order to find a region with $\alpha \approx .05$. These are shown in the following table.

Rejection Region	α
$x = 0; x = 9$	.004
$x \leq 1; x \geq 8$	.040
$x \leq 2; x \geq 7$	.180

We choose to reject H_0 if $x \leq 1$ or $x \geq 8$ with $\alpha = .040$. Since $x = 2$, H_0 is not rejected. There is insufficient evidence to indicate a difference between the two stimuli.

b   The experiment has been designed in a paired manner, and the paired difference test is used. The differences are shown below.

$$d_i \quad -.9 \quad -1.1 \quad 1.5 \quad -2.6 \quad -1.8 \quad -2.9 \quad -2.5 \quad 2.5 \quad -1.4$$

The hypothesis to be tested is

$$H_0: \mu_1 - \mu_2 = 0 \qquad H_a: \mu_1 - \mu_2 \neq 0$$

Calculate

$$\bar{d} = \frac{\Sigma d_i}{n} = \frac{-9.2}{9} = -1.022 \qquad s_d^2 = \frac{\Sigma d_i^2 - \frac{(\Sigma d_i)^2}{n}}{n-1} = \frac{37.14 - 9.404}{8} = 3.467$$

and the test statistic is

$$t = \frac{\bar{d}}{\sqrt{\frac{s_d^2}{n}}} = \frac{-1.022}{\sqrt{\frac{3.467}{9}}} = -1.646.$$

The rejection region with $\alpha = .05$ and 8 df is $|t| > 2.306$ and H_0 is not rejected.

15.59 The data, with corresponding ranks, are shown in the following table.

A (1)	B (2)
6.1 (1)	9.1 (16)
9.2 (17)	8.2 (8)
8.7 (12)	8.6 (11)
8.9 (13.5)	6.9 (2)
7.6 (5)	7.5 (4)
7.1 (3)	7.9 (7)
9.5 (18)	8.3 (9.5)
8.3 (9.5)	7.8 (6)
9.0 (1.5)	8.9 (13.5)
$T_1 = 94$	

The difference in the brightness levels using the two processes can be tested using the nonparametric Wilcoxon rank sum test, or the parametric two-sample t test.

1   To test the null hypothesis that the two population distributions are identical, calculate

$$T_1 = 1 + 17 + \cdots + 1.5 = 94$$
$$T_1^* = n_1(n_1 + n_2 + 1) - T_1 = 9(18 + 1) - 94 = 77$$

The test statistic is

$$T = min(T_1, T_1^*) = 77$$

With $n_1 = n_2 = 9$, the two-tailed rejection region with $\alpha = .05$ is found in Table 7b to be $T_1^* \leq 62$. The observed value, $T = 77$, does not fall in the rejection region and H_0 is not rejected. We cannot conclude that the distributions of brightness measurements is different for the two processes.

2   To test the null hypothesis that the two population means are identical, calculate

$$\bar{x}_1 = \frac{\Sigma x_{1j}}{n_1} = \frac{74.4}{9} = 8.2667 \qquad \bar{x}_2 = \frac{\Sigma x_{2j}}{n_2} = \frac{73.2}{9} = 8.1333$$

$$s^2 = \frac{\Sigma x_{1j}^2 - \dfrac{(\Sigma x_{1j})^2}{n_1} + \Sigma x_{2j}^2 - \dfrac{(\Sigma x_{2j})^2}{n_2}}{n_1 + n_2 - 2} = \frac{625.06 - \dfrac{(74.4)^2}{9} + 599.22 - \dfrac{(73.2)^2}{9}}{16} = .8675$$

and the test statistic is

$$t = \frac{\bar{x}_1 - \bar{x}_2}{\sqrt{s^2\left(\dfrac{1}{n_1} + \dfrac{1}{n_2}\right)}} = \frac{8.27 - 8.13}{\sqrt{.8675\left(\dfrac{2}{9}\right)}} = .304$$

The rejection region with $\alpha = .05$ and 16 degrees of freedom is $|t| > 1.746$ and H_0 is not rejected. There is insufficient evidence to indicate a difference in the average brightness measurements for the two processes.

Notice that the nonparametric and parametric tests reach the same conclusions.

15.61   Since this is a paired experiment, you can choose either the sign test, the Wilcoxon signed rank test, or the parametric paired t test. Since the tenderizers have been scored on a scale of 1 to 10, the parametric test is not applicable. Start by using the easiest of the two nonparametric tests—the sign test. Define $p =$ P(tenderizer A exceeds B for a given cut) and $x =$ number of times that A exceeds B. The hypothesis to be tested is

$$H_0: p = 1/2 \quad \text{versus} \quad H_a: p \neq 1/2$$

using the sign test with x as the test statistic. Notice that for this exercise $n = 8$ (there are two ties), and the observed value of the test statistic is $x = 2$.
p-value approach: For the observed value $x = 2$, calculate

$$p\text{-value} = 2P(x \leq 2) = 2(.145) = .290$$

Since the p-vallue is greater than .10, H_0 is not rejected. There is insufficient evidence to indicate a difference between the two tenderizers.

If you use the Wilcoxon signed rank test, you will find $T^+ = 7$ and $T^- = 29$ which will not allow rejection of H_0 at the 5% level of signficance. The results are the same.

15.65 The hypothesis to be tested is

H_0: population distributions 1 and 2 are identical

H_a: the distributions differ in location

and the test statistic is T, the rank sum of the positive (or negative) differences. The ranks are obtained by ordering the differences according to their absolute value. Define d_i to be the difference between a pair in populations 1 and 2 (i.e., $x_{1i} - x_{2i}$). The differences, along with their ranks (according to absolute magnitude), are shown in the following table.

d_i	-31	-31	-6	-11	-9	-7	7
Rank $|d_i|$	14.5	14.5	4.5	12.5	10.5	7	7

d_i	-11	7	-9	-2	-8	-1	-6	-3
Rank $|d_i|$	12.5	7	10.5	2	9	1	4.5	3

The rank sum for positive differences is $T^+ = 14$ and the rank sum for negative differences is $T^- = 106$ with $n = 15$. Consider the smaller rank sum and determine the appropriate lower portion of the two-tailed rejection region. Indexing $n = 15$ and $\alpha = .05$ in Table 8, the rejection region is $T \leq 25$ and H_0 is rejected. We conclude that there is a difference between math and art scores.

15.69 **a-b** Since the experiment is a completely randomized design, the Kruskal Wallis H test is used. The combined ranks are shown below.

Plant	Ranks					T_i
A	9	12	5	1	7	34
B	11	15	4	19	14	63
C	3	13	2	9	6	33
D	20	17	9	16	18	80

The test statistic, based on the rank sums, is

$$H = \frac{12}{n(n+1)} \sum \frac{T_i^2}{n_i} - 3(n+1)$$

$$= \frac{12}{20(21)} \left[\frac{(34)^2}{5} + \frac{(63)^2}{5} + \frac{(33)^2}{5} + \frac{(80)^2}{5} \right] - 3(21) = 9.08$$

With $df = k - 1 = 3$, the observed value $H = 9.08$ is between $\chi_{.025}$ and $\chi_{.05}$ so that $.025 < p$-value $< .05$. The null hypothesis is rejected and we conclude that there is a difference among the four plants.

c   From Exercise 11.58, $F = 5.20$ and H_0 is rejected. The results are the same.

Appendix I

Tables

Table 1 Cumulative Binomial Probabilities

Tabulated values are $P(x \leq k) = p(0) + p(1) + \ldots + p(k)$

(Computations are rounded at the third decimal place.)

$n = 2$

k	.01	.05	.10	.20	.30	.40	.50	.60	.70	.80	.90	.95	.99	k
							p							
0	.980	.902	.810	.640	.490	.360	.250	.160	.090	.040	.010	.002	.000	0
1	1.000	.998	.990	.960	.910	.840	.750	.640	.510	.360	.190	.098	.020	1
2	1.000	1.000	1.000	1.000	1.000	1.000	1.000	1.000	1.000	1.000	1.000	1.000	1.000	2

$n = 3$

k	.01	.05	.10	.20	.30	.40	.50	.60	.70	.80	.90	.95	.99	k
							p							
0	.970	.857	.729	.512	.343	.216	.125	.064	.027	.008	.001	.000	.000	0
1	1.000	.993	.972	.896	.784	.648	.500	.352	.216	.104	.028	.007	.000	1
2	1.000	1.000	.999	.992	.973	.936	.875	.784	.657	.488	.271	.143	.030	2
3	1.000	1.000	1.000	1.000	1.000	1.000	1.000	1.000	1.000	1.000	1.000	1.000	1.000	3

$n = 4$

k	.01	.05	.10	.20	.30	.40	.50	.60	.70	.80	.90	.95	.99	k
							p							
0	.961	.815	.656	.410	.240	.130	.062	.026	.008	.002	.000	.000	.000	0
1	.999	.986	.948	.819	.652	.475	.312	.179	.084	.027	.004	.000	.000	1
2	1.000	1.000	.996	.937	.916	.821	.688	.525	.348	.181	.052	.014	.001	2
3	1.000	1.000	1.000	.998	.992	.974	.938	.870	.760	.590	.344	.185	.039	3
4	1.000	1.000	1.000	1.000	1.000	1.000	1.000	1.000	1.000	1.000	1.000	1.000	1.000	4

(continued)

Table 1 *(Continued)*

$n = 5$

							p							
k	.01	.05	.10	.20	.30	.40	.50	.60	.70	.80	.90	.95	.99	k
0	.951	.774	.590	.328	.168	.078	.031	.010	.002	.000	.000	.000	.000	0
1	.999	.977	.919	.737	.528	.337	.188	.087	.031	.007	.000	.000	.000	1
2	1.000	.999	.991	.942	.837	.683	.500	.317	.163	.058	.009	.001	.000	2
3	1.000	1.000	1.000	.993	.969	.913	.812	.663	.472	.263	.081	.023	.001	3
4	1.000	1.000	1.000	1.000	.998	.990	.969	.922	.832	.672	.410	.226	.049	4
5	1.000	1.000	1.000	1.000	1.000	1.000	1.000	1.000	1.000	1.000	1.000	1.000	1.000	5

$n = 6$

							p							
k	.01	.05	.10	.20	.30	.40	.50	.60	.70	.80	.90	.95	.99	k
0	.941	.735	.531	.262	.118	.047	.016	.004	.001	.000	.000	.000	.000	0
1	.999	.967	.886	.655	.420	.233	.109	.041	.011	.002	.000	.000	.000	1
2	1.000	.998	.984	.901	.744	.544	.344	.179	.070	.017	.001	.000	.000	2
3	1.000	1.000	.999	.983	.930	.821	.656	.456	.256	.099	.016	.002	.000	3
4	1.000	1.000	1.000	.998	.989	.959	.891	.767	.580	.345	.114	.033	.001	4
5	1.000	1.000	1.000	1.000	.999	.996	.984	.953	.882	.738	.469	.265	.059	5
6	1.000	1.000	1.000	1.000	1.000	1.000	1.000	1.000	1.000	1.000	1.000	1.000	1.000	6

$n = 7$

							p							
k	.01	.05	.10	.20	.30	.40	.50	.60	.70	.80	.90	.95	.99	k
0	.932	.698	.478	.210	.082	.028	.008	.002	.000	.000	.000	.000	.000	0
1	.998	.956	.850	.577	.329	.159	.062	.019	.004	.000	.000	.000	.000	1
2	1.000	.996	.974	.852	.647	.420	.227	.096	.029	.005	.000	.000	.000	2
3	1.000	1.000	.997	.967	.874	.710	.500	.290	.126	.033	.003	.000	.000	3
4	1.000	1.000	1.000	.995	.971	.904	.773	.580	.353	.148	.026	.004	.000	4
5	1.000	1.000	1.000	1.000	.996	.981	.938	.841	.671	.423	.150	.044	.002	5
6	1.000	1.000	1.000	1.000	1.000	.998	.992	.972	.918	.790	.522	.302	.068	6
7	1.000	1.000	1.000	1.000	1.000	1.000	1.000	1.000	1.000	1.000	1.000	1.000	1.000	7

(continued)

Table 1 (*Continued*)

n = 8

k	.01	.05	.10	.20	.30	.40	.50	.60	.70	.80	.90	.95	.99	k
0	.923	.663	.430	.168	.058	.017	.004	.001	.000	.000	.000	.000	.000	0
1	.997	.943	.813	.503	.255	.106	.035	.009	.001	.000	.000	.000	.000	1
2	1.000	.994	.962	.797	.552	.315	.145	.050	.011	.001	.000	.000	.000	2
3	1.000	1.000	.995	.944	.806	.594	.363	.174	.058	.010	.000	.000	.000	3
4	1.000	1.000	1.000	.990	.942	.826	.637	.406	.194	.056	.005	.000	.000	4
5	1.000	1.000	1.000	.999	.989	.950	.855	.685	.448	.203	.038	.006	.000	5
6	1.000	1.000	1.000	1.000	.999	.991	.965	.894	.745	.497	.187	.057	.003	6
7	1.000	1.000	1.000	1.000	1.000	.999	.996	.983	.942	.832	.570	.337	.077	7
8	1.000	1.000	1.000	1.000	1.000	1.000	1.000	1.000	1.000	1.000	1.000	1.000	1.000	8

n = 9

k	.01	.05	.10	.20	.30	.40	.50	.60	.70	.80	.90	.95	.99	k
0	.914	.630	.387	.134	.040	.010	.002	.000	.000	.000	.000	.000	.000	0
1	.997	.929	.775	.436	.196	.071	.020	.004	.000	.000	.000	.000	.000	1
2	1.000	.992	.947	.738	.463	.232	.090	.025	.004	.000	.000	.000	.000	2
3	1.000	.999	.992	.914	.730	.483	.254	.099	.025	.003	.000	.000	.000	3
4	1.000	1.000	.999	.980	.901	.733	.500	.267	.099	.020	.001	.000	.000	4
5	1.000	1.000	1.000	.997	.975	.901	.746	.517	.270	.086	.008	.001	.000	5
6	1.000	1.000	1.000	1.000	.996	.975	.910	.768	.537	.262	.053	.008	.000	6
7	1.000	1.000	1.000	1.000	1.000	.996	.980	.929	.804	.564	.225	.071	.003	7
8	1.000	1.000	1.000	1.000	1.000	1.000	.998	.990	.960	.866	.613	.370	.086	8
9	1.000	1.000	1.000	1.000	1.000	1.000	1.000	1.000	1.000	1.000	1.000	1.000	1.000	9

n = 10

k	.01	.05	.10	.20	.30	.40	.50	.60	.70	.80	.90	.95	.99	k
0	.904	.599	.349	.107	.028	.006	.001	.000	.000	.000	.000	.000	.000	0
1	.996	.914	.736	.376	.149	.046	.011	.002	.000	.000	.000	.000	.000	1
2	1.000	.988	.930	.678	.383	.167	.055	.012	.002	.000	.000	.000	.000	2
3	1.000	.999	.987	.879	.650	.382	.172	.055	.011	.001	.000	.000	.000	3
4	1.000	1.000	.998	.967	.850	.633	.377	.166	.047	.006	.000	.000	.000	4
5	1.000	1.000	1.000	.994	.953	.834	.623	.367	.150	.033	.002	.000	.000	5
6	1.000	1.000	1.000	.999	.989	.945	.828	.618	.350	.121	.013	.001	.000	6
7	1.000	1.000	1.000	1.000	.998	.988	.945	.833	.617	.322	.070	.012	.000	7
8	1.000	1.000	1.000	1.000	1.000	.998	.989	.954	.851	.624	.264	.086	.004	8
9	1.000	1.000	1.000	1.000	1.000	1.000	.999	.994	.972	.893	.651	.401	.096	9
10	1.000	1.000	1.000	1.000	1.000	1.000	1.000	1.000	1.000	1.000	1.000	1.000	1.000	10

(*continued*)

Table 1 (*Continued*)

n = 11

k	.01	.05	.10	.20	.30	.40	.50	.60	.70	.80	.90	.95	.99	k
0	.895	.569	.314	.086	.020	.004	.000	.000	.000	.000	.000	.000	.000	0
1	.995	.898	.697	.322	.113	.030	.006	.001	.000	.000	.000	.000	.000	1
2	1.000	.985	.910	.617	.313	.119	.033	.006	.001	.000	.000	.000	.000	2
3	1.000	.998	.981	.839	.570	.296	.113	.029	.004	.000	.000	.000	.000	3
4	1.000	1.000	.997	.950	.790	.533	.274	.099	.022	.002	.000	.000	.000	4
5	1.000	1.000	1.000	.988	.922	.754	.500	.246	.078	.012	.000	.000	.000	5
6	1.000	1.000	1.000	.998	.978	.901	.726	.467	.210	.050	.003	.000	.000	6
7	1.000	1.000	1.000	1.000	.996	.971	.887	.704	.430	.161	.019	.002	.000	7
8	1.000	1.000	1.000	1.000	.999	.994	.967	.881	.687	.383	.090	.015	.000	8
9	1.000	1.000	1.000	1.000	1.000	.999	.994	.970	.887	.678	.303	.102	.005	9
10	1.000	1.000	1.000	1.000	1.000	1.000	1.000	.996	.980	.914	.686	.431	.105	10
11	1.000	1.000	1.000	1.000	1.000	1.000	1.000	1.000	1.000	1.000	1.000	1.000	1.000	11

n = 12

k	.01	.05	.10	.20	.30	.40	.50	.60	.70	.80	.90	.95	.99	k
0	.886	.540	.282	.069	.014	.002	.000	.000	.000	.000	.000	.000	.000	0
1	.994	.882	.659	.275	.085	.020	.003	.000	.000	.000	.000	.000	.000	1
2	1.000	.980	.889	.558	.253	.083	.019	.003	.000	.000	.000	.000	.000	2
3	1.000	.998	.974	.795	.493	.225	.073	.015	.002	.000	.000	.000	.000	3
4	1.000	1.000	.996	.927	.724	.438	.194	.057	.009	.001	.000	.000	.000	4
5	1.000	1.000	.999	.981	.882	.665	.387	.158	.039	.004	.000	.000	.000	5
6	1.000	1.000	1.000	.996	.961	.842	.613	.335	.118	.019	.001	.000	.000	6
7	1.000	1.000	1.000	.999	.991	.943	.806	.562	.276	.073	.004	.000	.000	7
8	1.000	1.000	1.000	1.000	.998	.985	.927	.775	.507	.205	.026	.002	.000	8
9	1.000	1.000	1.000	1.000	1.000	.997	.981	.917	.747	.442	.111	.020	.000	9
10	1.000	1.000	1.000	1.000	1.000	1.000	.997	.980	.915	.725	.341	.118	.006	10
11	1.000	1.000	1.000	1.000	1.000	1.000	1.000	.998	.986	.931	.718	.460	.114	11
12	1.000	1.000	1.000	1.000	1.000	1.000	1.000	1.000	1.000	1.000	1.000	1.000	1.000	12

(*continued*)

Table 1 (*Continued*)

n = 15

k	.01	.05	.10	.20	.30	.40	p .50	.60	.70	.80	.90	.95	.99	k
0	.860	.463	.206	.035	.005	.000	.000	.000	.000	.000	.000	.000	.000	0
1	.990	.829	.549	.167	.035	.005	.000	.000	.000	.000	.000	.000	.000	1
2	1.000	.964	.816	.398	.127	.027	.004	.000	.000	.000	.000	.000	.000	2
3	1.000	.995	.944	.648	.297	.091	.018	.002	.000	.000	.000	.000	.000	3
4	1.000	.999	.987	.836	.515	.217	.059	.009	.001	.000	.000	.000	.000	4
5	1.000	1.000	.998	.939	.722	.403	.151	.034	.004	.000	.000	.000	.000	5
6	1.000	1.000	1.000	.982	.869	.610	.304	.095	.015	.001	.000	.000	.000	6
7	1.000	1.000	1.000	.996	.950	.787	.500	.213	.050	.004	.000	.000	.000	7
8	1.000	1.000	1.000	.999	.985	.905	.696	.390	.131	.018	.000	.000	.000	8
9	1.000	1.000	1.000	1.000	.996	.966	.849	.597	.278	.061	.002	.000	.000	9
10	1.000	1.000	1.000	1.000	.999	.991	.941	.783	.485	.164	.013	.001	.000	10
11	1.000	1.000	1.000	1.000	1.000	.998	.982	.909	.703	.352	.056	.005	.000	11
12	1.000	1.000	1.000	1.000	1.000	1.000	.996	.973	.873	.602	.184	.036	.000	12
13	1.000	1.000	1.000	1.000	1.000	1.000	1.000	.995	.965	.833	.451	.171	.010	13
14	1.000	1.000	1.000	1.000	1.000	1.000	1.000	1.000	.995	.965	.794	.537	.140	14
15	1.000	1.000	1.000	1.000	1.000	1.000	1.000	1.000	1.000	1.000	1.000	1.000	1.000	15

n = 20

k	.01	.05	.10	.20	.30	.40	p .50	.60	.70	.80	.90	.95	.99	k
0	.818	.358	.122	.012	.001	.000	.000	.000	.000	.000	.000	.000	.000	0
1	.983	.736	.392	.069	.008	.001	.000	.000	.000	.000	.000	.000	.000	1
2	.999	.925	.677	.206	.035	.004	.000	.000	.000	.000	.000	.000	.000	2
3	1.000	.984	.867	.411	.107	.016	.001	.000	.000	.000	.000	.000	.000	3
4	1.000	.997	.957	.630	.238	.051	.006	.000	.000	.000	.000	.000	.000	4
5	1.000	1.000	.989	.804	.416	.126	.021	.002	.000	.000	.000	.000	.000	5
6	1.000	1.000	.998	.913	.608	.250	.058	.006	.000	.000	.000	.000	.000	6
7	1.000	1.000	1.000	.968	.772	.416	.132	.021	.001	.000	.000	.000	.000	7
8	1.000	1.000	1.000	.990	.887	.596	.252	.057	.005	.000	.000	.000	.000	8
9	1.000	1.000	1.000	.997	.952	.755	.412	.128	.017	.001	.000	.000	.000	9
10	1.000	1.000	1.000	.999	.983	.872	.588	.245	.048	.003	.000	.000	.000	10
11	1.000	1.000	1.000	1.000	.995	.943	.748	.404	.113	.010	.000	.000	.000	11
12	1.000	1.000	1.000	1.000	.999	.979	.868	.584	.228	.032	.000	.000	.000	12
13	1.000	1.000	1.000	1.000	1.000	.994	.942	.750	.392	.087	.002	.000	.000	13
14	1.000	1.000	1.000	1.000	1.000	.998	.979	.874	.584	.196	.011	.000	.000	14
15	1.000	1.000	1.000	1.000	1.000	1.000	.994	.949	.762	.370	.043	.003	.000	15
16	1.000	1.000	1.000	1.000	1.000	1.000	.999	.984	.893	.589	.133	.016	.000	16
17	1.000	1.000	1.000	1.000	1.000	1.000	1.000	.996	.965	.794	.323	.075	.001	17
18	1.000	1.000	1.000	1.000	1.000	1.000	1.000	.999	.992	.931	.608	.264	.017	18
19	1.000	1.000	1.000	1.000	1.000	1.000	1.000	1.000	.999	.988	.878	.642	.182	19
20	1.000	1.000	1.000	1.000	1.000	1.000	1.000	1.000	1.000	1.000	1.000	1.000	1.000	20

(*continued*)

Table 1 (*Continued*)

$n = 25$

							p							
k	.01	.05	.10	.20	.30	.40	.50	.60	.70	.80	.90	.95	.99	k
0	.778	.277	.072	.004	.000	.000	.000	.000	.000	.000	.000	.000	.000	0
1	.974	.642	.271	.027	.002	.000	.000	.000	.000	.000	.000	.000	.000	1
2	.998	.873	.537	.098	.009	.000	.000	.000	.000	.000	.000	.000	.000	2
3	1.000	.966	.764	.234	.033	.002	.000	.000	.000	.000	.000	.000	.000	3
4	1.000	.993	.902	.421	.090	.009	.000	.000	.000	.000	.000	.000	.000	4
5	1.000	.999	.967	.617	.193	.029	.002	.000	.000	.000	.000	.000	.000	5
6	1.000	1.000	.991	.780	.341	.074	.007	.000	.000	.000	.000	.000	.000	6
7	1.000	1.000	.998	.891	.512	.154	.022	.001	.000	.000	.000	.000	.000	7
8	1.000	1.000	1.000	.953	.677	.274	.054	.004	.000	.000	.000	.000	.000	8
9	1.000	1.000	1.000	.983	.811	.425	.115	.013	.000	.000	.000	.000	.000	9
10	1.000	1.000	1.000	.994	.902	.586	.212	.034	.002	.000	.000	.000	.000	10
11	1.000	1.000	1.000	.998	.956	.732	.345	.078	.006	.000	.000	.000	.000	11
12	1.000	1.000	1.000	1.000	.983	.846	.500	.154	.017	.000	.000	.000	.000	12
13	1.000	1.000	1.000	1.000	.994	.922	.655	.268	.044	.002	.000	.000	.000	13
14	1.000	1.000	1.000	1.000	.998	.966	.788	.414	.098	.006	.000	.000	.000	14
15	1.000	1.000	1.000	1.000	1.000	.987	.885	.575	.189	.017	.000	.000	.000	15
16	1.000	1.000	1.000	1.000	1.000	.996	.946	.726	.323	.047	.000	.000	.000	16
17	1.000	1.000	1.000	1.000	1.000	.999	.978	.846	.488	.109	.002	.000	.000	17
18	1.000	1.000	1.000	1.000	1.000	1.000	.993	.926	.659	.220	.009	.000	.000	18
19	1.000	1.000	1.000	1.000	1.000	1.000	.998	.971	.807	.383	.033	.001	.000	19
20	1.000	1.000	1.000	1.000	1.000	1.000	1.000	.991	.910	.579	.098	.007	.000	20
21	1.000	1.000	1.000	1.000	1.000	1.000	1.000	.998	.967	.766	.236	.034	.000	21
22	1.000	1.000	1.000	1.000	1.000	1.000	1.000	1.000	.991	.902	.463	.127	.002	22
23	1.000	1.000	1.000	1.000	1.000	1.000	1.000	1.000	.998	.973	.729	.358	.026	23
24	1.000	1.000	1.000	1.000	1.000	1.000	1.000	1.000	1.000	.996	.928	.723	.222	24
25	1.000	1.000	1.000	1.000	1.000	1.000	1.000	1.000	1.000	1.000	1.000	1.000	1.000	25

Table 2 Cumulative Poisson Probabilities

Tabulated values are $P(x \leq k) = p(0) + p(1) + p(2) + \ldots + p(k)$

(Computations are rounded at the third decimal place.)

						μ					
k	.1	.2	.3	.4	.5	.6	.7	.8	.9	1.0	1.5
0	.905	.819	.741	.670	.607	.549	.497	.449	.407	.368	.223
1	.995	.982	.963	.938	.910	.878	.844	.809	.772	.736	.558
2	1.000	.999	.996	.992	.986	.977	.966	.953	.937	.920	.809
3		1.000	1.000	.999	.998	.997	.994	.991	.987	.981	.934
4				1.000	1.000	1.000	.999	.999	.998	.996	.981
5							1.000	1.000	1.000	.999	.996
6										1.000	.999
7											1.000

(*continued*)

Table 2 (*Continued*)

k	2.0	2.5	3.0	3.5	4.0	4.5	5.0	5.5	6.0	6.5	7.0
0	.135	.082	.050	.030	.018	.011	.007	.004	.003	.002	.001
1	.406	.287	.199	.136	.092	.061	.040	.027	.017	.011	.007
2	.677	.544	.423	.321	.238	.174	.125	.088	.062	.043	.030
3	.857	.758	.647	.537	.433	.342	.265	.202	.151	.112	.082
4	.947	.891	.815	.725	.629	.532	.440	.358	.285	.224	.173
5	.983	.958	.916	.858	.785	.703	.616	.529	.446	.369	.301
6	.995	.986	.966	.935	.889	.831	.762	.686	.606	.563	.450
7	.999	.996	.988	.973	.949	.913	.867	.809	.744	.673	.599
8	1.000	.999	.996	.990	.979	.960	.932	.894	.847	.792	.729
9		1.000	.999	.997	.992	.983	.968	.946	.916	.877	.830
10			1.000	.999	.997	.993	.986	.975	.957	.933	.901
11				1.000	.999	.998	.995	.989	.980	.966	.947
12					1.000	.999	.998	.996	.991	.984	.973
13						1.000	.999	.998	.996	.993	.987
14							1.000	.999	.999	.997	.994
15								1.000	.999	.999	.998
16									1.000	1.000	.999
17											1.000

(*continued*)

Table 2 (*Continued*)

k	7.5	8.0	8.5	9.0	9.5	10.0	12.0	15.0	20.0
					μ				
0	.001	.000	.000	.000	.000	.000	.000	.000	.000
1	.005	.003	.002	.001	.001	.000	.000	.000	.000
2	.020	.014	.009	.006	.004	.003	.001	.000	.000
3	.059	.042	.030	.021	.015	.010	.002	.000	.000
4	.132	.100	.074	.055	.040	.029	.008	.001	.000
5	.241	.191	.150	.116	.089	.067	.020	.003	.000
6	.378	.313	.256	.207	.165	.130	.046	.008	.000
7	.525	.453	.386	.324	.269	.220	.090	.018	.001
8	.662	.593	.523	.456	.392	.333	.155	.037	.002
9	.776	.717	.653	.587	.522	.458	.242	.070	.005
10	.862	.816	.763	.706	.645	.583	.347	.118	.011
11	.921	.888	.849	.803	.752	.697	.462	.185	.021
12	.957	.936	.909	.876	.836	.792	.576	.268	.039
13	.978	.966	.949	.926	.898	.864	.682	.363	.066
14	.990	.983	.973	.959	.940	.917	.772	.466	.105
15	.995	.992	.986	.978	.967	.951	.844	.568	.157
16	.998	.996	.993	.989	.982	.973	.899	.664	.221
17	.999	.998	.997	.995	.991	.986	.937	.749	.297
18	1.000	.999	.999	.998	.996	.993	.963	.819	.381
19		1.000	.999	.999	.998	.997	.979	.875	.470
20			1.000	1.000	.999	.998	.988	.917	.559
21					1.000	.999	.994	.947	.644
22						1.000	.997	.967	.721
23							.999	.981	.787
24							.999	.989	.843
25							1.000	.994	.888
26								.997	.922
27								.998	.948
28								.999	.966
29								1.000	.978
30									.987
31									.992
32									.995
33									.997
34									.999
35									.999
36									1.000

Table 3   Normal Curve Areas

z	.00	.01	.02	.03	.04	.05	.06	.07	.08	.09
.0	.0000	.0040	.0080	.0120	.0160	.0199	.0239	.0279	.0319	.0359
.1	.0398	.0438	.0478	.0517	.0557	.0596	.0636	.0675	.0714	.0753
.2	.0793	.0832	.0871	.0910	.0948	.0987	.1026	.1064	.1103	.1141
.3	.1179	.1217	.1255	.1293	.1331	.1368	.1406	.1443	.1480	.1517
.4	.1554	.1591	.1628	.1664	.1700	.1736	.1772	.1808	.1844	.1879
.5	.1915	.1950	.1985	.2019	.2054	.2088	.2123	.2157	.2190	.2224
.6	.2257	.2291	.2324	.2357	.2389	.2422	.2454	.2486	.2517	.2549
.7	.2580	.2611	.2642	.2673	.2704	.2734	.2764	.2794	.2823	.2852
.8	.2881	.2910	.2939	.2967	.2995	.3023	.3051	.3078	.3106	.3133
.9	.3159	.3186	.3212	.3238	.3264	.3289	.3315	.3340	.3365	.3389
1.0	.3413	.3438	.3461	.3485	.3508	.3531	.3554	.3577	.3599	.3621
1.1	.3643	.3665	.3686	.3708	.3729	.3749	.3770	.3790	.3810	.3830
1.2	.3849	.3869	.3888	.3907	.3925	.3944	.3962	.3980	.3997	.4015
1.3	.4032	.4049	.4066	.4082	.4099	.4115	.4131	.4147	.4162	.4177
1.4	.4192	.4207	.4222	.4236	.4251	.4265	.4279	.4292	.4306	.4319
1.5	.4332	.4345	.4357	.4370	.4382	.4394	.4406	.4418	.4429	.4441
1.6	.4452	.4463	.4474	.4484	.4495	.4505	.4515	.4525	.4535	.4545
1.7	.4554	.4564	.4573	.4582	.4591	.4599	.4608	.4616	.4625	.4633
1.8	.4641	.4649	.4656	.4664	.4671	.4678	.4686	.4693	.4699	.4706
1.9	.4713	.4719	.4726	.4732	.4738	.4744	.4750	.4756	.4761	.4767
2.0	.4772	.4778	.4783	.4788	.4793	.4798	.4803	.4808	.4812	.4817
2.1	.4821	.4826	.4830	.4834	.4838	.4842	.4846	.4850	.4854	.4857
2.2	.4861	.4864	.4868	.4871	.4875	.4878	.4881	.4884	.4887	.4890
2.3	.4893	.4896	.4898	.4901	.4904	.4906	.4909	.4911	.4913	.4916
2.4	.4918	.4920	.4922	.4925	.4927	.4929	.4931	.4932	.4934	.4936
2.5	.4938	.4940	.4941	.4943	.4945	.4946	.4948	.4949	.4951	.4952
2.6	.4953	.4955	.4956	.4957	.4959	.4960	.4961	.4962	.4963	.4964
2.7	.4965	.4966	.4967	.4968	.4969	.4970	.4971	.4972	.4973	.4974
2.8	.4974	.4975	.4976	.4977	.4977	.4978	.4979	.4979	.4980	.4981
2.9	.4981	.4982	.4982	.4983	.4984	.4984	.4985	.4985	.4986	.4986
3.0	.4987	.4987	.4987	.4988	.4988	.4989	.4989	.4989	.4990	.4990

Source: This table is abridged from Table 1 of *Statistical Tables and Formulas* by A. Hald (New York: Wiley, 1952). Reproduced by permission of A. Hald and the publisher, John Wiley & Sons, Inc.

Table 4 Critical Values of t

df	$t_{.100}$	$t_{.050}$	$t_{.025}$	$t_{.010}$	$t_{.005}$	df
1	3.078	6.314	12.706	31.821	63.657	1
2	1.886	2.920	4.303	6.965	9.925	2
3	1.638	2.353	3.182	4.541	5.841	3
4	1.533	2.132	2.776	3.747	4.604	4
5	1.476	2.015	2.571	3.365	4.032	5
6	1.440	1.943	2.447	3.143	3.707	6
7	1.415	1.895	2.365	2.998	3.499	7
8	1.397	1.860	2.306	2.896	3.355	8
9	1.383	1.833	2.262	2.821	3.250	9
10	1.372	1.812	2.228	2.764	3.169	10
11	1.363	1.796	2.201	2.718	3.106	11
12	1.356	1.782	2.179	2.681	3.055	12
13	1.350	1.771	2.160	2.650	3.012	13
14	1.345	1.761	2.145	2.624	2.977	14
15	1.341	1.753	2.131	2.602	2.947	15
16	1.337	1.746	2.120	2.583	2.921	16
17	1.333	1.740	2.110	2.567	2.898	17
18	1.330	1.734	2.101	2.552	2.878	18
19	1.328	1.729	2.093	2.539	2.861	19
20	1.325	1.725	2.086	2.528	2.845	20
21	1.323	1.721	2.080	2.518	2.831	21
22	1.321	1.717	2.074	2.508	2.819	22
23	1.319	1.714	2.069	2.500	2.807	23
24	1.318	1.711	2.064	2.492	2.797	24
25	1.316	1.708	2.060	2.485	2.787	25
26	1.315	1.706	2.056	2.479	2.779	26
27	1.314	1.703	2.052	2.473	2.771	27
28	1.313	1.701	2.048	2.467	2.763	28
29	1.311	1.699	2.045	2.462	2.756	29
inf.	1.282	1.645	1.960	2.326	2.576	∞

Source: From "Table of Percentage Points of the t-Distribution," *Biometrika* 32 (1941): 300. Reproduced by permission of the *Biometrika* Trustees.

Table 5 Critical Values of Chi-square

df	$\chi^2_{.995}$	$\chi^2_{.990}$	$\chi^2_{.975}$	$\chi^2_{.950}$	$\chi^2_{.900}$	$\chi^2_{.100}$	$\chi^2_{.050}$	$\chi^2_{.025}$	$\chi^2_{.010}$	$\chi^2_{.005}$
1	.0000393	.0001571	.0009821	.0039321	.0157908	2.70554	3.84146	5.02389	6.63490	7.87944
2	.0100251	.0201007	.0506356	.102587	.210720	4.60517	5.99147	7.37776	9.21034	10.5966
3	.0717212	.114832	.215795	.351846	.584375	6.25139	7.81473	9.34840	11.3449	12.8381
4	.206990	.297110	.484419	.710721	1.063623	7.77944	9.48773	11.1433	13.2767	14.8602
5	.411740	.554300	.831211	1.145476	1.61031	9.23635	11.0705	12.8325	15.0863	16.7496
6	.675727	.872085	1.237347	1.63539	2.20413	10.6446	12.5916	14.4494	16.8119	18.5476
7	.989265	1.239043	1.68987	2.16735	2.83311	12.0170	14.0671	16.0128	18.4753	20.2777
8	1.344419	1.646482	2.17973	2.73264	3.48954	13.3616	15.5073	17.5346	20.0902	21.9550
9	1.734926	2.087912	2.70039	3.32511	4.16816	14.6837	16.9190	19.0228	21.6660	23.5893
10	2.15585	2.55821	3.24697	3.94030	4.86518	15.9871	18.3070	20.4831	23.2093	25.1882
11	2.60321	3.05347	3.81575	4.57481	5.57779	17.2750	19.6751	21.9200	24.7250	26.7569
12	3.07382	3.57056	4.40379	5.22603	6.30380	18.5494	21.0261	23.3367	26.2170	28.2995
13	3.56503	4.10691	5.00874	5.89186	7.04150	19.8119	23.3621	24.7356	27.6883	29.8194
14	4.07468	4.66043	5.62872	6.57063	7.78953	21.0642	23.6848	26.1190	29.1413	31.3193
15	4.60094	5.22935	6.26214	7.26094	8.54675	22.3072	24.9958	27.4884	30.5779	32.8013
16	5.14224	5.81221	6.90766	7.96164	9.31223	23.5418	26.2962	28.8454	31.9999	34.2672
17	5.69724	6.40776	7.56418	8.67176	10.0852	24.7690	27.5871	30.1910	33.4087	35.7185
18	6.26481	7.01491	8.23075	9.39046	10.8649	25.9894	28.8693	31.5264	34.8053	37.1564
19	6.84398	7.63273	8.90655	10.1170	11.6509	27.2036	30.1435	32.8523	36.1908	38.5822
20	7.43386	8.26040	9.59083	10.8508	12.4426	28.4120	31.4104	34.1696	37.5662	39.9968
21	8.03366	8.89720	10.28293	11.5913	13.2396	29.6151	32.6705	35.4789	38.9321	41.4010
22	8.64272	9.54249	10.9823	12.3380	14.0415	30.8133	33.9244	36.7807	40.2894	42.7956
23	9.26042	10.19567	11.6885	13.0905	14.8479	32.0069	35.1725	38.0757	41.6384	44.1813
24	9.88623	10.8564	12.4011	13.8484	15.6587	33.1963	36.4151	39.3641	42.9798	45.5585

Source: From "Tables of the Percentage Points of the χ^2-Distribution," *Biometrika Tables for Statisticians* 1, 3rd ed. (1966). Reproduced by permission of the *Biometrika* Trustees.

(*continued*)

Table 5 (*Continued*)

df	$\chi^2_{.995}$	$\chi^2_{.990}$	$\chi^2_{.975}$	$\chi^2_{.950}$	$\chi^2_{.900}$	$\chi^2_{.100}$	$\chi^2_{.050}$	$\chi^2_{.025}$	$\chi^2_{.010}$	$\chi^2_{.005}$
25	10.5197	11.5240	13.1197	14.6114	16.4734	34.3816	37.6525	40.6465	44.3141	46.9278
26	11.1603	12.1981	13.8439	15.3791	17.2919	35.5631	38.8852	41.9232	45.6417	48.2899
27	11.8076	12.8786	14.5733	16.1513	18.1138	36.7412	40.1133	43.1944	46.9630	49.6449
28	12.4613	13.5648	15.3079	16.9279	18.9392	37.9159	41.3372	44.4607	48.2782	50.9933
29	13.1211	14.2565	16.0471	17.7083	19.7677	39.0875	42.5569	45.7222	49.5879	52.3356
30	13.7867	14.9535	16.7908	18.4926	20.5992	40.2560	43.7729	46.9792	50.8922	53.6720
40	20.7065	22.1643	24.4331	26.5093	29.0505	51.8050	55.7585	59.3417	63.6907	66.7659
50	27.9907	29.7067	32.3574	34.7642	37.6886	63.1671	67.5048	71.4202	76.1539	79.4900
60	35.5346	37.4848	40.4817	43.1879	46.4589	74.3970	79.0819	83.2976	88.3794	91.9517
70	43.2752	45.4418	48.7576	51.7393	55.3290	85.5271	90.5312	95.0231	100.425	104.215
80	51.1720	53.5400	57.1532	60.3915	64.2778	96.5782	101.879	106.629	112.329	116.321
90	59.1963	61.7541	65.6466	69.1260	73.2912	107.565	113.145	118.136	124.116	128.299
100	67.3276	70.0648	74.2219	77.9295	82.3581	118.498	124.342	129.561	135.807	140.169

Table 6   Percentage Points of the F Distribution

df_2	a	df_1 1	2	3	4	5	6	7	8	9
1	.100	39.86	49.50	53.59	55.83	57.24	58.20	58.91	59.44	59.86
	.050	161.4	199.5	215.7	224.6	230.2	234.0	236.8	238.9	240.5
	.025	647.8	799.5	864.2	899.6	921.8	937.1	948.2	956.7	963.3
	.010	4052	4999.5	5403	5625	5764	5859	5928	5982	6022
	.005	16211	20000	21615	22500	23056	23437	23715	23925	24091
2	.100	8.53	9.00	9.16	9.24	9.29	9.33	9.35	9.37	9.38
	.050	18.51	19.00	19.16	19.25	19.30	19.33	19.35	19.37	19.38
	.025	38.51	39.00	39.17	39.25	39.30	39.33	39.36	39.37	39.39
	.010	98.50	99.00	99.17	99.25	99.30	99.33	99.36	99.37	99.39
	.005	198.5	199.0	199.2	199.2	199.3	199.3	199.4	199.4	199.4
3	.100	5.54	5.46	5.39	5.34	5.31	5.28	5.27	5.25	5.24
	.050	10.13	9.55	9.28	9.12	9.01	8.94	8.89	8.85	8.81
	.025	17.44	16.04	15.44	15.10	14.88	14.73	14.62	14.54	14.47
	.010	34.12	30.82	29.46	28.71	28.24	27.91	27.67	27.49	27.35
	.005	55.55	49.80	47.47	46.19	45.39	44.84	44.43	44.13	43.88
4	.100	4.54	4.32	4.19	4.11	4.05	4.01	3.98	3.95	3.94
	.050	7.71	6.94	6.59	6.39	6.26	6.16	6.09	6.04	6.00
	.025	12.22	10.65	9.98	9.60	9.36	9.20	9.07	8.98	8.90
	.010	21.20	18.00	16.69	15.98	15.52	15.21	14.98	14.80	14.66
	.005	31.33	26.28	24.26	23.15	22.46	21.97	21.62	21.35	21.14
5	.100	4.06	3.78	3.62	3.52	3.45	3.40	3.37	3.34	3.32
	.050	6.61	5.79	5.41	5.19	5.05	4.95	4.88	4.82	4.77
	.025	10.01	8.43	7.76	7.39	7.15	6.98	6.85	6.76	6.68
	.010	16.26	13.27	12.06	11.39	10.97	10.67	10.46	10.29	10.16
	.005	22.78	18.31	16.53	15.56	14.94	14.51	14.20	13.96	13.77

Source: A portion of "Tables of Percentage Points of the Inverted Beta (E) Distribution" *Biometrika*, vol. 33 (1943) by M. Merrington and C. M. Thompson and from Table 18 of *Biometrika Tables for Statisticians*, vol. 1. Cambridge University Press, 1954, edited by E. S. Pearson and H. O. Hartley. Reproduced with permission of the authors, editors, and *Biometrika* trustees.

(*continued*)

Table 6 (*Continued*)

df_2	a	1	2	3	4	5	6	7	8	9
6	.100	3.78	3.46	3.29	3.18	3.11	3.05	3.01	2.98	2.96
	.050	5.99	5.14	4.76	4.53	4.39	4.28	4.21	4.15	4.10
	.025	8.81	7.26	6.60	6.23	5.99	5.82	5.70	5.60	5.52
	.010	13.75	10.92	9.78	9.15	8.75	8.47	8.26	8.10	7.98
	.005	18.63	14.54	12.92	12.03	11.46	11.07	10.79	10.57	10.39
7	.100	3.59	3.26	3.07	2.96	2.88	2.83	2.78	2.75	2.72
	.050	5.59	4.74	4.35	4.12	3.97	3.87	3.79	3.73	3.68
	.025	8.07	6.54	5.89	5.52	5.29	5.12	4.99	4.90	4.82
	.010	12.25	9.55	8.45	7.85	7.46	7.19	6.99	6.84	6.72
	.005	16.24	12.40	10.88	10.05	9.52	9.16	8.89	8.68	8.51
8	.100	3.46	3.11	2.92	2.81	2.73	2.67	2.62	2.59	2.56
	.050	5.32	4.46	4.07	3.84	3.69	3.58	3.50	3.44	3.39
	.025	7.57	6.06	5.42	5.05	4.82	4.65	4.53	4.43	4.36
	.010	11.26	8.65	7.59	7.01	6.63	6.37	6.18	6.03	5.91
	.005	14.69	11.04	9.60	8.81	8.30	7.95	7.69	7.50	7.34
9	.100	3.36	3.01	2.81	2.69	2.61	2.55	2.51	2.47	2.44
	.050	5.12	4.26	3.86	3.63	3.48	3.37	3.29	3.23	3.18
	.025	7.21	5.71	5.08	4.72	4.48	4.32	4.20	4.10	4.03
	.010	10.56	8.02	6.99	6.42	6.06	5.80	5.61	5.47	5.35
	.005	13.61	10.11	8.72	7.96	7.47	7.13	6.88	6.69	6.54
10	.100	3.29	2.92	2.73	2.61	2.52	2.46	2.41	2.38	2.35
	.050	4.96	4.10	3.71	3.48	3.33	3.22	3.14	3.07	3.02
	.025	6.94	5.46	4.83	4.47	4.24	4.07	3.95	3.85	3.78
	.010	10.04	7.56	6.55	5.99	5.64	5.39	5.20	5.06	4.94
	.005	12.83	9.43	8.08	7.34	6.87	6.54	6.30	6.12	5.97
11	.100	3.23	2.86	2.66	2.54	2.45	2.39	2.34	2.30	2.27
	.050	4.84	3.98	3.59	3.36	3.20	3.09	3.01	2.95	2.90
	.025	6.72	5.26	4.63	4.28	4.04	3.88	3.76	3.66	3.59
	.010	9.65	7.21	6.22	5.67	5.32	5.07	4.89	4.74	4.63
	.005	12.23	8.91	7.60	6.88	6.42	6.10	5.86	5.68	5.54
12	.100	3.18	2.81	2.61	2.48	2.39	2.33	2.28	2.24	2.21
	.050	4.75	3.89	3.49	3.26	3.11	3.00	2.91	2.85	2.80
	.025	6.55	5.10	4.47	4.12	3.89	3.73	3.61	3.51	3.44
	.010	9.33	6.93	5.95	5.41	5.06	4.82	4.64	4.50	4.39
	.005	11.75	8.51	7.23	6.52	6.07	5.76	5.52	5.35	5.20

(*continued*)

Table 6 (*Continued*)

					df_1						
10	12	15	20	24	30	40	60	120	∞	a	df_2
60.19	60.71	60.22	61.74	62.00	62.26	62.53	62.79	63.06	63.33	.100	1
241.9	243.9	245.9	248.0	249.1	250.1	251.2	252.2	253.3	254.3	.050	
968.6	976.7	984.9	993.1	997.2	1001	1006	1010	1014	1018	.025	
6056	6106	6157	6209	6235	6261	6287	6313	6339	6366	.010	
24224	24426	24630	24836	24940	25044	25148	25253	25359	25465	.005	
9.39	9.41	9.42	9.44	9.45	9.46	9.47	9.47	9.48	9.49	.100	2
19.40	19.41	19.43	19.45	19.45	19.46	19.47	19.48	19.49	19.50	.050	
39.40	39.41	39.43	39.45	39.46	39.46	39.47	39.48	39.49	39.50	.025	
99.40	99.42	99.43	99.45	99.46	99.47	99.47	99.48	99.49	99.50	.010	
199.4	199.4	199.4	199.4	199.5	199.5	199.5	199.5	199.5	199.5	.005	
5.23	5.22	5.20	5.18	5.18	5.17	5.16	5.15	5.14	5.13	.100	3
8.79	8.74	8.70	8.66	8.64	8.62	8.59	8.57	8.55	8.53	.050	
14.42	14.34	14.25	14.17	14.12	14.08	14.04	13.99	13.95	13.90	.025	
27.23	27.05	26.87	26.69	26.60	26.50	26.41	26.32	26.22	26.13	.010	
43.69	43.39	43.08	42.78	42.62	42.47	42.31	42.15	41.99	41.83	.005	
3.92	3.90	3.87	3.84	3.83	3.82	3.80	3.79	3.78	3.76	.100	4
5.96	5.91	5.86	5.80	5.77	5.75	5.72	5.69	5.66	5.63	.050	
8.84	8.75	8.66	8.56	8.51	8.46	8.41	8.36	8.31	8.26	.025	
14.55	14.37	14.20	14.02	13.93	13.84	13.75	13.65	13.56	13.46	.010	
20.97	20.70	20.44	20.17	20.03	19.89	19.75	19.61	19.47	19.32	.005	
3.30	3.27	3.24	3.21	3.19	3.17	3.16	3.14	3.12	3.10	.100	5
4.74	4.68	4.62	4.56	4.53	4.50	4.46	4.43	4.40	4.36	.050	
6.62	6.52	6.43	6.33	6.28	6.23	6.18	6.12	6.07	6.02	.025	
10.05	9.89	9.72	9.55	9.47	9.38	9.29	9.20	9.11	9.02	.010	
13.62	13.38	13.15	12.90	12.78	12.66	12.53	12.40	12.27	12.14	.005	
2.94	2.90	2.87	2.84	2.82	2.80	2.78	2.76	2.74	2.72	.100	6
4.06	4.00	3.94	3.87	3.84	3.81	3.77	3.74	3.70	3.67	.050	
5.46	5.37	5.27	5.17	5.12	5.07	5.01	4.96	4.90	4.85	.025	
7.87	7.72	7.56	7.40	7.31	7.23	7.14	7.06	6.97	6.88	.010	
10.25	10.03	9.81	9.59	9.47	9.36	9.24	9.12	9.00	8.88	.005	
2.70	2.67	2.63	2.59	2.58	2.56	2.54	2.51	2.49	2.47	.100	7
3.64	3.57	3.51	3.44	3.41	3.38	3.34	3.30	3.27	3.23	.050	
4.76	4.67	4.57	4.47	4.42	4.36	4.31	4.25	4.20	4.14	.025	
6.62	6.47	6.31	6.16	6.07	5.99	5.91	5.82	5.74	5.65	.010	
8.38	8.18	7.97	7.75	7.65	7.53	7.42	7.31	7.19	7.08	.005	

(*continued*)

Table 6 (*Continued*)

10	12	15	20	24	30	40	60	120	∞	a	df_2
					df_1						
2.54	2.50	2.46	2.42	2.40	2.38	2.36	2.34	2.32	2.29	.100	8
3.35	3.28	3.22	3.15	3.12	3.08	3.04	3.01	2.97	2.93	.050	
4.30	4.20	4.10	4.00	3.95	3.89	3.84	3.78	3.73	3.67	.025	
5.81	5.67	5.52	5.36	5.28	5.20	5.12	5.03	4.95	4.86	.010	
7.21	7.01	6.81	6.61	6.50	6.40	6.29	6.18	6.06	5.95	.005	
2.42	2.38	2.34	2.30	2.28	2.25	2.23	2.21	2.18	2.16	.100	9
3.14	3.07	3.01	2.94	2.90	2.86	2.83	2.79	2.75	2.71	.050	
3.96	3.87	3.77	3.67	3.61	3.56	3.51	3.45	3.39	3.33	.025	
5.26	5.11	4.96	4.81	4.73	4.65	4.57	4.48	4.40	4.31	.010	
6.42	6.23	6.03	5.83	5.73	5.62	5.52	5.41	5.30	5.19	.005	
2.32	2.28	2.24	2.20	2.18	2.16	2.13	2.11	2.08	2.06	.100	10
2.98	2.91	2.85	2.77	2.74	2.70	2.66	2.62	2.58	2.54	.050	
3.72	3.62	3.52	3.42	3.37	3.31	3.26	3.20	3.14˙	3.08	.025	
4.85	4.71	4.56	4.41	4.33	4.25	4.17	4.08	4.00	3.91	.010	
5.85	5.66	5.47	5.27	5.17	5.07	4.97	4.86	4.75	4.64	.005	
2.25	2.21	2.17	2.12	2.10	2.08	2.05	2.03	2.00	1.97	.100	11
2.85	2.79	2.72	2.65	2.61	2.57	2.53	2.49	2.45	2.40	.050	
3.53	3.43	3.33	3.23	3.17	3.12	3.06	3.00	2.94	2.88	.025	
4.54	4.40	4.25	4.10	4.02	3.94	3.86	3.78	3.69	3.60	.010	
5.42	5.24	5.05	4.86	4.76	4.65	4.55	4.44	4.34	4.23	.005	
2.19	2.15	2.10	2.06	2.04	2.01	1.99	1.96	1.93	1.90	.100	12
2.75	2.69	2.62	2.54	2.51	2.47	2.43	2.38	2.34	2.30	.050	
3.37	3.28	3.18	3.07	3.02	2.96	2.91	2.85	2.79	2.72	.025	
4.30	4.16	4.01	3.86	3.78	3.70	3.62	3.54	3.45	3.36	.010	
5.09	4.91	4.72	4.53	4.43	4.33	4.23	4.12	4.01	3.90	.005	

(*continued*)

Table 6 (*Continued*)

df_2	a	1	2	3	4	5	6	7	8	9
						df_1				
13	.100	3.14	2.76	2.56	2.43	2.35	2.28	2.23	2.20	2.16
	.050	4.67	3.81	3.41	3.18	3.03	2.92	2.83	2.77	2.71
	.025	6.41	4.97	4.35	4.00	3.77	3.60	3.48	3.39	3.31
	.010	9.07	6.70	5.74	5.21	4.86	4.62	4.44	4.30	4.19
	.005	11.37	8.19	6.93	6.23	5.79	5.48	5.25	5.08	4.94
14	.100	3.10	2.73	2.52	2.39	2.31	2.24	2.19	2.15	2.12
	.050	4.60	3.74	3.34	3.11	2.96	2.85	2.76	2.70	2.65
	.025	6.30	4.86	4.24	3.89	3.66	3.50	3.38	3.29	3.21
	.010	8.86	6.51	5.56	5.04	4.69	4.46	4.28	4.14	4.03
	.005	11.06	7.92	6.68	6.00	5.56	5.26	5.03	4.86	4.72
15	.100	3.07	2.70	2.49	2.36	2.27	2.21	2.16	2.12	2.09
	.050	4.54	3.68	3.29	3.06	2.90	2.79	2.71	2.64	2.59
	.025	6.20	4.77	4.15	3.80	3.58	3.41	3.29	3.20	3.12
	.010	8.68	6.36	5.42	4.89	4.56	4.32	4.14	4.00	3.89
	.005	10.80	7.70	6.48	5.80	5.37	5.07	4.85	4.67	4.54
16	.100	3.05	2.67	2.46	2.33	2.24	2.18	2.13	2.09	2.06
	.050	4.49	3.63	3.24	3.01	2.85	2.74	2.66	2.59	2.54
	.025	6.12	4.69	4.08	3.73	3.50	3.34	3.22	3.12	3.05
	.010	8.53	6.23	5.29	4.77	4.44	4.20	4.03	3.89	3.78
	.005	10.58	7.51	6.30	5.64	5.21	4.91	4.69	4.52	4.38
17	.100	3.03	2.64	2.44	2.31	2.22	2.15	2.10	2.06	2.03
	.050	4.45	3.59	3.20	2.96	2.81	2.70	2.61	2.55	2.49
	.025	6.04	4.62	4.01	3.66	3.44	3.28	3.16	3.06	2.98
	.010	8.40	6.11	5.18	4.67	4.34	4.10	3.93	3.79	3.68
	.005	10.38	7.35	6.16	5.50	5.07	4.78	4.56	4.39	4.25
18	.100	3.01	2.62	2.42	2.29	2.20	2.13	2.08	2.04	2.00
	.050	4.41	3.55	3.16	2.93	2.77	2.66	2.58	2.51	2.46
	.025	5.98	4.56	3.95	3.61	3.38	3.22	3.10	3.01	2.93
	.010	8.29	6.01	5.09	4.58	4.25	4.01	3.84	3.71	3.60
	.005	10.22	7.21	6.03	5.37	4.96	4.66	4.44	4.28	4.14
19	.100	2.99	2.61	2.40	2.27	2.18	2.11	2.06	2.02	1.98
	.050	4.38	3.52	3.13	2.90	2.74	2.63	2.54	2.48	2.42
	.025	5.92	4.51	3.90	3.56	3.33	3.17	3.05	2.96	2.88
	.010	8.18	5.93	5.01	4.50	4.17	3.94	3.77	3.63	3.52
	.005	10.07	7.09	5.92	5.27	4.85	4.56	4.34	4.18	4.04

(*continued*)

Table 6 (*Continued*)

df_2	a	df_1								
		1	2	3	4	5	6	7	8	9
20	.100	2.97	2.59	2.38	2.25	2.16	2.09	2.04	2.00	1.96
	.050	4.35	3.49	3.10	2.87	2.71	2.60	2.51	2.45	2.39
	.025	5.87	4.46	3.86	3.51	3.29	3.13	3.01	2.91	2.84
	.010	8.10	5.85	4.94	4.43	4.10	3.87	3.70	3.56	3.46
	.005	9.94	6.99	5.82	5.17	4.76	4.47	4.26	4.09	3.96
21	.100	2.96	2.57	2.36	2.23	2.14	2.08	2.02	1.98	1.95
	.050	4.32	3.47	3.07	2.84	2.68	2.57	2.49	2.42	2.37
	.025	5.83	4.42	3.82	3.48	3.25	3.09	2.97	2.87	2.80
	.010	8.02	5.78	4.87	4.37	4.04	3.81	3.64	3.51	3.40
	.005	9.83	6.89	5.73	5.09	4.68	4.39	4.18	4.01	3.88
22	.100	2.95	2.56	2.35	2.22	2.13	2.06	2.01	1.97	1.93
	.050	4.30	3.44	3.05	2.82	2.66	2.55	2.46	2.40	2.34
	.025	5.79	4.38	3.78	3.44	3.22	3.05	2.93	2.84	2.76
	.010	7.95	5.72	4.82	4.31	3.99	3.76	3.59	3.45	3.35
	.005	9.73	6.81	5.65	5.02	4.61	4.32	4.11	3.94	3.81
23	.100	2.94	2.55	2.34	2.21	2.11	2.05	1.99	1.95	1.92
	.050	4.28	3.42	3.03	2.80	2.64	2.53	2.44	2.37	2.32
	.025	5.75	4.35	3.75	3.41	3.18	3.02	2.90	2.81	2.73
	.010	7.88	5.66	4.76	4.26	3.94	3.71	3.54	3.41	3.30
	.005	9.63	6.73	5.58	4.95	4.54	4.26	4.05	3.88	3.75
24	.100	2.93	2.54	2.33	2.19	2.10	2.04	1.98	1.94	1.91
	.050	4.26	3.40	3.01	2.78	2.62	2.51	2.42	2.36	2.30
	.025	5.72	4.32	3.72	3.38	3.15	2.99	2.87	2.78	2.70
	.010	7.82	5.61	4.72	4.22	3.90	3.67	3.50	3.36	3.26
	.005	9.55	6.66	5.52	4.89	4.49	4.20	3.99	3.83	3.69
25	.100	2.92	2.53	2.32	2.18	2.09	2.02	1.97	1.93	1.89
	.050	4.24	3.39	2.99	2.76	2.60	2.49	2.40	2.34	2.28
	.025	5.69	4.29	3.69	3.35	3.13	2.97	2.85	2.75	2.68
	.010	7.77	5.57	4.68	4.18	3.85	3.63	3.46	3.32	3.22
	.005	9.48	6.60	5.46	4.84	4.43	4.15	3.94	3.78	3.64
26	.100	2.91	2.52	2.31	2.17	2.08	2.01	1.96	1.92	1.88
	.050	4.23	3.37	2.98	2.74	2.59	2.47	2.39	2.32	2.27
	.025	5.66	4.27	3.67	3.33	3.10	2.94	2.82	2.73	2.65
	.010	7.72	5.53	4.64	4.14	3.82	3.59	3.42	3.29	3.18
	.005	9.41	6.54	5.41	4.79	4.38	4.10	3.89	3.73	3.60

(*continued*)

Table 6 (*Continued*)

10	12	15	20	24	30	40	60	120	∞	a	df_2
					df_1						
2.14	2.10	2.05	2.01	1.98	1.96	1.93	1.90	1.88	1.85	.100	13
2.67	2.60	2.53	2.46	2.42	2.38	2.34	2.30	2.25	2.21	.050	
3.25	3.15	3.05	2.95	2.89	2.84	2.78	2.72	2.66	2.60	.025	
4.10	3.96	3.82	3.66	3.59	3.51	3.43	3.34	3.25	3.17	.010	
4.82	4.64	4.46	4.27	4.17	4.07	3.97	3.87	3.76	3.65	.005	
2.10	2.05	2.01	1.96	1.94	1.91	1.89	1.86	1.83	1.80	.100	14
2.60	2.53	2.46	2.39	2.35	2.31	2.27	2.22	2.18	2.13	.050	
3.15	3.05	2.95	2.84	2.79	2.73	2.67	2.61	2.55	2.49	.025	
3.94	3.80	3.66	3.51	3.43	3.35	3.27	3.18	3.09	3.00	.010	
4.60	4.43	4.25	4.06	3.96	3.86	3.76	3.66	3.55	3.44	.005	
2.06	2.02	1.97	1.92	1.90	1.87	1.85	1.82	1.79	1.76	.100	15
2.54	2.48	2.40	2.33	2.29	2.25	2.20	2.16	2.11	2.07	.050	
3.06	2.96	2.86	2.76	2.70	2.64	2.59	2.52	2.46	2.40	.025	
3.80	3.67	3.52	3.37	3.29	3.21	3.13	3.05	2.96	2.87	.010	
4.42	4.25	4.07	3.88	3.79	3.69	3.58	3.48	3.37	3.26	.005	
2.03	1.99	1.94	1.89	1.87	1.84	1.81	1.78	1.75	1.72	.100	16
2.49	2.42	2.35	2.28	2.24	2.19	2.15	2.11	2.06	2.01	.050	
2.99	2.89	2.79	2.68	2.63	2.57	2.51	2.45	2.38	2.32	.025	
3.69	3.55	3.41	3.26	3.18	3.10	3.02	2.93	2.84	2.75	.010	
4.27	4.10	3.92	3.73	3.64	3.54	3.44	3.33	3.22	3.11	.005	
2.00	1.96	1.91	1.86	1.84	1.81	1.78	1.75	1.72	1.69	.100	17
2.45	2.38	2.31	2.23	2.19	2.15	2.10	2.06	2.01	1.96	.050	
2.92	2.82	2.72	2.62	2.56	2.50	2.44	2.38	2.32	2.25	.025	
3.59	3.46	3.31	3.16	3.08	3.00	2.92	2.83	2.75	2.65	.010	
4.14	3.97	3.79	3.61	3.51	3.41	3.31	3.21	3.10	2.98	.005	
1.98	1.93	1.89	1.84	1.81	1.78	1.75	1.72	1.69	1.66	.100	18
2.41	2.34	2.27	2.19	2.15	2.11	2.06	2.02	1.97	1.92	.050	
2.87	2.77	2.67	2.56	2.50	2.44	2.38	2.32	2.26	2.19	.025	
3.51	3.37	3.23	3.08	3.00	2.92	2.84	2.75	2.66	2.57	.010	
4.03	3.86	3.68	3.50	3.40	3.30	3.20	3.10	2.99	2.87	.005	
1.96	1.91	1.86	1.81	1.79	1.76	1.73	1.70	1.67	1.63	.100	19
2.38	2.31	2.23	2.16	2.11	2.07	2.03	1.98	1.93	1.88	.050	
2.82	2.72	2.62	2.51	2.45	2.39	2.33	2.27	2.20	2.13	.025	
3.43	3.30	3.15	3.00	2.92	2.84	2.76	2.67	2.58	2.49	.010	
3.93	3.76	3.59	3.40	3.31	3.21	3.11	3.00	2.89	2.78	.005	

(*continued*)

Table 6 (*Continued*)

					df_1							
10	12	15	20	24	30	40	60	120	∞	a	df_2
1.94	1.89	1.84	1.79	1.77	1.74	1.71	1.68	1.64	1.61	.100	20
2.35	2.28	2.20	2.12	2.08	2.04	1.99	1.95	1.90	1.84	.050	
2.77	2.68	2.57	2.46	2.41	2.35	2.29	2.22	2.16	2.09	.025	
3.37	3.23	3.09	2.94	2.86	2.78	2.69	2.61	2.52	2.42	.010	
3.85	3.68	3.50	3.32	3.22	3.12	3.02	2.92	2.81	2.69	.005	
1.92	1.87	1.83	1.78	1.75	1.72	1.69	1.66	1.62	1.59	.100	21
2.32	2.25	2.18	2.10	2.05	2.01	1.96	1.92	1.87	1.81	.050	
2.73	2.64	2.53	2.42	2.37	2.31	2.25	2.18	2.11	2.04	.025	
3.31	3.17	3.03	2.88	2.80	2.72	2.64	2.55	2.46	2.36	.010	
3.77	3.60	3.43	3.24	3.15	3.05	2.95	2.84	2.73	2.61	.005	
1.90	1.86	1.81	1.76	1.73	1.70	1.67	1.64	1.60	1.57	.100	22
2.30	2.23	2.15	2.07	2.03	1.98	1.94	1.89	1.84	1.78	.050	
2.70	2.60	2.50	2.39	2.33	2.27	2.21	2.14	2.08	2.00	.025	
3.26	3.12	2.98	2.83	2.75	2.67	2.58	2.50	2.40	2.31	.010	
3.70	3.54	3.36	3.18	3.08	2.98	2.88	2.77	2.66	2.55	.005	
1.89	1.84	1.80	1.74	1.72	1.69	1.66	1.62	1.59	1.55	.100	23
2.27	2.20	2.13	2.05	2.01	1.96	1.91	1.86	1.81	1.76	.050	
2.67	2.57	2.47	2.36	2.30	2.24	2.18	2.11	2.04	1.97	.025	
3.21	3.07	2.93	2.78	2.70	2.62	2.54	2.45	2.35	2.26	.010	
3.64	3.47	3.30	3.12	3.02	2.92	2.82	2.71	2.60	2.48	.005	
1.88	1.83	1.78	1.73	1.70	1.67	1.64	1.61	1.57	1.53	.100	24
2.25	2.18	2.11	2.03	1.98	1.94	1.89	1.84	1.79	1.73	.050	
2.64	2.54	2.44	2.33	2.27	2.21	2.15	2.08	2.01	1.94	.025	
3.17	3.03	2.89	2.74	2.66	2.58	2.49	2.40	2.31	2.21	.010	
3.59	3.42	3.25	3.06	2.97	2.87	2.77	2.66	2.55	2.43	.005	
1.87	1.82	1.77	1.72	1.69	1.66	1.63	1.59	1.56	1.52	.100	25
2.24	2.16	2.09	2.01	1.96	1.92	1.87	1.82	1.77	1.71	.050	
2.61	2.51	2.41	2.30	2.24	2.18	2.12	2.05	1.98	1.91	.025	
3.13	2.99	2.85	2.70	2.62	2.54	2.45	2.36	2.27	2.17	.010	
3.54	3.37	3.20	3.01	2.92	2.82	2.72	2.61	2.50	2.38	.005	
1.86	1.81	1.76	1.71	1.68	1.65	1.61	1.58	1.54	1.50	.100	26
2.22	2.15	2.07	1.99	1.95	1.90	1.85	1.80	1.75	1.69	.050	
2.59	2.49	2.39	2.28	2.22	2.16	2.09	2.03	1.95	1.88	.025	
3.09	2.96	2.81	2.66	2.58	2.50	2.42	2.33	2.23	2.13	.010	
3.49	3.33	3.15	2.97	2.87	2.77	2.67	2.56	2.45	2.33	.005	

(*continued*)

Table 6 (*Continued*)

df_2	a	1	2	3	4	5	6	7	8	9
						df_1				
27	.100	2.90	2.51	2.30	2.17	2.07	2.00	1.95	1.91	1.87
	.050	4.21	3.35	2.96	2.73	2.57	2.46	2.37	2.31	2.25
	.025	5.63	4.24	3.65	3.31	3.08	2.92	2.80	2.71	2.63
	.010	7.68	5.49	4.60	4.11	3.78	3.56	3.39	3.26	3.15
	.005	9.34	6.49	5.36	4.74	4.34	4.06	3.85	3.69	3.56
28	.100	2.89	2.50	2.29	2.16	2.06	2.00	1.94	1.90	1.87
	.050	4.20	3.34	2.95	2.71	2.56	2.45	2.36	2.29	2.24
	.025	5.61	4.22	3.63	3.29	3.06	2.90	2.78	2.69	2.61
	.010	7.64	5.45	4.57	4.07	3.75	3.53	3.36	3.23	3.12
	.005	9.28	6.44	5.32	4.70	4.30	4.02	3.81	3.65	3.52
29	.100	2.89	2.50	2.28	2.15	2.06	1.99	1.93	1.89	1.86
	.050	4.18	3.33	2.93	2.70	2.55	2.43	2.35	2.28	2.22
	.025	5.59	4.20	3.61	3.27	3.04	2.88	2.76	2.67	2.59
	.010	7.60	5.42	4.54	4.04	3.73	3.50	3.33	3.20	3.09
	.005	9.23	6.40	5.28	4.66	4.26	3.98	3.77	3.61	3.48
30	.100	2.88	2.49	2.28	2.14	2.05	1.98	1.93	1.88	1.85
	.050	4.17	3.32	2.92	2.69	2.53	2.42	2.33	2.27	2.21
	.025	5.57	4.18	3.59	3.25	3.03	2.87	2.75	2.65	2.57
	.010	7.56	5.39	4.51	4.02	3.70	3.47	3.30	3.17	3.07
	.005	9.18	6.35	5.24	4.62	4.23	3.95	3.74	3.58	3.45
40	.100	2.84	2.44	2.23	2.09	2.00	1.93	1.87	1.83	1.79
	.050	4.08	3.23	2.84	2.61	2.45	2.34	2.25	2.18	2.12
	.025	5.42	4.05	3.46	3.13	2.90	2.74	2.62	2.53	2.45
	.010	7.31	5.18	4.31	3.83	3.51	3.29	3.12	2.99	2.89
	.005	8.83	6.07	4.98	4.37	3.99	3.71	3.51	3.35	3.22

(*continued*)

Table 6 (*Continued*)

df_2	a	1	2	3	4	5	6	7	8	9
						df_1				
60	.100	2.79	2.39	2.18	2.04	1.95	1.87	1.82	1.77	1.74
	.050	4.00	3.15	2.76	2.53	2.37	2.25	2.17	2.10	2.04
	.025	5.29	3.93	3.34	3.01	2.79	2.63	2.51	2.41	2.33
	.010	7.08	4.98	4.13	3.65	3.34	3.12	2.95	2.82	2.72
	.005	8.49	5.79	4.73	4.14	3.76	3.49	3.29	3.13	3.01
120	.100	2.75	2.35	2.13	1.99	1.90	1.82	1.77	1.72	1.68
	.050	3.92	3.07	2.68	2.45	2.29	2.17	2.09	2.02	1.96
	.025	5.15	3.80	3.23	2.89	2.67	2.52	2.39	2.30	2.22
	.010	6.85	4.79	3.95	3.48	3.17	2.96	2.79	2.66	2.56
	.005	8.18	5.54	4.50	3.92	3.55	3.28	3.09	2.93	2.81
∞	.100	2.71	2.30	2.08	1.94	1.85	1.77	1.72	1.67	1.63
	.050	3.84	3.00	2.60	2.37	2.21	2.10	2.01	1.94	1.63
	.025	5.02	3.69	3.12	2.79	2.57	2.41	2.29	2.19	2.11
	.010	6.63	4.61	3.78	3.32	3.02	2.80	2.64	2.51	2.41
	.005	7.88	5.30	4.28	3.72	3.35	3.09	2.90	2.74	2.62

(*continued*)

Table 6 (*Continued*)

10	12	15	20	24	30	40	60	120	∞	a	df_2
					df_1						
1.85	1.80	1.75	1.70	1.67	1.64	1.60	1.57	1.53	1.49	.100	27
2.20	2.13	2.06	1.97	1.93	1.88	1.84	1.79	1.73	1.67	.050	
2.57	2.47	2.36	2.25	2.19	2.13	2.07	2.00	1.93	1.85	.025	
3.06	2.93	2.78	2.63	2.55	2.47	2.38	2.29	2.20	2.10	.010	
3.45	3.28	3.11	2.93	2.83	2.73	2.63	2.52	2.41	2.29	.005	
1.84	1.79	1.74	1.69	1.66	1.63	1.59	1.56	1.52	1.48	.100	28
2.19	2.12	2.04	1.96	1.91	1.87	1.82	1.77	1.71	1.65	.050	
2.55	2.45	2.34	2.23	2.17	2.11	2.05	1.98	1.91	1.83	.025	
3.03	2.90	2.75	2.60	2.52	2.44	2.35	2.26	2.17	2.06	.010	
3.41	3.25	3.07	2.89	2.79	2.69	2.59	2.48	2.37	2.25	.005	
1.83	1.78	1.73	1.68	1.65	1.62	1.58	1.55	1.51	1.47	.100	29
2.18	2.10	2.03	1.94	1.90	1.85	1.81	1.75	1.70	1.64	.050	
2.53	2.43	2.32	2.21	2.15	2.09	2.03	1.96	1.89	1.81	.025	
3.00	2.87	2.73	2.57	2.49	2.41	2.33	2.23	2.14	2.03	.010	
3.38	3.21	3.04	2.86	2.76	2.66	2.56	2.45	2.33	2.21	.005	
1.82	1.77	1.72	1.67	1.64	1.61	1.57	1.54	1.50	1.46	.100	30
2.16	2.09	2.01	1.93	1.89	1.84	1.79	1.74	1.68	1.62	.050	
2.51	2.41	2.31	2.20	2.14	2.07	2.01	1.94	1.87	1.79	.025	
2.98	2.84	2.70	2.55	2.47	2.39	2.30	2.21	2.11	2.01	.010	
3.34	3.18	3.01	2.82	2.73	2.63	2.52	2.42	2.30	2.18	.005	
1.76	1.71	1.66	1.61	1.57	1.54	1.51	1.47	1.42	1.38	.100	40
2.08	2.00	1.92	1.84	1.79	1.74	1.69	1.64	1.58	1.51	.050	
2.39	2.29	2.18	2.07	2.01	1.94	1.88	1.80	1.72	1.64	.025	
2.80	2.66	2.52	2.37	2.29	2.20	2.11	2.02	1.92	1.80	.010	
3.12	2.95	2.78	2.60	2.50	2.40	2.30	2.18	2.06	1.93	.005	

(*continued*)

Table 6 (*Continued*)

					df_1							
10	12	15	20	24	30	40	60	120	∞	a	df_2
1.71	1.66	1.60	1.54	1.51	1.48	1.44	1.40	1.35	1.29	.100	60
1.99	1.92	1.84	1.75	1.70	1.65	1.59	1.53	1.47	1.39	.050	
2.27	2.17	2.06	1.94	1.88	1.82	1.74	1.67	1.58	1.48	.025	
2.63	2.50	2.35	2.20	2.12	2.03	1.94	1.84	1.73	1.60	.010	
2.90	2.74	2.57	2.39	2.29	2.19	2.08	1.96	1.83	1.69	.005	
1.65	1.60	1.55	1.48	1.45	1.41	1.37	1.32	1.26	1.19	.100	120
1.91	1.83	1.75	1.66	1.61	1.55	1.50	1.43	1.35	1.25	.050	
2.16	2.05	1.94	1.82	1.76	1.69	1.61	1.53	1.43	1.31	.025	
2.47	2.34	2.19	2.03	1.95	1.86	1.76	1.66	1.53	1.38	.010	
2.71	2.54	2.37	2.19	2.09	1.98	1.87	1.75	1.61	1.43	.005	
1.60	1.55	1.49	1.42	1.38	1.34	1.30	1.24	1.17	1.00	.100	∞
1.83	1.75	1.67	1.57	1.52	1.46	1.39	1.32	1.22	1.00	.050	
2.05	1.94	1.83	1.71	1.64	1.57	1.48	1.39	1.27 ·	1.00	.025	
2.32	2.18	2.04	1.88	1.79	1.70	1.59	1.47	1.32	1.00	.010	
2.52	2.36	2.19	2.00	1.90	1.79	1.67	1.53	1.36	1.00	.005	

Table 7 Critical Values of T for the Wilcoxon Rank Sum Test, $n_1 \leq n_2$

Table 7(a). 5% Left-Tailed Critical Values

							n_1							
n_2	2	3	4	5	6	7	8	9	10	11	12	13	14	15
3	—	6												
4	—	6	11											
5	3	7	12	19										
6	3	8	13	20	28									
7	3	8	14	21	29	39								
8	4	9	15	23	31	41	51							
9	4	10	16	24	33	43	54	66						
10	4	10	17	26	35	45	56	69	82					
11	4	11	18	27	37	47	59	72	86	100				
12	5	11	19	28	38	49	62	75	89	104	120			
13	5	12	20	30	40	52	64	78	92	108	125	142		
14	6	13	21	31	42	54	67	81	96	112	129	147	166	
15	6	13	22	33	44	56	69	84	99	116	133	152	171	192

Source: Adapted from "An Extended Table of Critical Values for the Mann-Whitney (Wilcoxon) Two-Sample Statistics" by Roy C. Milton, *Journal of the American Statistical Association*, Volume 59, Number 307 (September, 1964). Reproduced with the permission of the Editor, *Journal of the American Statistical Association*.

Table 7(b). 2.5% Left-Tailed Critical Values

							n_1							
n_2	2	3	4	5	6	7	8	9	10	11	12	13	14	15
4	—	—	10											
5	—	6	11	17										
6	—	7	12	18	26									
7	—	7	13	20	27	36								
8	3	8	14	21	29	38	49							
9	3	8	14	22	31	40	51	62						
10	2	9	15	23	32	42	53	65	78					
11	3	9	16	24	34	44	55	68	81	96				
12	4	10	17	26	35	46	58	71	84	99	115			
13	4	10	18	27	37	48	60	73	88	103	119	136		
14	4	11	19	28	38	50	62	76	91	106	123	141	160	
15	4	11	20	29	40	52	65	79	94	110	127	145	164	184

Table 7(c). 1% Left-Tailed Critical Values

n_2	\multicolumn{14}{c}{n_1}
	2	3	4	5	6	7	8	9	10	11	12	13	14	15
3	—	—												
4	—	—	—											
5	—	—	10	16										
6	—	—	11	17	24									
7	—	6	11	18	25	34								
8	—	6	12	19	27	35	45							
9	—	7	13	20	28	37	47	59						
10	—	7	13	21	29	39	49	61	74					
11	—	7	14	22	30	40	51	63	77	91				
12	—	8	15	23	32	42	53	66	79	94	109			
13	3	8	15	24	33	44	56	68	82	97	113	130		
14	3	8	16	25	34	45	58	71	85	100	116	134	152	
15	3	9	17	26	36	47	60	73	88	103	120	138	156	176

Table 7(d). 0.5% Left-Tailed Critical Values

n_2	\multicolumn{13}{c}{n_1}
	3	4	5	6	7	8	9	10	11	12	13	14	15
3	—												
4	—	—											
5	—	—	15										
6	—	10	16	23									
7	—	10	16	24	32								
8	—	11	17	25	34	42							
9	6	11	18	26	35	45	56						
10	6	12	19	27	37	47	58	71					
11	6	12	20	28	38	49	61	73	87				
12	6	13	21	30	40	51	63	76	90	105			
13	6	13	22	31	41	53	65	79	83	109	125		
14	6	14	22	32	43	54	67	81	96	112	129	147	
15	8	15	23	33	44	56	69	84	99	115	133	151	171

Table 7(e). 0.1% Left-Tail Critical Values of the Rank Sum for the Smaller Sample, $n_1 \leq n_2$

n_2	n_1											
	4	5	6	7	8	9	10	11	12	13	14	15
3												
4	—											
5	—	—										
6	—	—	—									
7	—	—	21	29								
8	—	15	22	30	40							
9	—	16	23	31	41	52						
10	10	16	24	33	42	53	65					
11	10	17	25	34	44	55	67	81				
12	10	17	25	35	45	57	69	83	98			
13	11	18	26	36	47	59	72	86	101	117		
14	11	18	27	37	48	60	74	88	103	120	137	
15	11	19	28	38	50	62	76	90	106	123	141	160

Table 8 Critical Values of T in the Wilcoxon Signed-Rank Test; $n = 5(1)50$

One-sided	Two-sided	$n = 5$	$n = 6$	$n = 7$	$n = 8$	$n = 9$	$n = 10$
$\alpha = .05$	$\alpha = .10$	1	2	4	6	8	11
$\alpha = .025$	$\alpha = .05$		1	2	4	6	8
$\alpha = .01$	$\alpha = .02$			0	2	3	5
$\alpha = .005$	$\alpha = .01$				0	2	3

One-sided	Two-sided	$n = 11$	$n = 12$	$n = 13$	$n = 14$	$n = 15$	$n = 16$
$\alpha = .05$	$\alpha = .10$	14	17	21	26	30	36
$\alpha = .025$	$\alpha = .05$	11	14	17	21	25	30
$\alpha = .01$	$\alpha = .02$	7	10	13	16	20	24
$\alpha = .005$	$\alpha = .01$	5	7	10	13	16	19

One-sided	Two-sided	$n = 17$	$n = 18$	$n = 19$	$n = 20$	$n = 21$	$n = 22$
$\alpha = .05$	$\alpha = .10$	41	47	54	60	68	75
$\alpha = .025$	$\alpha = .05$	35	40	46	52	59	66
$\alpha = .01$	$\alpha = .02$	28	33	38	43	49	56
$\alpha = .005$	$\alpha = .01$	23	28	32	37	43	49

One-sided	Two-sided	$n = 23$	$n = 24$	$n = 25$	$n = 26$	$n = 27$	$n = 28$
$\alpha = .05$	$\alpha = .10$	83	92	101	110	120	130
$\alpha = .025$	$\alpha = .05$	73	81	90	98	107	117
$\alpha = .01$	$\alpha = .02$	62	69	77	85	93	102
$\alpha = .005$	$\alpha = .01$	55	68	68	76	84	92

One-sided	Two-sided	$n = 29$	$n = 30$	$n = 31$	$n = 32$	$n = 33$	$n = 34$
$\alpha = .05$	$\alpha = .10$	141	152	163	175	188	201
$\alpha = .025$	$\alpha = .05$	127	137	148	159	171	183
$\alpha = .01$	$\alpha = .02$	111	120	130	141	151	162
$\alpha = .005$	$\alpha = .01$	100	109	118	128	138	149

One-sided	Two-sided	$n = 35$	$n = 36$	$n = 37$	$n = 38$	$n = 39$	$n = 40$
$\alpha = .05$	$\alpha = .10$	214	228	242	256	271	287
$\alpha = .025$	$\alpha = .05$	195	208	222	235	250	264
$\alpha = .01$	$\alpha = .02$	174	186	198	211	224	238
$\alpha = .005$	$\alpha = .01$	160	171	183	195	208	221

One-sided	Two-sided	$n = 41$	$n = 42$	$n = 43$	$n = 44$	$n = 45$	$n = 46$
$\alpha = .05$	$\alpha = .10$	303	319	336	353	371	389
$\alpha = .025$	$\alpha = .05$	279	295	311	327	344	361
$\alpha = .01$	$\alpha = .02$	252	267	281	297	313	329
$\alpha = .005$	$\alpha = .01$	234	248	262	277	292	307

One-sided	Two-sided	$n = 47$	$n = 48$	$n = 49$	$n = 50$
$\alpha = .05$	$\alpha = .10$	408	427	446	466
$\alpha = .025$	$\alpha = .05$	379	397	415	434
$\alpha = .01$	$\alpha = .02$	345	362	380	398
$\alpha = .005$	$\alpha = .01$	323	339	356	373

Source: From "Some Rapid Approximate Statistical Procedures" (1964) 28, by F. Wilcoxon and R. A. Wilcox. Reproduced with the kind permission of Lederle Laboratories, a division of American Cyanamid Company.

Table 9 Critical Values of Spearman's Rank Correlation Coefficient for a One-tailed Test

n	$\alpha = .05$	$\alpha = .025$	$\alpha = .01$	$\alpha = .005$
5	.900	—	—	—
6	.829	.886	.943	—
7	.714	.786	.893	—
8	.643	.738	.833	.881
9	.600	.683	.783	.833
10	.564	.648	.745	.794
11	.523	.623	.736	.818
12	.497	.591	.703	.780
13	.475	.566	.673	.745
14	.457	.545	.646	.716
15	.441	.525	.623	.689
16	.425	.507	.601	.666
17	.412	.490	.582	.645
18	.399	.476	.564	.625
19	.388	.462	.549	.608
20	.377	.450	.534	.591
21	.368	.438	.521	.576
22	.359	.428	.508	.562
23	.351	.418	.496	.549
24	.343	.409	.485	.537
25	.336	.400	.475	.526
26	.329	.392	.465	.515
27	.323	.385	.456	.505
28	.317	.377	.448	.496
29	.311	.370	.440	.487
30	.305	.364	.432	.478

Source: From "Distribution of Sums of Squares of Rank Differences for Small Samples" by E. G. Olds, *Annals of Mathematical Statistics* 9 (1938). Reproduced with the kind permission of the Editor, *Annals of Mathematical Statistics.*

Table 10(a) Percentage points of the Studentized range, $q(k, df)$; upper 5% points

df	2	3	4	5	6	7	8	9	10	11
1	17.97	26.98	32.82	37.08	40.41	43.12	45.40	47.36	49.07	50.59
2	6.08	8.33	9.80	10.88	11.74	12.44	13.03	13.54	13.99	14.39
3	4.50	5.91	6.82	7.50	8.04	8.48	8.85	9.18	9.46	9.72
4	3.93	5.04	5.76	6.29	6.71	7.05	7.35	7.60	7.83	8.03
5	3.64	4.60	5.22	5.67	6.03	6.33	6.58	6.80	6.99	7.17
6	3.46	4.34	4.90	5.30	5.63	5.90	6.12	6.32	6.49	6.65
7	3.34	4.16	4.68	5.06	5.36	5.61	5.82	6.00	6.16	6.30
8	3.26	4.04	4.53	4.89	5.17	5.40	5.60	5.77	5.92	6.05
9	3.20	3.95	4.41	4.76	5.02	5.24	5.43	5.59	5.74	5.87
10	3.15	3.88	4.33	4.65	4.91	5.12	5.30	5.46	5.60	5.72
11	3.11	3.82	4.26	4.57	4.82	5.03	5.20	5.35	5.49	5.61
12	3.08	3.77	4.20	4.51	4.75	4.95	5.12	5.27	5.39	5.51
13	3.06	3.73	4.15	4.45	4.69	4.88	5.05	5.19	5.32	5.43
14	3.03	3.70	4.11	4.41	4.64	4.83	4.99	5.13	5.25	5.36
15	3.01	3.67	4.08	4.37	4.60	4.78	4.94	5.08	5.20	5.31
16	3.00	3.65	4.05	4.33	4.56	4.74	4.90	5.03	5.15	5.26
17	2.98	3.63	4.02	4.30	4.52	4.70	4.86	4.99	5.11	5.21
18	2.97	3.61	4.00	4.28	4.49	4.67	4.82	4.96	5.07	5.17
19	2.96	3.59	3.98	4.25	4.47	4.65	4.79	4.92	5.04	5.14
20	2.95	3.58	3.96	4.23	4.45	4.62	4.77	4.90	5.01	5.11
24	2.92	3.53	3.90	4.17	4.37	4.54	4.68	4.81	4.92	5.01
30	2.89	3.49	3.85	4.10	4.30	4.46	4.60	4.72	4.82	4.92
40	2.86	3.44	3.79	4.04	4.23	4.39	4.52	4.63	4.73	4.82
60	2.83	3.40	3.74	3.98	4.16	4.31	4.44	4.55	4.65	4.73
120	2.80	3.36	3.68	3.92	4.10	4.24	4.36	4.47	4.56	4.64
∞	2.77	3.31	3.63	3.86	4.03	4.17	4.29	4.39	4.47	4.55

(*continued*)

Table 10(a) (*Continued*)

					k				
12	13	14	15	16	17	18	19	20	*df*
51.96	53.20	54.33	55.36	56.32	57.22	58.04	58.83	59.56	1
14.75	15.08	15.38	15.65	15.91	16.14	16.37	16.57	16.77	2
9.95	10.15	10.35	10.52	10.69	10.84	10.98	11.11	11.24	3
8.21	8.37	8.52	8.66	8.79	8.91	9.03	9.13	9.23	4
7.32	7.47	7.60	7.72	7.83	7.93	8.03	8.12	8.21	5
6.79	6.92	7.03	7.14	7.24	7.34	7.43	7.51	7.59	6
6.43	6.55	6.66	6.76	6.85	6.94	7.02	7.10	7.17	7
6.18	6.29	6.39	6.48	6.57	6.65	6.73	6.80	6.87	8
5.98	6.09	6.19	6.28	6.36	6.44	6.51	6.58	6.64	9
5.83	5.93	6.03	6.11	6.19	6.27	6.34	6.40	6.47	10
5.71	5.81	5.90	5.98	6.06	6.13	6.20	6.27	6.33	11
5.61	5.71	5.80	5.88	5.95	6.02	6.09	6.15	6.21	12
5.53	5.63	5.71	5.79	5.86	5.93	5.99	6.05	6.11	13
5.46	5.55	5.64	5.71	5.79	5.85	5.91	5.97	6.03	14
5.40	5.49	5.57	5.65	5.72	5.78	5.85	5.90	5.96	15
5.35	5.44	5.52	5.59	5.66	5.73	5.79	5.84	5.90	16
5.31	5.39	5.47	5.54	5.61	5.67	5.73	5.79	5.84	17
5.27	5.35	5.43	5.50	5.57	5.63	5.69	5.74	5.79	18
5.23	5.31	5.39	5.46	5.53	5.59	5.65	5.70	5.75	19
5.20	5.28	5.36	5.43	5.49	5.55	5.61	5.66	5.71	20
5.10	5.18	5.25	5.32	5.38	5.44	5.49	5.55	5.59	24
5.00	5.08	5.15	5.21	5.27	5.33	5.38	5.43	5.47	30
4.90	4.98	5.04	5.11	5.16	5.22	5.27	5.31	5.36	40
4.81	4.88	4.94	5.00	5.06	5.11	5.15	5.20	5.24	60
4.71	4.78	4.84	4.90	4.95	5.00	5.04	5.09	5.13	120
4.62	4.68	4.74	4.80	4.85	4.89	4.93	4.97	5.01	∞

Source: From *Biometrika Tables for Statisticians*, Vol. 1, 3rd ed., edited by E. S. Pearson and H. O. Hartley (Cambridge University Press, 1966). Reproduced by permission of the Biometrika Trustees.

Table 10(b) Percentage points of the Studentized range, $q(k, df)$; upper 1% points

df	2	3	4	5	6	7	8	9	10	11
					k					
1	90.03	135.0	164.3	185.6	202.2	215.8	227.2	237.0	245.6	253.2
2	14.04	19.02	22.29	24.72	26.63	28.20	29.53	30.68	31.69	32.59
3	8.26	10.62	12.17	13.33	14.24	15.00	15.64	16.20	16.69	17.13
4	6.51	8.12	9.17	9.96	10.58	11.10	11.55	11.93	12.27	12.57
5	5.70	6.98	7.80	8.42	8.91	9.32	9.67	9.97	10.24	10.48
6	5.24	6.33	7.03	7.56	7.97	8.32	8.61	8.87	9.10	9.30
7	4.95	5.92	6.54	7.01	7.37	7.68	7.94	8.17	8.37	8.55
8	4.75	5.64	6.20	6.62	6.96	7.24	7.47	7.68	7.86	8.03
9	4.60	5.43	5.96	6.35	6.66	6.91	7.13	7.33	7.49	7.65
10	4.48	5.27	5.77	6.14	6.43	6.67	6.87	7.05	7.21	7.36
11	4.39	5.15	5.62	5.97	6.25	6.48	6.67	6.84	6.99	7.13
12	4.32	5.05	5.50	5.84	6.10	6.32	6.51	6.67	6.81	6.94
13	4.26	4.96	5.40	5.73	5.98	6.19	6.37	6.53	6.67	6.79
14	4.21	4.89	5.32	5.63	5.88	6.08	6.26	6.41	6.54	6.66
15	4.17	4.84	5.25	5.56	5.80	5.99	6.16	6.31	6.44	6.55
16	4.13	4.79	5.19	5.49	5.72	5.92	6.08	6.22	6.35	6.46
17	4.10	4.74	5.14	5.43	5.66	5.85	6.01	6.15	6.27	6.38
18	4.07	4.70	5.09	5.38	5.60	5.79	5.94	6.08	6.20	6.31
19	4.05	4.67	5.05	5.33	5.55	5.73	5.89	6.02	6.14	6.25
20	4.02	4.64	5.02	5.29	5.51	5.69	5.84	5.97	6.09	6.19
24	3.96	4.55	4.91	5.17	5.37	5.54	5.69	5.81	5.92	6.02
30	3.89	4.45	4.80	5.05	5.24	5.40	5.54	5.65	5.76	5.85
40	3.82	4.37	4.70	4.93	5.11	5.26	5.39	5.50	5.60	5.69
60	3.76	4.28	4.59	4.82	4.99	5.13	5.25	5.36	5.45	5.53
120	3.70	4.20	4.50	4.71	4.87	5.01	5.12	5.21	5.30	5.37
∞	3.64	4.12	4.40	4.60	4.76	4.88	4.99	5.08	5.16	5.23

(*continued*)

Table 10(b) (*Continued*)

12	13	14	15	16	17	18	19	20	df
260.0	266.2	271.8	277.0	281.8	286.3	290.0	294.3	298.0	1
33.40	34.13	34.81	35.43	36.00	36.53	37.03	37.50	37.95	2
17.53	17.89	18.22	18.52	18.81	19.07	19.32	19.55	19.77	3
12.84	13.09	13.32	13.53	13.73	13.91	14.08	14.24	14.40	4
10.70	10.89	11.08	11.24	11.40	11.55	11.68	11.81	11.93	5
9.48	9.65	9.81	9.95	10.08	10.21	10.32	10.43	10.54	6
8.71	8.86	9.00	9.12	9.24	9.35	9.46	9.55	9.65	7
8.18	8.31	8.44	8.55	8.66	8.76	8.85	8.94	9.03	8
7.78	7.91	8.03	8.13	8.23	8.33	8.41	8.49	8.57	9
7.49	7.60	7.71	7.81	7.91	7.99	8.08	8.15	8.23	10
7.25	7.36	7.46	7.56	7.65	7.73	7.81	7.88	7.95	11
7.06	7.17	7.26	7.36	7.44	7.52	7.59	7.66	7.73	12
6.90	7.01	7.10	7.19	7.27	7.35	7.42	7.48	7.55	13
6.77	6.87	6.96	7.05	7.13	7.20	7.27	7.33	7.39	14
6.66	6.76	6.84	6.93	7.00	7.07	7.14	7.20	7.26	15
6.56	6.66	6.74	6.82	6.90	6.97	7.03	7.09	7.15	16
6.48	6.57	6.66	6.73	6.81	6.87	6.94	7.00	7.05	17
6.41	6.50	6.58	6.65	6.72	6.79	6.85	6.91	6.97	18
6.34	6.43	6.51	6.58	6.65	6.72	6.78	6.84	6.89	19
6.28	6.37	6.45	6.52	6.59	6.65	6.71	6.77	6.82	20
6.11	6.19	6.26	6.33	6.39	6.45	6.51	6.56	6.61	24
5.93	6.01	6.08	6.14	6.20	6.26	6.31	6.36	6.41	30
5.76	5.83	5.90	5.96	6.02	6.07	6.12	6.16	6.21	40
5.60	5.67	5.73	5.78	5.84	5.89	5.93	5.97	6.01	60
5.44	5.50	5.56	5.61	5.66	5.71	5.75	5.79	5.83	120
5.29	5.35	5.40	5.45	5.49	5.54	5.57	5.61	5.65	∞

Source: From *Biometrika Tables for Statisticians*, Vol. 1, 3rd ed., edited by E. S. Pearson and H. O. Hartley (Cambridge University Press, 1966). Reproduced by permission of the Biometrika Trustees.